Lecture Notes in Computer Science 9462

Commenced Publication in 1973
Founding and Former Series Editors:
Gerhard Goos, Juris Hartmanis, and Jan van Leeuwen

More information about this series at http://www.springer.com/series/7410

Alex Biryukov · Vipul Goyal (Eds.)

Progress in Cryptology – INDOCRYPT 2015

16th International Conference on Cryptology in India
Bangalore, India, December 6–9, 2015
Proceedings

 Springer

Editors
Alex Biryukov
Université du Luxembourg
Luxembourg
Luxembourg

Vipul Goyal
Microsoft Research India
Bangalore
India

ISSN 0302-9743   ISSN 1611-3349   (electronic)
Lecture Notes in Computer Science
ISBN 978-3-319-26616-9   ISBN 978-3-319-26617-6   (eBook)
DOI 10.1007/978-3-319-26617-6

Library of Congress Control Number: 2015954620

LNCS Sublibrary: SL4 – Security and Cryptology

Springer International Publishing AG Switzerland is part of Springer Science+Business Media
(www.springer.com)

Preface

We are pleased to present the proceedings of INDOCRYPT 2015, held during December 6–9, 2015, in Bangalore, India. This was the 16th edition of the INDO-CRYPT series organized under the aegis of the Cryptology Research Society of India (CRSI).

The INDOCRYPT series of conferences began in 2000 under the leadership of Prof. Bimal Roy of Indian Statistical Institute.

The submissions for INDOCRYPT 2015 were due on July 20, 2015. We received 60 submissions from which, after a careful review and discussion process, 19 were selected for the conference proceedings.

The review process was conducted in two stages: In the first stage, most papers were reviewed by at least three committee members. In the second phase, which lasted for about two weeks, online discussion took place in order to decide on the acceptance of the submissions.

During the review process the Program Committee was helped by a team of 65 external reviewers.

We would like to thank the Program Committee members and the external reviewers for sharing their expertise and giving every paper a fair assessment. The review process was done with EasyChair, which greatly simplified the process.

We were delighted that Itai Dinur, Sanjam Garg, Seny Kamara, Alon Rosen, and Palash Sarkar agreed to deliver invited talks on several interesting topics of relevance to INDOCRYPT.

We were also pleased to have Yevgeniy Dodis and Manoj Prabhakaran deliver two tutorials as part of the conference.

We thank the General Chairs Satya Lokam and Sanjay Burman as well as the teams DRDO and the National Mathematics Initiative at the Indian Institute of Science, Bangalore, for their hard work and taking care of all the local organization matters for the conference. We are especially grateful to our sponsors for their generous support of the conference.

We acknowledge Springer for their active cooperation and timely production of the proceedings. Finally we thank all the authors who submitted papers to the INDO-CRYPT 2015, and all the attendees. We hope you enjoy the proceedings of this year's INDOCRYPT conference.

December 2015

Alex Biryukov
Vipul Goyal

Organization

Program Committee

Alex Biryukov	University of Luxembourg (Program Co-chair), Luxembourg
Vipul Goyal	Microsoft Research India (Program Co-chair), India
Adam O'Neill	Georgetown University, USA
Frederik Armknecht	University of Mannheim, Germany
Arpita Patra	IISc Bangalore, India
Rishiraj Bhattacharyya	R.C. Bose Centre for Cryptology and Security, Indian Statistical Institute, India
Bhavana Kanukurthi	IISc Bangalore, India
Céline Blondeau	Aalto University, School of Science and Technology, Finland
Anne Canteaut	Inria, France
Itai Dinur	Ecole Normale Superieure, Paris, France
Divya Gupta	UCLA, USA
Pierre-Alain Fouque	Université de Rennes, France
David Galindo	Scytl Secure Electronic Voting, Spain
Goutam Paul	Cryptology and Security Research Unit, Indian Statistical Institute, Kolkata, India
Hemanta Maji	Purdue University, USA
Charanjit Jutla	IBM Research, USA
Kannan Srinathan	IIIT Hyderabad, India
Gregor Leander	Horst Gortz University for IT Security, Ruhr-Universität Bochum, Germany
Gaëtan Leurent	Inria, France
Stefan Mangard	IAIK, Graz University of Technology, Austria
Willi Meier	FHNW, Switzerland
Michael Naehrig	Microsoft Research Redmond, USA
David Naccache	Ecole Normale Superieure, Paris, France
Ivica Nikolic	Nanyang Technological University, Singapore
Nishanth Chandran	Microsoft Research, India
Omkant Pandey	University of California, Berkeley, USA
Pandu Rangan	IIT Madras, India
Josef Pieprzyk	Queensland University of Technology, Australia
Pratyay Mukherjee	Aarhus University, Denmark
Emmanuel Prouff	ANSSI, France
Christian Rechberger	Technical University of Denmark, Denmark
Sanjit Chatterjee	IISc Bangalore, India

Palash Sarkar	Applied Statistics Unit, Indian Statistical Institute, India
Shweta Agrawal	IIT Delhi, India
Somitra Sanadhya	IIIT Delhi, India
Sourav Sen Gupta	R.C. Bose Centre for Cryptology and Security, Indian Statistical Institute, India
Francois-Xavier Standaert	CTEAM/ELEN/Crypto Group, Université catholique de Louvain, Belgium
Sushmita Ruj	Cryptology and Security Research Unit, Indian Statistical Institute, Kolkata, India
Ingrid Verbauwhede	KU Leuven ESAT/COSIC and iMind, Belgium
Hongjun Wu	Nanyang Technological University, Singapore

Additional Reviewers

Agrawal, Shashank
Badrinarayanan, Saikrishna
Balasch, Josep
Chakraborti, Avik
Chakraborty, Kaushik
Chattopadhyay, Anupam
Chattopadhyay, Eshan
Choudhury, Ashish
De Santis, Fabrizio
Deshpande, Apoorvaa
Dobraunig, Christoph
Dutta, Avijit
Feix, Benoit
Fuhr, Thomas
Fuller, Ben
Ghosh, Mohona
Gruss, Daniel
Jati, Arpan
Journault, Anthony
Järvinen, Kimmo
Karpman, Pierre
Keelveedhi, Sriram
Khalid, Ayesha
Khurana, Dakshita
Koelbl, Stefan

Koppula, Venkata
Maghrebi, Houssem
Mendel, Florian
Mikhalev, Vasily
Mukherjee, Imon
Müller, Christian
Poussier, Romain
Ramanna, Somindu C.
Rangan, C.
Regazzoni, Francesco
Roy, Arnab
Sarkar, Pinaki
Sengupta, Binanda
Sim, Siang Meng
Sinha Roy, Sujoy
Striecks, Christoph
Thillard, Adrian
Tiessen, Tyge
Varici, Kerem
Vasudevan, Prashant
Venelli, Alexandre
Vivek, Sree
Wentao, Zhang
Zhang, Guoyan

Abstracts of Invited Talks

On Randomness, Codes and Extractors in Cryptography

Yevgeniy Dodis[1]

Department of Computer Science, New York University, USA
dodis@cs.nyu.edu

Abstract. We survey several recent advances in information-theoretic cryptography, such as cryptography with imperfect randomness, randomness extractors, leftover hash lemma and non-malleable extractors/codes.

Secure Multi-Party Computation: A Tutorial

Manoj M. Prabhakaran

University of Illinois at Urbana-Champaign, USA

Abstract. Secure Multi-Party Computation (MPC) is a central problem in modern cryptography, that allows mutually distrusting parties to collaborate with each other on computational tasks, without compromising their private data (beyond what the output of the computation reveals). In this tutorial we shall cover some of the basic concepts behind MPC, informed by recent developments in the field.

The first half of the tutorial will introduce the concept of MPC and briefly present some of the classic constructions, including Yao's Garbled Circuits, the GMW protocol and the BGW protocol. We shall then see some *blackbox transformations* that can be applied to simpler protocols, to achieve higher security or efficiency goals.

The second half of the tutorial will deal with fundamental issues in the theory of MPC. These include definitions of security, classification of MPC tasks according to their cryptographic complexity (including characterization of tasks as possible or impossible to carry out), and questions regarding the communication complexity of MPC.

Encrypted Search:
Theory, Practice and Cryptanalysis

Seny Kamara

Microsoft Research

Abstract. Encrypted search is one of the most potentially impactful topics in cryptography research. Secure and practical encrypted search could fundamentally change how we store and process data, allowing us to design cloud services, databases and storage systems that are both end-to-end encrypted and usable. Research in encrypted search is now 15 years old and is more active and relevant than ever due to the emergence of cloud computing and to consumer, enterprise and government concerns over data privacy.

In this talk I will go over the evolution of encrypted search from its inception until now. I will describe the theoretical and practical advances that pushed the field forward and will discuss where research is headed. I will also survey the latest and most exciting directions including the design of inference attacks and the expansion of encrypted search techniques to handle graph and relational databases. Finally, I will highlight some of the most important theoretical and practical open problems in the area.

On the Cryptographic Hardness
of Finding a Nash Equilibrium

Alon Rosen

School of Computer Science
IDC Herzliya, Israel
alon.rosen@idc.ac.il

Abstract. The notion of *Nash equilibrium* (NE) is fundamental to game theory. While a mixed Nash equilibrium is guaranteed to exist in any game, there is no known polynomial-time algorithm for finding one. The tractability of the problem has received much attention in the past decade, in large part due to its theoretical and philosophical significance.

Prominent evidence for the hardness of finding a NE emerges from a line of works, originating in Papadimitriou and ultimately showing that the problem is complete for the complexity class **PPAD**. The class **PPAD** contains several other search problems that are not known to be tractable, such as finding a fixed point of the kind guaranteed by Brouwer's Theorem. Akin to the phenomenon of **NP**-completeness, this could be interpreted as evidence to computational difficulty. However, unlike in the case of **NP**, currently known problems in **PPAD** appear to be of fairly restricted nature, and carry similar flavor to one another.

In this talk I will show that finding a Nash equilibrium of a game is hard, assuming the existence of indistinguishability obfuscation and one-way functions with sub-exponential hardness. We do so by showing how these cryptographic primitives give rise to a hard computational problem that lies in **PPAD**. Previous proposals for basing **PPAD**-hardness on program obfuscation considered a strong "virtual black-box" notion that is subject to severe limitations and is unlikely to be realizable for the programs in question. In contrast, for indistinguishability obfuscation no such limitations are known, and recently, several candidate constructions of indistinguishability obfuscation were suggested.

Our result provides further evidence of the intractability of finding a Nash equilibrium, one that is extrinsic to the evidence presented so far.

The talk is based on joint work with Nir Bitansky (MIT) and Omer Paneth (BU). It was presented in FOCS'15.

New Advances in Program Obfuscation

Sanjam Garg

University of California, Berkeley
sanjamg@berkeley.edu

Abstract. Recent proposals for plausible candidate constructions of *obfuscation* have radically transformed what we imagined to be possible in cryptography. For over a decade cryptographers had been very skeptical about the existence of such objects. In this talk, I will first provide a very brief introduction to these results and some of their interesting consequences. Next I will present our recent progress towards basing obfuscation on weaker computational assumptions, and the challenges that remain.

The Power of Linear Algebra: Breaking Block Ciphers Using Linearization

Itai Dinur

Department of Computer Science, Ben-Gurion University, Beer-Sheva, Israel

Abstract. Linearization transforms a system of non-linear equations into a linear system using various operations such as replacing complex expressions with new variables. Despite its simplicity, linearization is a versatile and very power tool in cryptanalysis. In this talk, I will review attacks on recent block cipher proposals and emphasize the various roles that linearization plays in these attacks. The talk will consist of three parts, each part analyzing a different block cipher construction and demonstrating how to use lineararization to enhance a different cryptanalytic attack.

In the first part of the talk, I will analyze the security of the block cipher Zorro (presented by Gérard et al. at CHES 2013) using a tool that exploits linearization in order to enhance differential and linear cryptanalysis. The tool gives rise to devastating and practical attacks on Zorro, but also allows to repair it and prove the immunity of the fixed block cipher to these attacks.

The second part of the talk will focus on the LowMC family of block ciphers that was presented at EUROCRYPT 2015 by Albrecht et al. I will analyze the resistance of LowMC against the classical interpolation attack (due to Jakobsen and Knudsen) which uses linearization in order to recover the secret key in a meet-in-the-middle approach. While the LowMC instances proposed at EUROCRYPT 2015 seem to resist the original interpolation attack, I will show how to optimize it using new ideas in order to break their claimed security.

Finally, I will discuss the ASASA block cipher construction that was proposed by Biryukov et al. at ASIACRYPT 2014. A very recent attack on ASASA (presented by Minaud et al. at ASIACRYPT 2015) uses linearization in a novel way in order to recover the key by exploiting a high order differential distinguisher. Although the original attack applies to a subset of ASASA instances, I will show that it can be extended to all instances of this construction.

On the Appropriateness of (Normal) Approximations in Statistical Analysis of Attacks on Symmetric Ciphers

Palash Sarkar

Applied Statistics Unit
Indian Statistical Institute
203, B.T.Road, 700108 Kolkata, India
palash@isical.ac.in

Abstract. Statistical analysis of attacks on symmetric ciphers often require assuming that a test statistic follows the normal distribution. Typically such an assumption holds in an asymptotic sense. In contrast, we consider concrete versions of some important normal approximations that have been made in the literature. To do this, we use the Berry-Esséen theorem to derive explicit bounds on the approximation errors. Analysing these error bounds in several cryptanalytic contexts throws up several surprising results. One important implication is that this puts in doubt the applicability of the order statistics based approach for analysing key recovery attacks on block ciphers. This approach has been earlier used to obtain several results on the data complexities of (multiple) linear and differential cryptanalysis. The non-applicability of the order statistics based approach puts a question mark on the validity of data complexities obtained using this approach. Fortunately, it is possible to recover all of these results by utilising the hypothesis testing framework.

For analysing multiple linear and differential attacks, previous works had used the χ^2 and the log-likelihood ratio (LLR) based test statistics and had approximated their distributions using the normal distribution. The hypothesis testing framework that we consider also uses these statistics and their normal approximations. Detailed consideration of the error in such normal approximations, however, shows that there are serious implications for the applicability of these results.

A general message that we would like to convey is that all cryptanalytic attacks should properly derive and interpret the error bound for any normal (or other) approximation that is made. This will show that an attack is meaningful in a concrete setting rather than in an asymptotic sense.

The talk will be based on joint work with Subhabrata Samajder.

Contents

Side Channel Attacks

Information Theoretic Cryptography

Lightweight Cryptography

Public Key Encryption

Compact Attribute-Based Encryption and Signcryption for General Circuits from Multilinear Maps

Pratish Datta, Ratna Dutta, and Sourav Mukhopadhyay(⊠)

Department of Mathematics, Indian Institute of Technology Kharagpur,
Kharagpur 721302, India
{pratishdatta,ratna,sourav}@maths.iitkgp.ernet.in

Abstract. In this paper, we start by presenting a *key-policy attribute-based encryption* ABE supporting *general polynomial-size circuit* realizable decryption policies and featuring *compactness* in the sense that our ABE construction exhibits *short* ciphertexts and *shorter* decryption keys compared to existing similar works. We then design a *key-policy attribute-based signcryption* ABSC scheme which enjoys several interesting properties that were *never* achievable *before*. It supports signing and decryption policies representable as *arbitrary polynomial-size circuits*. Besides, it generates *short* ciphertext. Our constructions employ multilinear map and achieve *selective* security in the *standard model* under *standard* complexity assumptions. More interestingly, our key-policy constructions can be converted to the corresponding *ciphertext-policy* variants achieving *short* ciphertext by utilizing the technique of *universal circuits*.

Keywords: ABE · ABSC · Polynomial-size circuits · Multilinear map

1 Introduction

ABE: The recent advancements in online social networks and cloud technology have triggered an emerging trend among individuals and organizations to outsource potentially sensitive private data to external servers. This necessitates enforcing sophisticated access control while sharing the outsourced data with other individuals or organizations. *Attribute-based encryption* (ABE) offers a natural solution to the above scenario by enabling fine-grained management of decryption rights to encrypted data.

In 2013, the independent breakthrough works due to Garg et al. [7] and Gorbunov et al. [9] on ABE systems were able to realize decryption policies representable as *polynomial-size circuits* with *arbitrary fan-out*, following which a series of distinguished works have contributed in making ABE for general circuits more practical in terms of both efficiency and security [1,2,8]. Besides tackling the issue of complex access control, ABE for general polynomial-size

© Springer International Publishing Switzerland 2015
A. Biryukov and V. Goyal (Eds.): INDOCRYPT 2015, LNCS 9462, pp. 3–24, 2015.
DOI: 10.1007/978-3-319-26617-6_1

circuit realizable decryption policies has found countless applications in cryptography, most notably for publicly verifiable two message delegation scheme with a preprocessing phase, succinct one-query functional encryption, reuse garbled circuits, token-based obfuscation, and homomorphic encryption for Turing machines.

ABSC: *Attribute-based signcryption* (ABSC) is a logical mixture of attribute-based encryption (ABE) and *attribute-based signature* (ABS) into an unified cost-effective primitive. ABS aims to allow signers to preserve their anonymity while signing digital documents. ABSC resolves the issue of managing sophisticated authentication and decryption rights simultaneously in large distributed networks with better efficiency compared to a sequential implementation of ABE and ABS. For instance, in cloud-based data sharing systems, storing sensitive information securely to the cloud may not be sufficient. The data owner should also be able to prove its genuineness at the cloud as well as to the data recipients to avoid illegal data storage by the cloud server.

A desirable property of an ABSC scheme is *public verifiability* meaning that any party can verify the authenticity of a ciphertext even without the knowledge of the signcrypted message or a valid decryption key. This feature is especially appealing in real-life applications such as filtering out the spams in secure email systems. Here, a spam filter can check whether a signcrypted email is generated from a source with claimed credentials or not before sending to inbox, without knowing the original message. If an email does not satisfy the public verifiability mechanism, it can be treated as spam and can be sent to the spam folder.

Designing efficient ABSC schemes for highly expressive signing and decryption policies is a challenging task and have received considerable attention to the recent research community [5,11–13]. In all the aforementioned ABSC schemes the classes of admissible signing and decryption policies have been restricted to *circuits of fan-out one*.

Our Contribution: In this paper, we propose two attribute-based cryptographic constructions:

- A *key-policy* ABE scheme supporting *arbitrary polynomial-size circuits* with *short ciphertext* and *shorter decryption keys* compared to existing similar works under *standard* complexity assumption.
- The *first key-policy* ABSC scheme for *general polynomial-size circuits* achieving *public verifiability* and featuring *compact ciphertext* as well.

More precisely, similar to [1,2,7–9], our ABE construction permits circuits of arbitrary polynomial-size and unbounded fan-out with bounded depth and input sizes that are fixed at the setup. We develop our ABE scheme in current multilinear map setting [3,4,6] with multilinearity level $n + l + 1$, where n and l denote respectively the input length and depth of the decryption policy circuits. To realize short ciphertext, we adopt the technique of [10] in developing a full domain hash from multilinear map. The structure of our decryption keys is similar to that of [2,7] except that the key components corresponding to the input wires of the decryption policy circuits are suitably modified and are smaller than all previous multilinear map-based constructions [1,2,7,8].

We prove *selective* security of our ABE construction against *chosen plaintext attack* (CPA) under the *Multilinear Decisional Diffie-Hellman* assumption. This is a standard complexity assumption and one can securely instantiate schemes based on this assumption using the recent multilinear map candidate [4].

The second and more significant contribution of this paper is an ABSC scheme of the key-policy category. This scheme also supports signing and decryption policies realizable by polynomial-size circuits of arbitrary fan-out having bounded depths and input lengths. This scheme is developed by augmenting our ABE construction with an attribute-based authentication functionality. We utilize a multilinear map of multilinearity level $n + n' + l + 1$, where n, n', and l represent respectively the input length of decryption policy circuits, input size of signing policy circuits, and depth of both types of circuits.

Our ABSC construction is proven *selectively message confidential* against *chosen-plaintext attack* (CPA) and *selectively ciphertext unforgeability* against *chosen message attack* (CMA) under the *Multilinear Decisional Diffie-Hellman* and *Multilinear Computational Diffie-Hellman* assumption respectively. The number of group elements comprising our ABSC ciphertext is also constant – 3 to be exact.

Finally, an interesting aspect of our work is that using the technique of universal circuits, as in [1,7], both of our constructions can be utilized to realize their corresponding ciphertext-policy variants for arbitrary bounded-size circuits featuring short ciphertext as well.

2   Preliminaries

2.1   Circuit Notation

We adopt the same notations for circuits as in [7]. As pointed out in [7], without loss of generality we can consider only those circuits which are *monotone*, where gates are either OR or AND having fan-in two, and *layered*. Our circuits will have a single output gate. A circuit will be represented as a six-tuple $f = (n, q, l, \mathbb{A}, \mathbb{B}, \mathsf{GateType})$. Here, n and q respectively denote the length of the input and the number of gates, while l represents the depth of the circuit. We designate the set of input wires as $\mathsf{Input} = \{1, \ldots, n\}$, the set of gates as $\mathsf{Gates} = \{n + 1, \ldots, n + q\}$, the total set of wires in the circuit as $W = \mathsf{Input} \cup \mathsf{Gates} = \{1, \ldots, n + q\}$, and the wire $n + q$ to be the output wire. Let $\mathbb{A}, \mathbb{B} : \mathsf{Gates} \rightarrow W \backslash \{n + q\}$ be functions where for all $w \in \mathsf{Gates}$, $\mathbb{A}(w)$ and $\mathbb{B}(w)$ respectively identify w's first and second incoming wires. Finally, $\mathsf{GateType} : \mathsf{Gates} \rightarrow \{\mathsf{AND}, \mathsf{OR}\}$ defines a functions that identifies a gate as either an AND or an OR gate. We follow the convention that $w > \mathbb{B}(w) > \mathbb{A}(w)$ for any $w \in \mathsf{Gates}$.

We also define a function $\mathsf{depth} : W \rightarrow \{1, \ldots, l\}$ such that if $w \in \mathsf{Inputs}$, $\mathsf{depth}(w) = 1$, and in general $\mathsf{depth}(w)$ of wire w is equal to one plus the length of the shortest path from w to an input wire. Since our circuit is layered, we have, for all $w \in$ Gates, if $\mathsf{depth}(w) = j$ then $\mathsf{depth}(\mathbb{A}(w)) = \mathsf{depth}(\mathbb{B}(w)) = j-1$.

We will abuse notation and let $f(x)$ be the evaluation of the circuit f on input $x \in \{0,1\}^n$, and $f_w(x)$ be the value of wire w of the circuit f on input x.

2.2   The Notion of **ABE** for General Circuits

■ **Syntax of ABE for Circuits:** Consider a circuit family $\mathbb{F}_{n,l}$ that consists of all circuits f with input length n and depth l characterizing decryption rights. A key-policy attribute-based encryption (ABE) scheme for circuits in $\mathbb{F}_{n,l}$ with message space $\mathbb{M}$ consists of the following algorithms:

ABE.Setup$(1^\lambda, n, l)$: The trusted key generation center takes as input a security parameter 1^λ, the length n of Boolean inputs to decryption policy circuits, and the allowed depth l of the decryption policy circuits. It publishes the public parameters PP, while keeps the master secret key MK to itself.

ABE.KeyGen$(\mathsf{PP}, \mathsf{MK}, f)$: On input the public parameters PP, the master secret key MK, and the description of a decryption policy circuit $f \in \mathbb{F}_{n,l}$ from a decrypter, the key generation center provides a decryption key $\mathsf{SK}_f^{(\mathsf{DEC})}$ to the decrypter.

ABE.Encrypt(PP, x, M): Taking as input the public parameters PP, an encryption input string $x \in \{0,1\}^n$, and a message $M \in \mathbb{M}$, the encrypter prepares a ciphertext CT_x.

ABE.Decrypt$(\mathsf{PP}, \mathsf{CT}_x, \mathsf{SK}_f^{(\mathsf{DEC})})$: A decrypter takes as input the public parameters PP, a ciphertext CT_x encrypted for x, and its decryption key $\mathsf{SK}_f^{(\mathsf{DEC})}$ corresponding to circuit $f \in \mathbb{F}_{n,l}$. It succeeds to output the message $M \in \mathbb{M}$, if $f(x) = 1$; otherwise, it outputs the distinguished symbol $\bot$.

■ **Security Definition:** The *selective* security notion of ABE for circuits against chosen plaintext attack (CPA) is defined in terms of the following game between a probabilistic adversary $\mathcal{A}$ and a probabilistic challenger $\mathcal{B}$:

Init: $\mathcal{A}$ commits to a challenge encryption input string $x^* \in \{0,1\}^n$ that would be used by $\mathcal{B}$ to create the challenge ciphertext.

Setup: $\mathcal{B}$ performs ABE.Setup$(1^\lambda, n, l)$ to obtain PP, MK, and hands PP to $\mathcal{A}$.

Query Phase 1: $\mathcal{A}$ may adaptively make any polynomial number of decryption key queries for circuit description $f \in \mathbb{F}_{n,l}$ of its choice subject to the restriction that $f(x^*) = 0$. $\mathcal{B}$ returns the corresponding decryption keys $\mathsf{SK}_f^{(\mathsf{DEC})}$ to $\mathcal{A}$ by executing ABE.KeyGen$(\mathsf{PP}, \mathsf{MK}, f)$.

Challenge: $\mathcal{A}$ submits two equal length messages $M_0^*, M_1^* \in \mathbb{M}$. Then $\mathcal{B}$ flips a random coin $b \in \{0,1\}$, and computes the challenge ciphertext CT^* by running ABE.Encrypt(PP, x, M_b). The challenge ciphertext CT^* is given to $\mathcal{A}$.

Query Phase 2: $\mathcal{A}$ may continue adaptively to make decryption key queries as in **Query Phase 1** with the same restriction as above.

Guess: $\mathcal{A}$ eventually outputs a guess b' for b and wins the game if $b' = b$.

Definition 1. *An* ABE *scheme for circuits is defined to be selectively secure against* CPA *if the advantage of any probabilistic polynomial-time* (PPT) *adversaries* $\mathcal{A}$ *in the above game,* $\mathsf{Adv}_{\mathcal{A}}^{\mathsf{ABE,s\text{-}IND\text{-}CPA}}(\lambda) = |\Pr[b' = b] - 1/2|$, *is at most negligible.*

2.3   The Notion of **ABSC** for General Circuits

■ **Syntax of ABSC for Circuits:** Consider a circuit family $\mathbb{F}_{n,l}^{(\mathsf{DEC})}$ consisting of all circuits f with input length n and depth l expressing decryption access structures along with a circuit class $\mathbb{F}_{n',l}^{(\mathsf{SIG})}$ containing all circuits g of input length n' and depth l characterizing signing rights. A key-policy attribute-based signcryption (ABSC) scheme for circuits in $\mathbb{F}_{n,l}^{(\mathsf{DEC})}$ and $\mathbb{F}_{n',l}^{(\mathsf{SIG})}$ with message space $\mathbb{M}$ consists of the following algorithms:

ABSC.Setup($1^{\lambda}, n, n', l$): The trusted key generation center takes as input a security parameter 1^{λ}, the length n of Boolean inputs to decryption policy circuits, the length n' of Boolean inputs to signing policy circuits, and the common allowed depth l of both types of circuits. It publishes the public parameters PP and keeps the master secret key MK to itself.

ABSC.SKeyGen(PP, MK, g): On input the public parameters PP, the master secret key MK, and the description of a signing policy circuit $g \in \mathbb{F}_{n',l}^{(\mathsf{SIG})}$ from a signcrypter, the key generation center provides a signing key $\mathsf{SK}_g^{(\mathsf{SIG})}$ to the signcrypter.

ABSC.DKeyGen(PP, MK, f): Taking as input the public parameters PP, the master secret key MK, and the description a decryption policy circuit $f \in \mathbb{F}_{n,l}^{(\mathsf{DEC})}$ from a decrypter, the key generation center hands a decryption key $\mathsf{SK}_f^{(\mathsf{DEC})}$ to the decrypter.

ABSC.Signcrypt(PP, $\mathsf{SK}_g^{(\mathsf{SIG})}, x, y, M$): A signcrypter takes as input the public parameters PP, its signing key $\mathsf{SK}_g^{(\mathsf{SIG})}$ corresponding to some circuit $g \in \mathbb{F}_{n',l}^{(\mathsf{SIG})}$, an encryption input string $x \in \{0,1\}^n$ describing a set of legitimate decrypter, a signature input string $y \in \{0,1\}^{n'}$ such that $g(y) = 1$, and a message $M \in \mathbb{M}$. It outputs a ciphertext $\mathsf{CT}_{x,y}$.

ABSC.Unsigncrypt(PP, $\mathsf{CT}_{x,y}, \mathsf{SK}_f^{(\mathsf{DEC})}$): A decrypter takes as input the public parameters PP, a ciphertext $\mathsf{CT}_{x,y}$ signcrypted with x, y, and its decryption key $\mathsf{SK}_f^{(\mathsf{DEC})}$ corresponding to circuit $f \in \mathbb{F}_{n,l}^{(\mathsf{DEC})}$. It succeeds to output the message $M \in \mathbb{M}$ provided the ciphertext is valid, if $f(x) = 1$; otherwise, it outputs $\perp$ indicating that either the ciphertext is invalid or the ciphertext cannot be decrypted.

■ **Security Definitions:** An ABSC scheme for circuits has two security requirements, namely, (I) *message confidentiality* and (II) *ciphertext unforgeability* which are described below:

(I) **Message Confidentiality:** This security notion is defined on indistinguishability of ciphertexts under chosen plaintext attack (CPA) in the *selective* encryption input string model through an analogous game as in case of ABE. The details is omitted here due to page restriction.

(II) **Ciphertext Unforgeability:** This notion of security is defined on existential unforgeability under adaptive chosen message attack (CMA) in the *selective* signature input string model through the following game between a probabilistic adversary $\mathcal{A}$ and a probabilistic challenger $\mathcal{B}$.

Init: $\mathcal{A}$ declares a signature input string $y^* \in \{0,1\}^{n'}$ to $\mathcal{B}$ that will be used to forge a signcryption.

Setup: $\mathcal{B}$ runs ABSC.Setup($1^\lambda, n, n', l$) to obtain PP, MK and hands PP to $\mathcal{A}$.

Query Phase: $\mathcal{A}$ may adaptively make a polynomial number of queries of the following types to $\mathcal{B}$ and $\mathcal{B}$ provides the answer to them.

▷ *Signing key query:* Upon receiving a signing key query from $\mathcal{A}$ corresponding to a signing policy circuit $g \in \mathbb{F}_{n',l}^{(\mathsf{SIG})}$ subject to the constraint that $g(y^*) = 0$, $\mathcal{B}$ returns the $\mathsf{SK}_g^{(\mathsf{SIG})}$ to $\mathcal{A}$ by executing ABSC.SKeyGen(PP, MK, g).

▷ *Decryption key query:* When $\mathcal{A}$ queries a decryption key for a decryption policy circuit $f \in \mathbb{F}_{n,l}^{(\mathsf{DEC})}$, $\mathcal{B}$ gives $\mathsf{SK}_f^{(\mathsf{DEC})}$ to $\mathcal{A}$ by performing ABSC.DKeyGen(PP, MK, f).

▷ *Signcryption query:* $\mathcal{A}$ queries a signcryption of a message M for a signature input string $y(\neq y^*) \in \{0,1\}^{n'}$ along with an encryption input string $x \in \{0,1\}^n$. $\mathcal{B}$ samples a signing policy circuit $g \in \mathbb{F}_{n',l}^{(\mathsf{SIG})}$ such that $g(y) = 1$ and returns the ciphertext $\mathsf{CT}_{x,y}$ to $\mathcal{A}$ by performing ABSC.Signcrypt(PP, $\mathsf{SK}_g^{(\mathsf{SIG})}$, x, y, M), where $\mathsf{SK}_g^{(\mathsf{SIG})}$ is got from ABSC.SKeyGen(PP, MK, g) by $\mathcal{B}$.

▷ *Unsigncryption query:* In response to a unsigncryption query from $\mathcal{A}$ for a ciphertext $\mathsf{CT}_{x,y}$ under the decryption policy circuit $f \in \mathbb{F}_{n,l}^{(\mathsf{DEC})}$, $\mathcal{B}$ obtains the decryption key $\mathsf{SK}_f^{(\mathsf{DEC})}$ by running ABSC.DkeyGen(PP, MK, f) and sends output of ABSC.Unsigncrypt(PP, $\mathsf{CT}_{x,y}$, $\mathsf{SK}_f^{(\mathsf{DEC})}$) to $\mathcal{A}$.

Forgery: $\mathcal{A}$ eventually outputs a forgery CT^* for some message M^* with the signature input string y^* and an encryption input string x^*. $\mathcal{A}$ wins the game if the ciphertext CT^* is valid, i.e., $M^*(\neq \perp)$ is the output of ABSC.Unsigncrypt(PP, CT^*, $\mathsf{SK}_{f^*}^{(\mathsf{DEC})}$) for any $f^* \in \mathbb{F}_{n,l}^{(\mathsf{DEC})}$ satisfying $f^*(x^*) = 1$, and CT^* is not obtained from any signcryption query to $\mathcal{B}$.

Definition 2. *An* ABSC *scheme for circuits is defined to be selectively ciphertext unforgeable against* CMA *if the advantage of any* PPT *adversaries* $\mathcal{A}$ *in the above game,* $\mathsf{Adv}_{\mathcal{A}}^{\mathsf{ABSC,s\text{-}UF\text{-}CMA}}(\lambda) = \Pr[\mathcal{A} \ wins]$, *is at most negligible.*

2.4 Multilinear Maps and Complexity Assumption

A (leveled) multilinear map [3,4,6] consists of the following two algorithms:

(i) $\mathcal{G}(1^\lambda, k)$: It takes as input a security parameter 1^λ and a positive integer k indicating the number of allowed pairing operations. It outputs a sequence of groups $\vec{\mathbb{G}} = (\mathbb{G}_1, \ldots, \mathbb{G}_k)$ each of large prime order $p > 2^\lambda$ together with the canonical generators g_i of $\mathbb{G}_i$. We call $\mathbb{G}_1$ the source group, $\mathbb{G}_k$ the target group, and $\mathbb{G}_2, \ldots, \mathbb{G}_{k-1}$ intermediate groups.

(ii) $e_{i,j}(g, h)$ (for $i, j \in \{1, \ldots, k\}$ with $i + j \leq k$): On input two elements $g \in \mathbb{G}_i$ and $h \in \mathbb{G}_j$ with $i + j \leq k$, it outputs an element of $\mathbb{G}_{i+j}$ such that $e_{i,j}(g_i^a, g_j^b) = g_{i+j}^{ab}$ for $a, b \in \mathbb{Z}_p$. We often omit the subscripts and just write e. We can also generalize e to multiple inputs as $e(h^{(1)}, \ldots, h^{(t)}) = e(h^{(1)}, e(h^{(2)}, \ldots, h^{(t)}))$.

We refer g_i^a as a level-i encoding of $a \in \mathbb{Z}_p$. The scalar a itself is referred to as a level-0 encoding of a. Then the map e combines a level-i encoding of an element $a \in \mathbb{Z}_p$ and a level-j encoding of another element $b \in \mathbb{Z}_p$, and produces level-$(i + j)$ encoding of the product ab.

Assumption 1 (k-Multilinear Decisional Diffie-Hellman: k-MDDH [6]). *The k- Multilinear Decisional Diffie-Hellman (k-MDDH) assumption states that it is intractable for any* PPT *algorithm $\mathcal{B}$ to guess $\overline{b} \in \{0,1\}$ given $\varrho_{\overline{b}} = (\vec{\mathbb{G}}, g_1, \overline{S}, C_1, \ldots, C_k, T_{\overline{b}})$ generated by $\mathcal{G}_{\overline{b}}^{k\text{-MDDH}}(1^\lambda)$, where $\mathcal{G}_{\overline{b}}^{k\text{-MDDH}}(1^\lambda)$ works as follows:*

- *Run $\mathcal{G}(1^\lambda, k)$ to generate $\vec{\mathbb{G}} = (\mathbb{G}_1, \ldots, \mathbb{G}_k)$ with $g_1, \ldots, g_k$ of order p.*
- *Pick random $\overline{s}, c_1, \ldots, c_k \in \mathbb{Z}_p$ and compute $\overline{S} = g_1^{\overline{s}}, C_1 = g_1^{c_1}, \ldots, C_k = g_1^{c_k}$.*
- *Set $T_0 = g_k^{\overline{s}\prod_{j=1}^k c_j}$ while $T_1 = $ some random element in $\mathbb{G}_k$.*
- *Return $\varrho_{\overline{b}} = (\vec{\mathbb{G}}, g_1, \overline{S}, C_1, \ldots, C_k, T_{\overline{b}})$.*

Assumption 2 (k-Multilinear Computational Diffie-Hellman: k-MCDH [10]). *The k-multilinear computational Diffie-Hellman (k-MCDH) assumption states that it is intractable for any* PPT *algorithm $\mathcal{B}$ to output $T = g_{k-1}^{\prod_{i=1}^k c_i}$ given $\varrho = (\vec{\mathbb{G}}, g_1, C_1, \ldots, C_k)$ generated by $\mathcal{G}^{k\text{-MCDH}}(1^\lambda)$, where $\mathcal{G}^{k\text{-MCDH}}(1^\lambda)$ performs the following:*

- *Run $\mathcal{G}(1^\lambda, k)$ to generate $\vec{\mathbb{G}} = (\mathbb{G}_1, \ldots, \mathbb{G}_k)$ with $g_1, \ldots, g_k$ of order p.*
- *Pick random $c_1, \ldots, c_k \in \mathbb{Z}_p$ and compute $C_1 = g_1^{c_1}, \ldots, C_k = g_1^{c_k}$.*
- *Return $\varrho = (\vec{\mathbb{G}}, g_1, C_1, \ldots, C_k)$.*

3 Our ABE Scheme

The Construction

ABE.Setup$(1^\lambda, n, l)$: The trusted key generation center takes as input a security parameter 1^λ, the length of Boolean inputs n to the decryption policy circuits, and the allowed depth l of decryption policy circuits. It proceeds as follows:

1. It runs $\mathcal{G}(1^\lambda, k = n+l+1)$ to produce group sequence $\vec{\mathbb{G}} = (\mathbb{G}_1, \ldots, \mathbb{G}_k)$ of prime order $p > 2^\lambda$ with canonical generators $g_1, \ldots, g_k$.

2. It selects random $\alpha \in \mathbb{Z}_p$ together with random $(a_{1,0}, a_{1,1}), \ldots, (a_{n,0}, a_{n,1}) \in \mathbb{Z}_p^2$, and computes

$$H = g_{l+1}^\alpha, \ A_{i,\beta} = g_1^{a_{i,\beta}} \text{ for } i = 1, \ldots, n; \ \beta \in \{0,1\}.$$

3. It publishes the public parameters PP consisting of the group sequence description along with H and $\{A_{i,\beta}\}_{i=1,\ldots,n; \ \beta \in \{0,1\}}$. The master secret key MK $= g_l^\alpha$ is kept to itself.

ABE.KeyGen(PP, MK, f): The key generation center takes as input the public parameters PP, the master secret key MK, and the description $f = (n, q, l, \mathbb{A}, \mathbb{B}, \mathsf{GateType})$ of a decryption policy circuit from a decrypter. Our circuit has $n + q$ wires $\{1, \ldots, n + q\}$ where $\{1, \ldots, n\}$ are n input wires, $\{n + 1, \ldots, n + q\}$ are q gates (OR or AND gates), and the wire $n + q$ designated as the output wire. It performs the following steps:

1. It chooses random $r_1, \ldots, r_{n+q} \in \mathbb{Z}_p$ where we think of randomness r_w as being associated with wire $w \in \{1, \ldots, n + q\}$. It produces the "header" component

$$K = g_l^\alpha g_l^{-r_{n+q}} = g_l^{\alpha - r_{n+q}},$$

where g_l^α is obtained from MK.

2. Next, it generates key components for every wire w. The structure of the key component depends upon the category of w, i.e., whether w is an Input wire, an OR gate, or an AND gate. We describe how it generates the key components in each case.

   • <u>Input wire</u>: If $w \in \{1, \ldots, n\}$ then it corresponds to the w-th input. It computes the key component $\mathcal{K}_w = e(A_{w,1}, g_1)^{r_w} = g_2^{r_w a_{w,1}}$.

   • <u>OR gate</u>: Suppose that wire $w \in \mathsf{Gates}$, $\mathsf{GateType}(w) = \mathsf{OR}$, and $j = \mathsf{depth}(w)$. It picks random $b_w, d_w \in \mathbb{Z}_p$ and creates the key component

$$\mathcal{K}_w = \left(K_{w,1} = g_1^{b_w}, K_{w,2} = g_1^{d_w}, K_{w,3} = g_j^{r_w - b_w r_{\mathbb{A}(w)}}, K_{w,4} = g_j^{r_w - d_w r_{\mathbb{B}(w)}}\right).$$

   • <u>AND gate</u>: Let wire $w \in \mathsf{Gates}$, $\mathsf{GateType}(w) = \mathsf{AND}$, and $j = \mathsf{depth}(w)$. It selects random $b_w, d_w \in \mathbb{Z}_p$ and forms the key component

$$\mathcal{K}_w = \left(K_{w,1} = g_1^{b_w}, K_{w,2} = g_1^{d_w}, K_{w,3} = g_j^{r_w - b_w r_{\mathbb{A}(w)} - d_w r_{\mathbb{B}(w)}}\right).$$

3. It provides the decryption key $\mathsf{SK}_f^{(\mathsf{DEC})} = (f, \mathcal{K}, \{\mathcal{K}_w\}_{w \in \{1,\ldots,n+q\}})$ to the decrypter.

ABE.Encrypt(PP, x, M): Taking as input the public parameters PP, an encryption input string $x = x_1 \ldots x_n \in \{0,1\}^n$, and a message $M \in \mathbb{G}_k$, the encrypter forms the ciphertext as follows:

1. It picks random $s \in \mathbb{Z}_p$ and computes

$$C_M = e(h, A_{1,x_1}, \ldots, A_{n,x_n})^s M = g_{n+l+1}^{\alpha s \prod_{i=1}^n a_{i,x_i}} M = g_k^{\alpha s \delta(x)} M,$$

where we define $\delta(x) = \prod_{i=1}^n a_{i,x_i}$ for the ease of exposition. It also computes $C = g_1^s$.

2. It outputs the ciphertext $\mathsf{CT}_x = (x, C_M, C)$.

$\mathsf{ABE.Decrypt}(\mathsf{PP}, \mathsf{CT}_x, \mathsf{SK}_f^{(\mathsf{DEC})})$: A decrypter, on input the public parameters PP, a ciphertext $\mathsf{CT}_x = (x, C_M, C)$ encrypted for encryption input string $x = x_1 \ldots x_n \in \{0,1\}^n$, and its decryption key $\mathsf{SK}_f^{(\mathsf{DEC})} = (f, \mathcal{K}, \{\mathcal{K}_w\}_{w \in \{1,\ldots,n+q\}})$ corresponding to its decryption policy circuit $f = (n, q, l, \mathbb{A}, \mathbb{B}, \mathsf{GateType})$, outputs $\bot$, if $f(x) = 0$; otherwise, (i.e., if $f(x) = 1$) proceeds as follows:

1. First, there is a header computation, where it computes

$$D = e(A_{1,x_1}, \ldots, A_{n,x_n}) = g_n^{\delta(x)}$$

followed by $\widehat{E} = e(\mathcal{K}, D, C) = g_k^{(\alpha - r_{n+q})s\delta(x)}$

by extracting $\{A_{i,x_i}\}_{i=1,\ldots,n}$ from PP.

2. Next, it evaluates the circuit from the bottom up. For every wire w with corresponding $\mathsf{depth}(w) = j$, if $f_w(x) = 0$, nothing needs to be computed for that wire, otherwise (if $f_w(x) = 1$), it attempts to compute $E_w = g_{n+j+1}^{r_w s\delta(x)}$ as described below. The decrypter proceeds iteratively starting with computing E_1 and moves forward in order to finally compute E_{n+q}. Note that computing these values in order ensures that the computation on a wire w with $\mathsf{depth}(w) = j - 1$ that evaluates to 1 will be defined before computing for a wire w with $\mathsf{depth}(w) = j$. The computation procedure varies with the category of the wire, i.e., whether the wire is an Input wire, an OR gate, or an AND gate.

• <u>Input wire</u>: If $w \in \{1, \ldots, n\}$ then it corresponds to the w-th input. Suppose that $x_w = f_w(x) = 1$. The decrypter extracts $\{A_{i,x_i}\}_{i,\ldots,n}$ from PP and computes

$$E_w = e(\mathcal{K}_w, A_{1,x_1}, \ldots, A_{w-1,x_{w-1}}, A_{w+1,x_{w+1}}, \ldots, A_{n,x_n}, C) = g_{n+2}^{r_w s\delta(x)}.$$

• <u>OR gate</u>: Consider a wire $w \in \mathsf{Gates}$ with $\mathsf{GateType}(w) = \mathsf{OR}$ and $j = \mathsf{depth}(w)$. Assume that $f_w(x) = 1$. Then either $f_{\mathbb{A}(w)}(x) = 1$ or $f_{\mathbb{B}(w)}(x) = 1$. If $f_{\mathbb{A}(w)}(x) = 1$, i.e., the first input of gate w evaluates to 1, then the decrypter computes

$$E_w = e(E_{\mathbb{A}(w)}, K_{w,1})e(K_{w,3}, D, C) = g_{n+j+1}^{r_w s\delta(x)}.$$

Note that $E_{\mathbb{A}(w)}$ is already computed at this stage in the bottom-up circuit evaluation as $\mathsf{depth}(\mathbb{A}(w)) = j - 1$.
Alternatively, if $f_{\mathbb{A}(w)}(x) = 0$ but $f_{\mathbb{B}(w)}(x) = 1$, then it computes

$$E_w = e(E_{\mathbb{B}(w)}, K_{w,2})e(K_{w,4}, D, C) = g_{n+j+1}^{r_w s\delta(x)}.$$

• <u>AND gate</u>: Consider a wire $w \in \mathsf{Gates}$ with $\mathsf{GateType}(w) = \mathsf{AND}$ and $j = \mathsf{depth}(w)$. Suppose that $f_w(x) = 1$. Then $f_{\mathbb{A}(w)}(x) = f_{\mathbb{B}(w)}(x) = 1$. The decrypter computes

$$E_w = e(E_{\mathbb{A}(w)}, K_{w,1})e(E_{\mathbb{B}(w)}, K_{w,2})e(K_{w,3}, D, C) = g_{n+j+1}^{r_w s\delta(x)}.$$

In this process, the decrypter ultimately computes $E_{n+q} = g_k^{r_{n+q}s\delta(x)}$, as $f(x) = f_{n+q}(x) = 1$.

3. Finally, the decrypter computes $E = \widehat{E}E_{n+q} = g_k^{\alpha s\delta(x)}$ and retrieves the message by the computation $C_M E^{-1} = g_k^{\alpha s\delta(x)} M(g_k^{\alpha s\delta(x)})^{-1} = M$.

Security Analysis

Theorem 1 (Security of ABE). *The proposed* ABE *scheme supporting decryption policies expressable as arbitrary circuits of depth l and input length n achieves selective* CPA*-security as per the security model of Sect. 2.2 under the k-MDDH assumption where $k = n + l + 1$.*

Proof. Suppose that there is a PPT adversary $\mathcal{A}$ that breaks with non-negligible advantage the selective CPA security of the proposed ABE scheme supporting decryption policies representable as arbitrary circuits of depth l and input length n. We construct a PPT algorithm $\mathcal{B}$ that attempts to solve an instance of the k-MDDH problem, where $k = n + l + 1$, using $\mathcal{A}$ as a sub-routine. $\mathcal{B}$ is given an instance of the k-MDDH problem $\varrho_{\overline{5}} = (\overline{\mathbb{G}}, g_1, \overline{S}, C_1, \ldots, C_k, T_{\overline{5}})$ such that $\overline{S} = g_1^{\overline{s}}, C_1 = g_1^{c_1}, \ldots, C_k = g_1^{c_k}$. $\mathcal{B}$ plays the role of the challenger in the selective CPA-security game of Sect. 2.2 and interacts with $\mathcal{A}$ as follows:

Init: $\mathcal{A}$ declares the challenge encryption input string $x^* = x_1^* \ldots x_n^* \in \{0,1\}^n$ to $\mathcal{B}$.

Setup: $\mathcal{B}$ chooses random $z_1, \ldots, z_n \in \mathbb{Z}_p$ and sets $a_{i,\beta} = c_i$ implicitly, if $\beta = x_i^*$, while $a_{i,\beta} = z_i$, if $\beta \neq x_i^*$, for $i = 1, \ldots, n$; $\beta \in \{0,1\}$. This corresponds to setting $A_{i,\beta} = C_i = g_1^{c_i}$, if $\beta = x_i^*$, while $A_{i,\beta} = g_1^{z_i}$, if $\beta \neq x_i^*$, for $i = 1, \ldots, n$; $\beta \in \{0,1\}$. Observe that the values $A_{i,\beta}$ are distributed identically as in the real scheme. In addition $\mathcal{B}$ picks random $\xi \in \mathbb{Z}_p$ and *implicitly* sets $\alpha = \xi + \prod_{h=1}^{l+1} c_{n+h}$. For enhancing readability we define $\gamma(u,v) = \prod_{h=u}^{v} c_h$ for positive integers u and v. Then, $\mathcal{B}$'s view point of α is $\alpha = \xi + \gamma(n+1, n+l+1)$. $\mathcal{B}$ computes $H = e(C_{n+1}, \ldots, C_{n+l+1})g_{l+1}^{\xi} = g_{l+1}^{\alpha}$. $\mathcal{B}$ hands the public parameters PP consisting of the group sequence description together with $H, \{A_{i,\beta}\}_{i=1,\ldots,n;\ \beta \in \{0,1\}}$ to $\mathcal{A}$.

Query Phase 1 and Query Phase 2: Both the key query phases are executed in the same manner by $\mathcal{B}$. So, we describe them once here. $\mathcal{A}$ queries a decryption key for a circuit $f = (n, q, l, \mathbb{A}, \mathbb{B}, \mathsf{GateType})$ to $\mathcal{B}$ subject to the restriction that $f(x^*) = 0$. As in [7], we will think of the proof as having some invariant property on the depth of the wire we are looking at. Consider a wire w with $\mathsf{depth}(w) = j$ and $\mathcal{B}$'s view point (symbolically) of r_w. If $f_w(x^*) = 0$, then $\mathcal{B}$ will *implicitly* view r_w as the term $\gamma(n+1, n+j+1)$ plus some additional known randomization term. On the other hand, if $f_w(x^*) = 1$ then $\mathcal{B}$ will view r_w as 0 plus some additional known randomization term. Keeping this property intact for simulating the keys up the circuit, $\mathcal{B}$ will ultimately view r_{n+q} as $\gamma(n+1, n+l+1)$ plus some additional known randomization term since $f_{n+q}(x^*) = f(x^*) = 0$. As will be demonstrated shortly, this would allow $\mathcal{B}$ to simulate the header component $\mathcal{K}$ by cancelation.

The bottom up simulation of the key component for each wire w by $\mathcal{B}$ varies depending on whether w is an Input wire, an OR gate, or an AND gate.

- Input wire: Consider $w \in \{1, \ldots, n\}$, i.e., an input wire.
  - If $x_w^* = 1$, then $\mathcal{B}$ picks random $r_w \in \mathbb{Z}_p$ (as is done honestly) and sets the key component

$$\mathcal{K}_w = e(C_w, g_1)^{r_w} = g_2^{r_w a_{w,1}}.$$

  - Otherwise, if $x_w^* = 0$, then $\mathcal{B}$ *implicitly* lets $r_w = \gamma(n+1, n+2) + \eta_w$, where $\eta_w \in \mathbb{Z}_p$ is randomly selected by $\mathcal{B}$, and sets the key component

$$\mathcal{K}_w = (e(C_{n+1}, C_{n+2})g_2^{\eta_w})^{z_w} = g_2^{r_w a_{w,1}}.$$

- OR gate: Consider a wire $w \in$ Gates with GateType$(w) =$ OR and $j =$ depth(w).
  - If $f_w(x^*) = 1$, then $f_{\mathbb{A}(w)}(x^*) = 1$ or $f_{\mathbb{B}(w)}(x^*) = 1$. $\mathcal{B}$ chooses random $b_w, d_w, r_w \in \mathbb{Z}_p$ as in the real scheme, and forms the key component as

$$\mathcal{K}_w = \left(K_{w,1} = g_1^{b_w}, K_{w,2} = g_1^{d_w}, K_{w,3} = g_j^{r_w - b_w r_{\mathbb{A}(w)}}, K_{w,4} = g_j^{r_w - d_w r_{\mathbb{B}(w)}}\right).$$

    Observe that, due to the bottom up simulation, $r_{\mathbb{A}(w)}$ and $r_{\mathbb{B}(w)}$ are already selected or implicitly set by $\mathcal{B}$ according as the corresponding gates, i.e., $\mathbb{A}(w)$ and $\mathbb{B}(w)$, evaluate to 1 or 0 upon input x^*. Note that even if $\mathbb{A}(w)$ or $\mathbb{B}(w)$ gate evaluates to 0 upon input x^*, $\mathcal{B}$ can still simulate its corresponding component, i.e., $K_{w,3}$ or $K_{w,4}$ in $\mathcal{K}_w$ using multilinear map. For instance, $f_{\mathbb{A}(w)}(x^*) = 0$ implies $r_{\mathbb{A}(w)}$ has been implicitly set as $\gamma(n+1, n+j) + \eta_{\mathbb{A}(w)}$ by $\mathcal{B}$, as depth$(\mathbb{A}(w)) = j - 1$ for the reason that our circuit is layered. Thus, in this case $\mathcal{B}$ can create $K_{w,3}$ as $K_{w,3} = e(C_{n+1}, \ldots, C_{n+j})^{-b_w} g_j^{r_w} = g_j^{r_w - b_w r_{\mathbb{A}(w)}}$.
  - On the other hand, if $f_w(x^*) = 0$, then $f_{\mathbb{A}(w)}(x^*) = f_{\mathbb{B}(w)}(x^*) = 0$. $\mathcal{B}$ picks random $\psi_w, \phi_w, \eta_w \in \mathbb{Z}_p$, implicitly sets $b_w = c_{n+j+1} + \psi_w, d_w = c_{n+j+1} + \phi_w$, and $r_w = \gamma(n+1, n+j+1) + \eta_w$, and creates the decryption key component $\mathcal{K}_w = (K_{w,1}, K_{w,2}, K_{w,3}, K_{w,4})$ where

$$K_{w,1} = C_{n+j+1}g_1^{\psi_w} = g_1^{b_w}, K_{w,2} = C_{n+j+1}g_1^{\phi_w} = g_1^{d_w},$$

$$K_{w,3} = e(C_{n+j+1}, g_{j-1})^{-\eta_{\mathbb{A}(w)}} e(C_{n+1}, \ldots, C_{n+j})^{-\psi_w} g_j^{\eta_w - \psi_w \eta_{\mathbb{A}(w)}}$$
$$= g_j^{\eta_w - c_{n+j+1}\eta_{\mathbb{A}(w)} - \psi_w(\gamma(n+1, n+j) + \eta_{\mathbb{A}(w)})} = g_j^{r_w - b_w r_{\mathbb{A}(w)}},$$

$$K_{w,4} = e(C_{n+j+1}, g_{j-1})^{-\eta_{\mathbb{B}(w)}} e(C_{n+1}, \ldots, C_{n+j})^{-\phi_w} g_j^{\eta_w - \phi_w \eta_{\mathbb{B}(w)}}$$
$$= g_j^{\eta_w - c_{n+j+1}\eta_{\mathbb{B}(w)} - \phi_w(\gamma(n+1, n+j) + \eta_{\mathbb{B}(w)})} = g_j^{r_w - d_w r_{\mathbb{B}(w)}}.$$

Note that according to our bottom up simulation, $r_{\mathbb{A}(w)}$ has been implicitly set as $r_{\mathbb{A}(w)} = \gamma(n+1, n+j) + \eta_{\mathbb{A}(w)}$ by $\mathcal{B}$ and similarly for $r_{\mathbb{B}(w)}$. Therefore,

$$r_w - b_w r_{\mathbb{A}(w)}$$
$$= (\gamma(n+1, n+j+1) + \eta_w) - (c_{n+j+1} + \psi_w)(\gamma(n+1, n+j) + \eta_{\mathbb{A}(w)})$$
$$= \eta_w - c_{n+j+1}\eta_{\mathbb{A}(w)} - \psi_w(\gamma(n+1, n+j) + \eta_{\mathbb{A}(w)})$$

which enables $\mathcal{B}$ to simulate $K_{w,3}$ and analogously $K_{w,4}$ in this case.

- **AND gate:** Consider wire $w \in$ Gates with GateType$(w) =$ AND and $j =$ depth(w).
  - If $f_w(x^*) = 1$, then $f_{\mathbb{A}(w)}(x^*) = f_{\mathbb{B}(w)}(x^*) = 1$. $\mathcal{B}$ selects random $b_w, d_w, r_w \in \mathbb{Z}_p$ and forms the key component as

  $$\mathcal{K}_w = \left(K_{w,1} = g_1^{b_w}, K_{w,2} = g_1^{d_w}, K_{w,3} = g_j^{r_w - b_w r_{\mathbb{A}(w)} - d_w r_{\mathbb{B}(w)}} \right).$$

  Notice that since $f_{\mathbb{A}(w)}(x^*) = f_{\mathbb{B}(w)}(x^*) = 1$, $r_{\mathbb{A}(w)}$ and $r_{\mathbb{B}(w)}$ are random values which have already been chosen by $\mathcal{B}$ in the course of simulation.
  - Alternatively, if $f_w(x^*) = 0$, then $f_{\mathbb{A}(w)}(x^*) = 0$ or $f_{\mathbb{B}(w)}(x^*) = 0$. If $f_{\mathbb{A}(w)}(x^*) = 0$, then $\mathcal{B}$ selects random $\psi_w, \phi_w, \eta_w \in \mathbb{Z}_p$, *implicitly* lets $b_w = c_{n+j+1} + \psi_w, d_w = \phi_w$, and $r_w = \gamma(n+1, n+j+1) + \eta_w$, and forms the decryption key component as $\mathcal{K}_w = (K_{w,1}, K_{w,2}, K_{w,3})$ where

  $$K_{w,1} = C_{n+j+1}g_1^{\psi_w} = g_1^{b_w}, K_{w,2} = g_1^{\phi_w} = g_1^{d_w},$$

  $$K_{w,3} = e(C_{n+j+1}, g_{j-1})^{-\eta_{\mathbb{A}(w)}} e(C_{n+1}, \ldots, C_{n+j})^{-\psi_w} g_j^{\eta_w - \psi_w \eta_{\mathbb{A}(w)} - \phi_w r_{\mathbb{B}(w)}}$$

  $$= g_j^{\eta_w - c_{n+j+1}\eta_{\mathbb{A}(w)} - \psi_w(\gamma(n+1,n+j) + \eta_{\mathbb{A}(w)}) - \phi_w r_{\mathbb{B}(w)}} = g_j^{r_w - b_w r_{\mathbb{A}(w)} - d_w r_{\mathbb{B}(w)}}.$$

  Observe that $\mathcal{B}$ can form $K_{w,3}$ due to a similar cancelation as explained in case of OR gates since, the $\mathbb{A}(w)$ gate being evaluated to 0, $r_{\mathbb{A}(w)} = \gamma(n+1, n+j) + \eta_{\mathbb{A}(w)}$ has already been implicitly set by $\mathcal{B}$. Moreover, $g_j^{r_{\mathbb{B}(w)}}$ is always computable by $\mathcal{B}$ from the available information regardless of whether $f_{\mathbb{B}(w)}(x^*) = 1$, in which case $r_{\mathbb{B}(w)}$ is a random value chosen by $\mathcal{B}$ itself, or $f_{\mathbb{B}(w)}(x^*) = 0$, for which $r_{\mathbb{B}(w)}$ has been implicitly set to be $r_{\mathbb{B}(w)} = \gamma(c_{n+1}, c_{n+j}) + \eta_{\mathbb{B}(w)}$ by $\mathcal{B}$ and, hence, $\mathcal{B}$ can compute $e(C_{n+1}, \ldots, C_{n+j})g_j^{\eta_{\mathbb{B}(w)}} = g_j^{r_{\mathbb{B}(w)}}$. The case where $f_{\mathbb{B}(w)}(x^*) = 0$ and $f_{\mathbb{A}(w)}(x^*) = 1$ is performed in a symmetric manner, with the roles of b_w and d_w reversed.

Since $f(x^*) = f_{n+q}(x^*) = 0$, r_{n+q} at the output gate is *implicitly* set as $\gamma(n+1, n+l+1) + \eta_{n+q}$ by $\mathcal{B}$. This allows $\mathcal{B}$ to perform a final cancelation in computing the "header" component of the key as $\mathcal{K} = g_l^{\xi - \eta_{n+q}} = g_l^{\alpha - r_{n+q}}$. $\mathcal{B}$ provides the decryption key $\mathsf{SK}_f^{(\mathsf{DEC})} = (f, \mathcal{K}, \{\mathcal{K}_w\}_{w \in \{1, \ldots, n+q\}})$ to $\mathcal{A}$.

Challenge: $\mathcal{A}$ submits two challenge messages $M_0^*, M_1^* \in \mathbb{G}_k$ to $\mathcal{B}$. $\mathcal{B}$ flips a random coin $b \in \{0, 1\}$, *implicitly* views $\bar{s}$ as the randomness used in generating the challenge ciphertext, and sets challenge ciphertext

$$\mathsf{CT}^* = (x^*, C_M^* = T_{\bar{b}}e(\overline{S}, C_1, \ldots, C_n, g_l^{\xi})M_b^* = T_{\bar{b}}g_k^{\xi \bar{s}\gamma(1,n)}M_b^*, C^* = \overline{S}),$$

and gives it to $\mathcal{A}$.

Guess: $\mathcal{B}$ eventually receives back the guess $b' \in \{0, 1\}$ from $\mathcal{A}$. If $b = b'$, $\mathcal{B}$ outputs $\bar{b}' = 1$; otherwise, it outputs $\bar{b}' = 0$.

Note that if $\bar{b} = 0$, the challenge ciphertext CT^* is properly generated by $\mathcal{B}$, while if $\bar{b} = 1$ the challenge ciphertext is random. Hence the theorem. □

Table 1. Communication and storage comparison

ABE	Security	Complexity assumptions	k	$|PP|$	$|CT_x|$	$|SK_f^{(DEC)}|$
[7]	selective	MDDH	$l+1$	$n+1$	$n+2$	$2n+4q+1$
[2]	selective	MDHE	$l+1$	$n+1$	3	$n+n^2+4q+1$
[8]	adaptive	3 new non-standard assumptions	$n+2q+2$	$2n+4q+3$	$4q+3$	$4q+2$
[1]	adaptive	$SD_1, SD_2, EMDDH_1, EMDDH_2$	$3l$	$n+4$	$n+4$	$2n+4q+3$
Ours	selective	MDDH	$n+l+1$	$2n+1$	2	$n+4q+1$

Here, $MDDH, MDHE, SD_1, SD_2, EMDDH_1, EMDDH_2$ stand respectively for the Multilinear Decisional Diffie-Hellman [7], Multilinear Diffie-Hellman Exponent [2], two variants of Multilinear Subgroup Decision [1], and the two versions of the Expanded Multilinear Decisional Diffie-Hellman assumptions [1].

In this table, k denotes the maximum multilinearity level of the underlying multilinear maps, n, q, and l represent respectively the input length, number of gates, and depth of the decryption policy circuits, while $|PP|, |CT_x|$, and $|SK_f^{(DEC)}|$ stand respectively for the number of group elements comprising PP, CT_x, and $SK_f^{(DEC)}$.

Table 2. Comparison of multilinear operation count

ABE	ABE.Setup	ABE.KeyGen	ABE.Encrypt	ABE.Decrypt
[7]	$n+2$	$3n+4q+1$	$n+2$	$2n+3q+1$
[2]	$n+2$	$n^2+2n+4q+1$	3	$2n+3q+1$
Ours	$2n+2$	$2n+4q+1$	3	$n+3q+3$

In this table, n and q denote respectively the input size and number of gates in the decryption policy circuits.

Efficiency

Table 1 compares the communication and storage requirements of our proposed ABE scheme with previously known multilinear map-based ABE constructions supporting general circuits in terms of the number of group elements comprising the public parameters PP, ciphertext CT_x, and decryption key $SK_f^{(DEC)}$. As is clear from the table, the most significant achievement of our construction is that our ABE ciphertext involves only 2 (constant) group elements which is smaller than all earlier constructions. Also, our decryption key contains only a single group element corresponding to each input wire of the decryption policy circuits. In all existing constructions, number of group elements required for each input wire of the decryption policy circuits is strictly greater than one.

Regarding computational efficiency, notice that unlike traditional bilinear map setting, in current multilinear map candidates [3,4,6], exponentiation is also realized through multilinear operation. Since multilinear operations are costlier compared to group operations in multilinear groups, we consider the count of multilinear operations required in each algorithm of ABE scheme as a parameter for comparing computational cost. Table 2 demonstrates the number of multilinear operations involved in the setup, key generation, encryption and decryption algorithms of our ABE scheme in comparison to existing multilinear map-based selectively secure ABE constructions for arbitrary circuits. From the table it readily follows that the key generation, encryption, as well as decryption algorithms of our scheme requires the least number of multilinear operations among all the three schemes.

4 Our ABSC Scheme

The Construction

ABSC.Setup($1^\lambda, n, n', l$): The trusted key generation center takes as input a security parameter 1^λ, the length n of inputs to decryption policy circuits, the length n' of inputs to signing policy circuits, and the common allowed depth l of both types of circuits. It proceeds as follows:

1. It runs $\mathcal{G}(1^\lambda, k = n + n' + l + 1)$ to produce group sequence $\vec{\mathbb{G}} = (\mathbb{G}_1, \ldots, \mathbb{G}_k)$ of prime order $p > 2^\lambda$ with canonical generators $g_1, \ldots, g_k$.
2. It picks random $\alpha_1, \alpha_2 \in \mathbb{Z}_p$ along with random $(a_{1,0}, a_{1,1}), \ldots, (a_{n,0}, a_{n,1}), (b_{1,0}, b_{1,1}), \ldots, (b_{n',0}, b_{n',1}) \in \mathbb{Z}_p^2$, sets $\alpha = \alpha_1 + \alpha_2$, and computes $H = g_{l+1}^{\alpha_1}$, $A_{i,\beta} = g_1^{a_{i,\beta}}$, $B_{t,\beta} = g_1^{b_{t,\beta}}$ for $i = 1, \ldots, n$; $t = 1, \ldots, n'$; $\beta \in \{0,1\}$.
3. Additionally, it chooses random $\theta \in \mathbb{Z}_p$ and computes $\Theta = g_n^\theta, Y = g_{n+l+1}^{\theta\alpha_2}$.
4. It publishes the public parameters PP consisting of the group sequence description together with $\{A_{i,\beta}\}_{i=1,\ldots,n; \beta\in\{0,1\}}$, $\{B_{t,\beta}\}_{t=1,\ldots,n'; \beta\in\{0,1\}}$, H, Θ, Y, while keeps the master secret key MK $= (g_l^\alpha, g_l^{\alpha_2})$.

ABSC.SKeyGen(PP, MK, g): On input the public parameters PP, the master secret key MK, and the description $g = (n', q', l, \mathbb{A}, \mathbb{B}, \mathsf{GateType})$ of a signing policy circuit from a signcrypter, the key generation center forms a signing key as described below. Recall that the circuit g has $n' + q'$ wires $\{1, \ldots, n' + q'\}$ with n' input wires $\{1, \ldots, n'\}$, q' gates $\{n' + 1, \ldots, n' + q'\}$, and the wire $n' + q'$ designated as the output wire.

1. It chooses random $r_1', \ldots, r_{n'+q'-1}' \in \mathbb{Z}_p$ and sets $r_{n'+q'}' = \alpha_2$, where again we will think of the random value r_w' as being associated with the wire w.
2. It proceeds to generate the key components for every wire w. Here also the structure of the key component depends upon the category of the wire $w \in \{1, \ldots, n' + q'\}$, i.e., whether w is an Input wire, an OR gate, or an AND gate. We describe how it generates the key component in each case.

   • Input wire: If $w \in \{1, \ldots, n'\}$ then it corresponds to the w-th input. It computes the key component $\mathcal{K}_w' = e(B_{w,1}, g_1)^{r_w'} = g_2^{r_w' b_{w,1}}$.
   • OR gate: Suppose that wire $w \in \mathsf{Gates}$, $\mathsf{GateType}(w) = \mathsf{OR}$, and $j = \mathsf{depth}(w)$. It picks random $b_w', d_w' \in \mathbb{Z}_p$ and creates the key component

$$\mathcal{K}_w' = \left(\mathcal{K}_{w,1}' = g_1^{b_w'}, \mathcal{K}_{w,2}' = g_1^{d_w'}, \mathcal{K}_{w,3}' = g_j^{r_w' - b_w' r_{\mathsf{A}(w)}'}, \mathcal{K}_{w,4}' = g_j^{r_w' - d_w' r_{\mathsf{B}(w)}'}\right).$$

- **AND gate:** Let wire $w \in \mathsf{Gates}$, $\mathsf{GateType}(w) = \mathsf{AND}$, and $j = \mathsf{depth}(w)$. It selects random $b'_w, d'_w \in \mathbb{Z}_p$ and generates the key component

$$\mathcal{K}'_w = \left(K'_{w,1} = g_1^{b'_w}, K'_{w,2} = g_1^{d'_w}, K'_{w,3} = g_j^{r'_w - b'_w r'_{\mathbb{A}(w)} - d'_w r'_{\mathbb{B}(w)}} \right).$$

Notice that while computing the key component $\mathcal{K}'_{n'+q'}$ for the output gate $n' + q'$ which has depth l, the required $g_l^{r'_{n'+q'}} = g_l^{\alpha_2}$ is retrieved from MK.

3. It gives the signing key $\mathsf{SK}_g^{(\mathsf{SIG})} = (g, \{\mathcal{K}'_w\}_{w \in \{1,\dots,n'+q'\}})$ to the signcrypter.

$\mathsf{ABSC.DKeyGen}(\mathsf{PP}, \mathsf{MK}, f)$: Taking as input the public parameters PP, the master secret key MK, and the description $f = (n, q, l, \mathbb{A}, \mathbb{B}, \mathsf{GateType})$ of a decryption policy circuit from a decrypter, the key generation center creates a decryption key $\mathsf{SK}_f^{(\mathsf{DEC})} = (f, \mathcal{K}, \{\mathcal{K}_w\}_{w \in \{1,\dots,n+q\}})$ in the same manner as the $\mathsf{ABE.KeyGen}(\mathsf{PP}, \mathsf{MK}, f)$ algorithm described in Sect. 3 using $\{A_{i,\beta}\}_{i=1,\dots,n;\ \beta \in \{0,1\}}$ obtained from PP and g_l^{α} extracted from MK. We omit the details here. It hands the decryption key $\mathsf{SK}_f^{(\mathsf{DEC})}$ to the decrypter.

$\mathsf{ABSC.Signcrypt}(\mathsf{PP}, \mathsf{SK}_g^{(\mathsf{SIG})}, x, y, M)$: A signcrypter takes as input the public parameters PP, its signing key $\mathsf{SK}_g^{(\mathsf{SIG})} = (g, \{\mathcal{K}'_w\}_{w \in \{1,\dots,n'+q'\}})$ corresponding to some signing policy circuit $g = (n', q', l, \mathbb{A}, \mathbb{B}, \mathsf{GateType})$, an encryption input string $x = x_1 \dots x_n \in \{0,1\}^n$, a signature input string $y = y_1 \dots y_{n'} \in \{0,1\}^{n'}$ satisfying $g(y) = 1$, and a message $M \in \mathbb{G}_k$. It prepares the ciphertext as follows:

1. It first evaluates the signing policy circuit from the bottom up. As before, we define $\delta(x) = \prod_{i=1}^n a_{i,x_i}$ and $\delta'(y) = \prod_{t=1}^{n'} b_{t,y_t}$ for improving readability. It starts by computing

$$D' = e(B_{1,y_1}, \dots, B_{n',y_{n'}}) = g_{n'}^{\delta'(y)},$$

where $\{B_{t,y_t}\}_{t=1,\dots,n'}$ are extracted from PP. For every wire w in g with corresponding $\mathsf{depth}(w) = j$, if $g_w(y) = 0$ then nothing needs to be computed for that wire; on the other hand, if $g_w(y) = 1$ then it computes $E'_w = g_{n'+j}^{r'_w \delta'(y)}$ as described below. The signcrypter proceeds iteratively starting with computing E'_1 and moves forward in order to ultimately compute $E'_{n'+q'} = g_{n'+l}^{r'_{n'+q'} \delta'(y)} = g_{n'+l}^{\alpha_2 \delta'(y)}$. Note that $r'_{n'+q'}$ has been set to α_2 by the key generation center. Moreover, observe that computing the E'_w values in order ensures that the computation on a wire w with $\mathsf{depth}(w) = j - 1$ that evaluates to 1 will be defined before computing for a wire w with $\mathsf{depth}(w) = j$. The computation procedure varies with the category of the wire, i.e., Input wire, OR gate, or AND gate in this case as well.

- **Input wire:** If $w \in \{1, \dots, n'\}$ then it corresponds to the w-th input. Suppose that $y_w = 1$. The signcrypter computes

$$E'_w = e(\mathcal{K}'_w, B_{1,y_1}, \dots, B_{w-1,y_{w-1}}, B_{w+1,y_{w+1}}, \dots, B_{n',y_{n'}}) = g_{n'+1}^{r'_w \delta'(y)}.$$

- **OR gate:** Consider a wire $w \in$ Gates with GateType$(w) =$ OR and $j =$ depth(w). Assume that $g_w(y) = 1$. Then either $g_{\mathbb{A}(w)}(y) = 1$ or $g_{\mathbb{B}(w)}(y) = 1$. If $g_{\mathbb{A}(w)}(y) = 1$ then the signcrypter computes

$$E'_w = e(E'_{\mathbb{A}(w)}, K'_{w,1})e(K'_{w,3}, D') = g_{n'+j}^{r'_w \delta'(y)}.$$

Alternatively, if $g_{\mathbb{A}(w)}(y) = 0$ but $g_{\mathbb{B}(w)}(y) = 1$ then it computes

$$E'_w = e(E'_{\mathbb{B}(w)}, K'_{w,2})e(K'_{w,4}, D') = g_{n'+j}^{r'_w \delta'(y)}.$$

- **AND gate:** Consider a wire $w \in$ Gates with GateType$(w) =$ AND and $j =$ depth(w). Suppose that $g_w(y) = 1$. Hence $g_{\mathbb{A}(w)}(y) = g_{\mathbb{B}(w)}(y) = 1$. The signcrypter computes

$$E'_w = e(E'_{\mathbb{A}(w)}, K'_{w,1})e(E'_{\mathbb{B}(w)}, K'_{w,2})e(K'_{w,3}, D') = g_{n'+j}^{r'_w \delta'(y)}.$$

2. Next the signcrypter picks random $s \in \mathbb{Z}_p$ and computes

$$
\begin{aligned}
C_M &= \left(e(H, A_{1,x_1}, \ldots, A_{n,x_n}, D')e(E'_{n'+q'}, A_{1,x_1}, \ldots, A_{n,x_n}, g_1)\right)^s M \\
&= g_k^{\alpha s \delta(x) \delta'(y)} M,
\end{aligned}
$$
$$C = g_1^s, \quad C' = e(\Theta, E'_{n'+q'}) = g_{k-1}^{\theta \alpha_2 \delta'(y)}.$$

Here, H, Θ, and $\{A_{i,x_i}\}_{i=1,\ldots,n}$ are extracted from PP.

3. The signcrypter outputs the ciphertext $\mathsf{CT}_{x,y} = (x, y, C_M, C, C')$.

ABSC.Unsigncrypt$(\mathsf{PP}, \mathsf{CT}_{x,y}, \mathsf{SK}_f^{(\mathsf{DEC})})$: A decrypter takes as input the public parameters PP, a ciphertext $\mathsf{CT}_{x,y} = (x, y, C_M, C, C')$ signcrypted with an encryption input string $x = x_1 \ldots x_n \in \{0,1\}^n$ and a signature input string $y = y_1 \ldots y_{n'} \in \{0,1\}^{n'}$, as well as its decryption key $\mathsf{SK}_f^{(\mathsf{DEC})} = (f, \mathcal{K}, \{\mathcal{K}_w\}_{w \in \{1,\ldots,n+q\}})$ corresponding to its legitimate decryption circuit $f = (n, q, l, \mathbb{A}, \mathbb{B}, \mathsf{GateType})$. It performs the following steps:

1. It first computes $D' = e(B_{1,y_1}, \ldots, B_{n',y_{n'}}) = g_{n'}^{\delta'(y)}$ and checks the validity of the ciphertext as $e(C', g_1) = e(Y, D')$.
   Note that if the ciphertext is valid then both sides of the above equality should evaluate to $g_k^{\theta \alpha_2 \delta'(y)}$. If the above equation is invalid then it outputs $\bot$; otherwise, it proceeds to the next step.

2. If $f(x) = 0$ then it outputs $\bot$; on the other hand, if $f(x) = 1$ then it proceeds in the same way as in the case of ABE.Decrypt$(\mathsf{PP}, \mathsf{CT}_x, \mathsf{SK}_f^{(\mathsf{DEC})})$ algorithm of Sect. 3 to compute the header $\widehat{E} = g_{n+l+1}^{(\alpha - r_{n+q})s\delta(x)}$ followed by a computation of the circuit from the bottom up ultimately obtaining $E_{n+q} = g_{n+l+1}^{r_{n+q}s\delta(x)}$. In this computation it makes use of C obtained from $\mathsf{CT}_{x,y}$ and $\{A_{i,x_i}\}_{i=1,\ldots,n}$ extracted from PP along with its decryption key components. We omit the details here.

3. Finally, the decrypter retrieves the message by computing

$$
C_M \left[e(\widehat{E} E_{n+q}, D')\right]^{-1} = g_k^{\alpha s \delta(x) \delta'(y)} M \left[e(g_{n+l+1}^{(\alpha - r_{n+q})s\delta(x)} g_{n+l+1}^{r_{n+q}s\delta(x)}, g_{n'}^{\delta'(y)})\right]^{-1}
$$
$$= M.$$

Security Analysis

Theorem 2 (Message Confidentiality of ABSC). *The proposed* ABSC *construction supporting arbitrary decryption policy circuits of input length n and depth l, as well as, arbitrary signing policy circuits of input length n' and the same depth l achieves selective message confidentiality against* CPA *under the k-MDDH assumption, where $k = n + n' + l + 1$.*

The proof of Theorem 2 closely resembles that of Theorem 1 and is omitted due to page consideration.

Theorem 3 (Ciphertext Unforgeability of ABSC). *The proposed* ABSC *scheme supporting arbitrary decryption policy circuits of input length n and depth l, as well as, arbitrary signing policy circuits of input length n' and depth l achieves selective ciphertext unforgeability against* CMA *as per the security model of Sect. 2.3 under the k-MCDH assumption, where $k = n + n' + l + 1$.*

Proof. Assume that there is a PPT adversary $\mathcal{A}$ that breaks with non-negligible advantage the selective CMA ciphertext unforgeability of the proposed ABSC scheme supporting decryption policy circuits of input length n and depth l, as well as, signing policy circuits of input length n' and the same depth l. We construct a PPT algorithm $\mathcal{B}$ that attempts to solve an instance of the k-MCDH problem, where $k = n + n' + l + 1$, using $\mathcal{A}$ as a sub-routine. $\mathcal{B}$ is given an instance of the k-MCDH problem $\varrho = (\vec{\mathbb{G}}, g_1, C_1, \ldots, C_k)$ such that $C_1 = g_1^{c_1}, \ldots, C_k = g_1^{c_k}$. $\mathcal{B}$ plays the role of the challenger in the selective CMA ciphertext unforgeability game of Sect. 2.3 and interacts with $\mathcal{A}$ as follows:

Init: $\mathcal{A}$ declares a signature input string $y^* = y_1^* \ldots y_{n'}^* \in \{0,1\}^{n'}$ to $\mathcal{B}$ that will be used to forge a signcryption.

Setup: $\mathcal{B}$ picks random $a_{i,\beta} \in \mathbb{Z}_p$ and computes $A_{i,\beta} = g_1^{a_{i,\beta}}$ for $i = 1, \ldots, n$; $\beta \in \{0,1\}$ as is done in the original scheme. Further $\mathcal{B}$ selects random $z_1', \ldots, z_{n'}' \in \mathbb{Z}_p$ and *implicitly* sets $b_{t,\beta} = c_{n+t}$, if $\beta = y_t^*$, and $b_{t,\beta} = z_t'$, if $\beta \neq y_t^*$, for $t = 1, \ldots, n'$; $\beta \in \{0,1\}$. This corresponds to setting $B_{t,\beta} = C_{n+t}$, if $\beta = y_t^*$, while $B_{t,\beta} = g_1^{z_t'}$, if $\beta \neq y_t^*$, for $t = 1, \ldots, n'$; $\beta \in \{0,1\}$. Additionally, $\mathcal{B}$ selects random $\alpha \in \mathbb{Z}_p$, *implicitly* lets $\theta = \gamma(1, n), \alpha_1 = \alpha - \gamma(n+n'+1, n+n'+l+1), \alpha_2 = \gamma(n+n'+1, n+n'+l+1)$, where $\gamma(u,v) = \prod_{h=u}^{v} c_h$ for integers u, v, and sets $\Theta = e(C_1, \ldots, C_n) = g_n^\theta, H = e(C_{n+n'+1}, \ldots, C_{n+n'+l+1})^{-1} g_{l+1}^\alpha = g_{l+1}^{\alpha_1}, Y = e(C_1, \ldots, C_n, C_{n+n'+1}, \ldots, C_{n+n'+l+1}) = g_{n+l+1}^{\theta \alpha_2}$. $\mathcal{B}$ hands the public parameters PP consisting of the group sequence description plus $\{A_{i,\beta}\}_{i=1,\ldots,n;\beta\in\{0,1\}}$, $\{B_{t,\beta}\}_{t=1,\ldots,n'; \beta\in\{0,1\}}, H, \Theta, Y$ to $\mathcal{A}$. Note that all the simulated PP components are identically distributed as in the original scheme.

Query Phase: $\mathcal{A}$ issues a series of queries to which $\mathcal{B}$ answers as follows:

▷ *Signing key query:* $\mathcal{A}$ queries a signing key corresponding to a circuit $g = (n', q', l, \mathbb{A}, \mathbb{B}, \mathsf{GateType})$ subject to the constraint that $g(y^*) = 0$. $\mathcal{B}$ proceeds to generate the key components from the bottom up the circuit as described below. Here also we will think the simulation to have some invariant property on the depth of the wire we are looking at. Consider a wire w with $\mathsf{depth}(w) = j$. If $g_w(y^*) = 0$, then $\mathcal{B}$ will view r'_w as $\gamma(n + n' + 1, n + n' + j + 1)$ plus some additional known randomization term, while if $g_w(y^*) = 1$ then $\mathcal{B}$ will view r'_w as 0 plus some additional known randomization term. Keeping this property intact up the circuit, $\mathcal{B}$ will *implicitly* set $r'_{n'+q'} = \gamma(n+n'+1, n+n'+l+1) = \alpha_2$ as $g_{n'+q'}(y^*) = g(y^*) = 0$. We describe how $\mathcal{B}$ creates the signing key components for each wire w organizing the simulation into the following cases:

- **Input wire:** Suppose $w \in \{1, \dots, n'\}$, i.e., an input wire.

  - If $y^*_w = 1$, then $\mathcal{B}$ chooses random $r'_w \in \mathbb{Z}_p$ and computes

    $$\mathcal{K}'_w = e(C_{n+w}, g_1)^{r'_w} = g_2^{r'_w b_{w,1}}.$$

  - If $y^*_w = 0$, then $\mathcal{B}$ picks random $\eta''_w \in \mathbb{Z}_p$, *implicitly* lets $r'_w = \gamma(n + n' + 1, n + n' + 2) + \eta'_w$, and sets

    $$\mathcal{K}'_w = \left(e(C_{n+n'+1}, C_{n+n'+2}) g_2^{\eta'_w}\right)^{z_w} = g_2^{r'_w b_{w,1}}.$$

- **OR gate:** Consider a wire $w \in \mathsf{Gates}$ with $\mathsf{GateType}(w) = \mathsf{OR}$ and $j = \mathsf{depth}(w)$.

  - If $g_w(y^*) = 1$, then $g_{\mathbb{A}(w)}(y^*) = 1$ or $g_{\mathbb{B}(w)}(y^*) = 1$. $\mathcal{B}$ chooses random $b'_w, d'_w, r'_w \in \mathbb{Z}_p$ as in the real scheme, and creates the key component as

    $$\mathcal{K}'_w = \left(K'_{w,1} = g_1^{b'_w}, K'_{w,2} = g_1^{d'_w}, K'_{w,3} = g_j^{r'_w - b'_w r'_{\mathbb{A}(w)}}, K'_{w,4} = g_j^{r'_w - d'_w r'_{\mathbb{B}(w)}}\right).$$

    Observe that, due to the bottom up simulation, $r'_{\mathbb{A}(w)}$ and $r'_{\mathbb{B}(w)}$ are already selected or implicitly set by $\mathcal{B}$ according as the corresponding gates, i.e., $\mathbb{A}(w)$ and $\mathbb{B}(w)$, evaluate to 1 or 0 upon input y^*. Note that even if $\mathbb{A}(w)$ or $\mathbb{B}(w)$ gate evaluates to 0 upon input y^*, $\mathcal{B}$ can still simulate its corresponding component, i.e., $K'_{w,3}$ or $K'_{w,4}$ in $\mathcal{K}'_w$ using multilinear map in a similar fashion as in simulating the queried decryption key components for OR gates in analogous situation in proof of Theorem 1.

  - On the other hand, if $g_w(y^*) = 0$, then $g_{\mathbb{A}(w)}(y^*) = g_{\mathbb{B}(w)}(y^*) = 0$. $\mathcal{B}$ chooses random $\psi'_w, \phi'_w, \eta'_w \in \mathbb{Z}_p$, *implicitly* lets $b'_w = c_{n+n'+j+1} + \psi'_w$, $d'_w = c_{n+n'+j+1} + \phi'_w$, and $r'_w = \gamma(n + n' + 1, n + n' + j + 1) + \eta'_w$, and sets $\mathcal{K}'_w = (K'_{w,1}, K'_{w,2}, K'_{w,3}, K'_{w,4})$ where

    $$K'_{w,1} = C_{n+n'+j+1} g_1^{\psi'_w} = g_1^{b'_w}, K'_{w,2} = C_{n+n'+j+1} g_1^{\phi'_w} = g_1^{d'_w},$$

    $$K'_{w,3} = e(C_{n+n'+j+1}, g_{j-1})^{-\eta'_{\mathbb{A}(w)}} e(C_{n+n'+1}, \dots, C_{n+n'+j})^{-\psi'_w} g_j^{\eta'_w - \psi'_w \eta'_{\mathbb{A}(w)}}$$
    $$= g_j^{r'_w - b'_w r'_{\mathbb{A}(w)}},$$

    $$K'_{w,4} = e(C_{n+n'+j+1}, g_{j-1})^{-\eta'_{\mathbb{B}(w)}} e(C_{n+n'+1}, \dots, C_{n+n'+j})^{-\phi'_w} g_j^{\eta'_w - \phi'_w \eta'_{\mathbb{B}(w)}}$$
    $$= g_j^{r'_w - d'_w r'_{\mathbb{B}(w)}}.$$

Observe that $\mathcal{B}$ can form $K'_{w,3}$ and $K'_{w,4}$ due to a cancelation analogous to the simulation of the decryption key components corresponding to OR gates in similar situation in proof of Theorem 1, since both the $\mathbb{A}(w)$ and $\mathbb{B}(w)$ gates being evaluated to 0, $r'_{\mathbb{A}(w)} = \gamma(n+n'+1, n+n'+j) + \eta'_{\mathbb{A}(w)}$ and similarly $r'_{\mathbb{B}(w)}$ have already been implicitly set by $\mathcal{B}$ in course of the bottom up simulation.

- **AND gate:** Consider a wire $w \in$ Gates with GateType$(w) =$ AND and $j =$ depth(w).
  - If $g'_w(y^*) = 1$, then $g_{\mathbb{A}(w)}(y^*) = g_{\mathbb{B}(w)}(y^*) = 1$. $\mathcal{B}$ selects random $b'_w, d'_w, r'_w \in \mathbb{Z}_p$ and forms the key component as

$$\mathcal{K}'_w = \left(K'_{w,1} = g_1^{b'_w}, K'_{w,2} = g_1^{d'_w}, K'_{w,3} = g_j^{r'_w - b'_w r'_{\mathbb{A}(w)} - d'_w r'_{\mathbb{B}(w)}} \right).$$

Notice that since $g_{\mathbb{A}(w)}(y^*) = g_{\mathbb{B}(w)}(y^*) = 1$, $r'_{\mathbb{A}(w)}$ and $r'_{\mathbb{B}(w)}$ are random values which have already been chosen by $\mathcal{B}$ in the course of the bottom-up simulation.

  - Alternatively, if $g_w(y^*) = 0$, then $g_{\mathbb{A}(w)}(y^*) = 0$ or $g_{\mathbb{B}(w)}(y^*) = 0$. If $g_{\mathbb{A}(w)}(y^*) = 0$, then $\mathcal{B}$ picks $\psi'_w, \phi'_w, \eta'_w \in \mathbb{Z}_p$, *implicitly lets* $b'_w = c_{n+n'+j+1} + \psi'_w, d'_w = \phi'_w$, and $r'_w = \gamma(n+n'+1, n+n'+j+1) + \eta'_w$, and forms $\mathcal{K}'_w = (K'_{w,1}, K'_{w,2}, K'_{w,3})$ where

$$K'_{w,1} = c_{n+n'+j+1} g_1^{\psi'_w} = g_1^{b'_w}, \quad K'_{w,2} = g_1^{\phi'_w} = g_1^{d'_w},$$

$$K'_{w,3} = e(c_{n+n'+j+1}, g_{j-1})^{-\eta'_{\mathbb{A}(w)}} e(c_{n+n'+1}, \ldots, c_{n+n'+j})^{-\psi'_w}.$$

$$g_j^{\eta'_w - \psi'_w \eta'_{\mathbb{A}(w)} - \phi'_w r'_{\mathbb{B}(w)}} = g_j^{r'_w - b'_w r'_{\mathbb{A}(w)} - d'_w r'_{\mathbb{B}(w)}}.$$

Note that $\mathcal{B}$ can generate $K'_{w,3}$ due to a similar cancelation as in the simulation of the decryption key components for AND gates in analogous scenario in the proof of Theorem 1 since, the $\mathbb{A}(w)$ gate being evaluated to 0, $\mathcal{B}$ has already set $r'_{\mathbb{A}(w)} = \gamma(n+n'+1, n+n'+j) + \eta'_{\mathbb{A}(w)}$ implicitly during the bottom up simulation. Moreover, notice that $g_j^{r'_{\mathbb{B}(w)}}$ is always computable for $\mathcal{B}$ regardless of whether $g_{\mathbb{B}(w)}(y^*)$ evaluates to 0 or 1 as $g_j^{\gamma(n+n'+1, n+n'+j)}$ is computable using the multilinear map from the available information for $\mathcal{B}$. The case where $g_{\mathbb{B}(w)}(y^*) = 0$ and $g_{\mathbb{A}(w)}(y^*) = 1$ is executed in a symmetric manner with the roles of b'_w and d'_w reversed.

We mention that at the output gate $n' + q'$, $\mathcal{B}$ will take the additional randomness $\eta'_{n'+q'}$ to be zero while setting $r'_{n'+q'}$. Observe that this would not prevent the distribution of the simulated signing keys from being identical to that of the real scheme. $\mathcal{B}$ gives the signing key $\mathsf{SK}_g^{(\mathsf{SIG})} = (g, \{\mathcal{K}'_w\}_{w \in \{1, \ldots, n'+q'\}})$ to $\mathcal{A}$.

▷ *Decryption key query:* Note that $\mathcal{B}$ knows α, therefore, $\mathcal{B}$ can provide the decryption key $\mathsf{SK}_f^{(\mathsf{DEC})}$ corresponding to any decryption policy circuit $f = (n, q, l, \mathbb{A}, \mathbb{B}, \mathsf{GateType})$ queried by $\mathcal{A}$.

▷ *Signcryption query*: $\mathcal{A}$ queries the signcryption of a message M relative to a signature input string $y = y_1 \ldots y_{n'} (\neq y^*) \in \{0,1\}^{n'}$ and an encryption input string $x = x_1 \ldots x_n \in \{0,1\}^n$. $\mathcal{B}$ chooses random $s \in \mathbb{Z}_p$ and computes

$$C_M = e(g_{l+1}^\alpha, A_{1,x_1}, \ldots, A_{n,x_n}, B_{1,y_1}, \ldots, B_{n',y_{n'}})^s M = g_k^{\alpha s \delta(x) \delta'(y)} M, C = g_1^s,$$

where $\delta(x) = \prod_{i=1}^n a_{i,x_i}$, $\delta'(y) = \prod_{t=1}^{n'} b_{t,y_t}$. $\mathcal{B}$ also computes C' as described below. Since $y \neq y^*$, there exists some $t \in \{1, \ldots, n'\}$ such that $y_t \neq y_t^*$ and, hence, $B_{t,y_t} = g_1^{z_t'}$ as per the simulation. $\mathcal{B}$ computes

$$C' = e(\Theta, C_{n+n'+1}, \ldots, C_{n+n'+l+1}, B_{1,x_1}, \ldots, B_{t-1,y_{t-1}}, B_{t+1,y_{t+1}}, \ldots, B_{n',y_{n'}})^{z_t'}$$
$$= g_{n+n'+l}^{\theta \alpha_2 \delta'(y)}.$$

$\mathcal{B}$ gives the ciphertext $\mathsf{CT}_{x,y} = (x, y, C_M, C, C')$ to $\mathcal{A}$.

▷ *Unsigncryption query*: Note that $\mathcal{B}$ can create the decryption key $\mathsf{SK}_f^{(\mathsf{DEC})}$ corresponding to any decryption policy circuit f. Therefore, when $\mathcal{A}$ queries the unsigncryption of a ciphertext $\mathsf{CT}_{x,y}$ under a decryption policy circuit f, $\mathcal{B}$ first computes $\mathsf{SK}_f^{(\mathsf{DEC})}$ and then provides the result of $\mathsf{ABSC.Unsigncrypt}(\mathsf{PP}, \mathsf{CT}_{x,y}, \mathsf{SK}_f^{(\mathsf{DEC})})$ to $\mathcal{A}$.

Forgery: $\mathcal{A}$ eventually produces a valid forgery $\mathsf{CT}^* = (x^*, y^*, C_M^*, C^*, C'^*)$ for some message M^* with an encryption input string x^* and the committed signature input string y^*. Then $\mathcal{B}$ solves the k-MCDH problem by outputting C'^*.

Note that, since CT^* is a valid forgery, we have

$$C'^* = g_{n+n'+l}^{\theta \alpha_2 \prod_{i'=1}^{n'} b_{i',y_{i'}}} = g_{k-1}^{\gamma(1,n)\gamma(n+n'+1,n+n'+l+1)\gamma(n+1,n+n')} = g_{k-1}^{\gamma(1,k)}$$

which is the desired answer of the k-MCDH problem instance given to $\mathcal{B}$. The theorem follows. $\qquad\square$

Efficiency

Regarding communication and storage complexity of the proposed ABSC construction, the number of multilinear group elements comprising the public parameters PP, ciphertext $\mathsf{CT}_{x,y}$, decryption key $\mathsf{SK}_f^{(\mathsf{DEC})}$, and signing key $\mathsf{SK}_g^{(\mathsf{SIG})}$ are respectively $2n + 2n' + 3$, 3, $n + 4q + 1$, and $n' + 4q'$ where we have used a multilinear map with multilinearity level $k = n + n' + l + 1$, n, q being respectively the input length and number of gates of the decryption policy circuits, n', q' being the corresponding values for the signing policy circuits, and l being the allowed depth of both kinds of circuits. On the other hand, about computational cost, notice that the count of multilinear operations involved in the setup, signing key generation, decryption key generation, encryption, and decryption algorithms of our ABSC scheme are respectively $2n + 2n' + 3$, $2n' + 4q'$, $2n + 4q + 1$, $n' + 3q' + 6$, and $n + 3q + 5$. We emphasize that our ABSC construction supports arbitrary polynomial-size circuits of unbounded fan-out, whereas, all the earlier constructions could support at most circuits of fan-out one.

5   Conclusion

In this work, we designed an ABE scheme followed by an ABSC scheme both supporting general circuit realizable access policies. Our constructions were proven selectively secure under Multilinear Decisional Diffie-Hellman and Multilinear Computational Diffie-Hellman assumptions. The ciphertext sizes of both our constructions are very short. Most importantly, our ABSC scheme is the first to support signing and decryption policies representable as arbitrary polynomial-size circuits which are highly expressive.

References

1. Attrapadung, N.: Fully secure and succinct attribute based encryption for circuits from multi-linear maps. Tech. rep., IACR Cryptology ePrint Archive, 2014/772 (2014)
2. Boneh, D., Gentry, C., Gorbunov, S., Halevi, S., Nikolaenko, V., Segev, G., Vaikuntanathan, V., Vinayagamurthy, D.: Fully key-homomorphic encryption, arithmetic circuit ABE and compact garbled circuits. In: Nguyen, P.Q., Oswald, E. (eds.) EUROCRYPT 2014. LNCS, vol. 8441, pp. 533–556. Springer, Heidelberg (2014)
3. Coron, J.-S., Lepoint, T., Tibouchi, M.: Practical multilinear maps over the integers. In: Canetti, R., Garay, J.A. (eds.) CRYPTO 2013, Part I. LNCS, vol. 8042, pp. 476–493. Springer, Heidelberg (2013)
4. Coron, J.S., Lepoint, T., Tibouchi, M.: New multilinear maps over the integers. Tech. rep., IACR Cryptology ePrint Archive, 2015/162 (2015)
5. Gagné, M., Narayan, S., Safavi-Naini, R.: Threshold attribute-based signcryption. In: Garay, J.A., De Prisco, R. (eds.) SCN 2010. LNCS, vol. 6280, pp. 154–171. Springer, Heidelberg (2010)
6. Garg, S., Gentry, C., Halevi, S.: Candidate multilinear maps from ideal lattices. In: Johansson, T., Nguyen, P.Q. (eds.) EUROCRYPT 2013. LNCS, vol. 7881, pp. 1–17. Springer, Heidelberg (2013)
7. Garg, S., Gentry, C., Halevi, S., Sahai, A., Waters, B.: Attribute-based encryption for circuits from multilinear maps. In: Canetti, R., Garay, J.A. (eds.) CRYPTO 2013, Part II. LNCS, vol. 8043, pp. 479–499. Springer, Heidelberg (2013)
8. Garg, S., Gentry, C., Halevi, S., Zhandry, M.: Fully secure attribute based encryption from multilinear maps. Tech. rep., IACR Cryptology ePrint Archive, 2014/622 (2014)
9. Gorbunov, S., Vaikuntanathan, V., Wee, H.: Attribute-based encryption for circuits. In: Proceedings of the Forty-fifth Annual ACM Symposium on Theory of Computing, pp. 545–554. ACM (2013)
10. Hohenberger, S., Sahai, A., Waters, B.: Full domain hash from (leveled) multilinear maps and identity-based aggregate signatures. In: Canetti, R., Garay, J.A. (eds.) CRYPTO 2013, Part I. LNCS, vol. 8042, pp. 494–512. Springer, Heidelberg (2013)
11. Rao, Y.S., Dutta, R.: Expressive attribute based signcryption with constant-size ciphertext. In: Pointcheval, D., Vergnaud, D. (eds.) AFRICACRYPT. LNCS, vol. 8469, pp. 398–419. Springer, Heidelberg (2014)

12. Rao, Y.S., Dutta, R.: *Expressive* bandwidth-efficient attribute based signature and signcryption in standard model. In: Susilo, W., Mu, Y. (eds.) ACISP 2014. LNCS, vol. 8544, pp. 209–225. Springer, Heidelberg (2014)

13. Wang, C., Huang, J.: Attribute-based signcryption with ciphertext-policy and claim-predicate mechanism. In: Seventh International Conference on Computational Intelligence and Security-CIS 2011, pp. 905–909. IEEE (2011)

Dynamic Key-Aggregate Cryptosystem on Elliptic Curves for Online Data Sharing

Sikhar Patranabis[✉], Yash Shrivastava, and Debdeep Mukhopadhyay

Department of Computer Science and Engineering,
Indian Institute of Technology Kharagpur, Kharagpur, India
{sikhar.patranabis,yash.shrivastava,debdeep}@cse.iitkgp.ernet.in

Abstract. The recent advent of cloud computing and the IoT has made it imperative to have efficient and secure cryptographic schemes for online data sharing. Data owners would ideally want to store their data/files online in an encrypted manner, and delegate decryption rights for some of these to users with appropriate credentials. An efficient and recently proposed solution in this regard is to use the concept of aggregation that allows users to decrypt multiple classes of data using a single key of constant size. In this paper, we propose a secure and dynamic key aggregate encryption scheme for online data sharing that operates on elliptic curve subgroups while allowing dynamic revocation of user access rights. We augment this basic construction to a generalized two-level hierarchical structure that achieves optimal space and time complexities, and also efficiently accommodates extension of data classes. Finally, we propose an extension to the generalized scheme that allows use of efficiently computable bilinear pairings for encryption and decryption operations. Each scheme is formally proven to be semantically secure. Practical experiments have been conducted to validate all claims made in the paper.

Keywords: Key-aggregate cryptoystem · Online data sharing · Semantic security · Dynamic access rights

1 Introduction

The advent of cloud computing and the Internet of Things (IoT) has led to a massive rise in the demand for online data storage and data sharing services. Two very important paradigms that any data sharing service provider must ensure are privacy and flexibility. Since online data almost always resides in shared environments (for instance, multiple virtual machines running on the same physical device), ensuring privacy is a non trivial task. Current technology for secure data sharing comes in two major flavors - trusting a third party auditor [1] or using the user's own key to encrypt her data [2]. Figure 1 describes a realistic online data sharing set-up. Suppose a data owner stores multiple classes of encrypted data online with the intention of providing users decryption keys to one or more such ciphertext classes, based on their respective credentials. She might also wish to dynamically update the delegated access rights based on

© Springer International Publishing Switzerland 2015
A. Biryukov and V. Goyal (Eds.): INDOCRYPT 2015, LNCS 9462, pp. 25–44, 2015.
DOI: 10.1007/978-3-319-26617-6_2

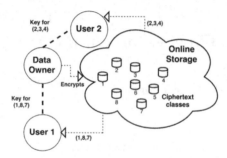

Fig. 1. Example of online data sharing

changes to the data/credibility issues. The challenge therefore is to provide her with a secure and efficient online data sharing scheme that allows updates to user access rights on the fly.

A näive (and extremely inefficient) solution is to have a different decryption key for each ciphertext class, and share them accordingly with users via secured channels. A more efficient proposition is the key-aggregate encryption (KAC) scheme proposed in [3] that combines the power of individual decryption keys, for ciphertext classes in a given subset, into a single key for that subset. This key is specific to the designated subset, meaning that it cannot be used to decrypt any ciphertext class outside that subset. KAC derives its roots from the seminal work by Boneh *et. al.* [4] that allows broadcasting of data (encrypted by the same public key) among multiple users, each of whom possess their own private keys for decryption. Both these schemes make use of bilinear mappings on multiplicative cyclic groups.

Contributions: In this paper, we propose a basic key-aggregate scheme on additive elliptic subgroups that delegate decryption rights to multiple ciphertext classes using a single constant sized key. The scheme is dynamic in nature, that is, it allows the data owner to revoke access rights of users without having to change the entire set-up, unlike in the existing KAC scheme. We then generalize this scheme into a two-level construction that allows flexible public key extension and maintains constant ciphertext size, while avoiding many of the pitfalls of earlier hierarchical schemes. We provide a formal proof of semantic security for the generalized scheme. We further extend the generalized scheme to allow using popular and efficiently implementable elliptic curve pairing schemes. We compare the time and space requirements of the proposed generalized scheme under various operating configurations. We also compare the performance of our proposed scheme, in terms of key size and resource utilization, with that of other existing schemes in literature.

Organization: The rest of the paper is organized as follows. Section 2 provides a brief overview of state of the art data sharing schemes. Section 3 introduces the notion of key aggregate cryptosystem, and provides a description of the complexity assumptions used to prove the semantic security of our proposed schemes. Our basic dynamic key-aggregate scheme is presented in Sect. 4. We follow up with a more generalized two-tiered construction of the scheme for efficient public key extension in Sect. 5, and prove its semantic security. A further extension for the generalized scheme that allows using efficiently implementable pairings is introduced and proved semantically secure in Sect. 6. Experimental results using Tate pairings based implementations of the extended scheme are presented in Sect. 7. Finally Sect. 8 concludes the paper.

2 Related Work

In this section we present a brief overview of public and private key cryptographic schemes in literature for secure online data sharing. While many of them focus on key aggregation in some form or the other, very few have the ability to provide constant size keys to decrypt an arbitrary number of encrypted entities. One of the most popular techniques for access control in online data storage is to use a pre-defined hierarchy of secret keys [5] in the form of a tree-like structure, where access to the key corresponding to any node implicitly grants access to all the keys in the subtree rooted at that node. A major disadvantage of hierarchical encryption schemes is that granting access to only a selected set of branches within a given subtree warrants an increase in the number of granted secret keys. This in turn blows up the size of the key shared. Compact key encryption for the symmetric key setting has been used in [6] to solve the problem of concisely transmitting large number of keys in the broadcast scenario. However, symmetric key sharing via a secured channel is costly and not always practically viable for many applications on the cloud. Proxy re-encryption is another technique to achieve fine-grained access control and scalable user revocation in unreliable clouds [7]. However, proxy re-encryption essentially transfers the responsibility for secure key storage from the delegatee to the proxy and is susceptible to collusion attacks. It is also important to ensure that the transformation key of the proxy is well protected, and every decryption would require a separate interaction with the proxy, which is inconvenient for applications on the cloud.

The authors of [3] proposes an efficient scheme, namely KAC, that allows secure and efficient sharing of data on the cloud. The scheme is a public-key cryptosystem that uses constant size ciphertexts such that efficient delegation of decryption rights for any set of ciphertexts are possible. When a user demands for a particular subset of the available classes of data, the data owner computes an aggregate key which integrates the power of the individual decryption keys corresponding to each class of data. KAC as proposed in [3] suffers from three major drawbacks, each of which we address in this paper. First of all, the security assumption of KAC seems to be the Bilinear Diffie Hellman Exponent (BDHE) assumption [8]; however no concrete proofs of semantic security are provided by the authors in [3]. Secondly, with respect to user access rights, KAC is a static scheme in the sense that once a user is in possession of the aggregate key corresponding to a subset of files from data owner, the owner cannot dynamically revoke the permission of the client for accessing one or more updated files. Since dynamic changes in access rights is extremely common in online data storage, this scenario needs to be tackled. Finally, the public key extension of KAC proposed in [3] is extremely cumbersome and resource consuming since registration of each new public key-private key pair requires the number of classes to be extended by the original number of classes.

3   Preliminaries

We begin by formally defining the Key Aggregate Cryptosystem (KAC), and stating the complexity assumptions used to prove the security of the encryption schemes proposed in this paper.

3.1   The Key Aggregate Cryptosystem (KAC)

A key aggregate cryptosystem is an ensemble of the following randomized algorithms:

1. **Setup**$(1^\lambda, n)$: Takes as input the number of ciphertext classes n and the group order parameter λ. Outputs the public parameter PK. Also computes a secret parameter t used for encryption which is not made public. It is only known to data owners with credentials to control client access rights.
2. **Keygen**(): Outputs the public and master-secret key pair:
   $(PK = \gamma P, msk = \gamma)$.
3. **Encrypt**(PK, i, m): Takes as input the public key parameter PK, the ciphertext class i and the message m. Outputs the ciphertext $\mathcal{C}$ corresponding to the message m belonging to class i.
4. **Extract**$(msk = \gamma, \mathcal{S})$: Takes as input the master secret key γ and a subset $\mathcal{S} \subset \{1, 2, \cdots, n\}$. Computes the aggregate key $K_\mathcal{S}$ and the dynamic access control parameter U. The tuple $(K_\mathcal{S}, U)$ is transmitted via a secure channel to users that have access rights to $\mathcal{S}$.
5. **Decrypt**$(K_\mathcal{S}, U, \mathcal{S}, i, \mathcal{C} = \{c_1, c_2, c_3\})$: Takes as input the aggregate key $K_\mathcal{S}$ corresponding to a subset $\mathcal{S} \subset \{1, 2, \cdots, n\}$, the dynamic access parameter U, the ciphertext class i and the ciphertext $\mathcal{C}$. Outputs the decrypted message m.

3.2   Semantic Security of KAC

We now define the semantic security of a key-aggregate encryption system against an adversary using the following game between an attack algorithm $\mathcal{A}$ and a challenger $\mathcal{B}$. Both $\mathcal{A}$ and $\mathcal{B}$ are given n, the total number of ciphertext classes, as input. The game proceeds through the following stages.

1. **Init:** Algorithm $\mathcal{A}$ begins by outputting a set $\mathcal{S} \subset \{1, 2, \cdots, n\}$ of receivers that it wishes to attack. For each ciphertext class $i \in \mathcal{S}$, challenger $\mathcal{B}$ performs the **SetUp-i**, **Challenge-i** and **Guess-i** steps. Note that the number of iterations is polynomial in $|S|$.

2. **SetUp-i**: Challenger $\mathcal{B}$ generates the public *param*, public key PK, the access parameter U, and provides them to $\mathcal{A}$. In addition, $\mathcal{B}$ also generates and furnishes $\mathcal{A}$ with the aggregate key $K_{\overline{S}}$ that allows $\mathcal{A}$ to decrypt any ciphertext class $j \notin \mathcal{S}$.

3. **Challenge-i**: Challenger $\mathcal{B}$ performs an encryption of the secret message m_i belonging to the ith class to obtain the ciphertext $\mathcal{C}$. Next, $\mathcal{B}$ picks a random $b \in (0,1)$. It sets $K_b = m_i$ and picks a random K_{1-b} from the set of possible plaintext messages. It then gives $(\mathcal{C}, K_0, K_1)$ to algorithm $\mathcal{A}$ as a challenge.

4. **Guess-i**: The adversary $\mathcal{A}$ outputs a guess b' of b. If $b' = b$, $\mathcal{A}$ wins and the challenger $\mathcal{B}$ loses. Otherwise, the game moves on to the next ciphertext class in $\mathcal{S}$ until all ciphertext classes in $\mathcal{S}$ are exhausted.

If the adversary $\mathcal{A}$ fails to predict correctly for all ciphertext classes in $\mathcal{S}$, only then $\mathcal{A}$ loses the game. Let $AdvKAC_{\mathcal{A},n}$ denote the probability that $\mathcal{A}$ wins the game when the challenger is given n as input. We say that a key-aggregate encryption system is (τ, ϵ, n) semantically secure if for all τ-time algorithms $\mathcal{A}$ we have that $|AdvKAC_{\mathcal{A},n} - \frac{1}{2}| < \epsilon$ where ϵ is a very small quantity. Note that the adversary $\mathcal{A}$ is non-adaptive; it chooses $\mathcal{S}$, and obtains the aggregate decryption key for all ciphertext classes outside of $\mathcal{S}$, before it even sees the public parameters *param* or the public key PK.

3.3   The Complexity Assumptions

We now introduce the complexity assumptions used in this paper. In this section, we make several references to bilinear non-degenerate mappings on elliptic curve sub-groups, popularly known in literature as pairings. For a detailed descriptions on pairings and their properties, refer [9].

The First Complexity Assumption: Our first complexity assumption is the l-BDHE problem [4] in a bilinear elliptic curve subgroup $\mathbb{G}$, defined as follows. Given a vector of $2l+1$ elements $(H, P, \alpha P, \alpha^2 P, \cdots, \alpha^l P, \alpha^{l+2} P \cdots, \alpha^{2l} P) \in \mathbb{G}^{2l+1}$ as input, and a bilinear pairing $\hat{e}' : \mathbb{G}_1 \times \mathbb{G}_1 \longrightarrow \mathbb{G}_T$ output $\hat{e}'(P, H)^{\alpha^{l+1}} \in \mathbb{G}_T$. Since $\alpha^{l+1} P$ is not an input, the bilinear pairing is of no real use in this regard. Using the shorthand $P_i = \alpha^i P$, an algorithm $\mathcal{A}$ is said to have an advantage ϵ in solving l-BDHE if $Pr[\mathcal{A}(H, P, P_1, P_2, \cdots, P_l, P_{l+2} \cdots, P_{2l}) = \hat{e}'(P_{l+1}, H)] \geq \epsilon$, where the probability is over the random choice of $H, P \in \mathbb{G}$, random choice of $\alpha \in \mathbb{Z}_q$ and random bits used by $\mathcal{A}$. The decisional version of l-BDHE for elliptic curve subgroups may be analogously defined. Let $Y_{(P,\alpha,l)} = (P_1, P_2, \cdots, P_l, P_{l+2} \cdots, P_{2l})$. An algorithm $\mathcal{B}$ that outputs $b \in \{0,1\}$ has advantage ϵ in solving decisional l-BDHE in $\mathbb{G}$ if $|Pr[\mathcal{B}(P, H, Y_{(P,\alpha,l)}, \hat{e}'(P_{l+1}, H)) = 0] - Pr[\mathcal{B}(P, H, Y_{(P,\alpha,l)}, T) = 0]| \geq \epsilon$, where the probability is over the random choice of $H, P \in \mathbb{G}$, random choice of $\alpha \in \mathbb{Z}_q$, random choice of $T \in \mathbb{G}_T$ and random bits used by $\mathcal{B}$. We refer to the left and right probability distributions as L-BDHE and R-BDHE respectively. Thus, it can be said that the decision (τ, ϵ, l)-BDHE assumption for elliptic curves holds in $\mathbb{G}$ if no τ-time algorithm has advantage ϵ in solving the decisional l-BDHE problem over elliptic curve subgroup $\mathbb{G}$.

The Second Complexity Assumption: We next define the (l, l)-BDHE problem over a pair of equi-prime order bilinear elliptic curve subgroups $\mathbb{G}_1$ with generator P and $\mathbb{G}_2$ with generator Q. Given a vector of $3l + 2$ elements $(H, P, Q, \alpha P, \alpha^2 P, \cdots, \alpha^l P, \alpha^{l+2} P \cdots, \alpha^{2l} P, \alpha Q, \alpha^2 Q, \cdots, \alpha^l Q)$ as input, where P and $\alpha^i P \in \mathbb{G}_1$ and $H, Q, \alpha^i Q \in \mathbb{G}_2$, along with a bilinear pairing $\hat{e}'' : \mathbb{G}_1 \times \mathbb{G}_2 \longrightarrow \mathbb{G}_T$, output $\hat{e}'(P, H)^{\alpha^{l+1}} \in \mathbb{G}_T$. Since $\alpha^{l+1} P$ is not an input, the bilinear pairing is of no real use in this regard. Using the shorthand $P_i = \alpha^i P$ and $Q_i = \alpha^i Q$, an algorithm $\mathcal{A}$ is said to have an advantage ϵ in solving (l, l)-BDHE if $Pr[\mathcal{A}(H, P, Q, P_1, P_2, \cdots, P_l, P_{l+2} \cdots, P_{2l}, Q_1, \cdots, Q_l) = \hat{e}'(P_{l+1}, H)] \geq \epsilon$ where the probability is over the random choice of $P \in \mathbb{G}_1$, $H, Q \in \mathbb{G}_2$, random choice of $\alpha \in \mathbb{Z}_q$ and random bits used by $\mathcal{A}$. We may also define the decisional (l, l)-BDHE problem over elliptic curve subgroup pairs as follows. Let $Y_{(P,\alpha,l)} = (P_1, P_2, \cdots, P_l, P_{l+2} \cdots, P_{2l})$ and $Y'_{(Q,\alpha,l)} = (Q_1, Q_2, \cdots, Q_l)$. Also let H be a random element in $\mathbb{G}_2$. An algorithm $\mathcal{B}$ that outputs $b \in \{0, 1\}$ has advantage ϵ in solving decisional (l, l)-BDHE if $|Pr[\mathcal{B}(P, Q, H, Y_{(P,\alpha,l)}, Y'_{(Q,\alpha,l)}, \hat{e}'(P_{l+1}, H)) = 0] - Pr[\mathcal{B}(P, Q, H, Y_{(P,\alpha,l)}, Y'_{(Q,\alpha,l)}, T) = 0]| \geq \epsilon$, where the probability is over the random choice of $P \in \mathbb{G}_1$, $H, Q \in \mathbb{G}_2$, random choice of $\alpha \in \mathbb{Z}_q$, random choice of $T \in \mathbb{G}_T$ and random bits used by $\mathcal{B}$. We refer to the left and right probability distributions as L'-BDHE and R'-BDHE respectively. Thus, it can be said that the decision (τ, ϵ, l, l)-BDHE assumption for elliptic curves holds in $(\mathbb{G}_1, \mathbb{G}_2)$ if no τ-time algorithm has advantage ϵ in solving the decisional (l, l)-BDHE problem over elliptic curve subgroups $\mathbb{G}_1$ and $\mathbb{G}_2$. To the best of our knowledge, the (l, l)-BDHE problem has not been introduced in literature before.

Proving the Validity of the Second Complexity Assumption: We prove here that the decision (τ, ϵ, l, l)-BDHE assumption for elliptic curves holds in equi-prime order subgroups $(\mathbb{G}_1, \mathbb{G}_2)$ if the decision (τ, ϵ, l)-BDHE assumption for elliptic curves holds in $\mathbb{G}_1$. Let $\hat{e}' : \mathbb{G}_1 \times \mathbb{G}_1 \longrightarrow \mathbb{G}_T$ and $\hat{e}'' : \mathbb{G}_1 \times \mathbb{G}_2 \longrightarrow \mathbb{G}_T$ be bilinear pairings. Also, let P and Q are the generators for $\mathbb{G}_1$ and $\mathbb{G}_2$ respectively. We first make the following observation.

Observation 1: Since G_1 and G_2 have the same prime order (say q), there exists a bijection $\varphi : \mathbb{G}_1 \longrightarrow \mathbb{G}_2$ such that $\varphi(aP) = aQ$ for all $a \in \mathbb{Z}_q$. Similarly, since $\mathbb{G}_T$ also has order q, there also exists a mapping $\phi : \mathbb{G}_T \longrightarrow \mathbb{G}_T$ such that $\phi(\hat{e}'(H_1, H_2)) = \hat{e}''(H_1, \varphi(H_2))$ for all $H_1, H_2 \in \mathbb{G}_1$.

Let $\mathcal{A}$ be a τ-time adversary that has advantage greater than ϵ in solving the decision (l, l)-BDHE problem over equi-prime order subgroups $(\mathbb{G}_1, \mathbb{G}_2)$. We build an algorithm $\mathcal{B}$ that has advantage at least ϵ in solving the l-BDHE problem in $\mathbb{G}_1$. Algorithm $\mathcal{B}$ takes as input a random l-BDHE challenge $(P, H, Y_{(P,\alpha,l)}, Z)$ where Z is either $\hat{e}'(P_{l+1}, H)$ or a random value in $\mathbb{G}_T$. $\mathcal{B}$ computes $Y'_{Q,\alpha,l}$ by setting $Q_i = \varphi(P_i)$ for $i = 1, 2, \cdots, l$. $\mathcal{B}$ also computes $H' = \varphi(H) \in \mathbb{G}_2$ and $Z' = \phi(Z) \in \mathbb{Z}$. Then randomly chooses a bit $b \in (0, 1)$ and sets T_b as Z' and T_{1-b} as a random element in $\mathbb{G}_T$. The challenge given to $\mathcal{A}$ is $((P, Q, H', Y_{(P,\alpha,l)}, Y'_{Q,\alpha,l}), T_0, T_1)$. Quite evidently, when $Z = \hat{e}'(P_{l+1}, H)$ (i.e. the input to $\mathcal{B}$ is a l-BDHE tuple), then $((P, Q, H', Y_{(P,\alpha,l)}, Y'_{Q,\alpha,l}), T_0, T_1)$ is a valid challenge to A. This is because in such a case, $T_b = Z' = \phi(Z) = \phi(\hat{e}'(P_{l+1}, H)) = \hat{e}''(P_{l+1}, H')$. On the other hand, if Z is a random element in $\mathbb{G}_T$ (i.e. the input to $\mathcal{B}$ is a random tuple), then T_0 and T_1 are just random independent elements of $\mathbb{G}_T$.

Now, A outputs a guess b' of b. If $b' = b$, $\mathcal{B}$ outputs 0 (indicating that $Z = \hat{e}'(P_{l+1}, H)$). Otherwise, it outputs 1 (indicating that Z is random in $\mathbb{G}_T$). A simple analysis reveals that if $(P, H, Y_{(P,\alpha,l)}, Z)$ is sampled from R-BDHE, $Pr[\mathcal{B}(G, H, Y_{(P,\alpha,l)}, Z) = 0] = \frac{1}{2}$, while if $(P, H, Y_{(P,\alpha,l)}, Z)$ is sampled from L-BDHE, $|Pr[\mathcal{B}(G, H, Y_{(P,\alpha,l)}, Z)] - \frac{1}{2}| \geq \epsilon$. So, the probability that $\mathcal{B}$ outputs correctly is at least ϵ, which in turn implies that $\mathcal{B}$ has advantage at least ϵ in solving the l-BDHE problem. This concludes the proof.

4   The Proposed Dynamic Key-Aggregate Cryptosystem: The Basic Case

In this section, we present the design of our proposed dynamic key-aggregate storage scheme on additive elliptic curve subgroups assuming that there are n ciphertext classes. Our scheme ensures that the ciphertext and aggregate key are of constant size, while the public parameter size is linear in the number of ciphertext classes. Unlike the scheme proposed in [3], the proposed scheme allows dynamic revocation of user access rights without having to massively change the system parameters. We also present a proof of security for the proposed scheme.

4.1   The Basic Construction of Dynamic KAC

Let $\mathbb{G}$ be an additive cyclic elliptic curve subgroup of prime order q, where $2^\lambda \leq q \leq 2^{\lambda+1}$, such that the point P is a generator for $\mathbb{G}$. Also, let $\mathbb{G}_T$ be a multiplicative group of order q with identity element 1. We assume that there exists an efficiently computable bilinear pairing $\hat{e}' : \mathbb{G} \times \mathbb{G} \longrightarrow \mathbb{G}_T$. We now present the basic construction of our proposed key-aggregate encryption scheme.

The scheme consists of the following five phases.

1. **Setup**$(1^\lambda, n)$: Randomly pick $\alpha \in \mathbb{Z}_q$. Compute $P_i = \alpha^i P \in \mathbb{G}$ for $i = 1, \cdots, n, n+2, \cdots, 2n$. Output the system parameter as
   $param = (P, P_1, \cdots, P_n, P_{n+2}, \cdots, P_{2n})$. The system also randomly chooses a secret parameter $t \in \mathbb{Z}_q$ which is not made public. It is only known to data owners with credentials to control client access rights.
2. **Keygen**(): Pick $\gamma \in \mathbb{Z}_q$, output the public and master-secret key pair: $(PK = \gamma P, msk = \gamma)$.
3. **Encrypt**(PK, i, m): For a message $m \in \mathbb{G}_T$ and an index $i \in \{1, 2, \cdots, n\}$, randomly choose $r \in \mathbb{Z}_q$ and let $t' = t + r \in \mathbb{Z}_q$. Then the ciphertext is computed as
   $\mathcal{C} = (rP, t'(PK + P_i), m.\hat{e}'(P_n, t'P_1)) = (c_1, c_2, c_3)$

4. **Extract**$(msk = \gamma, \mathcal{S})$: For the set $\mathcal{S}$ of indices j the aggregate key is computed as
$K_{\mathcal{S}} = \sum_{j \in \mathcal{S}} \gamma P_{n+1-j} = \sum_{j \in \mathcal{S}} \alpha^{n+1-j} PK$
and the dynamic access control parameter U is computed as tP. Thus the net aggregate key is $(K_{\mathcal{S}}, U)$ which is transmitted via a secure channel to users that have access rights to $\mathbb{S}$.

5. **Decrypt**$(K_{\mathcal{S}}, U, \mathcal{S}, i, \mathcal{C} = \{c_1, c_2, c_3\})$: If $i \notin \mathcal{S}$, output $\perp$. Otherwise return the message $\hat{m} = c_3 \hat{e'}(K_{\mathcal{S}} + \sum_{j \in \mathcal{S}, j \neq i} P_{n+1-j+i}, U + c_1)/(\hat{e'}(\sum_{j \in \mathcal{S}} P_{n+1-j}, c_2))$.

The proof of correctness of this scheme is presented below.

$$\hat{m} = c_3 \frac{\hat{e'}(K_{\mathcal{S}} + \sum_{j \in \mathcal{S}, j \neq i} P_{n+1-j+i}, U + c_1)}{\hat{e'}(\sum_{j \in \mathcal{S}} P_{n+1-j}, c_2)}$$

$$= c_3 \frac{\hat{e'}(\sum_{j \in \mathcal{S}} \gamma P_{n+1-j}, t'P) \hat{e'}(\sum_{j \in \mathcal{S}} (P_{n+1-j+i}) - P_{n+1}, t'P)}{\hat{e'}(\sum_{j \in \mathcal{S}} P_{n+1-j}, t'PK) \hat{e'}(\sum_{j \in \mathcal{S}} P_{n+1-j}, t'P_i))}$$

$$= c_3 \frac{\hat{e'}(\sum_{j \in \mathcal{S}} P_{n+1-j+i}, t'P)}{\hat{e'}(P_{n+1}, t'P) \hat{e'}(\sum_{j \in \mathcal{S}} P_{n+1-j+i}, t'P))}$$

$$= m$$

4.2   Dynamic Access Control

An important aspect of the proposed scheme is the fact that it allows the data owner to dynamically update user access permissions. In KAC [3], once the data owner issues an aggregate key corresponding to a set of ciphertext classes to a user, revoking the user's access permissions to the same is not possible without changing the master secret key. However, changing the master secret key each time an user's access privileges to a ciphertext class need to be updated, is a very expensive option and may not be practically feasible. Our scheme, on the other hand, offers a solution to this problem by allowing the data owner to dynamically update user access privileges.

We achieve this by making the parameter $U = tP$ a part of the aggregate key in our proposed scheme and not a part of the ciphertext. The user must have the correct value of U in possession to be able to decrypt any encrypted ciphertext class in the subset $\mathcal{S}$. Now suppose the data owner wishes to alter the access rights to the subset $\mathcal{S}$. She can simply re-encrypt all ciphertexts in that class using a different random element $\hat{t} \in \mathbb{Z}_q$, and then provide the updated dynamic access parameter $\hat{U} = \hat{t}P$ to only those users who she wishes to delegate access to. The decrypted value will give the correct message m only if the same t is used for both encryption and decryption. This is a major difference between our scheme and the scheme proposed in [3], where the knowledge of the random parameter was only embedded as part of the ciphertext itself, and could not be used to control access rights of users. Moreover, since U is of constant size and needs to be transmitted only when changed (and not for every encryption), there is no significant degradation in performance.

4.3   Performance and Efficiency

The decryption time for any subset of ciphertext classes S is essentially dominated by the computation of $W_S = \sum_{j \in S} P_{n+1-j+i}$. However, if a user has already computed $\sum_{j \in S'} P_{n+1-j+i}$ for a subset S' similar to S, then she can easily compute the desired value by at most $|S - S'|$ operations. For similar subsets S and S', this value is expected to be fairly small. A suggested in [4], for subsets of very large size $(n - r, r \ll n)$, an advantageous approach could be to pre-compute $\sum_{j=1}^{j=n} P_{n+1-j+i}$ corresponding to $i = 1$ to n, which would allow the user to decrypt using only r group operations, and would require only r elements of *param*. Similar optimizations would also hold for the encryption operation where pre-computation of $\sum_{j=1}^{j=n} P_{n+1-j}$ is useful for large subsets.

It is important to note that our proposed scheme fixes the number of ciphertext classes beforehand, thus limiting the scope for ciphertext class extension. The only way to increase the number of classes is to change the public key parameters, which would therefore require some kind of administrative privileges, and cannot be done by an user for her own purposes. However, in online data sharing environments, users may wish to register their own public key-private key pairs for new ciphertext classes according to their own requirements. Such an extension to the scheme would make extremely convenient and attractive to potential users. A proposal made in [3] recommends that the user be allowed to register new public-private key pairs, at the cost of increasing the number of ciphertext classes by n each time. This is both impractical and wasteful. In the next section, we present a two-tier generalization of our scheme that tackles this issue in a more economical fashion. We avoid a separate proof of semantic security for the base case presented here, since the proof is a special case of the proof for the generalized scheme presented in the next section.

5   A Generalized Version of Dynamic KAC

In this section, we focus on building an efficiently extensible version of our proposed scheme that allows an user to economically increase the number of ciphertext classes while registering a new public key-private key pair. We adopt the idea presented in [4] to develop a hierarchical structure that has multiple instances (say n_1) of the original scheme running in parallel. Each such instance in turn provides *locally aggregate keys* for n_2 ciphertext sub-classes. Each ciphertext class thus now has a double index (i_1, i_2) where $1 \leq i_1 \leq n_1$ and $1 \leq i_2 \leq n_2$. This allows the overall setup to handle $n = n_1 n_2$ classes. However, it is important to note that all the instances can use the same public parameters. This interaction among the instances helps to largely improve performance. We further point out that while in [4], the generalized construction offers a trade-off between the public parameter size and the ciphertext size, our generalized scheme actually reduces the public parameter size without compromising on the size of the ciphertext. Further, addition of a single new key increases the number of classes only by n_2 and not by n. Setting $n_2 \ll n$ thus achieves significant improvement in performance over the existing proposal.

5.1 The Construction of the Generalized KAC

Let n_2 be a fixed positive integer. Our proposed n_2-generalized key-aggregate encryption scheme over elliptic curve subgroups is as described below. It may be noted that the bilinear additive elliptic curve sub-group $\mathbb{G}$ and the multiplicative group $\mathbb{G}_T$, as well as the pairing $\hat{e}'$ are the same as in the basic scheme. The algorithm sets up $n_1 = \lfloor n/n_2 \rfloor$ instances of the basic scheme, each of which handles n_2 ciphertext classes. The original scheme is thus a special case of the extended scheme with $n_1 = 1$ and $n_2 = n$.

1. **Setup**$(1^\lambda, n_2)$: Randomly pick $\alpha \in \mathbb{Z}_q$. Compute $P_i = \alpha^i P \in \mathbb{G}$ for $i = 1, \cdots,$ $n_2, n_2 + 2, \cdots, 2n_2$. Output the system parameter as $param = (P, P_1, \cdots, P_{n_2}, P_{n_2+2}, \cdots, P_{2n_2})$. The system randomly chooses a secret parameter $t \in \mathbb{Z}_q$ which is not made public. It is only known to data owners with credentials to control client access rights.
2. **Keygen**(): Pick $\gamma_1, \gamma_2, \cdots, \gamma_{n_1} \in \mathbb{Z}_q$, output the public and master-secret key pair:
   $(PK=(pk_1, pk_2, \cdots, pk_{n_1}) = (\gamma_1 P, \gamma_2 P, \cdots, \gamma_{n_1} P), msk=(\gamma_1, \gamma_2, \cdots, \gamma_{n_1}))$.
3. **Encrypt**$(pk_{i_1}, (i_1, i_2), m)$: For a message $m \in \mathbb{G}_T$ and an index $(i_1, i_2) \in \{1, 2, \cdots, n_1\} \times \{1, 2, \cdots, n_2\}$, randomly choose $r \in \mathbb{Z}_q$ and let $t' = t + r \in \mathbb{Z}_q$. Then compute the ciphertext $\mathcal{C}=(rP, t'(pk_{i_1} + P_{i_2}), m.\hat{e}'(P_{n_2}, t'P_1)) = (c_1, c_2, c_3)$.
4. **Extract**$(msk = \gamma, \mathcal{S})$: For the set $\mathcal{S}$ of indices (j_1, j_2) the aggregate key is computed as $K_\mathcal{S} = (k_\mathcal{S}^1, k_\mathcal{S}^2, \cdots, k_\mathcal{S}^{n_1}) =$
   $(\sum_{(1,j_2)\in\mathcal{S}} \gamma_1 P_{n_2+1-j_2}, \sum_{(2,j_2)\in\mathcal{S}} \gamma_2 P_{n_2+1-j_2}, \cdots, \sum_{(n_1,j_2)\in\mathcal{S}} \gamma_{n_1} P_{n_2+1-j_2})$
   and the dynamic access control parameter U is computed as tP. Thus the net aggregate key is $(K_\mathcal{S}, U)$ which is transmitted via a secure channel to users that have access rights to $\mathcal{S}$. Note that $k_\mathcal{S}^{j_1} = \sum_{(j_1,j_2)\in\mathcal{S}} \alpha^{n+1-j_2} pk_{j_1}$ for $j_1 = 1, 2, \cdots, n_1$.
5. **Decrypt**$(K_\mathcal{S}, U, \mathcal{S}, (i_1, i_2), \mathcal{C} = \{c_1, c_2, c_3\})$: If $(i_1, i_2) \notin \mathcal{S}$, output $\bot$. Otherwise return the message
   $$\hat{m} = c_3 \frac{\hat{e}'(k_\mathcal{S}^{i_1} + \sum_{(i_1,j_2)\in\mathcal{S}, j_2\neq i_2} P_{n_2+1-j_2+i_2}, U+c_1)}{\hat{e}'(\sum_{(i_1,j_2)\in\mathcal{S}} P_{n_2+1-j_2}, c_2)}.$$

The proof of correctness for the generalized scheme is very similar to that for the basic scheme.

5.2 Semantic Security of the Generalized KAC

The Reduced Generalized Scheme: We define a reduced version of the generalized encryption scheme. We note that the ciphertext $\mathcal{C} = (c_1, c_2, c_3)$ output by the *Encypt* operation essentially embeds the value of m in c_3 by multiplying it with $\hat{e}'(P_{n_2}, tP_1)$. Consequently, the security of our proposed scheme is equivalent to that of a *reduced* generalized key-aggregate encryption scheme that simply uses the reduced ciphertext (c_1, c_2), the aggregate key $K_\mathcal{S}$ and the dynamic access parameter U to successfully transmit and decrypt the value of $\hat{e}'(P_{n_2}, t'P_1) = \hat{e}'(P_{n_2+1}, t'P)$. We prove the semantic security of this *reduced* *scheme* parameterized with a given number of ciphertext classes n_2 for each instance, which also amounts to proving the semantic security of our original encryption scheme for the same number of ciphertext classes. Note that the proof of security is independent of the number of instances n_1 that run in parallel.

The Adversarial Model: We make the following assumptions about the adversary $\mathcal{A}$:

1. The adversary has the aggregate key that allows her to access any ciphertext class other than those in the target subset $\mathcal{S}$, that is, she possesses $K_{\overline{\mathcal{S}}}$.
2. The adversary has access to the public parameters *param* and PK, and also possesses the dynamic access parameter U.

The Security Proof: The security proof presented here uses the first complexity assumption stated in Sect. 3.3 (The First Complexity Assumption). Let $\mathbb{G}$ be a bilinear elliptic curve subgroup of prime order q and G_T be a multiplicative group of order q. Let $\hat{e}' : \mathbb{G} \times \mathbb{G} \longrightarrow \mathbb{G}_T$ be a bilinear non-degenerate pairing. For any pair of positive integers $n_2, n'(n' > n_2)$ our proposed n_2-generalized reduced key-aggregate encryption scheme over elliptic curve subgroups is (τ, ϵ, n') semantically secure if the decision (τ, ϵ, n_2)-BDHE assumption holds in $\mathbb{G}$. We now prove this statement below.

Proof: Let for a given input n', $\mathcal{A}$ be a τ-time adversary that has advantage greater than ϵ for the *reduced scheme* parameterized with a given n_2. We build an algorithm $\mathcal{B}$ that has advantage at least ϵ in solving the n_2-BDHE problem in $\mathbb{G}$. Algorithm $\mathcal{B}$ takes as input a random n_2-BDHE challenge $(P, H, Y_{(P,\alpha,n_2)}, Z)$ where Z is either $\hat{e}'(P_{n_2+1}, H)$ or a random value in $\mathbb{G}_T$. Algorithm $\mathcal{B}$ proceeds as follows.

1. **Init:** Algorithm $\mathcal{B}$ runs $\mathcal{A}$ and receives the set $\mathcal{S}$ of ciphertext classes that $\mathcal{A}$ wishes to be challenged on. For each ciphertext class $(i_1, i_2) \in \mathcal{S}$, $\mathcal{B}$ performs the **SetUp-(i_1, i_2)**, **Challenge-(i_1, i_2)** and **Guess-(i_1, i_2)** steps. Note that the number of iterations is polynomial in $|\mathcal{S}|$.
2. **SetUp-(i_1, i_2):** $\mathcal{B}$ should generate the public *param*, public key PK, the access parameter U, and the aggregate key $K_{\overline{\mathcal{S}}}$. For the iteration corresponding to ciphertext class (i_1, i_2), they are generated as follows.
   - *param* is set as (P, Y_{P,α,n_2}).
   - Randomly generate $u_1, u_2, \cdots, u_{n_1} \in \mathbb{Z}_q$. Then, set $PK = (pk_1, pk_2, \cdots, pk_{n_1})$, with $pk_{j_1} = u_{j_1}P - P_{i_2}$ for $j_1 = 1, 2, \cdots, n_1$.
   - Set $K_{\overline{\mathcal{S}}} = (k_{\overline{\mathcal{S}}}^1, k_{\overline{\mathcal{S}}}^2, \cdots, k_{\overline{\mathcal{S}}}^{n_1})$, where $k_{\overline{\mathcal{S}}}^{j_1}$ is set as $\sum_{(j_1,j_2) \notin \mathcal{S}} (u_{j_1} P_{n_2+1-j_2} - (P_{n_2+1-j_2+i_2}))$. Then, $k_{\overline{\mathcal{S}}}^{j_1} = \sum_{(j_1,j_2) \notin \mathcal{S}} \alpha^{n_2+1-j_2} pk_{j_1}$, which is as per the scheme specification. Note that $\mathcal{B}$ knows that $(i_1, i_2) \notin \overline{\mathcal{S}}$, and hence has all the resources to compute this aggregate key for $\overline{\mathcal{S}}$.
   - U is set as some random element in $\mathbb{G}$.
   Note that since P, α, U and the u_{j_1} values are chosen uniformly at random, the public key has an identical distribution to that in the actual construction.

3. **Challenge-(i_1, i_2):** To generate the challenge for the ciphertext class (i_1, i_2), $\mathcal{B}$ computes (c_1, c_2) as $(H - U, u_{i_1} H)$. It then randomly chooses a bit $b \in (0, 1)$ and sets K_b as Z and K_{1-b} as a random element in $\mathbb{G}_T$. The challenge given to $\mathcal{A}$ is $((c_1, c_2), K_0, K_1)$.

   We claim that when $Z = \hat{e'}(P_{n_2+1}, H)$ (i.e. the input to $\mathcal{B}$ is a n_2-BDHE tuple), then $((c_1, c_2), K_0, K_1)$ is a valid challenge to A. We prove this claim here. we point out that P is a generator of $\mathbb{G}$ and so $H = t'P$ for some $t' \in \mathbb{Z}_q$. Putting H as $t'P$ gives us the following:

   - $U = tP$ for some $t \in \mathbb{Z}_q$
   - $c_1 = H - U = (t' - t)P = rP$ for $r = t' - t$
   - $c_2 = u_{i_1} H = (u_{i_1})t'P = t'(u_{i_1} P) = t'(u_{i_1} P - P_{i_2} + P_{i_2}) = t'(pk_{i_1} + P_{i_2})$
   - $K_b = Z = \hat{e'}(P_{n_2+1}, H) = \hat{e'}(P_{n_2+1}, t'P)$

   On the other hand, if Z is a random element in $\mathbb{G}_T$ (i.e. the input to $\mathcal{B}$ is a random tuple), then K_0 and K_1 are just random independent elements of $\mathbb{G}_T$.

4. **Guess-(i_1, i_2):** The adversary $\mathcal{A}$ outputs a guess b' of b. If $b' = b$, $\mathcal{B}$ outputs 0 (indicating that $Z = \hat{e'}(P_{n_2+1}, H)$), and terminates. Otherwise, it goes for the next ciphertext class in $\mathcal{S}$.

If after $|\mathcal{S}|$ iterations, $b' \neq b$ for each ciphertext class $(i_1, i_2) \in \mathcal{S}$, the algorithm $\mathcal{B}$ outputs 1 (indicating that Z is random in $\mathbb{G}_T$). We now analyze the probability that $\mathcal{B}$ gives a correct output. If $(P, H, Y_{(P,\alpha,n_2)}, Z)$ is sampled from R-BDHE, $Pr[\mathcal{B}(G, H, Y_{(P,\alpha,n_2)}, Z) = 0] = \frac{1}{2}$, while if $(P, H, Y_{(P,\alpha,n_2)}, Z)$ is sampled from L-BDHE, $|Pr[\mathcal{B}(G, H, Y_{(P,\alpha,n_2)}, Z)] - \frac{1}{2}| \geq \epsilon$. So, the probability that $\mathcal{B}$ outputs correctly is at least $1 - (\frac{1}{2} - \epsilon)^{|\mathcal{S}|} \geq \frac{1}{2} + \epsilon$. Thus $\mathcal{B}$ has advantage at least ϵ in solving the n_2-BDHE problem. This concludes the proof. *Note that the instance of this proof with $n_1 = 1$ and $n_2 = n$ serves as the proof of security for the basic KAC scheme proposed in Sect. 4.*

Performance Trade Off with the Basic Scheme: We compare the various parameter sizes for the proposed original and extended schemes in Table 1. We note that *SetUp* and *KeyGen* are both one-time operations, and for a given subset $\mathcal{S}$, the *Extract* operation is also performed once to generate the corresponding aggregate key $K_{\mathcal{S}}$. The most important advantage that the generalized scheme provides is the user's ability to efficiently extend the number of ciphertext classes. As far as encryption and decryption are concerned, encryption should ideally take the same time for both schemes, while decryption is actually expected to be faster for the generalized construction as $n_2 \leq n$.

5.3   A Flexible Extension Policy

If a user needs to classify her ciphertexts into more that n classes, she can register for additional key pairs $(pk_{n_1+1}, msk_{n_1+1}), \cdots, (pk_{n_1+l}, msk_{n_1+l})$ as per her requirements. Each new key registration increases the number of classes by n_2,

Table 1. Comparison between the basic and generalized schemes

Item	Nature of computation	Original scheme	Generalized scheme
$param$(SetUp)	One-time	$\mathcal{O}(n)$	$\mathcal{O}(n_2)$
PK(KeyGen)	One-time	$\mathcal{O}(1)$	$\mathcal{O}(n_1)$
K_S(Extract)	One-time	$\mathcal{O}(1)$	$\mathcal{O}(n_1)$
$\mathcal{C}$	One per message	$\mathcal{O}(1)$	$\mathcal{O}(1)$
Encrypt	One per message	$\mathcal{O}(1)$	$\mathcal{O}(1)$
Decrypt	One per message	$\mathcal{O}(|\mathcal{S}|)$	$\mathcal{O}(|\mathcal{S}|)$

where $n_2 \leq n$. The idea of under-utilization stems from the fact that registration of each public-private key pair increases the number of classes by n_2. However, it is not necessary that all the existing classes are utilized at any given point of time. For instance, a user may at any point of time want to register l new private-public key pairs, however she will in all probability not use up all ln_2 additional classes of messages that could be encrypted using the newly registered keys. We stress here is that, unlike in the public key extension scheme proposed in [3] where the values of n_1 and n_2 are fixed to 1 and n respectively, our generalized construction *provides a choice* of n_1 and n_2 so that the system administrator could choose pair of values suited to their requirements.

We propose a metric to quantify the under-utilization of ciphertext classes for a given configuration of the system. Let us assume that at some instance of time, there are $n_1 + l$ private-public key pairs registered in the system, and c_i classes corresponding to each key are being utilized. We define the utilization coefficient as $\frac{1}{1+\xi}$, where $\xi = -\frac{1}{n_1}\sum_{i=1 c_i \neq 0}^{n_1} \log(\frac{c_i}{n_2})$. An efficient scheme tries to minimize the value of ξ to achieve good utilization of the existing set of classes. The value is maximum when $c_i = n_2 \forall i = 1, 2, \cdots, n_2$. Note that $c_i = 0$ implies that no subclasses under the given key pk_i are being utilized, which is equivalent to not registering the key at all.

To stress the importance of the flexible extension policy, we provide a simplified example here. We consider two possible configurations of the extended scheme. In the first configuration, $n_1 = 1$ and $n_2 = n$, which is essentially identical to the public key extension scheme proposed in [3]. The other configuration has $n_1 > 1$ and $n_2 < n$. Now assume that before extension, both schemes utilized c ciphertext classes out of the n possible classes, equally distributed across all key pairs. Now suppose a situation arises where an user needs to register l more key pairs, and utilizes $z < n_2$ classes corresponding to each key. In the first configuration, we have $\xi_1 = -\frac{1}{l+1}(l\log(\frac{z}{n}) + \log(\frac{c}{n}))$, while for the second configuration, $\xi_2 = -\frac{1}{l+n_1}(l(\log(\frac{z}{n_2})) + n_1 \log(\frac{c}{n}))$. Now for $l > (\frac{n_1}{\log n_1} - 1)\log(\frac{z}{c}) - 1$, $\xi_2 < \xi_1$. Thus for any value of (n_1, n_2) other than $(1, n)$, there exists a value of l for which the scheme achieves better utilization coefficient. Since l is expected to increase in a dynamic scenario, our public key extension scheme eventually performs better than the scheme suggested in [3].

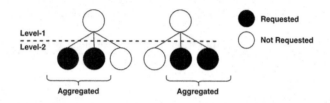

Fig. 2. A practical request scenario in the hierarchical setting

5.4 Advantage over Hierarchical Encryption Based Schemes

Although the generalized scheme has a two level hierarchy (with each of the n_1 parallely executing instances of the basic scheme representing a node in the top level and the actual ciphertext classes representing nodes in the lower level), it avoids the pitfalls of existing hierarchical encryption based schemes [5,10]. In standard tree based hierarchical systems, granting access to the key corresponding to any node implicitly grants access to all the keys in the subtree rooted at that node. This means granting access to a selected set of nodes in a given subtree would blow up the key-size to be the same as the number of nodes. This is avoided in our generalized scheme, since any number of nodes (ciphertext classes) that belong to the same instance may be aggregated into a single key. Figure 2 summarizes this phenomenon. In the situation depicted, a tree-based hierarchy system would require 4 decryption keys, while our scheme would require only 2. In this respect, our scheme has similar advantages to that of [3].

6   Extending the Generalized KAC for Efficient Pairings on Elliptic Curve Subgroups

The encryption schemes proposed so far use the assumption that the elliptic curve pairing bilinear pairing $\hat{e}' : \mathbb{G}_1 \times \mathbb{G}_1 \longrightarrow \mathbb{G}_T$ satisfies the property $\hat{e}'(P,P) \neq 1$, where P is the generator for $\mathbb{G}_1$. In this section, we propose an extension to the generalized n_2-scheme that allows using pairings of the form $\hat{e}'' : \mathbb{G}_1 \times \mathbb{G}_2 \longrightarrow \mathbb{G}_T$, where $\mathbb{G}_1$ and $\mathbb{G}_2$ are two elliptic curve subgroups of the same prime order. The motivation behind this extension is that many popular pairing algorithms such as the Tate [11], Eta [12], and Ate [13] pairings are defined over two distinct elliptic curve subgroups $\mathbb{G}_1$ and $\mathbb{G}_2$ of the same order. Many efficient implementations of such pairings on sensor nodes such as TinyTate [14] have been proposed in literature. This motivates us to modify our scheme in a manner that allows using such well-known pairings. The modified encryption scheme described below allows using a pairing $\hat{e}'' : \mathbb{G}_1 \times \mathbb{G}_2 \longrightarrow \mathbb{G}_T$ .with P generator of $\mathbb{G}_1$ and Q generator of $\mathbb{G}_2$.

6.1   Construction of the Extended KAC

1. **Setup$(1^\lambda, n_2)$:** Randomly pick $\alpha \in \mathbb{Z}_q$. Compute $P_i = \alpha^i P \in \mathbb{G}_1$ for $i = 1, \cdots, n_2, n_2 + 2, \cdots, 2n_2$ and $Q_i = \alpha^i Q \in \mathbb{G}_2$ for $i = 1, \cdots, n_2$.

Output the system parameter as $param = (P, P_1, \cdots, P_{n_2}, P_{n_2+2}, \cdots, P_{2n_2}, Q, Q_1, \cdots, Q_{n_2})$. The system also randomly chooses secret parameters $t \in \mathbb{Z}_q$ which is not made public. It is only transferred through a secure channel to data owners with credentials to control client access rights.

2. **Keygen()**: Pick $\gamma_1, \gamma_2, \cdots, \gamma_{n_1} \in \mathbb{Z}_q$, output the public and master-secret key tuple:
$(PK^1 = (pk^1_1, pk^1_2, \cdots, pk^1_{n_1}) = (\gamma_1 P, \gamma_2 P, \cdots, \gamma_{n_1} P),\ PK^2 = (pk^2_1, pk^2_2, \cdots, pk^2_{n_1}) = (\gamma_1 Q, \gamma_2 Q, \cdots, \gamma_{n_1} Q),\ msk = (\gamma_1, \gamma_2, \cdots, \gamma_{n_1}))$.

3. **Encrypt**$(pk_{i_1}, (i_1, i_2), m)$: For a message $m \in \mathbb{G}_T$ and an index $(i_1, i_2) \in \{1, 2, \cdots, n_1\} \times \{1, 2, \cdots, n_2\}$, randomly choose $r \in \mathbb{Z}_q$ and let $t' = t + r \in \mathbb{Z}_q$. Then compute the ciphertext as
$\mathcal{C} = (rQ, t'(pk^2_{i_1} + Q_{i_2}), m.\hat{e}''(P_{n_2}, t'Q_1)) = (c_1, c_2, c_3)$.

4. **Extract**$(msk = \gamma, \mathcal{S})$: For the set $\mathcal{S}$ of indices (j_1, j_2) the aggregate key is computed as $K_\mathcal{S} = (k^1_\mathcal{S}, k^2_\mathcal{S}, \cdots, k^{n_1}_\mathcal{S}) = (\sum_{(1,j_2)\in\mathcal{S}} \gamma_1 P_{n_2+1-j_2}, \sum_{(2,j_2)\in\mathcal{S}} \gamma_2 P_{n_2+1-j_2}, \cdots, \sum_{(n_1,j_2)\in\mathcal{S}} \gamma_{n_1} P_{n_2+1-j_2})$ and the dynamic access control parameter U is computed as tQ. Thus the net aggregate key is $(K_\mathcal{S}, U)$ which is transmitted via a secure channel to users that have access rights to $\mathcal{S}$. Note that $k^{j_1}_\mathcal{S} = \sum_{(j_1,j_2)\in\mathcal{S}} \alpha^{n+1-j_2} pk^1_{j_1}$ for $j_1 = 1, 2, \cdots, n_1$.

5. **Decrypt**$(K_\mathcal{S}, U, \mathcal{S}, (i_1, i_2), \mathcal{C} = \{c_1, c_2, c_3\})$: If $(i_1, i_2) \notin \mathcal{S}$, output $\perp$. Otherwise return the message
$$\hat{m} = c_3 \frac{\hat{e}''(k^{i_1}_\mathcal{S} + \sum_{(i_1,j_2)\in\mathcal{S}, j_2\neq i_2} P_{n_2+1-j_2+i_2}, U+c_1)}{\hat{e}''(\sum_{(i_1,j_2)\in\mathcal{S}} P_{n_2+1-j_2}, c_2)}.$$

The proof of correctness of this scheme is presented below.

$$\hat{m} = c_3 \frac{\hat{e}''(k^{i_1}_\mathcal{S} + \sum_{(i_1,j_2)\in\mathcal{S}, j_2\neq i_2} P_{n_2+1-j_2+i_2}, U + c_1)}{\hat{e}''(\sum_{(i_1,j_2)\in\mathcal{S}} P_{n_2+1-j_2}, c_2)}$$

$$= c_3 \frac{\hat{e}''(\sum_{(i_1,j_2)\in\mathcal{S}} \gamma_{i_1} P_{n_2+1-j_2}, t'Q)\hat{e}''(\sum_{(i_1,j_2)\in\mathcal{S}}(P_{n_2+1-j_2+i_2}) - P_{n_2+1}, t'Q)}{\hat{e}''(\sum_{(i_1,j_2)\in\mathcal{S}} P_{n_2+1-j_2}, \gamma_{i_1}(t'Q))\hat{e}''(\sum_{(i_1,j_2)\in\mathcal{S}} P_{n_2+1-j_2}, \alpha^{i_2}(t'Q))}$$

$$= c_3 \frac{\hat{e}''(\sum_{(i_1,j_2)\in\mathcal{S}} P_{n_2+1-j_2+i_2}, t'Q)}{\hat{e}''(P_{n_2+1}, t'Q)\hat{e}''(\sum_{(i_1,j_2)\in\mathcal{S}} P_{n_2+1-j_2+i_2}, t'Q)}$$

$$= m$$

6.2   Semantic Security of the Extended KAC

The proof of security uses a reduced version of the extended KAC scheme, analogous to the reduced scheme used for proving the security of the generalized KAC. The adversarial model is also the assumed to be the same as for the generalized KAC. The proof uses the (l, l)-BDHE assumption proposed in Sect. 3.3 (The Second Complexity Assumption). Let $\mathbb{G}_1$ and $\mathbb{G}_2$ be additive elliptic curve subgroups of prime order q, and G_T be a multiplicative group of order q. Let $\hat{e}'' : \mathbb{G}_1 \times \mathbb{G}_2 \longrightarrow \mathbb{G}_T$ be a bilinear non-degenerate pairing. We claim that for any pair of positive integers $n_2, n'(n' > n_2)$ our proposed extension to the

n_2-generalized reduced key-aggregate encryption scheme over elliptic curve subgroups is (τ, ϵ, n') semantically secure if the decision $(\tau, \epsilon, n_2, n_2)$-BDHE assumption holds in $(\mathbb{G}_1, \mathbb{G}_2)$. We prove the claim below.

Proof: Let for a given input n', $\mathcal{A}$ be a τ-time adversary that has advantage greater than ϵ for the *reduced scheme* parameterized with a given n_2. We build an algorithm $\mathcal{B}$ that has advantage at least ϵ in solving the (n_2, n_2)-BDHE problem in $\mathbb{G}$. Algorithm $\mathcal{B}$ takes as input a random (n_2, n_2)-BDHE challenge $(P, Q, H, Y_{(P,\alpha,n_2)}, Y'_{Q,\alpha,n_2}, Z)$ where Z is either $\hat{e}''(P_{n_2+1}, H)$ or a random value in $\mathbb{G}_T$. Algorithm $\mathcal{B}$ proceeds as follows.

1. **Init:** Algorithm $\mathcal{B}$ runs $\mathcal{A}$ and receives the set $\mathcal{S}$ of ciphertext classes that $\mathcal{A}$ wishes to be challenged on. For each ciphertext class $(i_1, i_2) \in \mathcal{S}$, $\mathcal{B}$ performs the **SetUp-(i_1, i_2)**, **Challenge-(i_1, i_2)** and **Guess-(i_1, i_2)** steps. Note that the number of iterations is polynomial in $|S|$.
2. **SetUp-(i_1, i_2)**: $\mathcal{B}$ should generate the public *param*, public keys PK^1, PK^2, the access parameter U, and the aggregate key $K_{\overline{\mathcal{S}}}$. For the iteration corresponding to ciphertext class (i_1, i_2), they are generated as follows.
   - *param* is set as $(P, Q, Y_{P,\alpha,n_2}, Y'_{Q,\alpha,n_2})$.
   - Randomly generate $u_1, u_2, \cdots, u_{n_1} \in \mathbb{Z}_q$. Then, set
     $PK^1 = (pk^1_{\ 1}, pk^1_{\ 2}, \cdots, pk^1_{\ n_1})$, where $pk^1_{\ j_1}$ is set as $u_{j_1} P - P_{i_2}$ for $j_1 = 1, 2, \cdots, n_1$, and set
     $PK^2 = (pk^2_{\ 1}, pk^2_{\ 2}, \cdots, pk^2_{\ n_1})$, where $pk^2_{\ j_1}$ is set as $u_{j_1} Q - Q_{i_2}$ for $j_1 = 1, 2, \cdots, n_1$
   - $K_{\overline{\mathcal{S}}}$ is set as $(k^1_{\overline{\mathcal{S}}}, k^2_{\overline{\mathcal{S}}}, \cdots, k^{n_1}_{\overline{\mathcal{S}}})$ where $k^{j_1}_{\overline{\mathcal{S}}}$
     $= \sum_{(j_1,j_2) \notin \mathcal{S}} (u_{j_1} P_{n_2+1-j_2} - (P_{n_2+1-j_2+i_2}))$ for $j_1 = 1, 2, \cdots, n_1$. Note that this implies $k^{j_1}_{\overline{\mathcal{S}}} = \sum_{(j_1,j_2) \notin \mathcal{S}} \alpha^{n_2+1-j_2} pk^1_{\ j_1}$, as is supposed to be as per the scheme specification. Note that $\mathcal{B}$ knows that $(i_1, i_2) \notin \overline{\mathcal{S}}$, and hence has all the resources to compute this aggregate key for $\overline{\mathcal{S}}$.
   - U is set as some random element in $\mathbb{G}_2$.

   Note that since P, Q, α, U and the u_{j_1} values are chosen uniformly at random, the public key has an identical distribution to that in the actual construction.
3. **Challenge-(i_1, i_2)**: To generate the challenge for the ciphertext class (i_1, i_2), $\mathcal{B}$ computes (c_1, c_2) as $(H - U, u_{i_1} H)$. It then randomly chooses a bit $b \in (0, 1)$ and sets K_b as Z and K_{1-b} as a random element in $\mathbb{G}_T$. The challenge given to $\mathcal{A}$ is $((c_1, c_2), K_0, K_1)$.
   We claim that when $Z = \hat{e}''(P_{n_2+1}, H)$ (i.e. the input to $\mathcal{B}$ is a n_2-BDHE tuple), then $((c_1, c_2), K_0, K_1)$ is a valid challenge to $\mathcal{A}$. We prove this claim here. we point out that Q is a generator of $\mathbb{G}_2$ and so $H = t'P$ for some $t' \in \mathbb{Z}_q$. Putting H as $t'Q$ gives us the following:
   - $U = tQ$ for some $t \in \mathbb{Z}_q$
   - $c_1 = H - U = (t' - t)Q = rQ$ where $r = t' - t$
   - $c_2 = u_{i_1} H = (u_{i_1}) t'Q = t'(u_{i_1} Q) = t'(u_{i_1} Q - Q_{i_2} + Q_{i_2}) = t'(pk^2_{\ i_1} + Q_{i_2})$
   - $K_b = Z = \hat{e}''(P_{n_2+1}, H) = \hat{e}''(P_{n_2+1}, t'Q)$

On the other hand, if Z is a random element in $\mathbb{G}_T$ (i.e. the input to $\mathcal{B}$ is a random tuple), then K_0 and K_1 are just random independent elements of $\mathbb{G}_T$.

4. **Guess-(i_1, i_2):** The adversary $\mathcal{A}$ outputs a guess b' of b. If $b' = b$, $\mathcal{B}$ outputs 0 (indicating that $Z = \hat{e}''(P_{n+1}, H)$), and terminates. Otherwise, it goes for the next ciphertext class in $\mathcal{S}$.

If after $|\mathcal{S}|$ iterations, $b' \neq b$ for each ciphertext class $(i_1, i_2) \in \mathcal{S}$, the algorithm $\mathcal{B}$ outputs 1 (indicating that Z is random in $\mathbb{G}_T$). We now analyze the probability that $\mathcal{B}$ gives a correct output. If $(P, H, Y_{(P,\alpha,n_2)}, Z)$ is sampled from R'-BDHE, $Pr[\mathcal{B}(G, H, Y_{(P,\alpha,n_2)}, Z) = 0] = \frac{1}{2}$, while if $(P, H, Y_{(P,\alpha,n_2)}, Z)$ is sampled from L'-BDHE, $|Pr[\mathcal{B}(G, H, Y_{(P,\alpha,n_2)}, Z)] - \frac{1}{2}| \geq \epsilon$. So, the probability that $\mathcal{B}$ outputs correctly is at least $1 - (\frac{1}{2} - \epsilon)^{|\mathcal{S}|} \geq \frac{1}{2} + \epsilon$. Thus $\mathcal{B}$ has advantage at least ϵ in solving the (n_2, n_2)-BDHE problem. This concludes the proof.

7   Experimental Results Using Tate Pairings

In this section we present experimental results from our implementations of the extended generalized scheme using Tate pairings on BN-curves using 256 bit primes [15]. All our experiments have been carried out on an AMD Opteron (TM) Processor 6272 × 16 with a clock frequency 1.4 GHz.

7.1   Space and Time Complexities

Table 2 summarizes the space requirements for various parameters of the scheme for different values of (n_1, n_2). The results have been averaged over 100 randomly chosen subsets of the $n = 100$ ciphertext classes. Table 3 summarizes the time complexity for various operations of the scheme for different values of (n_1, n_2). The results have been averaged over 100 randomly chosen subsets of the $n = 100$ ciphertext classes. The encryption and decryption operation complexities are further averaged over 10 message transmissions corresponding to each subset. We point out that both the overall space and time requirements are minimum for $n_1 = n_2 = 10 = \sqrt{n}$, which proves the usefulnesss of the generaalization.

7.2   Comparison with Hierarchy Based Schemes

Next, we compare specifically the key size required for the proposed extended scheme, for different values of n_1 and n_2 (again corresponding to $n = 100$), with that required for a hierarchical encryption construction [16]. Since our scheme uses a hierarchy depth of 2, we use the same for the hierarchical construction as well, with n_1 nodes in level 0, and n_2 level 1 nodes in the subtree rooted at each level 0 node. Figure 3 summarizes the findings. Evidently, lower the value of n_1, better the key aggregation, hence lower the ratio.

Table 2. Space complexities

n_1	n_2	$param$ (in bytes)	PK (in bytes)	msk (in bytes)	K_S (in bytes)	U (in bytes)	Total (in KB)
1	100	16112	144	40	72	64	16.046875
2	50	8112	240	56	120	64	8.390625
4	25	4112	432	88	216	64	4.796875
5	20	3312	528	104	264	64	4.171875
10	**10**	**1712**	**1008**	**184**	**504**	**64**	**3.390625**
20	5	912	1968	344	984	64	4.171875
25	4	752	2448	424	1224	64	4.796875
50	2	432	4848	824	2424	64	8.390625
100	1	272	9648	1624	4824	64	16.046875

Table 3. Time complexities

n_1	n_2	$SetUp$ (in clock cycles)	$KeyGen$ (in clock cycles)	$Encrypt$ (in clock cycles)	$Extract$ (in clock cycles)	$Decrypt$ (in clock cycles)	Total (in clock cycles)
1	100	2920000	10000	7932000	47000	16095000	27004100
2	50	1410000	30000	8065000	53000	16110000	25668000
4	25	690000	60000	8130000	81000	16284000	25245000
5	20	590000	70000	8091000	96000	16379000	25226000
10	**10**	**280000**	**140000**	**7957000**	**170000**	**16049000**	**25136000**
20	5	130000	270000	8070000	320000	16361000	25151000
25	4	120000	350000	8256000	370000	16239000	25836000
50	2	50000	680000	8265000	712000	16398000	26105000
100	1	30000	1360000	8201000	1315000	16142000	27048000

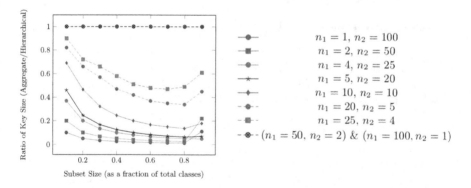

Fig. 3. Key size ratio - proposed aggregate scheme vs hierarchical scheme

7.3 Utilization Coefficient Comparison

Finally we compare the utilization-coefficient of the extended scheme for various values of n_1 and n_2 (corresponding to $n = 100$) with increase in the number of registered key pairs l, where each key pair increases the number of classes by n_2. We leave out the configuration $n_1 = n, n_2 = 1$ because that always leads to an utilization coefficient of 1 but is impractical due to huge space requirements. Figure 4 demonstrates that beyond a certain value of l, the combination $(1, n)$

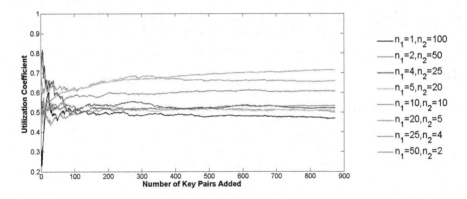

Fig. 4. Utilization coefficient vs newly registered keys

proposed in [3] has a lower utilization coefficient that all other combinations of (n_1, n_2) for a given n. This emphasizes the advantage of making the choice of (n_1, n_2) flexible.

8   Conclusions and Future Work

In this paper, we have proposed a secure and dynamic key aggregate encryption scheme for online data sharing. Our scheme allows data owners to delegate users with access rights to multiple ciphertext classes using a single decryption key that combines the decrypting power of individual keys corresponding to each ciphertext class. Unlike existing key aggregate schemes that are static in their access right delegation policies, our scheme allows data owners to dynamically revoke one or more users' access rights without having to change either the public or the private parameters/keys. The use of bilinear pairings over additive elliptic curve subgroups in our scheme helps achieve massive reductions in key and ciphertext sizes over existing schemes that use multiplicative groups. We pointed out that a possible criticism of this scheme is that the number of classes is predefined to some fixed n. To deal with this issue, we next proposed a generalized two-level construction of the basic scheme that runs n_1 instances of the basic scheme in parallel, with each instance handling key aggregation for n_2 ciphertext classes. This scheme provides two major advantages. First of all, it allows dynamic extension of ciphertext classes by registering of new public key-private key pairs without affecting other system parameters. Secondly, it provides a wide range of choices for n_1 and n_2 that allows efficient utilization of ciphertext classes while also achieving optimum space and time complexities. Finally, we extend the generalized scheme to allow the use of popular and efficiently implementable bilinear pairings in literature such as Tate Pairings that operate on multiple elliptic curve subgroups instead of one. Each of the three proposed schemes have been proven to be semantically secure. Experimental studies have demonstrated the superiority of our proposed scheme over existing ones in terms of key size as well as efficient utilization of ciphertext classes. A possible future work is to make the proposed schemes secure against chosen ciphertext attacks.

References

1. Wang, C., Chow, S.S.M., Wang, Q., Ren, K., Lou, W.: Privacy-preserving public auditing for secure cloud storage. Cryptology ePrint Archive, Report 2009/579 (2009). http://eprint.iacr.org/
2. Chow, S.S.M., Chu, C.-K., Huang, X., Zhou, J., Deng, R.H.: Dynamic secure cloud storage with provenance. In: Naccache, D. (ed.) Cryphtography and Security: From Theory to Applications. LNCS, vol. 6805, pp. 442–464. Springer, Heidelberg (2012)
3. Chu, C.-K., Chow, S.S.M., Tzeng, W.-G., Zhou, J., Deng, R.H.: Key-aggregate cryptosystem for scalable data sharing in cloud storage. IEEE Trans. Parallel Distrib. Syst. **25**(2), 468–477 (2014)
4. Boneh, D., Gentry, C., Waters, B.: Collusion resistant broadcast encryption with short ciphertexts and private keys. In: Shoup, V. (ed.) CRYPTO 2005. LNCS, vol. 3621, pp. 258–275. Springer, Heidelberg (2005)
5. Ateniese, G., De Santis, A., Ferrara, A.L., Masucci, B.: Provably-secure time-bound hierarchical key assignment schemes. J. Cryptology **25**(2), 243–270 (2012)
6. Benaloh, J., Chase, M., Horvitz, E., Lauter, K.: Patient controlled encryption: ensuring privacy of electronic medical records. In: Proceedings of the 2009 ACM Workshop on Cloud Computing Security, pp. 103–114. ACM (2009)
7. Ateniese, G., Kevin, F., Green, M., Hohenberger, S.: Improved proxy re-encryption schemes with applications to secure distributed storage. ACM Trans. Inf. Syst. Secur. (TISSEC) **9**(1), 1–30 (2006)
8. Miller, V.S.: Use of elliptic curves in cryptography. In: Williams, H.C. (ed.) CRYPTO 1985. LNCS, vol. 218, pp. 417–426. Springer, Heidelberg (1986)
9. Silverman, J.H.: Advanced Topics in the Arithmetic of Elliptic Curves, vol. 151. Springer, New York (1994)
10. Akl, S.G., Taylor, P.D.: Cryptographic solution to a problem of access control in a hierarchy. ACM Trans. Comput. Syst. (TOCS) **1**(3), 239–248 (1983)
11. Frey, G., Rück, H.-G.: A remark concerning-divisibility and the discrete logarithm in the divisor class group of curves. Math. Comput. **62**(206), 865–874 (1994)
12. Hess, F., Smart, N.P., Vercauteren, F.: The eta pairing revisited. IEEE Trans. Inf. Theor. **52**(10), 4595–4602 (2006)
13. Zhao, C.-A., Zhang, F., Huang, J.: A note on the ate pairing. Int. J. Inf. Secur. **7**(6), 379–382 (2008)
14. Oliveira, L.B., Aranha, D.F., Morais, E., Daguano, F., López, J., Dahab, R.: Tinytate: computing the tate pairing in resource-constrained sensor nodes. In: 2007 Sixth IEEE International Symposium on Network Computing and Applications, NCA 2007, pp. 318–323. IEEE (2007)
15. Ghosh, S., Mukhopadhyay, D., Roychowdhury, D.: Secure dual-core cryptoprocessor for pairings over barreto-naehrig curves on FPGA platform. IEEE Trans. Very Large Scale Integr. (VLSI) Syst. **21**(3), 434–442 (2013)
16. Sandhu, R.S.: Cryptographic implementation of a tree hierarchy for access control. Inf. Process. Lett. **27**(2), 95–98 (1988)

Lite-Rainbow: Lightweight Signature Schemes Based on Multivariate Quadratic Equations and Their Secure Implementations

Kyung-Ah Shim[1]($\boxtimes$), Cheol-Min Park[1], and Yoo-Jin Baek[2]

[1] Division of Mathematical Modeling, National Institute for Mathematical Sciences,
Daejeon, Republic of Korea
{kashim,mpcm}@nims.re.kr
[2] Department of Information Security, Woosuk University,
Wanju-gun, Republic of Korea
yoojin.baek@gmail.com

Abstract. Rainbow signature scheme based on multivariate quadratic equations is one of alternatives to guarantee secure communications in the post-quantum world. Its speed is about dozens of times faster than classical public-key signatures, RSA and ECDSA, while its key size is much heavier than those of the classical ones. We propose lightweight variants of Rainbow, Lite-Rainbow-0 and Lite-Rainbow-1, for constrained devices. By replacing some parts of a public key or a secret key with small random seeds via a pseudo-random number generator, we reduce a public key in Lite-Rainbow-1 and a secret key in Lite-Rainbow-0 by factors 71 % and 99.8 %, respectively, compared to Rainbow. Although our schemes require additional costs for key recovery processes, they are still highly competitive in term of performance. We also prove unforgeability of our scheme with special parameter sets in the random oracle model under the hardness assumption of the multivariate quadratic polynomial solving problem. Finally, we propose countermeasures of Rainbow-like schemes against side channel attacks such as power analysis for their secure implementations.

Keywords: Multivariate quadratic equations · Key size reduction · Pseudo-random number generator · Rainbow

1 Introduction

In 1995, Shor [24] presented a quantum algorithm to solve the Integer Factorization problem (IFP) and the Discrete Logarithm problem (DLP) in polynomial time. As a result, the existence of a sufficiently large quantum computer would be a real-world threat to break RSA, Diffie-Hellman key exchange, DSA, and ECDSA, the most widely used public-key cryptography [27]. Thus, there is an increasing demand in investigating possible alternatives. Such classes of so-called Post-Quantum Cryptography (PQ-PKC) are lattice-based (NTRU [11]),

© Springer International Publishing Switzerland 2015
A. Biryukov and V. Goyal (Eds.): INDOCRYPT 2015, LNCS 9462, pp. 45–63, 2015.
DOI: 10.1007/978-3-319-26617-6_3

code-based (McEliece [16]), hash-based (Merkle's hash-tree signature [17]), and multivariate quadratic equations-based (UOV [12], Rainbow [7]). All of these systems are believed to resist classical computers and quantum computers.

Table 1. Current MQ-PKCs vs. classical ones at an 80-bit security level.

Scheme	Public key	Secret key	Verify	Sign
RSA(1024)	128 B	1024 B	22.4 μs	813.5 μs
ECDSA(160)	40 B	60 B	409.2 μs	357.8 μs
Rainbow($\mathbb{F}_{31}$, 24, 20, 20)	57 KB	150 KB	17.7 μs	70.6 μs

Multivariate quadratic public-key cryptography (MQ-PKC) are cryptosystems based on the NP-hard problem of solving random systems of quadratic equations over finite fields, known as the MQ-problem. Since Mastsumoto-Imai scheme [15], there have been proposed a number of MQ-PKCs. However, nearly all MQ-encryption schemes and most of the MQ-signature schemes have been broken due to uncertainty of the Isomorphism of Polynomials problem. There are only very few exceptions like the signature schemes, Unbalanced Oil and Vinegar (UOV) [12], and Rainbow [7]. Recently, two of the most important standardization bodies in the world, NIST [19] and ETSI [8] have started initiatives for developing cryptographic standards with long-term security resistant to quantum algorithms.

Beyond the supposed resistance against quantum computers, one of MQ-PKCs' advantages is a fast speed, especially for signatures. Chen *et al.* [5] presented performances of MQ-schemes and classical ones on an Intel Core 2 Quad Q9550 at 2.833 GHz, summarized in Table 1. According to Table 1, Rainbow is 23 times and 5 times faster than ECDSA in verification and signing, respectively. MQ-PKCs require simplicity of operations (matrices and vectors) and small fields avoid multiple-precision arithmetic. Classical PKCs need coprocessors in smart cards with low flexibility for use or optimizations, so operating on units hundreds of bits long is prohibitively expensive for embedded devices without a co-processor. Despite these advantages, the biggest problem of MQ-PKCs is a relatively large key size. Petzoldt *et al.* [23] proposed CyclicRainbow to reduce a public key size in Rainbow by up to 62 % using a circulant matrix. However, the use of this circulant matrix provides a limited randomness of the quadratic parts of a public key. If the key size is reduced to an adequate level then MQ-PKCs are the most competitive PKC for constrained devices. Side channel attacks (SCAs) are a serious threat to constrained small devices. If implementations of cryptographic algorithms on such a device are careless, attackers can recover secret keys. Recently, Hashimoto *et al.* [10] proposed general fault attacks on MQ-PKCs to reduce complexity of key recovery attacks by causing faults on the central map or on ephemeral random values. It has never been proposed countermeasures of Rainbow-like schemes against other SCAs such as power analysis. In this paper, we provide solutions to these problems.

Our Contributions. We construct two lightweight variants of Rainbow, Lite-Rainbow-0 and Lite-Rainbow-1. Our contributions are as follows:

- **Public/Secret Key Size Reduction.** We utilize a pseudo-random number generator (PRNG) to replace some quadratic parts, linear and constant parts of a public key or a secret key with small random seeds for key size reduction. To do it, we use "*Recover then Sign*" and "*Recover then Verify*" methodologies: after recovering some parts of the secret key and public key, signing and verification are performed. As a result, we reduce a public key in Lite-Rainbow-1 and a secret key in Lite-Rainbow-0 up to 71 % and 99.9 %, respectively, compared to Rainbow.
- **Provable Security of Lite-Rainbow.** Unlike RSA and ECC, security of MQ-schemes is not related to the MQ-problem only. This causes the difficulty of their security proofs against unforgeability in formal security model. We prove unforgeability of our scheme with special parameter sets against adaptive chosen-message attacks and direct attacks in the random oracle model under the hardness assumption of the MQ-problem.
- **Countermeasures of Rainbow-Like Schemes Against SCAs and Implementations.** We propose countermeasures of Rainbow-like schemes against power analysis, SPA and DPA. To randomize signing, we use "*Splitting technique*" which divides a solution of the central map into three parts to leave one part of them after performing a binary operation. To the best of our knowledge, this is the first work that deals with these SCAs on them. We provide a direct comparison of the implementation results for our schemes and classical ones, unprotected ones and protected ones for secure and optimal parameters at a 128-bit security level on the same platform.

Organization. The rest of the paper is organized as follows. In Sect. 2, we describe underlying hard problems of MQ-PKCs and Rainbow signature scheme. In Sect. 3, we construct lightweight variants of Rainbow, Lite-Rainbow-0 and Lite-Rainbow-1 and provide their security proofs. In Sect. 4, we propose countermeasures of Rainbow-like schemes against SCAs and implement protected and unprotected ones. Concluding remarks are given in Sect. 5.

2 Preliminaries

Here, we describe underlying hard problems of MQ-PKCs, and Rainbow [12].

2.1 Underlying Hard Problems

First, we describe the Solving Polynomial System problem and a variant of Isomorphism of Polynomials problem.

- **Solving Polynomial System (SPS) Problem:** Given a system $\mathcal{P} = (p^{(1)}, \cdots, p^{(m)})$ of m nonlinear polynomial equations defined over a finite field K with degree of d in variables $x_1, \cdots, x_n$ and $\mathbf{y} \in K^m$, find values $(x'_1, \cdots, x'_n) \in K^n$ such that $p^{(1)}(x'_1, \cdots, x'_n) = \cdots = p^{(m)}(x'_1, \cdots, x'_n) = \mathbf{y}$.

- **IP2S (Isomorphism of Polynomials with 2 Secrets) Problem:** Given nonlinear multivariate systems A and B such that $B = T \circ A \circ S$ for linear or affine maps S and T, find two maps S' and T' such that $B = T' \circ A \circ S'$.

The SPS problem is proven to be NP-complete [9]. For efficiency, MQ-PKCs restrict to quadratic polynomials. The SPS problem with all polynomials $p^{(1)}, \cdots, p^{(m)}$ of degree 2 is called the MQ-Problem for multivariate quadratic. The IP problem was first described by Patarin at Eurocrypt'96 [21], in contrast to the MQ-problem, there is not much known about the difficulty of the IP problem.

Definition 1. We say that algorithm has advantage $\epsilon(\lambda)$ in solving the MQ-problem over a finite field K if for a sufficiently large λ,

$$Adv_K^{\mathcal{A}}(t) = Pr \left[\begin{array}{c} \mathcal{A}(K, \mathcal{P}(\mathbf{x}), \mathbf{y}) = \mathbf{s} \in K^n \ s.t. \ \mathcal{P}(\mathbf{s}) = \mathbf{y} \\ | \ \mathcal{P}(\mathbf{x}) \leftarrow \mathcal{MQ}(K), \ \mathbf{y} \leftarrow K^m \end{array} \right] \geq \varepsilon(\lambda),$$

where $\mathcal{MQ}(K)$ is a set of all systems of quadratic equations over K. We say that the MQ-problem is (t, ϵ)-hard if no t-time algorithm has advantage at least ϵ in solving the MQ-problem.

2.2 Rainbow Signature Scheme

Ding and Schmidt [7] proposed an MQ-signature scheme, Rainbow, based on Unbalanced Oil and Vinegar (UOV) [12]. A system $\mathcal{P} = (p^{(1)}, \cdots, p^{(m)})$ of multivariate quadratic polynomials with m equations and n variables over a finite field K is defined by

$$p^{(k)}(x_1, \cdots, x_n) = \sum_{i=1}^{n} \sum_{j=i}^{n} p_{ij}^{(k)} x_i x_j + \sum_{i=1}^{n} p_i^{(k)} x_i + p_0^{(k)},$$

for $k = 1, \cdots, m$. Let $v_1, \ldots, v_{u+1}$ $(u \geq 1)$ be integers such that $0 < v_1 < v_2 < \cdots < v_u < v_{u+1} = n$. Define sets of integers $V_i = \{1, \cdots, v_i\}$ for $i = 1, \ldots, u$ and set $o_i = v_{i+1} - v_i$ and $O_i = \{v_i + 1, \ldots, v_{i+1}\}$ $(i = 1, \ldots, u)$. Then $|V_i| = v_i$ and $|O_i| = o_i$. For $k = v_1 + 1, \ldots, n$, we define multivariate quadratic polynomials in the n variables $x_1, \ldots, x_n$ by

$$f^{(k)}(\mathbf{x}) = \sum_{i \in O_l, j \in V_l} \alpha_{ij}^{(k)} x_i x_j + \sum_{i,j \in V_l, i \leq j} \beta_{ij}^{(k)} x_i x_j + \sum_{i \in V_l \cup O_l} \gamma_i^{(k)} x_i + \eta^{(k)},$$

where l is the only integer such that $k \in O_l$ and $\mathbf{x} = (x_1, \cdots, x_n)$. Note that these are Oil and Vinegar polynomials with x_i $(i \in V_l)$ being Vinegar variables and x_j $(j \in O_l)$ being Oil variables. The map $\mathcal{F}(\mathbf{x}) = (f^{(v_1+1)}(\mathbf{x}), \cdots, f^{(n)}(\mathbf{x}))$ can be inverted by using Oil-Vinegar method. To hide the structure of $\mathcal{F}$, one composes it with two invertible affine maps $\mathcal{S} : K^m \rightarrow K^m$ and $\mathcal{T} : K^n \rightarrow K^n$. A public key is given as $\mathcal{P} = \mathcal{S} \circ \mathcal{F} \circ \mathcal{T} : K^n \rightarrow K^m$ and a secret key is $(\mathcal{S}, \mathcal{F}, \mathcal{T})$ which allows to invert the public key. Rainbow is denoted by Rainbow$(K, v_1, o_1, \cdots, o_u)$. For $u = 1$, we get the original UOV scheme. We use Rainbow with two layers, Rainbow(K, v_1, o_1, o_2). Let $H : \{0,1\}^* \rightarrow K^m$ be a collision-resistant hash function. Here, $m = o_1 + o_2$ and $n = v_1 + o_1 + o_2$.

■ **Rainbow**

- **KeyGen**(1^λ). For a security parameter λ, a public key is $PK = \mathcal{P}$ and a secret key is $SK = (\mathcal{S}, \mathcal{F}, \mathcal{T})$ such that $\mathcal{P} = \mathcal{S} \circ \mathcal{F} \circ \mathcal{T}$.
- **Sign**($SK, \mathbf{m}$). Given a message $\mathbf{m}$, compute $\mathbf{h} = h(\mathbf{m}) \in K^m$ and recursively compute $\alpha = \mathcal{S}^{-1}(\mathbf{h}), \beta = \mathcal{F}^{-1}(\alpha)$ and $\gamma = \mathcal{T}^{-1}(\beta)$. To compute $\beta = \mathcal{F}^{-1}(\alpha)$, i.e., $\mathcal{F}(\beta) = \alpha$,
  - First choose $(s_1, \ldots, s_{v_1})$ at random, substitute $(s_1, \ldots, s_{v_1})$ into o_1 polynomials $f^{(k)}$ $(v_1 + 1 \le k \le v_2 = v_1 + o_1)$ and get $(s_{v_1+1}, \ldots, s_{v_2})$ by solving a system of o_1 linear equations o_1 unknowns $x_{v_1+1}, \ldots, x_{v_2}$ using the Gaussian Elimination.
  - Next, substitute $(s_1, \ldots, s_{v_2})$ into o_2 polynomials $f^{(k)}$ $(v_2 + 1 \le k \le n)$ and get $(s_{v_2+1}, \ldots, s_n)$ by solving a system of o_2 linear equations with o_2 unknowns $x_{v_2+1}, \cdots, x_n$ using the Gaussian Elimination.
  - Then $\beta = (s_1, \cdots, s_n)$.

  If one of the linear systems do not have a solution, choose other values of $x_1, \cdots, x_{v_1}$ and try again. Then, γ is a signature of $\mathbf{m}$.
- **Verify**(PK, σ). Given $(\gamma, \mathbf{m})$, check $\mathcal{P}(\gamma) = h(\mathbf{m})$. If it holds, accept the signature, otherwise reject it.

3   Lite-Rainbow

Now, we construct Lite-Rainbow-0 and Lite-Rainbow-1 based on Rainbow.

3.1   Properties of Lite-Rainbow Keys

Define three integers D_1, D_2 and D as:

- $D_1 := \frac{v_1 \cdot (v_1 + 1)}{2} + o_1 \cdot v_1$ be the number of nonzero quadratic terms in the central polynomials of the first layer.
- $D_2 := \frac{v_2 \cdot (v_2 + 1)}{2} + o_2 \cdot v_2$ be the number of nonzero quadratic terms in the central polynomials of the second layer.
- $D := \frac{n \cdot (n + 1)}{2}$ be the number of quadratic terms in the public polynomials.

We use a special blockwise ordering of monomials and the lexicographical ordering for inside the blocks as in [22]. A public key is $\mathcal{P} = \mathcal{S} \circ \mathcal{F} \circ \mathcal{T}$, where $\mathcal{F}$ is a central map and two invertible linear maps $T : K^n \to K^n$ and $S : K^m \to K^m$ given by $n \times n$ matrix $T = (t_{ij})$ and $m \times m$ matrix $S = (s_{ij})$, respectively. For simplicity, we use two invertible linear maps S and T instead of invertible affine maps. This replacement doesn't affect security of MQ-schemes due to the following Lemma:

Lemma 1 [26]. The IP2S problem using bijective affine maps $\mathcal{S}$ and $\mathcal{T}$ is polynomial reducible to the IP2S problem using $\mathcal{S} \in \mathbb{GL}_m(K)$ and $\mathcal{T} \in \mathbb{GL}_n(K)$, where $\mathbb{GL}_n(K)$ is a general linear group of invertible $n \times n$ matrices over K.

We denote $\mathcal{Q} = \mathcal{F} \circ T$ and get $\mathcal{P} = S \circ \mathcal{Q}$. Let $q_{ij}^{(k)}$ be the coefficients of the monomial $x_i x_j$ in the k-th component of $\mathcal{Q}$, respectively $(1 \leq k \leq m)$. We get the following equations:

$$q_{ij}^{(k)} = \sum_{r=1}^{n} \sum_{s=r}^{n} \alpha_{ij}^{rs} \cdot f_{rs}^{(k)} \quad (1 \leq k \leq m), \tag{1}$$

with

$$\alpha_{ij}^{rs} = \begin{cases} t_{ri} \cdot t_{si} & (i = j) \\ t_{ri} \cdot t_{sj} + t_{rj} \cdot t_{si} & \text{otherwise.} \end{cases}$$

Due to the special structure of the central map $\mathcal{F}$, we can reduce the number of terms in Eq. (1) as

$$q_{ij}^{(k)} = \begin{cases} \sum_{r=1}^{v_1} \sum_{s=r}^{v_2} \alpha_{ij}^{rs} \cdot f_{rs}^{(k)} & (1 \leq k \leq o_1) \\ \sum_{r=1}^{v_2} \sum_{s=r}^{n} \alpha_{ij}^{rs} \cdot f_{rs}^{(k)} & (o_1 + 1 \leq k \leq m). \end{cases}$$

We define a transformation matrix $\hat{A}$ which is $D \times D$ matrix as

$$\hat{A} = \begin{cases} (\alpha_{ij}^{rs}) \ (1 \leq r \leq v_2,\ r \leq s \leq n \text{ for the rows}) \\ (\alpha_{ij}^{rs}) \ (1 \leq i \leq j \leq n \text{ for the columns}). \end{cases} \tag{2}$$

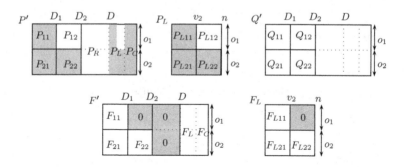

Fig. 1. Layout of the matrices P', Q' and F' for Lite-Rainbow-1

We divide the matrix $\hat{A}$ into six parts as $\begin{pmatrix} A_{11} & A_{12} & A_{13} \\ A_{21} & A_{22} & A_{23} \\ A_{31} & A_{32} & A_{33} \end{pmatrix}$. Let $A = \begin{pmatrix} A_{11} & A_{12} \\ A_{21} & A_{22} \end{pmatrix}$ be the matrix consisting of the first D_2 rows, columns of $\hat{A}$. We write down the coefficients of $\mathcal{P}, \mathcal{Q}$ and $\mathcal{F}$ (according to the monomial ordering) into three matrices P', Q' and F' and divide these matrices as in Fig. 1. We define P, Q and F to be matrices consisting of the first D_2 columns of P', Q' and F', respectively. We also define P'' and F'' to be the matrices consisting of the first D columns of P' and F', respectively. Then we have $P'' = S \cdot F'' \cdot \hat{A}$.

Finally, we get the following relations between the three matrices P, Q and F:

$$P = S \cdot Q, \ i.e., \ Q = \tilde{S} \cdot P = \begin{pmatrix} S'_{11} & S'_{12} \\ S'_{21} & S'_{22} \end{pmatrix} \begin{pmatrix} P_{11} & P_{12} \\ P_{21} & P_{22} \end{pmatrix}, \tag{3}$$

where $\tilde{S} = S^{-1}$, P is $m \times D_2$ matrix, $\tilde{S}$ is $m \times m$ matrix, S'_{11} is $o_1 \times o_1$ submatrix of $\tilde{S}$ and Q is $m \times D_2$ matrix.

$$Q = F \cdot A = \begin{pmatrix} Q_{11} & Q_{12} \\ Q_{21} & Q_{22} \end{pmatrix} = \begin{pmatrix} F_{11} & 0 \\ F_{21} & F_{22} \end{pmatrix} \begin{pmatrix} A_{11} & A_{21} \\ A_{12} & A_{22} \end{pmatrix}, \tag{4}$$

where F is $m \times D_2$ matrix, F_{11} is $o_1 \times D_1$ matrix, A is $D_2 \times D_2$ submatrix of $\hat{A}$, and A_{11} is $D_1 \times D_1$ submatrix of A. We have the same relations the matrices P_L, Q_L and F_L, where P_L, Q_L and F_L are linear parts of P', Q' and F', respectively.

$$P_L = S \cdot Q_L, i.e., Q_L = \tilde{S} \cdot P_L = \begin{pmatrix} S'_{11} & S'_{12} \\ S'_{21} & S'_{22} \end{pmatrix} \begin{pmatrix} P_{L11} & P_{L12} \\ P_{L21} & P_{L22} \end{pmatrix} = \begin{pmatrix} Q_{L11} & Q_{L12} \\ Q_{L21} & Q_{L22} \end{pmatrix},$$

$$Q_L = F_L \cdot T = \begin{pmatrix} Q_{L11} & Q_{L12} \\ Q_{L21} & Q_{L22} \end{pmatrix} = \begin{pmatrix} F_{L11} & 0 \\ F_{L21} & F_{L22} \end{pmatrix} \cdot \begin{pmatrix} T_{11} & T_{12} \\ T_{21} & T_{22} \end{pmatrix}$$

According to these notations, $P' = (P, P_R, P_L, P_C)$ and $F' = (F, F_L, F_C)$ are representations of $\mathcal{P}$ and $\mathcal{F}$, respectively.

3.2   Lite-Rainbow-0

In Lite-Rainbow-0, we use a random seed se' to get the entire secret key $(S, \mathcal{F}, T)$. So, unlike Rainbow, it needs to recover the entire secret key for signing. Its key generation is given in Algorithm 1. Let $\mathcal{G} : \{0,1\}^\lambda \to \{0,1\}^k$ be a secure PRNG which outputs a sequence of random numbers, for $(k, \lambda) \in (\mathbb{Z}^+)^2$.

Algorithm 1. Key Generation for Lite-Rainbow-0.

Input: A security parameter λ, system parameters (K, v_1, o_1, o_2), and a PRNG $\mathcal{G}$
Output: $(\mathcal{P}, se')$
  1: Choose a λ-bit random seed se' and compute $\mathcal{G}(se') = (\tilde{S}, \mathcal{F}, \tilde{T})$. If neither $\tilde{S}$ nor $\tilde{T}$ is invertible then choose a new random seed again, where $\tilde{T} = T^{-1}$.
  2: Compute $\mathcal{P}$ from $\mathcal{P} = S \circ \mathcal{F} \circ T$.
  3: **return** $(\mathcal{P}, se')$

■ Lite-Rainbow-0

- **KeyGen**(1^λ). After performing **Algorithm 1**, set $PK = \mathcal{P}$, and $SK = se'$.
- **Sign**$(SK, \mathbf{m})$. Given a message $\mathbf{m}$:
    • **Secret Key Recovery.** Compute $\mathcal{G}(se') = (\tilde{S}, \mathcal{F}, \tilde{T})$ from SK.
    • **Rainbow Sign.** Generate a Rainbow signature σ on $\mathbf{m}$ using $(\tilde{S}, \mathcal{F}, \tilde{T})$.
- **Verify**(PK, σ). Given a signature σ on $\mathbf{m}$ and $PK = \mathcal{P}$, check $\mathcal{P}(\sigma) = h(\mathbf{m})$. If it holds, accept σ, otherwise, reject it.

3.3   Lite-Rainbow-1

In Lite-Rainbow-1, we use a random seed se to get three blocks P_{11}, P_{21}, P_{22} for the quadratic part, $P_{L11}, P_{L21}, P_{L22}$ for the linear part, and P_C for the constant part which appear in gray parts of the matrix P' in Fig. 1. The goal of key generation in Algorithm 2 is to compute all the coefficients of the central map F' and the remainder P_{12}, P_R of P and P_{L12} of P_L. For these, we need the following condition (*):

- The matrix A is invertible and the $D_1 \times D_1$ submatrix A_{11} of A is invertible.
- The $o_1 \times o_1$ submatrix S'_{11} of $\tilde{S}$ is invertible.
- The $v_2 \times v_2$ submatrix T_{11} of T is invertible.

Algorithm 2. Key Generation for Lite-Rainbow-1.

Input: A security parameter λ, system parameters (K, v_1, o_1, o_2), a random seed se, and a PRNG $\mathcal{G} : \{0,1\}^\lambda \to \{0,1\}^k$

Output: (P_{12}, P_R, P_{L12}), $(\text{se}_1, F, F_L, F_C)$

1: Compute $\mathcal{G}(\text{se}) = (P_{11}, P_{21}, P_{22}, P_{L11}, P_{L21}, P_{L22}, P_C)$ and take P_{11}, P_{21}, P_{22} from the designated parts of $\mathcal{G}(\text{se})$.

2: Choose a λ-bit random seed se_1 and compute $\mathcal{G}(\text{se}_1) = (\tilde{S}, \tilde{T})$. If neither $(\tilde{S}, S'_{11})$ nor $(\tilde{T}, T_{11})$ is invertible then choose a new random seed again.

3: Compute A using Eq. (2). If neither A nor A_{11} is invertible then go back to line 2.

4: Compute Q_{11} and Q_{21} using $\begin{pmatrix} Q_{11} \\ Q_{21} \end{pmatrix} = \tilde{S} \cdot \begin{pmatrix} P_{11} \\ P_{21} \end{pmatrix}$.

5: Compute F_{11} using $F_{11} = Q_{11} \cdot (A_{11}^{-1})$.

6: Compute Q_{12} using $Q_{12} = F_{11} \cdot A_{12}$.

7: Compute P_{12} using $P_{12} = S'^{-1}_{11} \cdot (Q_{12} - S'_{12} P_{22})$.

8: Compute Q_{22} using $Q_{22} = S'_{21} P_{12} + S'_{22} P_{22}$

9: Compute F_{21} and F_{22} using $(F_{21} \| F_{22}) = (Q_{21} \| Q_{22}) \cdot (A^{-1})$.

10: Compute P_R from $P_R = S \cdot F \cdot \begin{pmatrix} A_{13} \\ A_{23} \end{pmatrix}$.

11: Compute $\mathcal{G}(\text{se}) = (P_{11}, P_{21}, P_{22}, P_{L11}, P_{L21}, P_{L22}, P_C)$ and take $P_{L11}, P_{L21}, P_{L22}$ from the designated parts of $\mathcal{G}(\text{se})$.

12: Compute Q_{L11} and Q_{L21} using $\begin{pmatrix} Q_{L11} \\ Q_{L21} \end{pmatrix} = \tilde{S} \cdot \begin{pmatrix} P_{L11} \\ P_{L21} \end{pmatrix}$.

13: Compute F_{L11} using $F_{L11} = Q_{L11} \cdot (T_{11}^{-1})$.

14: Compute Q_{L12} using $Q_{L12} = F_{L11} \cdot T_{12}$.

15: Compute P_{L12} using $P_{L12} = S'^{-1}_{11} \cdot (Q_{L12} - S'_{12} P_{L22})$ and Q_{L22} using $Q_{L22} = S'_{21} P_{L12} + S'_{22} P_{L22}$.

16: Compute F_{L21} and F_{L22} using $(F_{L21} \| F_{L22}) = (Q_{L21} \| Q_{L22}) \cdot (T^{-1})$.

17: Take P_C from the designated parts of $g(\text{se})$.

18: Compute F_C from $F_C = \tilde{S} \cdot P_C$.

19: **return** (P_{12}, P_R, P_{L12}), $(\text{se}_1, F, F_L, F_C)$

Proposition 1. Given $P_{*1} = (\frac{P_{11}}{P_{21}})$, P_{22}, $P_{L*1} = (\frac{P_{L11}}{P_{L21}})$, P_{L22} and P_C, and two invertible linear maps (S, T) such that (S, T, A) satisfies the condition (*), it is possible to construct an MQ-signature scheme with a public/secret key pair $< \mathcal{P}/(S, \mathcal{F}, T) >$ such that $\mathcal{P} = S \circ \mathcal{F} \circ T$.

Proof. Under the condition (*), for the quadratic part, Q, P_{12} and F are uniquely determined from (3) and (4) given (P_{*1}, P_{22}). Then we can compute P_{12}. For the linear part, (P_{L12}, F_L) is uniquely determined from (P_{L*1}, P_{L22}). At last, for the constant part, F_C is uniquely determined from P_C. Therefore, an MQ-signature scheme is determined by a public/secret key pair $< \mathcal{P}/(S, \mathcal{F}, T) >$. $\qquad\square$

Remark 1. The part P_{12} is minimal, i.e., one cannot choose the lager part generated by a seed in advance, as the size of P_{12} is uniquely determined by the size of a submatrix which consists of only zeros in F, i.e., $F_{12} = (0)$. The seed se for some parts of a public key is a public information, while se' for $(\tilde{S}, \tilde{T})$ must be kept secret.

In Lite-Rainbow-1, before signing and verification, the recovery processes of the entire secret key and some parts of public key are required, respectively.

■ Lite-Rainbow-1

- **KeyGen**(1^λ). After performing Algorithm 2, set $PK = (\mathsf{se}, P_{12}, P_R, P_{L12})$ and $SK = (\mathsf{se}_1, \mathcal{F} = (F, F_L, F_C))$.
- **Sign**(SK, m). Given a message m:
  - **Secret Key Recovery.** Compute $\mathcal{G}(\mathsf{se}_1) = (\tilde{S}, \tilde{T})$ from SK.
  - **Rainbow Sign.** Using $(\tilde{S}, \mathcal{F}, \tilde{T})$, generate a Rainbow signature σ on m.
- **Verify**(SK, σ). Given a signature σ on m and PK,
  - **Public Key Recovery.** compute $\mathcal{G}(\mathsf{se}) = (P_{11}, P_{21}, P_{22}, P_{L11}, P_{L21}, P_{L22}, P_C)$ from PK to recover the entire public key $\mathcal{P}$.
  - **Rainbow Verify.** Check $\mathcal{P}(\sigma) = h(\mathsf{m})$. If it holds, accept σ, otherwise, reject it.

3.4   Security Analysis and Proof of Lite-Rainbow

Here, we prove unforgeability of Lite-Rainbow-1 in the random oracle model (ROM) under the hardness assumption of the MQ-problem. It is believed that breaking RSA (resp., ECC) is as hard as factoring (resp., the DLP in a group of points of an elliptic curve). However, security of Rainbow-like schemes is not related to the MQ-problem only: its security is based on the MQ-problem, the IP2S problem and MinRank problem. This causes difficulty of their security proofs against unforgeability in formal security model. Known attacks of Rainbow-like schemes can be divided into two classes:

- **Direct Attack.** Given $\mathbf{y} \in \mathbb{F}_q^m$, find a solution $\mathbf{x} \in \mathbb{F}_q^n$ of $\mathcal{P}(\mathbf{x}) = \mathbf{y}$.
- **Key Recovery Attack (KRA).** Given $\mathcal{P} = S \circ \mathcal{F} \circ T$, find secret linear maps S and T.

The direct attack is related to the MQ-problem, and the latter which contains rank-based attacks and KRAs using equivalent keys and good keys is related to the MinRank problem and the IP2S problem. In Lite-Rainbow(K, v_1, o_1, o_2), security against direct attacks and the KRAs using good keys depends on the selection of (m, n), while security against the KRAs realted to the rank-based attacks and the Kipnis-Shamir attack depends on the selection of (v_1, o_1, o_2) associated to a security parameter λ.

We use a PRNG for the key size reduction and randomness of the quadratic part for the public key depends on security of the PRNG. Now, we prove unforgeability of our scheme with special parameter sets in the ROM. Security notion for public-key signature scheme is described in Appendix. We deal with a hash function, H, and a PRNG, $\mathcal{G}$, as random oracles. Theorem 1 guarantees the security of our scheme against direct attacks.

Theorem 1. If the MQ-problem is (t', ε')-hard, Lite-Rainbow-1(K, v_1, o_1, o_2) is $(t, q_H, q_S, q_{\mathcal{G}}, \varepsilon)$-existential unforgeable against an adaptively chosen message attack, for any t and ε satisfying

$$\varepsilon \geq \mathsf{e} \cdot (q_S + 1) \cdot \varepsilon', \ t' \geq t + q_H \cdot c_V + q_S \cdot c_S + c_{\mathcal{G}},$$

where e is the base of the natural logarithm, and c_S, c_V and $c_{\mathcal{G}}$ are time for a signature generation, a signature verification and a $\mathcal{G}$ evaluation to recover some parts of a public key, respectively, provided that the parameter set (K, v_1, o_1, o_2) is chosen to be resistant to the KRAs.

Proof. See Appendix.

A main difference between Rainbow and Lite-Rainbow-1 is the key generation method by randomizing some quadratic parts of a public key resulting in key size reduction. After recovering a secret key and a public key from seeds, the rest of Lite-Rainbow for signing and verification is the same as Rainbow. Security analysis of Lite-Rainbow against known attacks except direct attacks is the same as that of Rainbow. According to [4], security of MQ-scheme against direct attacks is based on one assumption that appears to be quite reasonable due to considerable empirical evidence gathered by the community of polynomial system solving.

Assumption. Solving a random quadratic system with m equations and n variables is as hard as solving a quadratic system with m equations and n variables with a completely random quadratic part.

This assumption deals with Gröbner bases techniques and other general techniques for solving the MQ-problem. Since these techniques are general, it is quite plausible to assume that the complexity is determined mainly by the quadratic part. Proposition 1 in §3 says that the quadratic part $\mathcal{R} = (P_{11}, P_{21}, P_{22})$ for a public key $\mathcal{P}$ is the largest part we can randomly choose. Thus, security of Lite-Rainbow-1 against direct attacks depends on the randomness of this quadratic part $\mathcal{R}$ of $\mathcal{P}$. The complexity of direct attacks is the complexity of the MQ-Problem determined by that of the HybridF5 algorithm [2] which is currently the fastest algorithm to solve systems of multivariate nonlinear equations.

Table 2. Running time (Second) for solving quadratic systems on $\mathbb{F}_{2^8}$

(v_1, o_1, o_2)	(5,4,5)	(6,5,5)	(7,5,6)	(8,6,6)	(9,6,7)	(10,7,7)
Random system	1.55	11.59	87.35	718.25	5565.15	43187.33
Lite-Rainbow-1	1.55	11.60	87.25	699.36	5411.12	43084.24

We compare experimental results of solving quadratic systems derived from a public key in Lite-Rainbow-1 with random quadratic systems on $\mathbb{F}_{2^8}$ using F4 algorithm (the details of F5 algorithm are not publicly known) with MAGMA in Table 2. We use as a hardware Intel Xeon E5-2687W CPU 3.1 GHz. According to these results, it makes little difference in complexities for solving two types of quadratic systems.

4 Secure Implementations of Lite-Rainbow

Now, we propose how to implement our schemes in a secure manner.

4.1 Countermeasures of Rainbow Against Side-Channel Attacks

A side-channel attack (SCA) tries to extract secret information from physical implementations of cryptographic algorithms on constrained small devices. Side-channel information obtained during cryptographic operations includes operation time, power consumption profile, electro-magnetic signal and faulty output. Since Kocher *et al.* [13] introduced timing attacks on classical PKCs running on smart cards, various SCAs and their countermeasures were proposed. Power analysis (PA) [14] which analyzes the power consumption patterns of cryptographic devices has two basic forms, simple power analysis (SPA) and differential power analysis (DPA). SPA observes power signals for executions of cryptographic operations to distinguish cryptographic primitive operations. DPA collects a number of power consumption signals and uses some sophisticated statistical tools to obtain some useful information from the data. As a countermeasure, various randomization techniques including random exponent blinding and random message blinding were proposed [6,14]. Fault injection is another very powerful cryptanalytic tool, its basic idea is to induce some faults inside a device by analyzing faulty outputs to get some meaningful information [3]. Here, we mainly concern how to randomize Rainbow to prevent the PA. Basic operations of Rainbow, multiplications and matrix multiplications in finite fields, may be vulnerable to various PA methods. Certainly, many PAs to AES are focusing on the S-box computation which consists of finite field inversions [18]. Since a finite field inversion can, in turn, be implemented with finite field multiplications, this implies feasibility of the PA to Rainbow. In [20,25], the authors proposed DPAs on SHA-1 in an MQ-signature, SFLASH, which was chosen as one of the final selection of the NESSIE project in 2003. They showed that if implementation of SHA-1 in SFLASH is careless, one can recover a secret key Δ, a random seed used for SHA-1. However, their attack cannot be applied to Rainbow-like schemes. Hence, to overcome these susceptibilities and get some robustness to the PA, it needs to randomize Rainbow.

We assume that K ia a binary field, but our method can easily be extended to a prime field case. Our strategy is that all the operations and intermediate data are randomized during signing. Rainbow consists of three basic operations, two matrix multiplications by S^{-1} and T^{-1}, and a central map inversion. To randomize the multiplication by S^{-1} and its input data h $(= h(\mathfrak{m}))$ together, we use the following identity:

$$S^{-1} \cdot h = (S^{-1} \oplus R)(h \oplus r) \oplus (S^{-1} \oplus R)r \oplus R \cdot h,$$

for a random vector $r \in K^m$ and a random $m \times m$ matrix R. Three quantities, $(S^{-1} \oplus R)(h \oplus r)$, $(S^{-1} \oplus R)r$ and $R \cdot h$ will subsequently be used in randomizing the central map inversion. In the matrix multiplication by T^{-1}, its input data t is already randomized, as it is an output of the central map inversion $\mathcal{F}^{-1}$, which introduces some randomness by choosing a random vector $(t_1, \cdots, t_{v_1})$ for Vinegar variables. Thus, it can be randomized by using

$$T^{-1} \cdot t = (T^{-1} \oplus R)t \oplus R \cdot t,$$

for an $n \times n$ random matrix R. Finally, to randomize the central map inversion of Rainbow, we explain how to randomize the UOV central map inversion, since the Rainbow central map is a two-layer UOV central map [12]. Given $s = (s_{v+1}, \cdots, s_n) \in K^o$, the UOV central map inversion is to find a solution $t \in K^n$ with $\mathcal{F}(t) = s$. To do so, we first choose a random vector $(t_1, \cdots, t_v)$ and try to find a solution $(t_{v+1}, \cdots, t_n)$ such that $f^{(k)}(t_1, \cdots, t_n) = s_k$ for component functions $f^{(k)}$ of $\mathcal{F}$ $(k = v+1, \cdots, n)$. We denote the solution $t = (t_1, \cdots, t_n)$ as $t = \mathcal{F}^{-1}_{(t_1, \cdots, t_v)}(s)$ for emphasizing its connection with a random vector $(t_1, \cdots, t_v)$. Suppose that, instead of the original input s, three values $s^{(1)}, s^{(2)}, s^{(3)}$ with $s = s^{(1)} \oplus s^{(2)} \oplus s^{(3)}$ are plugged into the central map inversion. Then we can get $t^{(1)}, t^{(2)}, t^{(3)}$ such that $t^{(l)} = \mathcal{F}^{-1}_{(t_1, \cdots, t_v)}(s^{(l)})$ for $l = 1, 2, 3$. Note that we assume that all the solutions $t^{(1)}, t^{(2)}, t^{(3)}$ have the same random components $t_1, \cdots, t_v$. Now, using the following Proposition, we can show that $t = t^{(1)} \oplus t^{(2)} \oplus t^{(3)}$ is a solution for the central map inversion with the input s.

Proposition 2. For $s^{(1)}, s^{(2)}, s^{(3)}$ with $s = s^{(1)} \oplus s^{(2)} \oplus s^{(3)}$ and $t^{(1)}, t^{(2)}, t^{(3)}$ with $t = t^{(1)} \oplus t^{(2)} \oplus t^{(3)}$, if $t^{(l)} = \mathcal{F}^{-1}_{(t_1, \cdots, t_v)}(s^{(l)})$ for $l = 1, 2, 3$, then the vector t satisfies $\mathcal{F}(t) = s$.

Proof. Since $t^{(l)} = \mathcal{F}^{-1}_{(t_1, \cdots, t_v)}(s^{(l)})$ for $l = 1, 2, 3$, we can set $t^{(l)} = (t_1, \cdots, t_v, t^{(l)}_{v+1}, \cdots, t^{(l)}_n)$, for some $t^{(l)}_j \in K$ and $j = v+1, \cdots, n$. Hence, we have

$$t = t^{(1)} \oplus t^{(2)} \oplus t^{(3)} = (t_1, \cdots, t_v, t^{(1)}_{v+1} \oplus t^{(2)}_{v+1} \oplus t^{(3)}_{v+1}, \cdots, t^{(1)}_n \oplus t^{(2)}_n \oplus t^{(3)}_n).$$

Now, $t^{(l)} = \mathcal{F}^{-1}_{(t_1, \cdots, t_v)}(s^{(l)})$ means that,

$$\sum_{i \in V, j \in O} \alpha^{(k)}_{ij} t_i t^{(l)}_j + \sum_{i,j \in V, i \leq j} \beta^{(k)}_{ij} t_i t_j + \sum_{i \in V} \gamma^{(k)}_i t^{(l)}_i + \sum_{i \in O} \gamma^{(k)}_i t_i + \eta^{(k)} = s^{(l)}_k,$$

for all k, where $s_k^{(l)}$ denotes the k-th component of $s^{(l)}$, i.e., $s^{(l)} = (s_{v+1}^{(l)}, \cdots, s_n^{(l)})$. Hence, summing up the equations above for $l = 1, 2, 3$, we get

$$\sum_{i \in V, \, j \in O} \alpha_{ij}^{(k)} t_i (t_j^{(1)} \oplus t_j^{(2)} \oplus t_j^{(3)}) + \sum_{i,j \in V, \, i \leq j} \beta_{ij}^{(k)} t_i t_j +$$

$$\sum_{i \in V} \gamma_i^{(k)} (t_j^{(1)} \oplus t_j^{(2)} \oplus t_j^{(3)}) + \sum_{i \in O} \gamma_i^{(k)} t_i + \eta^{(k)} = s_k.$$

Thus, $t = (t_1, \cdots, t_v, t_{v+1}^{(1)} \oplus t_{v+1}^{(2)} \oplus t_{v+1}^{(3)}, \cdots, t_n^{(1)} \oplus t_n^{(2)} \oplus t_n^{(3)})$ is a solution of $\mathcal{F}(\mathbf{x}) = s$, i.e., $\mathcal{F}(t) = s$. $\qquad\square$

Combining all the arguments above, we get Algorithm 3 for randomizing signing of Rainbow. Its implementation result is given in Table 4.

Algorithm 3. Randomization for Rainbow Signing

Input: a secret key of Rainbow$(v_1, o_1, \cdots, o_u)$, $(S^{-1}, \mathcal{F}, T^{-1})$ and $h = h(\mathbf{m})$
Output: a signature σ on h
   $(S^{-1}$**-Computation)**
1: Randomly choose an $m \times m$ matrix R and $r \in K^m$.
2: Compute $s^{(1)} = (s_{v_1+1}^{(1)}, \cdots, s_n^{(1)})$ by $s^{(1)} = (S^{-1} \oplus R) \cdot (h \oplus r)$.
3: Compute $s^{(2)} = (s_{v_1+1}^{(2)}, \cdots, s_n^{(2)})$ by $s^{(2)} = (S^{-1} \oplus R) \cdot r$.
4: Compute $s^{(3)} = (s_{v_1+1}^{(3)}, \cdots, s_n^{(3)})$ by $s^{(3)} = R \cdot h$.
   (Inversion of the Central Map $\mathcal{F}$)
5: Randomly generate a vector $(t_1, \cdots, t_{v_1}) \in K^{v_1}$.
6: $(t_1^{(1)}, \cdots, t_{v_1}^{(1)}) \leftarrow (t_1, \cdots, t_{v_1})$;
7: $(t_1^{(2)}, \cdots, t_{v_1}^{(2)}) \leftarrow (t_1, \cdots, t_{v_1})$;
8: $(t_1^{(3)}, \cdots, t_{v_1}^{(3)}) \leftarrow (t_1, \cdots, t_{v_1})$;
9: For $i = 1$ to u do
   i). Let $\mathcal{F}_i = (f^{(v_i+1)}, \cdots, f^{(v_{i+1})})$.
   ii). Let $t^{(l)} = (t_1^{(l)}, \cdots, t_{v_i}^{(l)}, x_{v_i+1}, \cdots, x_{v_{i+1}})$, for $l = 1, 2, 3$,
   iii). For $l = 1, 2, 3$, solve the linear equation system $\mathcal{F}_i(t^{(l)}) = (s_{v_i+1}^{(l)}, \cdots, s_{v_{i+1}}^{(l)})$ to get a solution $(x_{v_i+1}^{(l)}, \cdots, x_{v_{i+1}}^{(l)}) = (t_{v_i+1}^{(l)}, \cdots, t_{v_{i+1}}^{(l)})$, if any.
   If there is no solution, go to the step 5) and repeat the process.
10: $t^{(1)} \leftarrow (t_1^{(1)}, \cdots, t_n^{(1)})$; $t^{(2)} \leftarrow (t_1^{(2)}, \cdots, t_n^{(2)})$; $t^{(3)} \leftarrow (t_1^{(3)}, \cdots, t_n^{(3)})$;
13: $t \leftarrow t^{(1)} \oplus t^{(2)} \oplus t^{(3)}$;
   $(T^{-1}$**-Computation)**
14: Choose a $n \times n$ random matrix R.
15: $\sigma^{(1)} \leftarrow (T^{-1} \oplus R) \cdot t$;
16: $\sigma^{(2)} \leftarrow R \cdot t$;
17: $\sigma \leftarrow \sigma^{(1)} \oplus \sigma^{(2)}$;
return σ

Table 3. Key Sizes (KB) of Rainbow and Lite-Rainbow.

Security level	Scheme	Public key	Secret key
80	Rainbow ($\mathbb{F}_{2^8}$,17,13,13)	25.1	19.1
	Lite-Rainbow-0 ($\mathbb{F}_{2^8}$,17,10,16)	25.1	**0.009766**
	Lite-Rainbow-1 ($\mathbb{F}_{2^8}$,17,10,16)	**8.38**	**16.79**
128	Rainbow($\mathbb{F}_{2^8}$,36,21,22)	136.1	102.5
	Lite-Rainbow-0 ($\mathbb{F}_{2^8}$,36,12,31)	136.1	**0.01563**
	Lite-Rainbow-1 ($\mathbb{F}_{2^8}$,36,12,31)	**39.56**	**96.54**

4.2 Selection of Secure and Optimal Parameters

In Rainbow, to find a good compromise between public and secret key size, $(\mathbb{F}_q, v_1, o_1, o_2) = (\mathbb{F}_{2^8}, 17, 13, 13)$ for an 80-bit security level and $(36, 21, 22)$ for a 128-bit security level are chosen. In Lite-Rainbow, secure parameter (m, n, v_1, o_1, o_2) for 80-bit and **128-bit** security level should be selected so that they defend the following five attacks [23]:

– Direct attack: $m \geq 26, \mathbf{43}$,
– Rainbow-Band-Separation attack: $n \geq 43, \mathbf{79}$,
– MinRank attack: $v_1 \geq 9, \mathbf{15}$, HighRank attack: $o_2 \geq 10, \mathbf{16}$,
– UOV attack (the Kipnis-Shamir attack): $n - 2o_2 \geq 11, \mathbf{17}$.

Finally, we can choose secure and optimal parameters with the smallest public key size for Lite-Rainbow-1 as $(\mathbb{F}_{2^8}, 17, 10, 16)$ for $\lambda = 80$ and $(\mathbb{F}_{2^8}, 36, 12, 31)$ for $\lambda = 128$. In Table 3, we compare the key sizes of Lite-Rainbow with Rainbow. Compared to Rainbow, the secret key size in Lite-Rainbow-0 and the public key size in Lite-Rainbow-1 are reduced by factors 99.8 % and 71 %, respectively.

4.3 Implementations

Now, we evaluate performance of Rainbow, Lite-Rainbow-0,-1, and their protected ones resistant to PA, and ECDSA, RSA for the same security level on the same platform. We use Rainbow, ECDSA and RSA based on the open source code in [1]. Details of RSA and ECDSA in our implementations are as follows:

– RSA-1024 and RSA-3072: 1024-bit and 3072-bit RSA signature with message recovery, respectively.
– ECDSA-160: ECDSA using the standard SECP160R1 elliptic curve, a curve modulo the prime $2^{160} - 2^{31} - 1$.
– ECDSA-283: ECDSA using the standard NIST B-283 elliptic curve, a curve over a field of size 2^{283}.

For consistent results, we use as a hardware Intel Xeon E5-2687W CPU 3.1 GHz. To speed up performance of CTR-PRNG using AES, we implement our schemes

Table 4. Implementation results (Cycles) of Rainbow, Lite-Rainbow-0, -1 and protected ones, and RSA, ECDSA.

λ	Signature scheme	Sign	Verify	Protected sign
80	RSA-1024	1,061,719	28,258	
	ECDSA-160	589,249	620,296	
	Rainbow ($\mathbb{F}_{2^8}$,17,13,13)	47,152	31,684	154,929
	Lite-Rainbow-0 ($\mathbb{F}_{2^8}$,17,10,16)	**74,179**	**32,103**	188,806
	Lite-Rainbow-1 ($\mathbb{F}_{2^8}$,17,10,16)	**48,165**	**59,178**	152,501
128	RSA-3072	12,730,848	105,648	
	ECDSA-283	1,523,052	2,924,292	
	Rainbow ($\mathbb{F}_{2^8}$,36,21,22)	162,821	132,419	539,338
	Lite-Rainbow-0 ($\mathbb{F}_{2^8}$,36,12,31)	**299,753**	**130,380**	628,552
	Lite-Rainbow-1 ($\mathbb{F}_{2^8}$,36,12,31)	**163,401**	**270,323**	514,609

on CPU supporting Advanced Encryption Standard New Instructions (AES-NI). We summarize implementation results of our schemes, RSA, and ECDSA, and unprotected ones and protected ones for optimal parameters in Table 4. Performance results given in Table 4 are averages of 1,000 measurements for each function using C++ with g++ compiler. Compared to Rainbow, when the total key size is decreased by 44 %, cost for signing and verification has increased by 34 % – 45 % in Lite-Rainbow-0 and Lite-Rainbow-1. Nevertheless, verification in Lite-Rainbow-0 and Lite-Rainbow-1 is about 22 times and 11 times faster than that in ECDSA, respectively, at the 128-bit security level. Signing in Lite-Rainbow-0 and Lite-Rainbow-1 is about 42 times and 78 times faster than that in RSA, respectively.

5 Conclusion

We have proposed lightweight variants of Rainbow, Lite-Rainbow-0 and Lite-Rainbow-1 for constrained devices. By replacing some parts of a public key, or a secret key with small random seeds, we reduce the secret key in Lite-Rainbow-0 and the public key size in Lite-Rainbow-1 by factors about 99.8 % and 71 %, respectively, compared to Rainbow. We have proved unforgeability of our scheme with special parameter sets (K, v_1, o_1, o_2) in the ROM under the hardness assumption of the MQ-problem. We have proposed a randomizing signing for Rainbow-like schemes resistant to the PA. Finally, we have provided a direct comparison of implementation results for our schemes, RSA, and ECDSA, and unprotected one and protected one for optimal parameters on the same platform. Modern microprocessors support an embedded AES accelerator for the PRNG and hash function, where AES operations are independently executed on the AES accelerator. We believe that our schemes are leading candidates for these constrained devices. Future works consist of optimizations for performance and investigation for practicability of our signature schemes on constrained devices such as an 8-bit microcontroller.

Acknowledgments. This research was supported by the National Institute for Mathematical Sciences funded by Ministry of Science, ICT and Future Planning of Korea (B21503-1).

Appendix

The most general security notion of a public-key signature (PKS) scheme is existential unforgeability under an adaptively chosen-message attack (EUF-acma). Its formal security model is defined as follows:

Unforgeability of Signature Schemes Against EUF-acma. An adversary $\mathcal{A}$'s advantage $Adv_{\mathcal{PKS},\mathcal{A}}$ is defined as its probability of success in the following game between a challenger $\mathcal{C}$ and $\mathcal{A}$:

- **Setup.** The challenger runs Setup algorithm and its resulting system parameters are given to $\mathcal{A}$.
- **Queries.** $\mathcal{A}$ issues the following queries:
  - **Sign Query:** Adaptively, $\mathcal{A}$ requests a signature on a message m_i, $\mathcal{C}$ returns a signature σ_i.
- **Output.** Eventually, $\mathcal{A}$ outputs σ^* on a message m^* and wins the game if (i) Verify$(\mathrm{m}^*, \sigma^*) = 1$ and ii) m^* has never requested to the Sign oracle.

Definition 2. A forger $\mathcal{A}(t, q_H, q_S, \epsilon)$-breaks a PKS scheme if $\mathcal{A}$ runs in time at most t, $\mathcal{A}$ makes at most q_H queries to the hash oracle, q_S queries to the signing oracle and $Adv_{\mathcal{PKS},\mathcal{A}}$ is at least ϵ. A PKS scheme is (t, q_E, q_S, ϵ)-EUF-acma if no forger (t, q_H, q_S, ϵ)-breaks it in the above game.

Theorem 1. If the MQ-problem is (t', ε')-hard, Lite-Rainbow-1(K, v_1, o_1, o_2) is $(t, q_H, q_S, q_{\mathcal{G}}, \varepsilon)$-existential unforgeable against an adaptively chosen message attack, for any t and ε satisfying

$$\varepsilon \geq \mathrm{e} \cdot (q_S + 1) \cdot \varepsilon', \ t' \geq t + q_H \cdot c_V + q_S \cdot c_S + c_{\mathcal{G}},$$

where e is the base of the natural logarithm, and c_S, c_V and $c_{\mathcal{G}}$ are time for a signature generation, a signature verification and a $\mathcal{G}$ evaluation to recover some parts of a public key, respectively, provided that the parameter set (K, v_1, o_1, o_2) is chosen to be resistant to the KRAs.

Proof. A random instance of the MQ-problem $(\mathcal{P}, \eta)$ is given. Suppose that $\mathcal{A}$ is a forger who breaks Lite-Rainbow-1(K, v_1, o_1, o_2) with the target public key $\mathcal{P}$. By using $\mathcal{A}$, we will construct an algorithm $\mathcal{B}$ which outputs a solution $\mathbf{x} \in K^n$ such that $\mathcal{P}(\mathbf{x}) = \eta$. Algorithm $\mathcal{B}$ performs the following simulation by interacting with $\mathcal{A}$.

Setup. Algorithm $\mathcal{B}$ chooses a random seed se and set $PK = (\mathrm{se}, P_{12}, P_R, P_{L12})$ as a public key, where P_{12}, P_R, P_{L12} are the corresponding parts of $\mathcal{P}$.

At any time, $\mathcal{A}$ can query random oracles, $\mathcal{G}$ and H, and Sign oracle. To answer these queries, $\mathcal{B}$ does the following:

$H-$**Queries.** To respond to H-queries, $\mathcal{B}$ maintains a list of tuples $(\mathbf{m}_i, c_i, \tau_i, \mathcal{P}(\tau_i))$ as explained below. We refer to this list as H-list. When $\mathcal{A}$ queries H at a point $\mathbf{m}_i \in \{0,1\}^*$,

1. If the query already appears on H-list in a tuple $(\mathbf{m}_i, c_i, \tau_i, \mathcal{P}(\tau_i))$ then $\mathcal{B}$ returns $H(\mathbf{m}_i) = \mathcal{P}(\tau_i)$.
2. Otherwise, $\mathcal{B}$ picks a random coin $c_i \in \{0,1\}$ with $Pr[c_i = 0] = \frac{1}{q_S+1}$.

- If $c_i = 1$ then $\mathcal{B}$ chooses a random $\tau_i \in K^n$, adds a tuple $(\mathbf{m}_i, c_i, \tau_i, \mathcal{P}(\tau_i))$ to H-list and returns $H(\mathbf{m}_i) = \mathcal{P}(\tau_i)$.
- If $c_i = 0$ then $\mathcal{B}$ adds $(\mathbf{m}_i, c_i, *, \eta)$ to H-list from the instance, and returns $H(\mathbf{m}_i) = \eta$.

Sign Queries. When $\mathcal{A}$ makes a Sign-query on $\mathbf{m}_i$, $\mathcal{B}$ finds the corresponding tuple $(\mathbf{m}_i, c_i, \tau_i, \mathcal{P}(\tau_i))$ from H-list.

- If $c_i = 1$ then $\mathcal{B}$ responds with τ_i.
- If $c_i = 0$ then $\mathcal{B}$ reports failure and terminates.

To verify a signature, $\mathcal{A}$ has to query the random oracle $\mathcal{G}$.

$\mathcal{G}-$**Queries.** To verify a signature, $\mathcal{A}$ must query the oracle $\mathcal{G}$. When $\mathcal{A}$ queries G at a point $\mathbf{se} \in \{0,1\}^\lambda$, $\mathcal{B}$ returns $\mathcal{G}(\mathbf{se}) = (P_{11}, P_{21}, P_{22}, P_{L11}, P_{L21}, P_{L22}, P_C)$, where $P_{11}, P_{21}, P_{22}, P_{L11}, P_{L21}, P_{L22}, P_C$ are the corresponding parts of $\mathcal{P}$.

All responses to Sign queries not aborted are valid. If $\mathcal{B}$ doesn't abort as a result of $\mathcal{A}$'s Sign query then $\mathcal{A}$'s view in the simulation is identical to its view in the real attack.

Output. Finally, $\mathcal{A}$ produces a signature $\mathbf{x}^*$ on a message $\mathbf{m}^*$. If it is not valid then $\mathcal{B}$ reports failure and terminates. Otherwise, a query on m^* already appears on H-list in a tuple $(\mathbf{m}^*, c^*, \tau^*, \mathcal{P}(\tau^*))$: if $c_* = 1$ then reports failure and terminates. Otherwise, $c^* = 0$, i.e., $(c^*, \mathbf{m}^*, *, \eta)$, then $\mathcal{P}(\mathbf{x}^*) = \eta$. Finally, $\mathcal{B}$ outputs $\mathbf{x}^*$ is a solution of $\mathcal{P}$.

To show that $\mathcal{B}$ solves the given instance with probability at least ε', we analyze four events needed for $\mathcal{B}$ to succeed:

- E_1: $\mathcal{B}$ doesn't abort as a result of $\mathcal{A}$'s Sign query.
- E_2: $\mathcal{A}$ generates a valid and nontrivial signature forgery σ on $\mathbf{m}_i$.
- E_3: Event E_2 occurs, $c_i = 0$ for the tuple containing $\mathbf{m}_i$ in the H-list.

Algorithm $\mathcal{B}$ succeeds if all of these events happen. The probability $Pr[E_1 \wedge E_3]$ is decomposed as

$$Pr[E_1 \wedge E_3] = Pr[E_1] \cdot Pr[E_2 \wedge E_1] \cdot Pr[E_3 | E_1 \wedge E_2] \cdot (*).$$

The probability that $\mathcal{B}$ doesn't abort as a result of $\mathcal{A}$'s Sign query is at least $(1 - \frac{1}{q_S+1})^{q_S}$ since $\mathcal{A}$ makes at most q_S queries to the Sign oracle. Thus, $Pr[E_1] \geq$

$(1 - \frac{1}{q_S+1})^{q_S}$. If $\mathcal{B}$ doesn't abort as a result of $\mathcal{A}$'s Sign query then $\mathcal{A}$'s view is identical to its view in the real attack. Hence, $Pr[E_1 \wedge E_2] \geq \varepsilon$. Given that events E_1, E_2 and E_3 happened, $\mathcal{B}$ will abort if $\mathcal{A}$ generates a forgery with $c_i = 1$. Thus, all the remaining c_i are independent of $\mathcal{A}$'s view. Since $\mathcal{A}$ could not have issued a signature query for the output we know that c is independent of $\mathcal{A}$'s current view and therefore $Pr[c = 0 | E_1 \wedge E_2] = \frac{1}{q_S+1}$. Then we get $Pr[E_3 | E_1 \wedge E_2] \geq \frac{1}{q_S+1}$. From $(*)$, $\mathcal{B}$ produces the correct answer with probability at least

$$(1 - \frac{1}{q_S + 1})^{q_S} \cdot \varepsilon \cdot \frac{1}{q_S + 1} \geq \frac{1}{\mathsf{e}} \cdot \frac{\varepsilon}{(q_S + 1)} \geq \varepsilon'.$$

Algorithm $\mathcal{B}$'s running time is the same as $\mathcal{A}$'s running time plus the time that takes to respond to q_H H-queries, and q_S Sign-queries. The H- and Sign-queries require a signature verification and a signature generation, respectively. We assume that a signature generation, a signature verification and a $\mathcal{G}$ evaluation to recover some parts of a public key take time c_S, c_V and $c_{\mathcal{G}}$, respectively. Hence, the total running time is at most $t' \geq t + q_H \cdot c_V + q_S \cdot c_S + c_{\mathcal{G}}$. □

References

1. Bernstein, D.J., Lange, T.: eBACS: ECRYPT benchmarking of cryptographic systems. http://bench.cr.yp.to
2. Bettale, L., Faugére, J.-C., Perret, L.: Hybrid approach for solving multivariate systems over finite fields. J. Math. Cryptol. **3**, 177–197 (2009)
3. Boneh, D., DeMillo, R.A., Lipton, R.J.: On the importance of checking cryptographic protocols for faults. In: Fumy, W. (ed.) EUROCRYPT 1997. LNCS, vol. 1233, pp. 37–51. Springer, Heidelberg (1997)
4. Bulygin, S., Petzoldt, A., Buchmann, J.: Towards Provable security of the unbalanced oil and vinegar signature scheme under direct attacks. In: Gong, G., Gupta, K.C. (eds.) INDOCRYPT 2010. LNCS, vol. 6498, pp. 17–32. Springer, Heidelberg (2010)
5. Chen, A.I.-T., Chen, M.-S., Chen, T.-R., Cheng, C.-M., Ding, J., Kuo, E.L.-H., Lee, F.Y.-S., Yang, B.-Y.: SSE implementation of multivariate PKCs on modern x86 CPUs. In: Clavier, C., Gaj, K. (eds.) CHES 2009. LNCS, vol. 5747, pp. 33–48. Springer, Heidelberg (2009)
6. Coron, J.-S.: Resistance against differential power analysis for elliptic curve cryptosystems. In: Koç, Ç.K., Paar, C. (eds.) CHES 1999. LNCS, vol. 1717, pp. 292–302. Springer, Heidelberg (1999)
7. Ding, J., Schmidt, D.: Rainbow, a new multivariable polynomial signature scheme. In: Ioannidis, J., Keromytis, A.D., Yung, M. (eds.) ACNS 2005. LNCS, vol. 3531, pp. 164–175. Springer, Heidelberg (2005)
8. ETSI: ETSI 2nd Quantum-Safe Crypto Workshop in partnership with the IQC. http://www.etsi.org/news-events/events/770-etsi-crypto-workshop-2014. (Accessed: September 2014)
9. Garey, M.R., Johnson, D.S.: Computers and Intractability: A Guide to the Theory of NP-Completeness. W.H. Freeman, New York (1979)
10. Hashimoto, Y., Takagi, T., Sakurai, K.: General Fault attacks on multivariate public key cryptosystems. In: Yang, B.-Y. (ed.) PQCrypto 2011. LNCS, vol. 7071, pp. 1–18. Springer, Heidelberg (2011)

11. Hoffstein, J., Pipher, J., Silverman, J.H.: NTRU: a ring-based public key cryptosystem. In: Buhler, J.P. (ed.) ANTS 1998. LNCS, vol. 1423, pp. 267–288. Springer, Heidelberg (1998)

12. Kipnis, A., Patarin, J., Goubin, L.: Unbalanced oil and vinegar signature schemes. In: Stern, J. (ed.) EUROCRYPT 1999. LNCS, vol. 1592, pp. 206–222. Springer, Heidelberg (1999)

13. Kocher, P.C.: Timing attacks on implementations of Diffie-Hellman, RSA, DSS, and other systems. In: Koblitz, N. (ed.) CRYPTO 1996. LNCS, vol. 1109, pp. 104–113. Springer, Heidelberg (1996)

14. Kocher, P.C., Jaffe, J., Jun, B.: Differential power analysis. In: Wiener, M. (ed.) CRYPTO 1999. LNCS, vol. 1666, pp. 388–397. Springer, Heidelberg (1999)

15. Matsumoto, T., Imai, H.: Public quadratic polynomial-tuples for efficient signature-verification and message-encryption. In: Günther, C.G. (ed.) EUROCRYPT 1988. LNCS, vol. 330, pp. 419–453. Springer, Heidelberg (1988)

16. McEliece, R.: A Public-Key Cryptosystem Based on Algebraic Coding Theory, DSN Progress Report 42–44. Jet Propulsion Laboratories, Pasadena (1978)

17. Merkle, R.C.: A digital signature based on a conventional encryption function. In: Pomerance, C. (ed.) CRYPTO 1987. LNCS, vol. 293, pp. 369–378. Springer, Heidelberg (1988)

18. Messerges, T.S., Dabbish, E.A., Sloan, R.H.: Investigations of Power analysis attacks on smartcards. In: USENIX 1999 (1999)

19. NIST: Workshop on Cybersecurity in a Post-Quantum World, post-quantum-crypto-workshop-2015.cfm. http://www.nist.gov/itl/csd/ct/. (Accessed: September 2014)

20. Okeya, K., Takagi, T., Vuillaume, C.: On the importance of protecting $\triangle$ in SFLASH against side channel attacks. IEICE Trans. **88-A**(1), 123–131 (2005)

21. Patarin, J.: Hidden fields equations (HFE) and isomorphisms of polynomials (IP): two new families of asymmetric algorithms. In: Maurer, U.M. (ed.) EUROCRYPT 1996. LNCS, vol. 1070, pp. 33–48. Springer, Heidelberg (1996)

22. Petzoldt, A., Bulygin, S., Buchmann, J.: Selecting parameters for the rainbow signature scheme. In: Sendrier, N. (ed.) PQCrypto 2010. LNCS, vol. 6061, pp. 218–240. Springer, Heidelberg (2010)

23. Petzoldt, A., Bulygin, S., Buchmann, J.: CyclicRainbow – a multivariate signature scheme with a partially cyclic public key. In: Gong, G., Gupta, K.C. (eds.) INDOCRYPT 2010. LNCS, vol. 6498, pp. 33–48. Springer, Heidelberg (2010)

24. Shor, P.W.: Polynomial-time algorithms for prime factorization and discrete logarithms on a quantum computer. SIAM J. Comput. **26**, 1484–1509 (1997)

25. Steinwandt, R., Geiselmann, W., Beth, T.: A theoretical DPA-based cryptanalysis of the NESSIE candidates FLASH and SFLASH. In: Davida, G.I., Frankel, Y. (eds.) ISC 2001. LNCS, vol. 2200. Springer, Heidelberg (2001)

26. Thomae, E.: About the Security of Multivariate Quadratic Public Key Schemes, Dissertation Thesis by Dipl. math. E. Thomae, RUB (2013)

27. Zalka, C.: Shor's algorithm with fewer (pure) qubits. arXiv:quant-ph/0601097

Lossy Projective Hashing and Its Applications

Haiyang Xue[1,2], Yamin Liu[1,2(✉)], Xianhui Lu[1,2], and Bao Li[1,2]

[1] Data Assurance and Communication Security Research Center,
Chinese Academy of Sciences, Beijing, China
[2] State Key Laboratory of Information Security,
Institute of Information Engineering, Chinese Academy of Sciences, Beijing, China
{hyxue12,ymliu,xhlu,lb}@is.ac.cn

Abstract. In this paper, we introduce a primitive called lossy projective hashing. It is unknown before whether smooth projective hashing (Cramer-Shoup, Eurocrypt'02) can be constructed from dual projective hashing (Wee, Eurocrypt'12). The lossy projective hashing builds a bridge between dual projective hashing and smooth projective hashing. We give instantiations of lossy projective hashing from DDH, DCR, QR and general subgroup membership assumptions (including 2^k-th residue, p-subgroup and higher residue assumptions). We also show how to construct lossy encryption and fully IND secure deterministic public key encryption from lossy projective hashing.

- We give a construction of lossy projective hashing via dual projective hashing. We prove that lossy projective hashing can be converted to smooth projective hashing via pairwise independent hash functions, which in turn yields smooth projective hashing from dual projective hashing.
- We propose a direct construction of lossy encryption via lossy projective hashing. Our construction is different from that given by Hemenway *et al.* (Eurocrypt 2011) via smooth projective hashing. In addition, we give a fully IND secure deterministic public key encryption via lossy projective hashing and one round UCE secure hash functions recently introduced by Bellare *et al.* (Crypto 2013).

Keywords: Lossy projective hashing · Dual projective hashing · Smooth projective hashing · Lossy encryption · Deterministic public key encryption

1 Introduction

Projective Hashing. Cramer and Shoup introduced smooth projective hashing [11] by generalizing their practical chosen ciphertext secure (CCA) encryption scheme under DDH assumption [10]. There are many other applications of smooth projective hashing beyond CCA secure PKE, such as password-based authenticated key exchange [6,20], oblivious transfer [15] and leakage resilient encryption [24,29]. A smooth projective hashing is a family of hash functions $\{H(k,x)\}$ where k is a hashing key and x is an instance from some "hard" language. The projective

© Springer International Publishing Switzerland 2015
A. Biryukov and V. Goyal (Eds.): INDOCRYPT 2015, LNCS 9462, pp. 64–84, 2015.
DOI: 10.1007/978-3-319-26617-6_4

property requires that for any YES instance x, the hash value $H(k, x)$ be determined by a projective map $\alpha(k)$ and x. The smoothness property requires that for any NO instance x, $\alpha(k)$ contain no information of $H(k, x)$. As a result, one can not guess $H(k, x)$ even knowing $\alpha(k)$ and the NO instance x.

Wee [31] proposed dual projective hashing in order to abstract various constructions of lossy trapdoor functions [28]. The dual projective hashing also contains a hash family $\{H(k, x)\}$ with a hashing key k and a language instance x. The difference is that, for any NO instance x, instead of the smoothness property, it requires invertibility, which means that there is a trapdoor to recover the hashing key k given the projected key $\alpha(k)$ and the hash value $H(k, x)$. The dual projective hashing can also be used to construct deterministic public key encryption (DPKE) schemes.

While the "smoothness" property concerns that the projected key and the instance tell nothing about the hash value, the "invertibility" property requires that the projected key and hash value reveal the hashing key via a trapdoor. The "smoothness" in smooth projective hashing provides security while the "invertibility" in dual projective hashing provides functionality.

It is unknown before whether smooth projective hashing can be constructed from dual projective hashing. One aim of this paper is to address this issue.

Lossy Encryption. In [27], Peikert *et al.* defined the dual-mode encryption with two modes. In the normal mode, the encryption and decryption algorithms behave as usual, and in the *lossy* mode, the encryption loses information of the message. In [3], Bellare *et al.* defined lossy encryption, extending the definition of dual-mode encryption in [27] and meaningful/meaningless encryption in [22]. Lossy encryption is useful for achieving selective opening security [3,16].

There are two types of methodology for constructing lossy encryption schemes. The first one is to embed the instance of the hard language into the randomness, such that for YES instances, the scheme provides normal functionality, and for NO instances, the randomness is extended to the message space and statistically cover the information of the message. The other one is to embed the instance of the hard language into the message, such that for NO instances, the scheme provides normal functionality, and for YES instances the message is lost into the randomness space. Consider QR based lossy encryption in [3,16] for instance in Table 1. Hemenway *et al.* [16] gave a general construction of lossy encryption based on smooth projective hashing which fits into the first methodology. For a smooth projective hashing with projective map $\alpha(k)$ and hash value $H(k, x)$, the ciphertext is $(\alpha(r), H(r, x) + m)$ where r is the randomness and m is the message. The language instance x is embedded into the randomness. For any YES instance x, the projective property provides the decryption functionality. For any NO instance x, $H(r, x)$ is statistically close to the uniform distribution, thus, $H(r, x) + m$ is statistically independent of the message m.

A natural question is that: whether there exists a general construction of lossy encryption from some hashing system which fits into the second methodology.

Table 1. Two methods of lossy encryption based on QR. We emphasize the key portion in boldface. In the ciphertext column, m is message and r is randomness.

	pk_{lossy} in lossy mode	pk in normal mode	Ciphertext
Method 1 [16]	$-1 \in$ QNR, **QR** $=<g>$	$-1 \in$ QNR, **QR∪QNR** $=<g>$	$(-1)^m g^r$
Method 2 [3]	$y \in$ **QNR**, QR $=<g>$	$y \in$ **QR**, QR $=<g>$	$y^m g^r$

To sum up, the motivations of this paper are to further clarify the relationship between the smooth projective hashing and the dual projective hashing and to give a general construction of lossy encryption from some hashing system.

1.1 Our Contributions

In this paper, we introduce the notion of lossy projective hashing. In lossy projective hashing, there is also a family of hash functions $\{H(k, x)\}$ with a hashing key k and a language instance x. The hashing key contains two components, k_1 and k_2. The projected key $\alpha(k_1, k_2)$ exactly loses the information of k_2 in both NO and YES instances. We also require the projective property, that is for any YES instance x, the hash value is determined by projected key and x. Instead of full invertibility, we require *partial invertibility*, which stipulates that there is a trapdoor, for the hash value $H(k_1, k_2, x)$ on a NO instance x, allowing us to efficiently recover k_2, the message part of the hashing keys.

There are two conceptual differences between dual projective hashing and dual projective hashing. The first one is that the dual projective hashing only requires the projective map to lose some information of the hashing key, while lossy projective hashing requires losing a specific part of the hashing keys, and the specific part is corresponding to the message in derived lossy encryption. The second one is that, on any NO instance, the dual projective hashing concerns that the projected key and hash value reveal the whole hashing key, while the lossy projective hashing is interested in that the projected key and hash value tell only the message part of hashing key which is lost by the projected key.

The difference between lossy projective hashing and smooth projective hashing is more obvious. On any NO instance, the "smoothness" property requires that the projected key and the hash value tell nothing about the hashing key, while lossy projective hashing concerns that the projected key and hash value tell the message part of hashing key.

From and to Other Projective Hashing. Interestingly, the lossy projective hashing builds a bridge between dual projective hashing and smooth projective hashing.

We give a construction of lossy projective hashing via dual projective hashing. Let $\alpha_d(k_1)$ be the projected key in dual projective hashing with lossiness of l bits, and h be a pairwise independent hash function with input k_1 and output length $l - 2\log(1/\epsilon)$. The average min-entropy of k_1 given $\alpha_d(k_1)$ is larger than the output length of h. For any key $k_2 \in \{0, 1\}^{l - 2\log(1/\epsilon)}$, $(\alpha_d(k_1), h(k_1) \oplus k_2)$ is the projected key in lossy projective hashing. $(H_d(k_1, x), h(k_1) \oplus k_2)$ is the hash value in the lossy projective hashing. The inversion algorithm of k_1, provides the possibility to recover k_2.

We then construct smooth projective hashing via lossy projective hashing and pairwise independent hash functions. Cramer and Shoup [11] showed the general construction of smooth projective hashing from universal projective hashing via pairwise independent hash functions. Here, we need only to prove that lossy projective hashing has the universal property. This result in turn yields a smooth projective hashing from dual projective hashing.

Instantiations. We give the instantiations of lossy projective hashing from DDH, DCR, QR and general subgroup membership (SGA) assumptions (including 2^k-th residue [14], p-subgroup [25] and higher residue [23] assumptions). The instantiation based on DDH is implied by the works of [17,18] on lossy encryption. Hemenway and Ostrovsky [17] proved that both QR and DCR assumptions imply the extended DDH assumption. We use the result given by Hemenway and Ostrovsky [17] in the instantiations based on QR and DCR assumptions (Fig. 1).

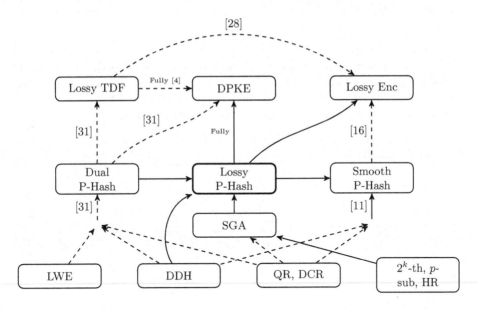

Fig. 1. Contributions of this paper. $\rightarrow$ Shown in this paper, $--\rightarrow$ Shown in previous papers

Lossy Encryption. We propose a direct construction of lossy encryption via lossy projective hashing. This generic construction is different from that given in [16] via smooth projective hashing. The instance of subset membership assumption is the public key. The NO instance is the public key in the normal mode while the YES instance is the public key in the lossy mode. The key pair (k_1, k_2)

play the roles of randomness and message respectively. The ciphertext of m is $(\alpha(r, m), H(r, m, x))$. The projective and lossy properties imply that, in the lossy mode, the ciphertext is statistically independent of the message m. The functionality of decryption is provided by the partial inversion algorithm.

Fully Secure DPKE. We also give a fully IND secure DPKE in the standard model via lossy projective hashing and UCE secure hash recently introduced by Bellare *et al.* [5]. Bellare *et al.* [4] gave the first fully IND secure DPKE by combining lossy trapdoor function and UCE secure hash. In their scheme, two rounds of UCE secure hash are required. We prove that lossy projective hashing + one round UCE secure hash is enough for fully IND secure DPKE.

1.2 Related Works

Previous Lossy Encryptions. It has been mentioned that most previous lossy encryption schemes [3,16,18,27] fall into the two methodologies. However, there are two exceptions. One is the RSA-OAEP scheme which is proved to be IND-CPA secure (in fact it is a lossy encryption) by Kiltz *et al.* [21] under the Φ-hiding assumption. The other is Rabin-OAEP which is proved to be a lossy encryption by Xue *et al.* [32] under the 2^k-$\Phi/4$-hiding assumption.

Fully Secure DPKE from LTDF+UCE. Previously, some standard model DPKE schemes [7,8,13,30,31] have been proposed. As noted by Bellare *et al.* in [4], those schemes only achieved security for block sources. Bellare *et al.* in [4] gave the first fully IND secure DPKE in the standard model via the LTDF + UCE method which employs two rounds of UCE secure hash.

2 Preliminaries

2.1 Notations

If S is a set, we denote by $|S|$ the cardinality of S, and denote by $x \leftarrow S$ the process of sampling x uniformly from S. If A is an algorithm, we denote by $z \leftarrow A(x, y, \cdots)$ the process of running A with input $x, y, \cdots$ and outputting z. For an integer n, we denote by $[n]$ the set of $\{0, 1, \cdots, n-1\}$. A function is *negligible* if for every $c > 0$ there exists a λ_c such that $f(\lambda) < 1/\lambda^c$ for all $\lambda > \lambda_c$. If A and B are distributions, $A =_s B$ means that the statistical distance between A and B is negligible.

2.2 Subset Membership Assumption

We first recall the definition of subset membership assumption. (Multiple versions of this assumption have appeared; we use the version in [31].)

Definition 1 (Subset Membership Assumption [31]). *Let PP be public parameters. Let Π_Y and Π_N be a pair of disjoint sets. The subset membership assumption states that the advantage of any PPT adversary A in distinguishing* the uniform distribution over Π_Y and Π_N is negligible even given PP. For any PPT adversary A, the advantage function

$$Adv_A^{SMA} = |Pr[A(PP, x) = 1 : x \leftarrow \Pi_N] - Pr[A(PP, x) = 1 : x \leftarrow \Pi_Y]|$$

is negligible.

3 Lossy Projective Hashing

In this section, we give the formal description of lossy projective hashing.

Definition 2 (Lossy Projective Hashing).

Setup. *On input the security parameter λ, output the public parameters PP and the hashing system $(H, X = \Pi_Y \cup \Pi_N, K = K_1 \times K_2, \Pi, \alpha, S)$. PP contains the subset membership problem associated with the language x, and Π_Y and Π_N correspond to YES and NO instances. There are a pair of sample algorithms respectively. SampYes(λ) outputs an uniform random x in Π_Y with corresponding witness w. SampNo(λ) outputs an uniform random x in Π_N with corresponding trapdoor t. $K = K_1 \times K_2$ is the key space with two parts. $H : K \times X \to \Pi$ is the hash map. $\alpha : K_1 \times K_2 \to S$ is a projective map of hashing keys.*

Lossiness. *For any $\widetilde{k}_2, \widehat{k}_2 \in K_2$, the statistical distance between the distributions $\{\alpha(K_1, \widetilde{k}_2)\}$ and $\{\alpha(K_1, \widehat{k}_2)\}$ is ϵ, which is negligible in λ. That is, for any $\widetilde{k}_2, \widehat{k}_2 \in K_2$,*

$$\{\alpha(k_1, \widetilde{k}_2)|k_1 \leftarrow K_1\} =_s \{\alpha(k_1, \widehat{k}_2)|k_1 \leftarrow K_1\}.$$

Projective Hashing. *For all $(x, w) \leftarrow$ SampYes(λ), and all $(k_1, k_2) \in K$, $\alpha(k_1, k_2)$ determines the hash value $H(k_1, k_2, x)$. Precisely, there exists an efficient algorithm Pub such that for all $(k_1, k_2) \in K$, and $(x, w) \leftarrow$ SampYes(λ),*

$$Pub(\alpha(k_1, k_2), x, w) = H(k_1, k_2, x).$$

Partial Invertibility. *For all $(x, t) \leftarrow$ SampNo(λ), and all $(k_1, k_2) \in K$, there exists an efficient inversion algorithm $TdInv$ such that*

$$TdInv(x, t, \alpha(k_1, k_2), H(k_1, k_2, x)) = k_2.$$

4 Dual and Smooth Projective Hashing

In this section, we investigate the relationship among lossy projective hashing, dual projective hashing, and smooth projective hashing. We construct lossy projective hashing from dual projective hashing in Sect. 4.1 and reduce smooth projective hashing to lossy projective hashing in Sect. 4.2, which implies that smooth projective hashing can be constructed from dual projective hashing.

4.1   From Dual Projective Hashing

We first recall the definition of dual projective hashing in [31], with which we construct lossy projective hashing.

Definition 3 (Dual Projective Hashing [31]). *Assume that the subset membership problem is over* (Π_Y, Π_N).

Setup. *Let* PP_d *be the public parameters,* $\alpha_d(k)$ *be the projective map of key. Let* K_d *be the key space and* $|K_d| = 2^n$. $(x, w) \leftarrow Samp\,Yes(\lambda)$ *and* $(x, t) \leftarrow SampNo(\lambda)$.

Lossiness. *For any* $k \in K_d$, *the image size of* $\alpha_d(k)$ *is at most* 2^{n-l}.

Projective Hashing. *For all* $x \in \Pi_Y$, *and all* $k \in K_d$, $\alpha_d(k)$ *determines the hash value* $H_d(k, x)$. *Precisely, there exists an efficient algorithm* Pub_d *such that for all* $\alpha_d(k)$, $(x, w) \leftarrow Samp\,Yes(\lambda)$,

$$Pub_d(\alpha_d(k), x, w) = H_d(k, x).$$

Invertibility. *For all* $(x, t) \leftarrow SampNo(\lambda)$, *and all* $k \in K_d$, *there exists an efficient inversion algorithm* $TdInv_d$ *such that*

$$TdInv_d(x, t, \alpha_d(k), H_d(k, x)) = k.$$

We now describe the construction of lossy projective hashing via dual projective hashing by adding another part of hashing key and a pairwise independent hash function. Assume that $(H, X = \Pi_Y \cup \Pi_N, K_d, \Pi, \alpha_d, S_d)$ is a dual projective hashing and α_d has lossiness of l bits. Let $h : K_d \rightarrow \{0,1\}^m$ be the pairwise independent hash function such that $m = l - 2\log(1/\epsilon)$ where ϵ is a negligible function of λ. The lossy projective hashing LPH is constructed as follows.

Setup. $PP = (PP_d, h)$. The subset membership problem is given by Π_Y, Π_N. $(x, w) \leftarrow$ SampYes(λ), $(x, t) \leftarrow SampNo(\lambda)$. $K = K_d \times \{0,1\}^m$ and $S = S_d \times \{0,1\}^m$.

Hashing. Given the input $(k_1, k_2) \in K_d \times \{0,1\}^m$, the projective map α is

$$\alpha(k) = (\alpha_d(k_1), h(k_1) \oplus k_2).$$

The hash value is given by

$$H(k, x) = (H_d(k_1, x), h(k_1) \oplus k_2).$$

Projective Hashing. On any YES instance x, the hash value can be computed from $\alpha(k) = (\alpha_1, \alpha_2)$ and the witness w by

$$Pub(\alpha(k), x, w) = (Pub_d(\alpha_1, x, w), \alpha_2).$$

Partial Inversion. On any NO instance x, given $\alpha(k) = (\alpha_1, \alpha_2)$ and the trapdoor t, compute $temp = TdInv_d(x, t, \alpha_1, H_d(k_1, x))$ firstly. Return $h(temp) \oplus \alpha_2$.

We need a generalized leftover hash lemma in order to prove the lossy property.

Lemma 1 (Lemma 2.4 in [12]). *Let X, Y be random variables such that the average min-entropy $\tilde{H}_\infty(X|Y) \geq l$. Let $\mathcal{H}$ be a family of universal hash family to $\{0,1\}^m$, where $m \leq l - 2\log(1/\varepsilon)$. It holds that for $h \leftarrow \mathcal{H}$ and $r \leftarrow \{0,1\}^l$,*

$$\Delta((h, h(X), Y)), (h, r, Y)) \leq \varepsilon.$$

Theorem 1. *Under the assumptions that $(H, X = \Pi_Y \cup \Pi_N, K_d, \Pi, \alpha_d, S_d)$ is a dual projective hashing, and $h : K_d \rightarrow \{0,1\}^m$ is a pairwise independent hash function, then LPH is a lossy projective hashing.*

Proof. The correctness of the Partial Inversion algorithm is trivial. Here, we prove the lossiness property. By the lossiness property of dual projective hashing, the average min-entropy $\tilde{H}_\infty(K_d|\alpha_d(K_d)) \geq n - (n - l) = l$. Since $m = l - 2\log(1/\varepsilon)$, according to Lemma 1, $(h, h(K_d) \oplus k_2, \alpha_d(K_d))$ is ε-close to $(h, U_m \in \{0,1\}^m, \alpha_d(K_d))$. For every $k_2^1, k_2^2 \in \{0,1\}^m$,

$$\Delta((\alpha_d(K_d), h, h(K) \oplus k_2^1), (\alpha_d(K_d), h, h(K) \oplus k_2^2) \leq 2\varepsilon.$$

Since the hash value is determined by projected key, the lossiness property holds.
□

4.2   To Smooth Projective Hashing

Cramer and Shoup [11] showed the general construction of smooth projective hashing from universal projective hashing via pairwise independent hash functions. In order to reduce smooth projective hashing to lossy projective hashing, we just need to prove that a lossy projective hashing is also a universal projective hashing. At first, we recall the definition of smooth and universal projective hashing systems given in [11].

Definition 4 (Smooth projective hashing [11]). *Let $(H_s, K_s, X_s, L, \Pi, \alpha_s, S_s)$ be a hash family where $L \subset X_s$, $H_s : K_s \times X_s \rightarrow \Pi$ and $\alpha_s : K_s \rightarrow S_s$. We say that this is a smooth projective hashing, if the following properties hold:*

Projetive. *For all $k \in K_s$ and all $x \in L$, the action of $H_s(k, x)$ is determined by $\alpha_s(k)$.*
Smoothness. *The following distributions are statistically indistinguishable:*

$$\{x, \alpha_s(k), H_s(k, x)\} =_s \{x, \alpha_s(k), \pi\},$$

where $k \in K_s$, $x \in X \setminus L$, and $\pi \in \Pi$ are chosen ramdomly.

Definition 5 (Universal projective hashing [11]). *Let $(H_u, K_u, X_u, L, \Pi, \alpha_u, S_u)$ be a hash family where $L \subset X_u$, $H_u : K_u \times X_u \rightarrow \Pi$ and $\alpha_u : K_u \rightarrow S_u$. We say that the hashing family is a universal projective hashing, if the following properties hold:*

Projetive. *For all $k \in K_u$ and all $x \in L$, the action of $H_u(k, x)$ is determined by $\alpha_u(k)$.*

ϵ-**universal.** *Consider the probability defined by choosing* $k \in K_u$ *at random. For all* $s \in S_u$, $x \in X_u \setminus L$, *and* $\pi \in \Pi$, *it holds that*

$$Pr[H(k, x) = \pi | \alpha_u(k) = s] \leq \epsilon.$$

Theorem 2. *Assume that* $(H, X = \Pi_Y \cup \Pi_N, K = K_1 \times K_2, \Pi, \alpha, S)$ *is a lossy projective hashing where* $K = K_1 \times K_2 = \{0,1\}^{n-l} \times \{0,1\}^l$, *then it is also a* $1/2^l$-*universal projective hashing.*

Proof. As the projective property is obvious, it is enough to prove that the lossy projective hashing satisfies the "universal" property. Let $(H, X = \Pi_Y \cup \Pi_N, K = K_1 \times K_2, \Pi, \alpha, S)$ be a lossy projective hashing where $K = K_1 \times K_2 = \{0,1\}^{n-l} \times \{0,1\}^l$. Consider the probability space defined by choosing $(k_1, k_2) \in K$ at random. Let $s \in S$, $x \in X \setminus L$, and $\pi \in \Pi$. Denote by k_2^* the output of partial inversion algorithm $TdInv(x, t, s, \pi)$. It holds that $Pr[k_2 = k_2^* | (H(k_1, k_2, x) = \pi \wedge \alpha(k_1, k_2) = s)] = 1$. According to the lossy property, $Pr[k_2 = k_2^* | \alpha(k_1, k_2) = s] = 1/2^l$. Then, consider the probability defined by choosing k_1, k_2 at random,

$$\begin{aligned}
&Pr[H(k_1, k_2, x) = \pi \wedge \alpha(k_1, k_2) = s] \\
=&Pr[H(k_1, k_2, x) = \pi \wedge \alpha(k_1, k_2) = s]Pr[k_2 = k_2^* | (H(k_1, k_2, x) = \pi \wedge \alpha(k_1, k_2) = s)] \\
=&Pr[k_2 = k_2^* \wedge H(k_1, k_2, x) = \pi \wedge \alpha(k_1, k_2) = s] \\
\leq&Pr[k_2 = k_2^* \wedge \alpha(k_1, k_2) = s] \\
=&Pr[k_2 = k_2^* | \alpha(k_1, k_2) = s]Pr[\alpha(k_1, k_2) = s] \\
=&1/2^l Pr[\alpha(k_1, k_2) = s].
\end{aligned}$$

Thus $Pr[H(k_1, k_2, x) = \pi | \alpha(k_1, k_2) = s] \leq 1/2^l$, and $(H, X = \Pi_Y \cup \Pi_N, K = K_1 \times K_2, \Pi, \alpha, S)$ is a $1/2^l$-universal projective hashing. $\square$

5   Instantiations

In this section, we present the instantiations of lossy projective hashing based on decisional Diffie-Hellman (DDH), quadratic residue (QR), decision composite residuosity (DCR) and general subgroup membership assumptions.

5.1   Instantiation from DDH Assumption

This construction is highly related to the lossy encryption given by Hemenway and Ostrovsky [18]. Let $\mathbb{G}$ be a group of prime order p, and g be the generator. Firstly, We describe the subset membership problem. Sample $s_1, s_2, \cdots, s_n$, $r_0, \cdots, r_n \in Z_p$. Set $v = (g^{r_0}, \cdots, g^{r_n})^T$,

$$U_1 = \begin{pmatrix} g^{r_0 s_1}, & g^{r_0 s_2}, & \cdots, & g^{r_0 s_n} \\ g^{r_1 s_1}, & g^{r_1 s_2}, & \cdots, & g^{r_1 s_n} \\ \vdots & & \ddots & \vdots \\ g^{r_n s_1}, & g^{r_n s_2}, & \cdots, & g^{r_n s_n} \end{pmatrix}, \text{ or } U_2 = \begin{pmatrix} g^{r_0 s_1}, & g^{r_0 s_2}, & \cdots, & g^{r_0 s_n} \\ g^{r_1 s_1}g, & g^{r_1 s_2}, & \cdots, & g^{r_1 s_n} \\ \vdots & & \ddots & \vdots \\ g^{r_n s_1}, & g^{r_n s_2}, & \cdots, & g^{r_n s_n}g \end{pmatrix}.$$

Note that the $(i+1, i)$-th entry of U_1 is $g^{r_i s_i}$ while that of U_2 is $g^{r_i s_i} g$ for $1 \leq i \leq n$. Then we define the exponentiation rules. For a group $\mathbb{G}$, an element $g \in \mathbb{G}$, a vector $\overrightarrow{x} = (x_1, \cdots, x_m)$, let $(g^{r_1}, \cdots, g^{r_m})^{\overrightarrow{x}} = g^{\sum_{i=1}^n r_i x_i}$. For an $s \times t$ matrix $U = (u_{ij})$ over $\mathbb{G}$, and a vector $\overrightarrow{x} = (x_1, \cdots, x_s)$, let $U^{\overrightarrow{x}} = (\prod_{j=1}^s u_{ji}^{x_j})_{1 \leq i \leq t}$. The lossy projective hashing LPH_{DDH} based on DDH assumption is constructed as follows.

Setup. $PP = (\mathbb{G}, g, v)$. The subset membership problem is given by $\Pi_Y = \{U_1\}$, and $\Pi_N = \{U_2\}$. Both YES and NO instances are efficiently samplable. $s = (s_i)_{1 \leq l \leq n}$ is the witness for a YES instance and the trapdoor for a NO instance.

Hashing. The input is $k = (k_1, k_2) = (k_1, (k_{21}, \cdots, k_{2n})) \in Z_p \times \{0, 1\}^n$. The projective map α is

$$\alpha(k) = v^k = g^{r_0 k_1 + \sum_{i=1}^n r_i k_{2i}}.$$

The hash value is given by $H(k, U) = U^k = (u_{1i}^{k_1} \prod_{j=2}^{n+1} u_{ji}^{k_{2(j-1)}})_{1 \leq i \leq n}$.

Projective Hashing. On any YES instance U_1, the hash value can be computed from $\alpha(k)$ and the witness s by

$$Pub(v^k, v, U_1, s) = ((v^k)^{s_i})_{1 \leq i \leq n}.$$

Partial Inversion. Given $c_0 = v^k$, $C = H(k, U_2) = (c_1, \cdots, c_n)$ and s, we can recover k_2 by computing $(g^{k_{21}}, g^{k_{22}}, \cdots, g^{k_{2n}}) = (c_1 c_0^{-s_1}, c_2 c_0^{-s_2}, \cdots, c_n c_0^{-s_n})$.

Theorem 3. *Under the DDH assumption the scheme LPH_{DDH} is a lossy projective hashing.*

Proof. As shown in [28], the subset membership assumption holds even given $(\mathbb{G}, g, v)$ under the DDH assumption. At first, we have $H(k, U_1) = U_1^k = (g^{(r_0 k_1 + \sum_{j=1}^n r_j k_{2j}) s_i})_{1 \leq j \leq n} = ((v^k)^{s_i})_{1 \leq i \leq n}$, for any YES instance. The Pub algorithm is correct on any YES instance, thus the projective hashing property holds. On NO instances, the correctness of the partial inversion algorithm is obvious.

Since $v^k = g^{r_0 k_1 + \prod_{i=1}^n r_i k_{2i}}$, when k_1 is randomly chosen from Z_p, v^k statistically loses the information of k_2. Thus $\alpha(k)$ is independent of k_2, and the lossy property holds. $\square$

5.2 Instantiation from QR

Let $N = PQ$, where P, Q are safe primes and $P, Q \equiv 3 \mod 4$. Let QR denote the subgroup of quadratic residues and J_N denote the subgroup with Jacobi symbol 1. We have that $-1 \in J_N \setminus QR$. Let g and h be the generators of the group QR. The lossy projective hashing LPH_{QR} based on QR assumption is constructed as follows.

Setup. $PP = (N, g, h)$. The subset membership problem is given by

$$\Pi_Y = \{g, h, g_1 = g^a, h_1 = h^a : a \leftarrow \lfloor \frac{N}{2} \rfloor\},$$

$$\Pi_N = \{g, h, g_1 = g^a, h_1 = (-1)h^a : a \leftarrow \lfloor \frac{N}{2} \rfloor\}.$$

Both YES and NO instances are efficiently samplable. Here, a is the witness for a YES instance and the trapdoor for a NO instance.

Hashing. Given the input $k = (k_1, k_2) \in Z_{\lfloor \frac{N}{2} \rfloor} \times \{0, 1\}$, the projective map α is

$$\alpha(k) = g^{k_1} h^{k_2}.$$

The hash value is given by $H(k, (g, h, g_1, h_1)) = g_1^{k_1} h_1^{k_2}$.

Projective Hashing. On any YES instance (g, h, g_1, h_1), the hash value can be computed from $\alpha(k)$ and the witness a by

$$Pub(\alpha(k), (g, h, g_1, h_1), a) = \alpha(k)^a = H(k, (g, h, g_1, h_1)).$$

Partial Inversion. On any NO instance (g, h, g_1, h_1), given $\alpha(k) = g^{k_1} h^{k_2}$ and the trapdoor a, we have that $H(k, (g, h, g_1, h_1)) = (-1)^{k_2} \alpha(k)^a$. The algorithm returns 0 if $H(k, (g, h, g_1, h_1)) = \alpha(k)^a$ and 1 otherwise.

Theorem 4. *Under the QR assumption the scheme LPH_{QR} is a lossy projective hashing.*

Proof. According to [17], the subset membership assumption given above is implied by the QR assumption. The projective property and the correctness of partial inversion algorithm are obvious.

When k_1 is randomly chosen from $Z_{\lfloor \frac{N}{2} \rfloor}$, $g^{k_1} h^{k_2}$ statistically loses the information of k_2. $\alpha(k)$ is independent of k_2, thus the lossy property holds. $\square$

5.3 Instantiation from DCR

Let $N = pq$ be the product of two large primes of roughly the same size. For an element $w \in Z_{N^2}^*$ there exists a unique pair $(x, y) \in Z_N \times Z_N^*$ such that $w = (1+N)^x y^N$. The DCR assumption states that it is difficult to decide if $x = 0$ or not. Let g and h be the generator of $2N$-th residues. The computations below are over $Z_{N^2}^*$. The lossy projective hashing LPH_{DCR} based on DCR assumption is constructed as follows.

Setup. $PP = (N, g, h)$. The subset membership problem is given by

$$\Pi_Y = \{g, h, g_1 = g^a, h_1 = h^a : a \leftarrow \lfloor \frac{N^2}{4} \rfloor\},$$

$$\Pi_N = \{g, h, g_1 = g^a, h_1 = (1+N)h^a : a \leftarrow \lfloor \frac{N^2}{4} \rfloor \}.$$

Both YES and NO instances are efficiently samplable. Here, a is the witness for a YES instance and the trapdoor for a NO instance.

Hashing. Given the input $k = (k_1, k_2) \in Z_{\lfloor \frac{N}{2} \rfloor} \times Z_N$, the projective map α is

$$\alpha(k) = g^{k_1} h^{k_2}.$$

The hash value is given by $H(k, (g, h, g_1, h_1)) = g_1^{k_1} h_1^{k_2}$.

Projective Hashing. On any YES instance (g, h, g_1, h_1), the hash value can be computed from $\alpha(k)$ and the witness a by

$$Pub(\alpha(k), (g, h, g_1, h_1), a) = \alpha(k)^a = H(k, (g, h, g_1, h_1)).$$

Partial Inversion. On any NO instance (g, h, g_1, h_1), given $\alpha(k)$ and the trapdoor a, we have that

$$H(k, (g, h, g_1, h_1)) = (1+N)^{k_2} \alpha(k)^a = (1 + k_2 N)\alpha(k)^a.$$

Thus, $k_2 = \frac{H(k,(g,h,g_1,h_1))/\alpha(k)^a - 1}{N}$.

Theorem 5. *Under the DCR assumption the scheme LPH_{DCR} is a lossy projective hashing.*

Proof. According to [17], the subset membership assumption given above is implied by the DCR assumption. The projective property and the correctness of partial inversion algorithm are obvious.

When k_1 is randomly chosen from $Z_{\lfloor \frac{N}{2} \rfloor}$, $g^{k_1} h^{k_2}$ statistically loses the information of k_2. $\alpha(k)$ is independent of k_2, thus the lossy property holds. □

5.4 Instantiation from Subgroup Membership Assumption

We first recall the definition of subgroup membership assumption (SGA). Then lossy projective hashing based on SGA is shown.

Gjøsteen [19] gave the definition of subgroup discrete logarithm problem which is a generalization of Paillier's [26] partial discrete logarithm problem.

Definition 6 (Subgroup Membership Assumption [19]). *Let $\mathbb{G}$ be a finite cyclic group with subgroups $\mathbb{H}$ and $\mathbb{K}$. $\mathbb{G} = \mathbb{H} \times \mathbb{K}$. Let g (resp. J) be a generator of group $\mathbb{K}$ (resp. $\mathbb{G}$). The subgroup membership problem $SM_{(\mathbb{G},\mathbb{K})}$ asserts that, for any PPT distinguisher D, the advantage*

$$Adv_D^{SM_{(\mathbb{G},\mathbb{K})}} = |\Pr[D(\mathbb{G}, \mathbb{K}, x) = 1 | x \leftarrow \mathbb{K}] - \Pr[D(\mathbb{G}, \mathbb{K}, x) = 1 | x \leftarrow \mathbb{G} \setminus \mathbb{K}]|.$$

is negligible, where the probability is taken over coin tosses of D.

If $\varphi : \mathbb{G} \to \mathbb{G}/\mathbb{K}$ is the canonical epimorphism, then the subgroup discrete logarithm problem $SDL_{(\mathbb{G},\mathbb{H})}$ is: given a random $x \in \mathbb{G}$, compute $\log_{\varphi(J)}(\varphi(x))$. We assume that given a trapdoor t the subgroup discrete logarithm problem is efficiently solvable. There are some SGAs satisfy the subgroup discrete logarithm assumption, such as QR, DCR, 2^k-th residue [14], p-subgroup [25] and higher residue [23] assumptions.

Hemenway and Ostrovsky [17] gave the definition of extended DDH assumption. It states that the following two sets are computational indistinguishable,

$$\Pi_Y = \{g, g^x, g^y, g^{xy} : g \text{ is a generator of } \mathbb{K}, x, y \leftarrow Z_{|\mathbb{G}|}\},$$

$$\Pi_N = \{g, g^x, g^y, g^{xy}h : g \text{ (resp.} h) \text{ is a generator of } \mathbb{K} \text{ (resp.} \mathbb{H}), x, y \leftarrow Z_{|\mathbb{G}|}\}.$$

We now give a sketch proof that the SGA implies the extended DDH assumption. The proof is an abstract of Theorem 2 on DCR case in [17]. By the SGA assumption, it is computationally indistinguishable if we replace g^x by a random element $X \in \mathbb{G}$ in both Π_Y and Π_N. As $g \in \mathbb{K}$, g^y only contains the information of $y \mod |\mathbb{K}|$. Then, if represent X by $X_1 X_2 \in \mathbb{H}\mathbb{K}$, $X^y = X_1^{y \mod |\mathbb{H}|} X_2^{y \mod |\mathbb{K}|}$, and $X_1^{y \mod |\mathbb{H}|}$ is definitely undetermined even given g^y. Thus, the statistical instance between $\{g, X, g^y, X^y\}$ and $\{g, X, g^y, X^y h\}$ is negligible. The extended DDH assumption holds if SGA holds.

Under SGA, it is difficult to decide the instance $\{g_1, g_2, g_3, g_4\}$ is randomly chosen from Π_Y or Π_N. The lossy projective hashing LPH_{SGA} is constructed as follows.

Setup. $PP = (N, g_1, g_2)$. The subset membership problem is given by Π_Y and Π_N. Both YES and NO instances are efficiently samplable. Here, (y, h) is the witness for a YES instance and the trapdoor for a NO instance.

Hashing. Given the input $k = (k_1, k_2) \in Z_{|\mathbb{K}|} \times Z_{|\mathbb{H}|}$, the projective map α is

$$\alpha(k) = g_1^{k_1} g_2^{k_2}.$$

The hash value is given by $H(k, (g_1, g_2, g_3, g_4)) = g_3^{k_1} g_4^{k_2}$.

Projective Hashing. On any YES instance (g_1, g_2, g_3, g_4), the hash value can be computed from $\alpha(k)$ and witness y by

$$Pub(\alpha(k), (g_1, g_2, g_3, g_4), (y, h)) = \alpha(k)^y.$$

Partial Inversion. On any NO instance (g_1, g_2, g_3, g_4), given $\alpha(k)$ and the trapdoor (y, h), we have that $H(k, (g_1, g_2, g_3, g_4)) = h^{k_2} \alpha(k)^y$. Then k_2 can be recovered by the subgroup discrete logarithm algorithm.

Theorem 6. *Under the subgroup membership assumption, the scheme LPH_{SGA} is a lossy projective hashing.*

Proof. On any YES instance (g, g^x, g^y, g^{xy}), $\alpha(k)^y = g^{k_1 + y k_2} = H(k, (g, g^x, g^y, g^{xy}))$, the projective property holds. When k_1 is randomly chosen from $Z_{|\mathbb{K}|}$, $g^{k_1} h^{k_2}$ statistically loses the information of k_2. $\alpha(k)$ is independent of k_2, thus the lossy property holds.   □

Remark 1. The instantiations in Sects. 5.2 and 5.3 are concrete examples of this construction. This construction implies lossy projective hashing based on 2^k-th residue [14], p-subgroup [25] and higher residue [23] assumptions.

6 Lossy Encryption

In this section, we deal with the second motivation of this paper. We first recall the definition of lossy encryption in [3], then give a generic construction from lossy projective hashing. This construction fits into the second methodology of constructing lossy encryption mentioned in the Introduction section.

Definition 7 (Lossy Encryption [3]). *A lossy encryption scheme is a tuple of probability polynomial time (PPT) algorithms (Gen_{inj}, Gen_{loss}, Enc, Dec).*

Gen_{inj}: *Output injective keys (pk, sk). The private key space is $\mathcal{K}$.*
Gen_{loss}: *Output lossy keys $(pk_{loss}, \perp)$.*
Enc: $\mathcal{M} \times \mathcal{R} \to \mathcal{C}$. *The message space is $\mathcal{M}$. The randomness string space is $\mathcal{R}$ and the ciphertext space is $\mathcal{C}$.*
Dec: $\mathcal{K} \times \mathcal{C} \to \mathcal{M}$.

These algorithms satisfy the following properties:

- *Correctness. For all $m \in \mathcal{M}$ and $r \in \mathcal{R}$, $Dec(sk, Enc(pk, m, r)) = m$.*
- *Indistinguishability of the injective key from the lossy key. The injective and lossy public keys are computationally indistinguishable:*

$$\{pk | (pk, sk) \leftarrow Gen_{inj}\} =_c \{pk | (pk, \perp) \leftarrow Gen_{loss}\}$$

- *Lossiness of encryption with lossy keys. For any lossy public key pk_{loss}, and any pair of message $m_0, m_1 \in \mathcal{M}$, there is*

$$\{Enc(pk_{loss}, m_0, r)) | r \in \mathcal{R}\} =_s \{Enc(pk_{loss}, m_1, r)) | r \in \mathcal{R}\}$$

Let $(H, X = \Pi_Y \cup \Pi_N, K = K_1 \times K_2, \Pi, \alpha, S)$ be a lossy projective hashing and $K = \{0,1\}^n \times \{0,1\}^l$. The lossy encryption LE is constructed below:

Gen_{inj}: It generates a NO instance $(x, t) \leftarrow SampNo(\lambda)$.

$$pk = (x, \alpha, H), sk = t.$$

Gen_{loss}: It generates a YES instance $(x, w) \leftarrow SampYes(\lambda)$.

$$pk = (x, \alpha, H).$$

Enc: On input a message $m \in \{0,1\}^l$, it randomly chooses $r \in \{0,1\}^n$. The ciphertext is $(c_1, c_2) = (\alpha(r, m), H(r, m, x))$.
Dec: On input $sk = t$, and a ciphertext $C = (c_1, c_2)$, it computes m using the partial inversion algorithm $TdInv$.

Theorem 7. *If $(H, X = \Pi_Y \cup \Pi_N, K = K_1 \times K_2, \Pi, \alpha, S)$ is a lossy projective hashing, then the scheme LE above is a lossy encryption scheme.*

Proof. We show that the above scheme satisfies the three properties of lossy encryption.

- *Correctness on real keys.* By the correctness of partial inversion algorithm, for all $m \in \{0,1\}^l$ and $r \in \{0,1\}^n$,

$$TdInv(x, t, \alpha(r, m), H(r, m, x)) = m.$$

- *Indistinguishability of the injective key from the lossy key.* The only difference between the injective and the lossy keys is the instance sample algorithm. The indistinguishability between them is guaranteed by the subset membership assumption. That is

$$Adv_I^{inj, loss} = \Pr[I(x \leftarrow \Pi_N, \alpha, H) = 1] - \Pr[I(x \leftarrow \Pi_Y, \alpha, H) = 1] \leq Adv_A^{SMA}.$$

- *Lossiness of encryption with lossy keys.* Since x is a YES instance, $H(r, m, x)$ is fully determined by $\alpha(r, m)$. For every $m_1, m_2 \in \mathcal{M}$, $\Delta(\alpha(\mathcal{R}, m_1), \alpha(\mathcal{R}, m_2)) \leq \epsilon$, since $\alpha(\cdot, \cdot)$ statically loses the information of its second input. Thus

$$\Delta((\alpha(\mathcal{R}, m_1), H(\mathcal{R}, m_1, x)), (\alpha(\mathcal{R}, m_2), H(\mathcal{R}, m_2, x))) \leq \epsilon.$$

$\square$

Remark 2. The derived DDH based lossy encryption is exactly the scheme given in [18], which is less efficient than the DDH based lossy encryption given in [3,16].

7 Fully IND Secure DPKE

In this section, we give the construction of a fully IND secure DPKE by combination of lossy projective hashing system and UCE secure hash functions. Deterministic public key encryption (DPKE) requires that the encryption algorithm, on input the public key and the message, deterministically return a ciphertext. There are PRIV formalization [1] and IND formalization [2] for the security of DPKE schemes. We use the full IND formalization in [4] which captures the case that the messages are individually unpredictable but may be correlated. We recall the definition of IND security for DPKE scheme $(D.kg, D.Enc, D.Dec)$ firstly.

$$\text{Game} IND_D^A(\lambda):$$
$$b \leftarrow \{0,1\}, (pk, sk) \leftarrow D.kg(\lambda);$$
$$(\mathbf{m}_0 = \{m_0^i\}_{i=1}^v, \mathbf{m}_1 = \{m_1^i\}_{i=1}^v) \leftarrow A_1(\lambda);$$
$$\text{For } i = 1 \text{ to } v \text{ do}$$
$$\quad c[i] \leftarrow D.Enc(\lambda, pk, m_b^i);$$
$$\mathbf{c} = \{c[i]\}_{1 \leq i \leq v};$$
$$b' \leftarrow A_2(\lambda, pk, \mathbf{c});$$
$$\text{Return}(b' = b).$$

It is required that m_0^i (resp. m_1^i) be distinct for $1 \leq i \leq v$. The guessing probability is

$$Guess_A = \max\{\Pr[m_b^i = m : (\mathbf{m}_0, \mathbf{m}_1) \leftarrow A_1]\},$$

for all $b \in \{0, 1\}$, $1 \leq i \leq v$, $m \in \{0, 1\}^*$. We say that the scheme is IND secure if $Adv_{DE,A}^{IND} = 2\Pr[IND_D^A(\lambda)] - 1$ is negligible for any PPT adversary $A = (A_1, A_2)$ with $Guess_A$ being negligible.

Bellare et al. [5] gave the definition of UCE secure hash functions. The UCE secure hash family is indistinguishable from the random oracle, for some kinds of source S. A source S, is statistically unpredictable if it is hard to guess the source's hash queries even given the leakage for unbounded algorithm. A source S, is computationally unpredictable if it is hard to guess the source's hash queries even given the leakage for a computational bounded algorithm. Brzuska et al. [9] proved that UCE secure hash functions for computationally unpredictable source is not achievable if indistinguishable obfuscation is possible. But they gave some evidence that their result does not work on statistically unpredictable sources.

Please refer [5] for formal definition of UCE secure hash functions for statistically unpredictable source.

We now give the fully IND secure DPKE from lossy projective hashing and UCE secure hash functions. Bellare et al. in [4] gave the first fully IND secure DPKE by combining lossy trapdoor function and UCE secure hash function. In their scheme, two rounds of UCE secure hash function is required. Here, we prove that lossy projective hashing + one round UCE secure function is enough for fully IND secure DPKE. Let $(H, \Pi_Y \cup \Pi_N, K, \Pi, \alpha, S)$ be a lossy projective hashing. The key space is $\{0, 1\}^s \times \{0, 1\}^t$. Let H_{uce} be the family of UCE secure hash for statistically unpredictable source with input length t and output length s.

D.kg: It generates a NO instance $(x, \tau) \leftarrow SampNo(\lambda)$. Choose a random h from H_{uce}.

$$pk = (x, \alpha, H, h), sk = (\tau, h).$$

D.Enc: On input a message $m \in \{0, 1\}^t$, it computes $h(m)$. The ciphertext is

$$(c_1, c_2) = (\alpha(h(m), m), H(h(m), m, x)).$$

D.Dec: On input $sk = (T, h)$, and a ciphertext $C = (c_1, c_2)$, it computes m using the partial inversion algorithm $TdInv$.

Theorem 8. *Let $(H, \Pi_Y \cup \Pi_N, K, \Pi, \alpha, S)$ be a lossy projective hashing and H_{uce} be the family of UCE secure hash function with statically unpredictable source. If there is an IND adversary A, then there is an adversary B for solving the subset membership assumption or a distinguisher D of UCE secure hash with respect to a statistically unpredictable source S, such that,*

$$Adv_A^{ind} \leq 2Adv_B^{SMA} + 2Adv_{S,D}^{uce} + v^2/2^{s-1},$$

$$Adv_{S,P}^{pred} \leq \frac{v^2}{2^s} + qv \cdot Guess_A(\cdot),$$

where q bounds the output size of the predictor P.

Proof. The correctness of the decryption algorithm is guaranteed by the correctness of the partial inversion algorithm. In the following, we prove the fully IND security by describing a sequence of experiments **Game 0**, **Game 1**, **Game 2**, and **Game 3**. Let $A = (A_1, A_2)$ be an adversary attacking the above scheme. Let $T_i, i = 0, 1, 2, 3$ denote the event that the **Game** i returns 1.

Game 0. This game is identical to the original IND game of DPKE. That is, the *SampNo* algorithm outputs x and trapdoor τ. The public key and encryption algorithms are identical to those of the original scheme.

Game 1. The only difference with Game 0 is the instance sample algorithm. The instance x is chosen by *SampYes* rather than *SampNo*. That is, $(x, w) \leftarrow SampYes(\lambda)$. Then we have

$$\Pr[T_0] - \Pr[T_1] = \mathrm{Adv}_B^{SMA},$$

where B is the adversary attacking the underly subset membership problem. On input a YES or NO instance x, B runs as follows.

$B(\lambda, x)$:
$\quad h \leftarrow H_{uce}, pk \leftarrow (x, \alpha, H, h), b \leftarrow \{0, 1\},$
$\quad (\mathbf{m}_0 = \{m_0^i\}_{i=1}^v, \mathbf{m}_1 = \{m_1^i\}_{i=1}^v) \leftarrow A_1(\lambda)$
$\quad$ For $i = 1$, to v,
$\quad\quad c[i] \leftarrow (\alpha(h(m_b^i), m_b^i), H(h(m_b^i), m_b^i, x)).$
$\quad \mathbf{c} = \{c[i]\}_{1 \leq i \leq v}.$
$\quad b' \leftarrow A_2(\lambda, pk, \mathbf{c}),$
$\quad$ Return$(b' = b).$

If $x \in \Pi_N$, then B is simulating Game 0 for A. Otherwise B is simulating Game 1.

Game 2. The only difference with Game 1 is that the UCE secure hash function h is replaced by a random oracle RO. Then we have $\Pr[T_1] - \Pr[T_2] = \mathrm{Adv}_{S,D}^{uce}$. Here, S is a statistically unpredictable source interacting with $HASH$ (h in Game 1, RO in Game 2), and D is the distinguisher of the hash function and the random oracle. Then we have that

$$\Pr[T_1] - \Pr[T_2] = \mathrm{Adv}_{S,D}^{uce}.$$

What is left is to prove that the source S is statistically unpredictable. The leakage of S is $(b, pk, \mathbf{c})$. b and pk are independent of the message. According to the lossy property, $\alpha(RO(m_b^i), m_b^i)$ is statistically independent of the message. Since the public key is on YES instance, $H(RO(m_b^i), m_b^i, x)$ is totally determined by $\alpha(RO(m_b^i), m_b^i)$, and contains no information of the message. The ciphertext is statistically independent of the message. Let P be the statistical predicator, and q be a polynomial that bounds the size of predictor's outputs. The chance that the predicator P guesses the messages is at most $qv\mathrm{Guess}_A$.

Source $S^{HASH}(\lambda)$:

$x \leftarrow SampYes(\lambda)$;

$pk \leftarrow (x, \alpha, H, h), b \leftarrow \{0, 1\}$;

$(\mathbf{m}_0, \mathbf{m}_1) \leftarrow A_1(\lambda)$;

For $i = 1$ to v do

$\quad r \leftarrow HASH(m_b^i)$;

$\quad c[i] \leftarrow (\alpha(r, m_b^i), H(r, m_b^i, x))$;

$\mathbf{c} = \{c[i]\}_{1 \le i \le v}$;

Return $(b, pk, \mathbf{c})$.

Distinguisher $D(\lambda, L, h)$:

$(b, pk, \mathbf{c}) \leftarrow L$;

$b' \leftarrow A_2(\lambda, pk, \mathbf{c})$;

Return $(b' = b)$.

Game 3. The only change is that the random oracle does not store a list of queried instance and returns a fresh random answer for every query. Since the strings m_b^i are distinct, the collision happens only when the random answer are repeated, which happens with probability at most $\frac{v^2}{2^s}$. Then we have

$$\Pr[T_2] - \Pr[T_3] = v^2/2^s.$$

Since $\alpha(RO(m_b^i), m_b^i)$ is statistically independent of b, what A_2 receives is independent of the challenge bit b, thus $\Pr[T_3] = 1/2$.

By adding up the sequence of results, we have

$$\mathrm{Adv}_A^{ind} = 2\Pr[T_0] - 1 \le 2\mathrm{Adv}_B^{SMA} + 2\mathrm{Adv}_{S,D}^{uce} + v^2/2^{s-1}.$$

$\square$

8 Conclusion

In this paper, we introduce the primitive of lossy projective hashing. It is similar to, but significantly different from, dual projective hashing and smooth projective hashing. We also provide constructions of lossy projective hashing from DDH, DCR, QR and general SGA assumption. The lossy projective hashing builds a bridge between dual projective hashing and smooth projective hashing. It is applicable to lossy encryption scheme. Finally, we give a fully IND secure deterministic public key encryption via lossy projective hashing and one round UCE secure hash functions.

Acknowledgement. Haiyang Xue and Bao Li are supported by the National Natural Science Foundation of China (No. 61379137). Yamin Liu is supported by the National Natural Science Foundation of China (No. 61502480). Xianhui Lu is supported by the National Natural Science Foundation of China (No. 61572495, No. 61272534) and IIE's Cryptography Research Project (No. Y4Z0061D03).

References

1. Bellare, M., Boldyreva, A., O'Neill, A.: Deterministic and efficiently searchable encryption. In: Menezes, A. (ed.) CRYPTO 2007. LNCS, vol. 4622, pp. 535–552. Springer, Heidelberg (2007)
2. Bellare, M., Fischlin, M., O'Neill, A., Ristenpart, T.: Deterministic encryption: definitional equivalences and constructions without random oracles. In: Wagner, D. (ed.) CRYPTO 2008. LNCS, vol. 5157, pp. 360–378. Springer, Heidelberg (2008)
3. Bellare, M., Hofheinz, D., Yilek, S.: Possibility and impossibility results for encryption and commitment secure under selective opening. In: Joux, A. (ed.) EURO-CRYPT 2009. LNCS, vol. 5479, pp. 1–35. Springer, Heidelberg (2009)
4. Bellare, M., Hoang, V.T.: UCE+LTDFs: efficient, subversion-resistant PKE in the standard model. Cryptology ePrint Archive, Report 2014/876 (2014)
5. Bellare, M., Hoang, V. T., Keelveedhi, S.: Instantiating random oracles via UCEs. Cryptology ePrint Archive, Report 2013/424 (2013). Preliminary version in CRYPTO 2013
6. Benhamouda, F., Blazy, O., Chevalier, C., Pointcheval, D., Vergnaud, D.: New techniques for SPHFs and efficient one-round PAKE protocols. In: Canetti, R., Garay, J.A. (eds.) CRYPTO 2013, Part I. LNCS, vol. 8042, pp. 449–475. Springer, Heidelberg (2013)
7. Boldyreva, A., Fehr, S., O'Neill, A.: On notions of security for deterministic encryption, and efficient constructions without random oracles. In: Wagner, D. (ed.) CRYPTO 2008. LNCS, vol. 5157, pp. 335–359. Springer, Heidelberg (2008)
8. Brakerski, Z., Segev, G.: Better security for deterministic public-key encryption: the auxiliary-input setting. In: Rogaway, P. (ed.) CRYPTO 2011. LNCS, vol. 6841, pp. 543–560. Springer, Heidelberg (2011)
9. Brzuska, C., Farshim, P., Mittelbach, A.: Indistinguishability obfuscation and UCEs: the case of computationally unpredictable sources. In: Garay, J.A., Gennaro, R. (eds.) CRYPTO 2014, Part I. LNCS, vol. 8616, pp. 188–205. Springer, Heidelberg (2014)
10. Cramer, R., Shoup, V.: A practical public key cryptosystem provably secure against adaptive chosen ciphertext attack. In: Krawczyk, H. (ed.) CRYPTO 1998. LNCS, vol. 1462, pp. 13–25. Springer, Heidelberg (1998)
11. Cramer, R., Shoup, V.: Universal hash proofs and a paradigm for adaptive chosen ciphertext secure public-key encryption. In: Knudsen, L.R. (ed.) EUROCRYPT 2002. LNCS, vol. 2332, pp. 45–64. Springer, Heidelberg (2002)
12. Dodis, Y., Ostrovsky, R., Reyzin, L., Smith, A.: Fuzzy extractors: how to generate strong keys from biometrics and other noisy data. SIAM J. Comput. **38**(1), 97–139 (2008)
13. Fuller, B., O'Neill, A., Reyzin, L.: A unified approach to deterministic encryption: new constructions and a connection to computational entropy. In: Cramer, R. (ed.) TCC 2012. LNCS, vol. 7194, pp. 582–599. Springer, Heidelberg (2012)

14. Joye, M., Libert, B.: Efficient cryptosystems from 2^k-th power residue symbols. In: Johansson, T., Nguyen, P.Q. (eds.) EUROCRYPT 2013. LNCS, vol. 7881, pp. 76–92. Springer, Heidelberg (2013)

15. Kalai, Y.T.: Smooth projective hashing and two-message oblivious transfer. In: Cramer, R. (ed.) EUROCRYPT 2005. LNCS, vol. 3494, pp. 78–95. Springer, Heidelberg (2005)

16. Hemenway, B., Libert, B., Ostrovsky, R., Vergnaud, D.: Lossy encryption: constructions from general assumptions and efficient selective opening chosen ciphertext security. In: Lee, D.H., Wang, X. (eds.) ASIACRYPT 2011. LNCS, vol. 7073, pp. 70–88. Springer, Heidelberg (2011)

17. Hemenway, B., Ostrovsky, R.: Extended-DDH and lossy trapdoor functions. In: Fischlin, M., Buchmann, J., Manulis, M. (eds.) PKC 2012. LNCS, vol. 7293, pp. 627–643. Springer, Heidelberg (2012)

18. Hemenway, B., Ostrovsky, R.: Building lossy trapdoor functions from lossy encryption. In: Sako, K., Sarkar, P. (eds.) ASIACRYPT 2013, Part II. LNCS, vol. 8270, pp. 241–260. Springer, Heidelberg (2013)

19. Gjøsteen, K.: Symmetric subgroup membership problems. In: Vaudenay, S. (ed.) PKC 2005. LNCS, vol. 3386, pp. 104–119. Springer, Heidelberg (2005)

20. Katz, J., Vaikuntanathan, V.: Smooth projective hashing and password-based authenticated key exchange from lattices. In: Matsui, M. (ed.) ASIACRYPT 2009. LNCS, vol. 5912, pp. 636–652. Springer, Heidelberg (2009)

21. Kiltz, E., O'Neill, A., Smith, A.: Instantiability of RSA-OAEP under chosen-plaintext attack. In: Rabin, T. (ed.) CRYPTO 2010. LNCS, vol. 6223, pp. 295–313. Springer, Heidelberg (2010)

22. Kol, G., Naor, M.: Cryptography and game theory: designing protocols for exchanging information. In: Canetti, R. (ed.) TCC 2008. LNCS, vol. 4948, pp. 320–339. Springer, Heidelberg (2008)

23. Naccache, D., Stern, J.: A new public key cryptosystem based on higher residues. In: ACM Conference on Computer and Communications Security, pp. 59–66 (1998)

24. Naor, M., Segev, G.: Public-key cryptosystems resilient to key leakage. In: Halevi, S. (ed.) CRYPTO 2009. LNCS, vol. 5677, pp. 18–35. Springer, Heidelberg (2009)

25. Okamoto, T., Uchiyama, S.: A new public-key cryptosystem as secure as factoring. In: Nyberg, K. (ed.) EUROCRYPT 1998. LNCS, vol. 1403, pp. 308–318. Springer, Heidelberg (1998)

26. Paillier, P., Pointcheval, D.: Efficient public-key cryptosystems provably secure against active adversaries. In: Lam, K.-Y., Okamoto, E., Xing, C. (eds.) ASIACRYPT 1999. LNCS, vol. 1716, pp. 165–179. Springer, Heidelberg (1999)

27. Peikert, C., Vaikuntanathan, V., Waters, B.: A framework for efficient and composable oblivious transfer. In: Wagner, D. (ed.) CRYPTO 2008. LNCS, vol. 5157, pp. 554–571. Springer, Heidelberg (2008)

28. Peikert, C., Waters, B.: Lossy trapdoor functions and their applications. In: STOC, pp. 187–196 (2008)

29. Qin, B., Liu, S.: Leakage-resilient chosen-ciphertext secure public-key encryption from hash proof system and one-time lossy filter. In: Sako, K., Sarkar, P. (eds.) ASIACRYPT 2013, Part II. LNCS, vol. 8270, pp. 381–400. Springer, Heidelberg (2013)

30. Raghunathan, A., Segev, G., Vadhan, S.: Deterministic public-key encryption for adaptively chosen plaintext distributions. In: Johansson, T., Nguyen, P.Q. (eds.) EUROCRYPT 2013. LNCS, vol. 7881, pp. 93–110. Springer, Heidelberg (2013)

31. Wee, H.: Dual projective hashing and its applications — lossy trapdoor functions and more. In: Pointcheval, D., Johansson, T. (eds.) EUROCRYPT 2012. LNCS, vol. 7237, pp. 246–262. Springer, Heidelberg (2012)
32. Xue, H., Li, B., Lu, X., Wang, K., Liu, Y.: On the lossiness of 2^k-th power and the instantiability of Rabin-OAEP. In: Gritzalis, D., Kiayias, A., Askoxylakis, I. (eds.) CANS 2014. LNCS, vol. 8813, pp. 34–49. Springer, Heidelberg (2014)

(De-)Constructing TLS 1.3

Markulf Kohlweiss[1], Ueli Maurer[2], Cristina Onete[3], Björn Tackmann[4(✉)],
and Daniele Venturi[5]

[1] Microsoft Research, Cambridge, UK
markulf@microsoft.com
[2] Department of Computer Science, ETH Zürich, Zurich, Switzerland
maurer@inf.ethz.ch
[3] INSA/IRISA, Rennes, France
cristina.onete@gmail.com
[4] Department of Computer Science and Engineering, UC San Diego, San Diego, USA
btackmann@eng.ucsd.edu
[5] Sapienza University of Rome, Rome, Italy
venturi@di.uniroma1.it

Abstract. SSL/TLS is one of the most widely deployed cryptographic
protocols on the Internet. It is used to protect the confidentiality and
integrity of transmitted data in various client-server applications. The
currently specified version is TLS 1.2, and its security has been analyzed
extensively in the cryptographic literature. The IETF working group is
actively developing a new version, TLS 1.3, which is designed to address
several flaws inherent to previous versions.

In this paper, we analyze the security of a slightly modified version
of the current TLS 1.3 draft. (We do not encrypt the server's certificate.) Our security analysis is performed in the constructive cryptography framework. This ensures that the resulting security guarantees are
composable and can readily be used in subsequent protocol steps, such
as password-based user authentication over a TLS-based communication
channel in which only the server is authenticated. Most steps of our
proof hold in the standard model, with the sole exception that the key
derivation function HKDF is used in a way that has a proof only in
the random-oracle model. Beyond the technical results on TLS 1.3, this
work also exemplifies a novel approach towards proving the security of
complex protocols by a modular, step-by-step decomposition, in which
smaller sub-steps are proved in isolation and then the security of the
protocol follows by the composition theorem.

1 Introduction

SSL/TLS is arguably one of the most widely-used cryptographic protocols securing today's Internet. It was introduced by Netscape [15] in the context of protecting connections between web browsers and web servers, but nowadays the protocol is also used for many other Internet protocols including, e.g., SMTP or IMAP

© Springer International Publishing Switzerland 2015
A. Biryukov and V. Goyal (Eds.): INDOCRYPT 2015, LNCS 9462, pp. 85–102, 2015.
DOI: 10.1007/978-3-319-26617-6_5

(for e-mail transmissions) and LDAP (for accessing directories). Flaws and insecurities in the original design required the protocol to be fixed repeatedly; the current version is TLS 1.2 [12]. A preliminary version of TLS 1.3, which deviates from prior versions considerably, is currently under development [13]. In this paper, we analyze the security of this latest (draft) version of TLS.

1.1 Our Contributions

We prove the security of (a slightly modified version of) the ephemeral Diffie-Hellman handshake of TLS 1.3 with unilateral authentication, that is, where only the server has a certificate. We expect that this mode will be used widely in practice, although recently other modes based on pre-shared keys or Diffie-Hellman with a certified group element have been added to the draft.

More precisely, we prove that TLS 1.3 in ephemeral Diffie-Hellman mode[1] constructs a unilaterally secure channel, that is, a channel where the client has the guarantee that it securely communicates with the intended server while the server has no comparable guarantee. The protocol assumes that an insecure network and a public-key infrastructure (PKI) are available. Our results for TLS 1.3 are in the standard model, with the sole exception that the key derivation function HKDF is used in a way for which security has so-far only been proved in the random-oracle model.[2]

We stress that our result guarantees composability, both in the sense that multiple sessions of the protocol can be used concurrently, and in the sense that the constructed channel can safely be used in applications that assume such a channel. In particular, adding password-based authentication for the client in the unilaterally secure channel immediately yields a mutually secure channel.

Our proof follows a modular approach, in which we decompose the protocol into thinner layers, with easier intermediary proofs. The security guarantee of the entire protocol then follows by composition. In particular:

- Each individual proof consists of a reduction from only a small number of assumptions,[3] and can be updated individually if the corresponding step of the protocol is altered.
- If a better proof is found for one of the smaller sub-steps, re-proving only this sub-step immediately results in an improved security statement of the complete protocol by virtue of the composition theorem.

Modification of the Protocol. While in the original draft [13] the server sends its (PKI) certificate encrypted under preliminarily established keys, we analyze a version of the protocol in which the certificate is sent in clear. Encrypting the certificate complicates the security analysis: on the one hand, the symmetric

---

[1] Subject to the modification described below.

[2] HKDF is used to extract from a Diffie-Hellman group element without a salt. The only proof of this that we know of relies on random oracles.

[3] The ultimate goal in such a modularization is that the proof of each step consist of only a single reduction, but TLS 1.3 does not allow for this.

keys are authenticated by the certificate (as the latter authenticates the server's key-exchange share); on the other hand, the certificate is protected with the symmetric keys. Our proof can be modified along the lines of a similar analysis of IPsec [16], but at the cost of a more complicated formalization.

Limitations of Our Analysis. Our proof does not cover the notion of (perfect) forward secrecy; the main reason is that no formalization of this property currently exists in the constructive cryptography framework we work in. Note that while our definitions do not model the adaptive corruption of parties, they *do* guarantee that the keys can be used in arbitrary applications, which traditional game-based notions model via so-called key-reveal oracle queries.

Our proof only applies to sessions with a fixed TLS 1.3 version and uses an abstract formulation of a PKI corresponding to the guarantee that (a) a client knows the identity of the server with which it communicates; and (b) only the honest server can get a certificate for this identity [22,24]. This means that some types of attacks are precluded from the model, such as version rollback (by assuming a fixed version) and Triple Handshake [5] (by assuming that the server be honest). This implies, in particular, that our results do not require the collision resistance of the hash function for the security of the key derivation (but only during the authentication); in other words, the additional security achieved by including the session hash into the key derivation is neither defined nor proven. Furthermore, our analysis does not cover session resumption.

Our analysis covers concurrent sessions, at the cost of some complexity in our intermediary proof steps. Indeed, the specific design of TLS makes many of these steps cumbersome by requiring us to model multiple sessions explicitly; this is an effect of TLS breaking natural module boundaries between different parts of the protocol, by explicitly using protocol data from lower levels (i.e., the transmitted messages) in higher-level parts of the protocol (hashed transcripts in the key derivation and the finished messages). Since some of the low-level data used in these computations, such as the server certificate, are correlated across multiple sessions of the same server, we cannot use generic composition to prove them in isolation. In a protocol designed from scratch, one can ensure that the separation of these sessions comes into full effect at a "lower" protocol level, simplifying the proofs for the "higher" levels. Indeed, our difficulties in the analysis encourages constructing protocols that are modular by design and can be analyzed by combining simple modular steps. We stress that even for TLS, we make heavy use of the composition theorem, not only to modularize our analysis, but also to lift the security we obtain for one server and multiple (anonymous) clients to the more standard multiple clients and servers setting, and for composition with arbitrary other protocols.

As most cryptographic work on TLS, we focus on the cryptographic aspects of TLS and many applied concerns are abstracted over. Moreover, as our work is in the constructive cryptography model, with notation yet unfamiliar to our audience, we focused in the body of our submission on the *beauty* (and elegance, within the limits of TLS' design characteristics), rather than the *weight* of our contribution. We invite the interested reader to find the technical details in the full version [17].

1.2   Related Work

On Provable Security. One aspect that is important in modeling and proving security especially of practical protocols is that of *composability*, as cryptographic protocols are rarely designed to be used in isolation. Indeed, a security guarantee in isolation does not necessarily imply security when a proven protocol is part of a larger scheme. While one can generally prove the security of a composite scheme by proving a reduction from breaking any of the component schemes to breaking the composite scheme, security frameworks that allow for *general/universal composition* result in security definitions that relieve one from explicitly proving such a reduction for each composite scheme. Such a reduction immediately follows from the security of the component schemes and the composition theorem.

For instance, suppose that one can prove that a given scheme (e.g. password-based authentication) achieves mutual authentication, assuming that a unilaterally authenticated secure channel already exists. Suppose also that one has several choices of how to construct this unilaterally secure channel, e.g., by RSA or DH-based key-exchange, relying on the existence of a PKI and an insecure network. In this case, the composition theorem implies that one only has to prove that the two candidate schemes construct the unilaterally secure channel; the security of the composition with the password-authentication scheme follows immediately. Frameworks which allow for generic composition are the universal composability framework (UC) due to Canetti [7], the reactive simulatability (RSIM) framework of Pfitzmann and Waidner [23], and the constructive cryptography framework (CC) of Maurer and Renner [20,21], which we use in this work. In particular, one advantage we see in using constructive cryptography is that it describes the way primitives are used within protocols with given resources, and makes explicit the guarantees that they provide in an application context. This provides an indication of how they can be used as part of more complex protocols.

Authenticated Key Exchange. Authenticated key-exchange (AKE) protocols allow two parties to agree on a session key that can be used to secure their communication. The "handshake" of the SSL/TLS protocol can be seen as an AKE. Beyond secure Internet communication, AKE has many other applications, e.g., in card-based identity and payment protocols. The security of AKE protocols was first defined by Bellare and Rogaway [4] as the indistinguishability of real session keys from random keys. However, neither the initial Bellare-Rogaway model, nor its modifications [2,3,8,10] are inherently composable. One special composition of AKE protocols with record-layer-type encryption was shown by Brzuska et al. [6]; however, AKE game-based security is not generally composable. Notions of key exchange in composable frameworks have been defined by Canetti and Krawczyk [9] and by Maurer, Tackmann, and Coretti [22], respectively.

TLS 1.2 vs. 1.3. As the TLS handshake is at present the most prominent AKE protocol, the analysis of its versions up to and including TLS 1.2 has been the subject of numerous papers in the literature. We note, however, that TLS 1.3 has a fundamentally different design from TLS 1.2, which has only been thoroughly analyzed in one publication so far [14]. While elegant and covering all modes in which the TLS 1.3 key derivation is done, this approach follows traditional game-based methods and is neither as modular as ours, nor generally composable. Several parts of the current protocol draft are adapted from work by Krawczyk and Wee [19], this includes the new key derivation scheme that we also describe in Sect. 4 and analyze in [17].

2 Our Approach — Description and Rationale

In constructive cryptography, the (security) guarantees provided to parties in a specific context are formalized in terms of *resources* available to the parties. In our analysis of TLS, resources are typically communication channels or shared secret keys with certain properties. Cryptographic protocols *construct* (desired) resources from assumed resources, and the composition theorem of this framework guarantees that the protocol (using the resources assumed by it) can be used whenever the constructed resource is required (as an assumed resource) in future constructions, i.e., several subsequent constructions can be combined into a single construction.

We model resources as discrete systems that provide one interface to each honest party, along with a specific interface that formalizes the capabilities of a potential attacker. Interfaces are labeled, such as C for a client, S for a server, or E for the attacker. Interfaces can have sub-interfaces (think of them as grouping related capabilities at the same interface for the sake of modularity); we write for instance S/sid for the server sub-interface for session *sid*. Protocols consist of one protocol engine or *converter* for each honest party. Compared with "traditional" game-based definitions, the *adversary model* corresponds to the capabilities offered via the E-interface at the *assumed resource* and the honest parties' interfaces at the *constructed resource*. For instance, interaction with an insecure network resource corresponds to an active attacker that is in full control of the network (i.e., a chosen-ciphertext attack). The fact that in a constructed channel the messages to be transmitted can be chosen by the distinguisher then corresponds to a chosen-plaintext attack. The *goal* of the game is reflected in the description of the *constructed resource*. The advantage of the *adversary* in game-based definitions corresponds to the advantage of the *distinguisher* in constructive definitions.

Notation. We use a term algebra to describe composite systems, where resources and converters are symbols, and they are composed via specific operations. We read a composed expression starting from the right-hand side resource, extended by systems on the left-hand side. If resource $\mathbf{R}$ has an interface A to which we "connect" a converter α, the resulting system $\alpha^A \mathbf{R}$ is the composition of the two systems, such that the converter connects to the A-interface of the resource **R**. For resources **R** and **S**, $[\mathbf{R}, \mathbf{S}]$ denotes the parallel composition of **R** and **S**. If we compose a family of resources $(\mathbf{R}_i)_{i \in \{1,\ldots,n\}}$ in parallel, we also write this as a product, e.g. $\bigotimes_{i=1}^{n} \mathbf{R}_i$. We introduce special notation for families of interfaces $\mathcal{L}$ and converters $\alpha_{\mathcal{L}} = (\alpha_\ell)_{\ell \in \mathcal{L}}$. To attach each π_ℓ to interface ℓ of a resource **R**, we write $(\alpha_{\mathcal{L}})^{\mathcal{L}} \mathbf{R}$.

Constructions. The construction notion is defined based on the distinguishing advantage between two resources **U** and **V**, which can be seen as a distance measure on the set of resources.[4] A distinguisher is a discrete system that connects to all the interfaces of a resource and outputs a single bit. The *distinguishing advantage of a distinguisher* **D** *on two systems* **U** *and* **V** is defined as

$$\Delta^{\mathbf{D}}(\mathbf{U}, \mathbf{V}) \coloneqq |\Pr(\mathbf{DU} = 1) - \Pr(\mathbf{DV} = 1)|. \tag{1}$$

The two main conditions defining a *construction* are: (1) *availability* (often called correctness), stipulating that the protocol using the assumed resource behaves as the constructed resource if no attacker is present at the E-interface; and (2) *security*, requiring that there exists a simulator, which, if connected at the E-interface of the constructed resource, achieves that the constructed resource with the simulator behaves like the protocol with the assumed resource (w.r.t. the distinguisher). For the availability condition, the "special converter" $\bot$ signals the attacker's absence; this is taken into account explicitly in the description of the resources. Formally a construction is defined as follows:

Definition 1. *Let ε_1 and ε_2 be two functions mapping each distinguisher* **D** *to a real number in $[0, 1]$. Let $\mathcal{L}$ be the interfaces of protocol participants. A protocol $\pi_{\mathcal{L}} = (\pi_\ell)_{\ell \in \mathcal{L}}$ constructs resource* **S** *from resource* **R** *with distance $(\varepsilon_1, \varepsilon_2)$ and with respect to the simulator σ, denoted*

$$\mathbf{R} \quad \overset{\pi_{\mathcal{L}}, \sigma, (\varepsilon_1, \varepsilon_2)}{\Longrightarrow} \quad \mathbf{S},$$

if, for all distinguishers **D***,*

$$\begin{cases} \Delta^{\mathbf{D}}\left((\pi_{\mathcal{L}})^{\mathcal{L}} \bot^{E} \mathbf{R}, \bot^{E} \mathbf{S}\right) \leq \varepsilon_1(\mathbf{D}) & (availability), \\ \Delta^{\mathbf{D}}\left((\pi_{\mathcal{L}})^{\mathcal{L}} \mathbf{R}, \sigma^{E} \mathbf{S}\right) \leq \varepsilon_2(\mathbf{D}) & (security). \end{cases}$$

Games. Several of our construction steps are proved by reductions to the security of underlying primitives, which are defined via game-based notions. A game can be seen as a system that, when connected to an adversary, determines a single bit W (denoting whether the game is won or lost). The success probability of an adversary **A** with respect to a game **G** is

$$\Gamma^{\mathbf{A}}(\mathbf{G}) \coloneqq \Pr^{\mathbf{AG}}(W = 1).$$

---

[4] The distinguishing advantage is in fact a pseudo-metric on the set of resources, that is, it is symmetric, the triangle inequality holds, and $d(x, x) = 0$ for all x. However, it may be that $d(x, y) = 0$ for $x \neq y$.

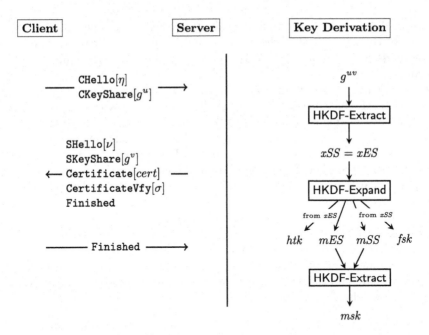

Fig. 1. The TLS 1.3 handshake and the key derivation in the case that the ephemeral and the static handshake secret coincide.

For games that are defined as distinguishing problems (such as IND-CPA security for encryption schemes), we use the notation from Eq. (1), that is, if the game is described by the pair $(\mathbf{G_0}, \mathbf{G_1})$, then we are interested in the advantage $\Delta^{\mathbf{A}}(\mathbf{G_0}, \mathbf{G_1})$. Both $\Gamma^{(\cdot)}(\mathbf{G})$ and $\Delta^{(\cdot)}(\mathbf{G_0}, \mathbf{G_1})$ define adversarial advantage functions $\epsilon(\cdot)$, such as $\varepsilon_{cr}(\mathbf{A}) = \Gamma^{\mathbf{A}}(\mathbf{G}^{cr})$ for the collision resistance of a hash function, $\varepsilon_{uf\text{-}cma}(\mathbf{A}) = \Gamma^{\mathbf{A}}(\mathbf{G}^{uf\text{-}cma})$ for the unforgeability of a signature, or $\varepsilon_{ddh}(\mathbf{A}) = \Delta^{\mathbf{A}}\big((g^A, g^B, g^{AB}), (g^A, g^B, g^C)\big)$ for the intractability of the DDH assumption.

3   TLS 1.3 and Unilaterally Secure Channels

The general structure of TLS 1.3 in (EC)DHE mode is depicted in Fig. 1 on the left. The client hello message includes a 32-byte nonce η; the client key share fixes an (elliptic curve) group $\mathbb{G}$ of order $q = |\mathbb{G}|$ (with some generator g), and an element g^u for some $u \leftarrow\!\!\$\ \{1, \dots, q\}$ in that group. The server verifies that the proposed group is in the list of acceptable groups; if so, it chooses a 32-bit nonce ν (the server hello message), and sends this, together with its key share g^v for $v \leftarrow\!\!\$\ \{1, \dots, q\}$, its certificate (in the initial draft, encrypted with the handshake keys, but in our case, without the encryption), and a certificate verify message, namely a signed session hash, also encrypted in the original draft. As a final message, the server sends an encryption (with its handshake transfer key htk) of the finished message.

The finished message is computed by evaluating a PRF keyed with the finished secret fsk on the session hash. If the signature and finished message verify, the client finished message is computed analogously and sent to the server, completing the handshake.

The current version of key derivation in TLS 1.3[5] uses HKDF[6] as a replacement of the TLS PRF construction that was the backbone of previous versions. This new key derivation, depicted in Fig. 1 on the right, follows a more stringent cryptographic design and adapts easily to various TLS handshake modes, such as the an as-yet underspecified zero round-trip time (0-RTT) mode, in which the client uses a previously-saved configuration to connect to a pre-known server.

While we leave a more technical, detailed description of the key-derivation steps to Sect. 4, note that we focus in this paper on one particular case of the key derivation in which the client and server calculate only one Diffie-Hellman value, obtained from the client and the server ephemeral key shares. The key derivation in the TLS draft is also prepared for cases in which the two parties compute *two* Diffie-Hellman values, one from the client share and the static server share, and another from the same client share and the ephemeral server share. In the case we consider, those two values are defined to be identical.

Unilaterally Secure Transmissions. The goal of TLS with server-only authentication is modeled by the following *unilateral* channel resource $\leftarrow\!\!\!-\rightarrow\!\!\bullet_n$. This resource is explicitly parametrized by the bound n on the number of sessions in which an attacker uses a specific *client* nonce (this parameter appears in the security bound). Parties input messages of length (at most) equal to TLS's maximum fragment size. We denote the set of all plaintexts as $\mathcal{PT}$.

$$\leftarrow\!\!\!-\rightarrow\!\!\bullet_n$$

No attacker present: Behave as a (multi-message) channel for messages in $\mathcal{PT}$ between interfaces C and $S/1$.

Attacker present:

- Upon the *first* input $(\texttt{allow}, e)$ with $e \in [n]$ at the E-interface (if e was not used before), provide a secure multiple-use (i.e., keep a buffer of undelivered messages) channel between C and S/e. In particular:
  - On input a message $m \in \mathcal{PT}$ at the C-interface, output $|m|$ at interface E.
  - On input $(\texttt{deliver}, \texttt{client})$ at the E-interface, deliver the next message at S/e.
  - On input a message $m \in \mathcal{PT}$ at the S/e-interface, output $|m|$ at interface E.
  - On input $(\texttt{deliver}, \texttt{server})$ at the E-interface, deliver the next message at C.
- After input $(\texttt{conquer}, e)$ with $e \in [n]$ at the E-interface (if e was not used before), forward messages in $\mathcal{PT}$ between the S/e- and E/e-interfaces in both directions.

---

[5] https://tools.ietf.org/id/draft-ietf-tls-tls13-07.txt.
[6] http://www.ietf.org/rfc/rfc5869.txt.

Intuitively, if no attacker is present, then the resource behaves like a direct channel between a client C and a server's $S/1$ sub-interface. If the attacker *is* present, then we have either a secure channel between the client and the server (first input $(\mathtt{allow}, e)$) or, if the attacker was the one performing the handshake (input $(\mathtt{conquer}, e)$), a channel between the attacker and the server.

The Assumed Resources. The resources we assume for the TLS protocol are: First, an insecure network NET (obtained by using the TCP/IP protocol over the Internet), where the attacker can also learn the contents of messages transmitted over the network, stop them, or inject arbitrary messages of his choice into the network. Second, a public-key infrastructure (PKI) resource, which we view as specific to a single server (whose identity we assume the client knows). This PKI resource allows the server to send one message (its signature verification key) authentically to all clients, thus capturing the guarantee that only the honest server can register a public key *relative to its own identity,* and the clients verify that the certificate is issued with respect to the expected identity. For simplicity, we consider a model where the PKI is local to the security statement; aspects of modeling a global PKI in composable security frameworks are discussed by Canetti el al. [11].

The Security Achieved by TLS 1.3. We show that TLS 1.3 constructs $\twoheadleftarrow\ \twoheadrightarrow\!\bullet_n$ from PKI and NET by sequential decomposition of the protocol in the main steps (right to left) shown in Fig. 2. At each step, the resources constructed in previous steps are used as assumed resources in order to construct a "new" resource, until we construct the unidirectional channel $\twoheadleftarrow\ \twoheadrightarrow\!\bullet_n$. We describe these steps in the rest of this paper.

Our reductions use the pseudorandomness of HMAC, as used internally by HKDF, the pseudorandomness of HKDF itself when seeded with seed 0, the unforgeability of signatures, the collision resistance for the hash function, the intractability of the DDH assumption, and the security of authenticated encryption. We write $\varepsilon_{\mathsf{hmac}}, \varepsilon_{\mathsf{kdf}}, \varepsilon_{\mathsf{uf\text{-}cma}}, \varepsilon_{\mathsf{cr}}, \varepsilon_{\mathsf{ddh}}, \varepsilon_{\mathsf{aead}}$ for their advantage functions.

Theorem 2. *Let $\mathcal{C}$ be a set of clients. The TLS 1.3 protocol constructs, for each client $C \in \mathcal{C}$, one unilaterally secure channel $\twoheadleftarrow\ \twoheadrightarrow\!\bullet_n$ from NET and PKI. Concretely, for the simulator σ and the adversaries $\mathbf{A}_1, \ldots, \mathbf{A}_{11}$ obtained from $\mathbf{D}$ by explicit reductions derived from those in the modular proof steps,*

$$[\mathsf{NET}, \mathsf{PKI}] \xLongrightarrow{(\mathsf{tls13c},\mathsf{tls13s}),\sigma,(\varepsilon_1,\varepsilon_2)} \bigotimes_{(I,J)\in\mathcal{P}} [\![\twoheadleftarrow\ \twoheadrightarrow\!\bullet_n]\!]^{(I,J)},$$

with:

$$\varepsilon_1(\mathbf{D}) := \binom{|\mathcal{C}|}{2} \cdot 2^{-256} + |\mathcal{C}|\left(\varepsilon_{\mathsf{ddh}}(\mathbf{A}_1) + 2\varepsilon_{\mathsf{prf}}(\mathbf{A}_2) + 2\varepsilon_{\mathsf{kdf}}(\mathbf{A}_3) + \varepsilon_{\mathsf{hmac}}(\mathbf{A}_4)\right)$$

and

$$\varepsilon_2(\mathbf{D}) := \left(\binom{n}{2} + \binom{|\mathcal{C}|}{2}\right) \cdot 2^{-256} + \varepsilon_{\mathsf{uf\text{-}cma}}(\mathbf{A}_5) + \varepsilon_{\mathsf{cr}}(\mathbf{A}_6) + n|\mathcal{C}| \cdot \varepsilon_{\mathsf{ddh}}(\mathbf{A}_7)$$
$$+ n|\mathcal{C}|\left(2\varepsilon_{\mathsf{prf}}(\mathbf{A}_8) + 2\varepsilon_{\mathsf{kdf}}(\mathbf{A}_9) + \varepsilon_{\mathsf{hmac}}(\mathbf{A}_{10})\right) + 2|\mathcal{C}| \cdot \varepsilon_{\mathsf{aead}}(\mathbf{A}_{11}).$$

This statement holds for all distinguishers **D**, *some injection* $\rho : \mathcal{C} \rightarrow \mathcal{N}$, *and* $\mathcal{P} \coloneqq \{(C, S/\rho(C)) : C \in \mathcal{C}\} \cup \{(E/\eta, S/\eta) : \eta \in \mathcal{N} \setminus \rho(\mathcal{C})\}$.

In the theorem, we construct the parallel composition $\bigotimes_{(I,J)\in\mathcal{P}} [\![{\leftarrow\!\!-\ \rightarrow\!\!\bullet}_n]\!]^{(I,J)}$ with interfaces (I, J) taken from the set $\mathcal{P}$. This models that the server can identify clients only by some value used in the handshake — we chose the random nonce $\rho(C) \in \mathcal{N}$ — and that the attacker can also interact with the server using "new" nonces, picked by none of the clients.

As a corollary and following a result by Tackmann [24], we model the use of password-based authentication to construct a bilaterally secure channel. We assume a password distribution with maximum guessing probability ϵ as an additional resource **Q**. Then the constructive corollary we postulate and prove in Sect. 5 is:

Lemma 3. *Sending and checking a password constructs from* $\leftarrow\!\!-\ \rightarrow\!\!\bullet_n$ *the channel* $\bullet\!\leftarrow\!\!-\ \rightarrow\!\!\bullet$, *for a distribution* **Q** *of passwords as described above. More formally, there is a simulator* σ *such that,*

$$[\leftarrow\!\!-\ \rightarrow\!\!\bullet_n, \mathbf{Q}] \xRightarrow{\mathsf{pwd}, \sigma, (0, \epsilon)} \bullet\!\leftarrow\!\!-\ \rightarrow\!\!\bullet.$$

4   De-Constructing TLS 1.3

This section acts as a stage-by-stage proof for Theorem 2. Our strategy is to prove that *individual parts* of the TLS protocol *construct* intermediate resources, which can be used as assumed resources for the next modular construction step. At the end, we use the composition theorem to show that the *entire* TLS 1.3 protocol constructs the $\leftarrow\!\!-\ \rightarrow\!\!\bullet_n$ channel shown in the previous section.

The structure of our proof follows Fig. 2, read from right to left. We begin by constructing a unique name resource, by choosing a random client nonce uniformly at random from the set of 32-byte strings. The unique name resource is then used to name client *sessions* on the insecure network NET; thus, from a constructive point of view, the nonce exchange at the beginning of the TLS protocol constructs from the resources NET and NAME the network-with-sessions resource SNET.

The subsequent two steps construct the handshake key resource DHKEY from the assumed PKI resource and the newly-constructed SNET resource. We proceed as follows: we first use these two resources to construct an authenticated network-transmission resource $\succ\!\!-\bullet$ (the corresponding TLS step is signing the server's first message; its ephemeral share). From this $\succ\!\!-\bullet$ resource, we construct the handshake key resource DHKEY by simply exchanging the client and server shares to calculate a Diffie-Hellman secret.

The next step is then to use the key derivation described in Fig. 1 to extract an (almost) uniformly random bit-string key from the Diffie-Hellman secret, and expand this to obtain all application keys required by the subsequent protocol steps.

The final step of the protocol is the actual payload protection, which begins by exchanging the finished messages computed using derived keys, and subsequently protecting plaintext messages using authenticated encryption.

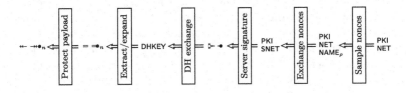

Fig. 2. The decomposition of TLS 1.3.

Session Naming. We formalize unique client naming by means of a resource NAME_ρ, parametrized by an injection ρ from the set $\mathcal{C}$ of honest clients to the set $\mathcal{N}$ of nonces; this resource returns to each client a unique nonce. NAME_ρ can be constructed from scratch: As a nonce contains 256 bits of randomness for TLS 1.3, choosing a nonce at random yields a unique nonce per client up to a loss of $\binom{|\mathcal{C}|}{2}2^{-256}$, where $|\mathcal{C}|$ is the total number of honest clients.

Naming Network Sessions. The client nonce η helps the server associate a session with some client C. Honest clients use distinct nonces, obtained from the NAME_ρ resource; however, an attacker can start many sessions with the same nonce (possibly generated by an honest client). Thus, we index sessions by pairs $sid = (\eta, e) \in \mathcal{N} \times \mathbb{N}$, where e differentiates sessions with the same η. The server's nonce ν for that session is chosen at random and sent to the client; this protocol constructs, from

---

Augmented Network Resource SNET$_{\rho,n}$

The resource is parametrized by an injective function $\rho : \mathcal{C} \to \mathcal{N}$, and $n \in \mathbb{N}$.

No attacker present: For each $C \in \mathcal{C}$ choose a nonce $\nu_{sid} \leftarrow^\$ \mathcal{N}$ with $sid = (\rho(C), 1)$, output $(\rho(C), \nu_{sid})$ at interface C and ν_{sid} at interface S/sid, and provide bidirectional channels between those two interfaces.

Attacker present: Initialize $e_\eta = 0$ for all $\eta \in \mathcal{N}$. For each $C \in \mathcal{C}$, output the nonce η at the interface C and forward all messages input at interface C to the interface E/C. Additionally:

- Upon input $(\mathbf{ack}, \eta)$ at interface E, if $e_\eta \leq n$, then for $sid = (\eta, e_\eta)$ choose a nonce ν_{sid} uniformly at random from $\mathcal{N} \setminus \{\nu_{(\eta,1)}, \ldots, \nu_{(\eta, e_\eta-1)}\}$, output ν_{sid} at interface E/sid and S/sid, and increase e_η. Then, forward all communication between the interfaces S/sid and E/sid.
- Upon input $(\mathbf{deliver}, C, \nu)$ with $C \in \mathcal{C}$, output the server's nonce ν at the interface C. Then, forward all messages between interfaces E/C and C.

Fig. 3. The network resource that additionally outputs nonces.

the resources NAME_ρ and the network resource NET, the resource SNET (the full details and description of the client and server converters, denoted hec and hes_n are left to the full version).

The resource SNET, described in Fig. 3, has interfaces labeled $C \in \mathcal{C}$ for the clients, a server interface S with one sub-interface for each pair (η, e), where $\eta \in \mathcal{N}$ is a nonce, not necessarily from an honest client, and $e \in [n]$ is a counter indicating how many sessions are initiated with nonce η, and an attacker's interface called E. To simplify further construction steps, we rule out collisions for *server* nonces in the SNET resource below, in sessions associated with the same nonce (i.e., $sid = (\eta, e)$ and $sid' = (\eta, e')$). Since the server nonce has the same structure as the client nonce, the security loss is analogous.

The following statement holds:

Lemma 4. *Let* $\mathcal{C} \subseteq \mathcal{A}$ *and let* $\rho : \mathcal{C} \to \mathcal{N}$ *be an injective mapping. The protocol* (hec, hes_n) *constructs the resource* $\mathsf{SNET}_{\rho,n}$ *from the resources* NET *and* NAME_ρ. *In more detail, for the simulator* σ *in the proof:*

$$[\mathsf{NET}, \mathsf{NAME}_\rho] \quad \overset{(\mathsf{hec},\mathsf{hes}_n),\sigma,(0,\varepsilon)}{\Longrightarrow} \quad \mathsf{SNET}_{\rho,n},$$

with $\varepsilon(\mathbf{D}) := \binom{n}{2} \cdot 2^{-256}$ *for all distinguishers* $\mathbf{D}$.

The Shared Key Resource. The next step is to construct the Diffie-Hellman key DHKEY; we decompose this step into two smaller steps, briefly described below (we refer to [17] for full details). We represent the DHKEY resource as a particular parametrization of the generic shared key resource $\mathsf{KEY}_{\rho,AUX,n,\mathcal{K}}$ detailed in Fig. 4, with a key space $\mathcal{K}$ that is the Diffie-Hellman group $\mathbb{G}$.

Our first step is to construct from the PKI and SNET resources an authenticated network resource $\succ\!\!-\!\!\bullet_{\rho,\mathfrak{F},SIG,n,h}$ using the certificate and the signature in the TLS certificate verify message. This resource allows the server to transmit one message in each session authentically; this is achieved by signing the message together with a hash of the handshake messages in order to bind it to the session. The reduction relies on the unforgeability of the signature scheme and the collision resistance in the handshake hash.

From $\succ\!\!-\!\!\bullet_{\rho,\mathfrak{F},SIG,n,h}$, we then construct, under the DDH assumption in $\mathbb{G}$, the resource DHKEY. Intuitively, the converters here are simply exchanging the Diffie-Hellman elements and perform the corresponding computation, where the transmission of the server's message relies on the authentication guarantees of the assumed resource. In particular, the signature computed and forwarded in the authentication step allows a client to abort an execution if the signature verification on the handshake hash fails. This is reflected in the second bullet point of the resource $\mathsf{KEY}_{\rho,AUX,n,\mathcal{K}}$.

The composition theorem allows us to combine the two intermediary steps in the following lemma, where we denote by hsc and hss the compositions of the two converters (protocol steps) outlined above:

Lemma 5. *The protocol* (hsc, hss) *constructs from the assumed resources* PKI *and* $\mathsf{SNET}_{\rho,n}$ *the resource* DHKEY, *given that: the signature scheme used in certification*

Shared Key Resource $\mathsf{KEY}_{\rho,AUX,n,\mathcal{K}}$

No attacker present: For each $C \in \mathcal{C}$, choose a key $\kappa_C \leftarrow_\$ \mathcal{K}$, a nonce $\nu_{sid} \leftarrow_\$ \mathcal{N}$ (with $sid = (\rho(C), 1)$), and auxiliary information $aux_C \leftarrow_\$ AUX$, and output $(\kappa_C, \rho(C), \nu_{sid}, aux_C)$ at interface C and $(\kappa_C, \nu_{sid}, aux_C)$ at the interface S/sid. Then, provide bidirectional channels between those two interfaces.

Attacker present: Initialize $e_\eta = 1$ for all $\eta \in \rho(\mathcal{C})$.

- Upon input $(\mathtt{ack}, \eta)$ at the E-interface, if $\eta \in \mathcal{N}$ and $e_\eta \leq n$, then choose ν_{sid} uniformly at random from $\mathcal{N} \setminus \{\nu_{(\eta,1)}, \dots, \nu_{(\eta,e_\eta-1)}\}$ for $sid = (\eta, e_\eta)$, output ν_{sid} at the E-interface and increase e_η.
- Upon input $(\mathtt{allow}, C, aux, \nu)$ at the E-interface with $aux \in \{0,1\}^*$, and $\nu \in \mathcal{N}$:
  1. If $\nu = \nu_{sid}$ and there was a previous $(\mathtt{server\text{-}allow}, \eta, e, aux)$ resulting in setting κ_{sid} for some $sid = (\eta, e)$ with $\eta = \rho(C)$ and $e < e_\eta$, then set $\kappa_C = \kappa_{sid}$; else, abort without generating keys.
  2. Output at the C-interface $(\kappa_C, \rho(C), \nu, aux)$.
  Afterward, if at the C-interface a message m is input, output m at the E-interface. Also, allow at the E-interface to inject messages m', to be output at the C-interface.
- Upon input $(\mathtt{server\text{-}allow}, \eta, e, aux)$ at the E-interface with $e < e_\eta$, with $sid = (\eta, e)$:
  1. If, for $C = \rho^{-1}(\eta)$, $(\mathtt{allow}, C, *, \nu_{sid})$ was input at the E-interface before (i.e., with the current server's nonce), then set $\kappa_{sid} = \kappa_C$; else draw κ_{sid} uniformly at random from $\mathcal{K}$.
  2. Output $(\kappa_{sid}, \nu_{sid}, aux)$ at interface S/sid.
- Upon input $(\mathtt{inject}, \eta, e, aux, \kappa)$ at the E-interface, if $e < e_\eta$, then for $sid = (\eta, e)$ output (κ, ν_{sid}, aux) at interface S/sid.

After either a $(\mathtt{server\text{-}allow}, \eta, e, *)$ or a $(\mathtt{inject}, \eta, e, *, *)$ message, if at interface S/sid (with $sid = (\eta, e)$) a message m is input, output m at interface E. Also, allow at interface E to inject messages m' to be output at interface S/sid.

Fig. 4. The shared key resource.

is unforgeable, the hash function is collision resistant, and the DDH assumption holds. More formally, for the simulator σ and the reductions $\mathbf{C}_1, \dots \mathbf{C}_4$ described in the proof,

$$[\mathsf{SNET}_{\rho,n}, \mathsf{PKI}_{\mathfrak{F}}] \stackrel{(\mathsf{hsc},\mathsf{hss}),\sigma,(\varepsilon_1,\varepsilon_2)}{\Longrightarrow} \mathsf{DHKEY}_{\rho,AUX,n},$$

such that for all distinguishers $\mathbf{D}$: $\varepsilon_1(\mathbf{D}) \coloneqq |\mathcal{C}| \cdot \varepsilon_{\mathsf{ddh}}(\mathbf{DC}_1)$, *and* $\varepsilon_2(\mathbf{D}) \coloneqq \varepsilon_{\mathsf{uf\text{-}cma}}(\mathbf{DC}_2) + \varepsilon_{\mathsf{CR}}(\mathbf{DC}_3) + n \cdot |\mathcal{C}| \cdot \varepsilon_{\mathsf{ddh}}(\mathbf{DC}_4)$.

Expanding the Key. The next step is to extract from the Diffie-Hellman secret and then expand the keys (following the scheme shown in Fig. 1). Finally, the finished messages used for key confirmation are computed. Interestingly, the only effect of the finished messages in our case is that the client and server detect mismatching keys before the first application data is accepted by the protocol. This does not exclude, however, that these messages serve a more crucial role in certain hand-shake modes or for proving specific security properties we do not consider in this paper.

The key derivation in the newest draft of TLS 1.3 differs considerably from that of TLS 1.2. From the Diffie-Hellman secret, several sets of session keys are derived for use in symmetric primitives: the application traffic keys atk for the protection of the payload data, handshake traffic keys htk used to protect some data packets in the handshake, the finished secret fsk used for the finished messages, and early-data keys used in the 0-RTT mode (the latter do not appear in our analysis). All computations are based on HKDF [18].

The key derivation can be described in several steps corresponding in our analysis to separate, simple construction steps that are composed via the composition theorem:

1. First, two keys xES and xSS are computed by calling HKDF.$extract(0, pmk)$, that is, evaluating the HKDF extraction with seed 0 on the Diffie-Hellman key pmk computed in the key exchange. This step assumes the security of HKDF as a computational extractor (therefore relying on a statement proven in the random-oracle model).
2. Using the expansion of HKDF, several keys are computed:
   (a) The finished secret $fsk \leftarrow$ HKDF.$expand(xSS,$ "finished"$, h)$ for the con-firmation messages, where h is the hash of the handshake messages,
   (b) the "static" master secret value $mSS \leftarrow$ HKDF.$expand(xSS,$ "static"$, h)$,
   (c) the "ephemeral" value $mES \leftarrow$ HKDF.$expand(xES,$ "ephemeral"$, h)$,
   (d) the handshake traffic keys $htk \leftarrow$ HKDF.$expand(xES,$ "handshake"$, h)$.
   This step assumes the security of the HKDF expansion as a pseudo-random function.
3. Then, compute the master secret key $msk \leftarrow$ HKDF.$extract(mSS, mES)$ by using HKDF to extract from mES using the seed mSS. This step relies only on the fact that the HKDF extraction is a pseudo-random function, as mSS is a good key — in fact a weak PRF is sufficient as mES is (pseudo) random.
4. Expand the application traffic keys atk by an HKDF expansion as follows: $atk \leftarrow$ HKDF.$expand(msk,$ "application"$, h)$. This step again relies on the HKDF expansion being a PRF.

In order to treat the expanded keys as separate resources for each client, we also incorporate the generation of the finished messages into the construction of those keys. Those messages are computed by evaluating HMAC with the key fsk on the session hash h and static labels. This requires that HMAC is a PRF. Since the expansion is the final step that explicitly relies on values that are consistent across several sessions (such as the server's certificate), the constructed expanded-key resource $= \Rightarrow\!\bullet_n$ can be described in a way that is single-client, as opposed to the more complicated $\mathsf{KEY}_{\rho,AUX,n,\mathcal{K}}$ resource. The resource $= \Rightarrow\!\bullet_n$ allows a single client and server session to compute the same keys and finished messages if the attacker did not establish that server session himself. Otherwise, the server and attacker share keys, as depicted in the description of $= \Rightarrow\!\bullet_n$ [17]. We describe the resource we want to obtain at key expansion by: $\bigotimes_{C \in \mathcal{C}} \llbracket = \twoheadrightarrow_n \rrbracket^{(C, S/\rho(C))}$, i.e. a parallel composition of such channels with appropriate interface labels.

The key-expansion steps yield the following constructive statement:

Lemma 6. *The protocol* (expc13, exps13) *constructs the parallel composition of keys* $\bigotimes_{(I,J) \in \mathcal{P}} \llbracket = \twoheadrightarrow_{n \, \text{cphs}, n} \rrbracket^{(I,J)}$ *from the secret key resource* DHKEY, *for* $\mathcal{P} := \{(C, S/\rho(C)) : C \in \mathcal{C}\} \cup \{(E/\eta, S/\eta) : \eta \notin \rho(\mathcal{C})\}$. *The construction holds under the assumptions that HKDF is a KDF with seed* 0, *and that HKDF expansion and HMAC are PRFs. In more detail, for the simulator* σ *and the reductions* $\mathbf{C}_1, \ldots, \mathbf{C}_5$ *described in the proof,*

$$\text{DHKEY}_{\rho, AUX, n} \quad \xRightarrow{\;(\text{expc13,exps13}), \sigma, (\varepsilon, \varepsilon')\;} \quad \bigotimes_{(I,J) \in \mathcal{P}} \llbracket = \twoheadrightarrow_n \rrbracket^{(I,J)}$$

where, for all distinguishers $\mathbf{D}$, $\varepsilon'(\mathbf{D}) = n \cdot \varepsilon(\mathbf{D})$ *and*

$$\varepsilon(\mathbf{D}) = |\mathcal{C}| \cdot \left(\varepsilon_{\text{kdf}}(\mathbf{DC}_1) + \varepsilon_{\text{prf}}(\mathbf{DC}_2) + \varepsilon_{\text{kdf}}(\mathbf{DC}_3) + \varepsilon_{\text{prf}}(\mathbf{DC}_4) + \varepsilon_{\text{hmac}}(\mathbf{DC}_5) \right).$$

The Record Layer. The authenticated key resource $= \twoheadrightarrow_n$ constructed in the previous step yields sets of keys (*htk*, *atk*, *fsk*) and the finished messages. The gap between the resource $= \twoheadrightarrow_n$ and our goal resource, i.e., the unilaterally-secure channel $\twoheadleftarrow \twoheadrightarrow_n$, is bridged by a pair of converters essentially exchanging and verifying the finished messages, then using authenticated encryption to protect messages. The key property of our constructed resource, $\twoheadleftarrow \twoheadrightarrow_n$, is notably that it allows for messages to be securely (confidentially and authentically) transmitted, either consistently between the server and the honest client, or between the server and the adversary (but never between the client and the adversary).

For TLS 1.3 the record-layer protocol is specified based on authenticated encryption with associated data (AEAD). This mode has been analyzed by Badertscher et al. [1] in recent work. Their result can be "imported" into our work. Thus, for the final step of the proof, we rely on the security of AEAD encryption, which is defined in terms of indistinguishability between two systems $\mathbf{G}_0^{\text{aead}}$ and $\mathbf{G}_1^{\text{aead}}$, formally detailed in the full version. In $\mathbf{G}_0^{\text{aead}}$, encryption and decryption queries to the scheme are answered by encryption and decryption using the given nonce and associated data. For $\mathbf{G}_1^{\text{aead}}$, encryption queries are answered with uniformly random strings of appropriate length, while decryption queries are answered either with a corresponding plaintext (if they were output by a previous encryption query) or by a special *invalid* symbol otherwise.

Lemma 7. *The protocol* (aeadc, aeads) *constructs from the authenticated key resource* $= \twoheadrightarrow_n$ *the unilaterally secure channel* $\twoheadleftarrow \twoheadrightarrow_n$, *under the assumption that the underlying AEAD cipher is secure. More formally, for the simulator* σ *and the reduction* $\mathbf{C}$ *described in the proof,*

$$= \twoheadrightarrow_n \quad \xRightarrow{\;(\text{aeadc,aeads}), \sigma, (0, \varepsilon)\;} \quad \twoheadleftarrow \twoheadrightarrow_n,$$

with $\varepsilon(\mathbf{D}) := 2 \cdot \varepsilon_{\text{aead}}(\mathbf{C})$ *for all distinguishers* $\mathbf{D}$.

Re-constructing TLS. At this point, using the composition theorem completes the proof of Theorem 2. In the full version, we also explain in detail how the composition of all the converters from the modular-steps yields the TLS protocol.

5   Composition with Password-Based Authentication

In prior work, Maurer et al. [22, 24] have discussed means of authenticating a unilaterally authenticated key by using password-based authentication. Thus, by starting from a unilateral key resource (similar to our $= \Rightarrow\bullet_n$ resource), one can use a password — a key with relatively low entropy — shared between a client and a server to obtain a key for which both client and server have authenticity guarantees, and which is sometimes denoted as $\bullet= \Rightarrow\bullet$ (the bullet on the left hand side indicates that the client is also authenticated). The resources $= \Rightarrow\bullet_n$ and $\bullet= \Rightarrow\bullet$ are different in that in $= \Rightarrow\bullet_n$ the attacker at the E-interface can also inject a key to be shared with the server (no client authentication). For $\bullet= \Rightarrow\bullet$ this is no longer possible.

We use the same ideas here, but our goal is to construct the fully secure channel $\bullet\twoheadleftarrow \twoheadrightarrow\bullet$ described below from the unilaterally secure bidirectional $\twoheadleftarrow \twoheadrightarrow\bullet_n$ and a password.

$$\bullet\twoheadleftarrow \twoheadrightarrow\bullet$$

No attacker present: Behave as a (multi-message) channel between interfaces C and S.

Attacker present: Provide a secure multiple-use (i.e., keep a buffer of undelivered messages) channel between C and S. In particular:

- On input a message $m \in \mathcal{PT}$ at the C-interface, output $|m|$ at interface E.
- On input (deliver, client) at the E-interface, deliver the next message at S.
- On input a message $m \in \mathcal{PT}$ at the S-interface, output $|m|$ at interface E.
- On input (deliver, server) at the E-interface, deliver the next message at C.

The protocol consists of two simple converters: sending the password (client) and verifying it (server), abbreviated as pwd $=$ (pwd.send, pwd.check). After the password exchange, the converters simply send and receive messages via the channel. For simplicity, we assume that the server accepts the same user password only once; this can be generalized along the lines of [24, Theorem 4.17]. We model a password distribution with maximum guessing probability ϵ as an additional resource $\mathbf{Q}$. The constructive statement we postulate is:

Lemma 3. *Sending and checking a password constructs from* $\twoheadleftarrow \twoheadrightarrow\bullet_n$ *the channel* $\bullet\twoheadleftarrow \twoheadrightarrow\bullet$, *for a distribution* $\mathbf{Q}$ *of passwords as described above. More formally, there is a simulator σ such that,*

$$[\twoheadleftarrow \twoheadrightarrow\bullet_n, \mathbf{Q}] \xRightarrow{\text{pwd},\sigma,(0,\epsilon)} \bullet\twoheadleftarrow \twoheadrightarrow\bullet.$$

Proof (Sketch). The availability condition follows since the client and the server obtain the same password. The simulator works as follows:

- the session between the honest client and the server is handled by (essentially) forwarding the communication between the E-interface of the constructed resource and the distinguisher,
- for all other sessions, the simulator simply drops all messages provided at its outside interface.

The only way for the distinguisher to be successful in distinguishing between the two cases is by guessing the correct password, since otherwise the behavior is the same in both cases. Since the server accepts a password only once, we can bound the overall success probability of the distinguisher by ϵ.   □

Acknowledgments. Ueli Maurer was supported by the Swiss National Science Foundation (SNF), project no. 200020-132794. Björn Tackmann was supported by the Swiss National Science Foundation (SNF) via Fellowship no. P2EZP2_155566 and the NSF grants CNS-1228890 and CNS-1116800. Daniele Venturi acknowledges support by the European Commission (Directorate General Home Affairs) under the GAINS project HOME/2013/CIPS/AG/4000005057, and by the European Union's Horizon 2020 research and innovation programme under grant agreement No 644666.

References

1. Badertscher, C., Matt, C., Maurer, U., Rogaway, P., Tackmann, B.: Augmented secure channels as the goal of the TLS record layer. In: Au, M.H., Miyaji, A. (eds.) Provable Security. LNCS, vol. 9451. Springer, Heidelberg (2015)
2. Bellare, M., Kohno, T., Namprempre, C.: Authenticated encryption in SSH: provably fixing the SSH binary packet protocol. ACM Trans. Inf. Syst. Secur. (TISSEC) **7**(2), 206–241 (2004)
3. Bellare, M., Pointcheval, D., Rogaway, P.: Authenticated key exchange secure against dictionary attacks. In: Preneel, B. (ed.) EUROCRYPT 2000. LNCS, vol. 1807, pp. 139–155. Springer, Heidelberg (2000)
4. Bellare, M., Rogaway, P.: Entity authentication and key distribution. In: Stinson, D.R. (ed.) CRYPTO 1993. LNCS, vol. 773, pp. 232–249. Springer, Heidelberg (1994)
5. Bhargavan, K., Delignat-Lavaud, A., Fournet, C., Pironti, A., Strub, P.Y.: Triple handshakes and cookie cutters: breaking and fixing authentication over TLS. In: IEEE Symposium on Security and Privacy (SP'14). IEEE (2014)
6. Brzuska, C., Fischlin, M., Smart, N., Warinschi, B., Williams, S.: Less is more: relaxed yet composable security notions for key exchange. Int. J. Inf. Secur. **12**(4), 267–297 (2013)
7. Canetti, R.: Universally composable security: A new paradigm for cryptographic protocols. Cryptology ePrint Archive, Report 2000/067, July 2013
8. Canetti, R., Krawczyk, H.: Analysis of key-exchange protocols and their use for building secure channels. In: Pfitzmann, B. (ed.) EUROCRYPT 2001. LNCS, vol. 2045, pp. 453–474. Springer, Heidelberg (2001)
9. Canetti, R., Krawczyk, H.: Universally composable notions of key exchange and secure channels. In: Knudsen, L.R. (ed.) EUROCRYPT 2002. LNCS, vol. 2332, pp. 337–351. Springer, Heidelberg (2002)

10. Krawczyk, H.: HMQV: a high-performance secure Diffie-Hellman protocol. In: Shoup, V. (ed.) CRYPTO 2005. LNCS, vol. 3621, pp. 546–566. Springer, Heidelberg (2005)
11. Canetti, R., Shahaf, D., Vald, M.: Universally composable authentication and key-exchange with global PKI. Cryptology ePrint Archive Report 2014/432, October 2014
12. Dierks, T., Rescorla, E.: The transport layer security (TLS) protocol version 1.2. RFC 5246, August 2008. http://www.ietf.org/rfc/rfc5246.txt
13. Dierks, T., Rescorla, E.: The transport layer security (TLS) protocol version 1.3. RFC draft, April 2015. http://tlswg.github.io/tls13-spec/
14. Dowling, B., Fischlin, M., Günther, F., Stebila, D.: A cryptographic analysis of the TLS 1.3 handshake protocol candidates. In: ACM Conference on Computer and Communications Security 2015 (2015)
15. Hickman, K.: The SSL protocol, February 1995. https://tools.ietf.org/html/draft-hickman-netscape-ssl-00 (internet draft)
16. Jost, D.: A Constructive Analysis of IPSec. Master's thesis, ETH Zürich, April 2014
17. Kohlweiss, M., Maurer, U., Onete, C., Tackmann, B., Venturi, D.: (De-)constructing TLS. Cryptology ePrint Archive, Report 020/2014 (2014)
18. Krawczyk, H.: Cryptographic extraction and key derivation: the HKDF scheme. In: Rabin, T. (ed.) CRYPTO 2010. LNCS, vol. 6223, pp. 631–648. Springer, Heidelberg (2010)
19. Krawczyk, H., Wee, H.: The OPTLS protocol and TLS 1.3. Manuscript, September 2015
20. Maurer, U.: Constructive cryptography – a new paradigm for security definitions and proofs. In: Mödersheim, S., Palamidessi, C. (eds.) TOSCA 2011. LNCS, vol. 6993, pp. 33–56. Springer, Heidelberg (2012)
21. Maurer, U., Renner, R.: Abstract cryptography. In: Innovations in Computer Science. Tsinghua University Press (2011)
22. Maurer, U., Tackmann, B., Coretti, S.: Key exchange with unilateral authentication: Composable security definition and modular protocol design. Cryptology ePrint Archive, Report 2013/555 (2013)
23. Pfitzmann, B., Waidner, M.: A model for asynchronous reactive systems and its application to secure message transmission. In: Proceedings of the 2001 IEEE Symposium on Security and Privacy, pp. 184–200. IEEE (2001)
24. Tackmann, B.: A Theory of Secure Communication. Ph.D. thesis, ETH Zürich (2014)

Cryptanalysis

Cryptanalysis of Variants of RSA with Multiple Small Secret Exponents

Liqiang Peng[1,2], Lei Hu[1,2](✉), Yao Lu[1,3], Santanu Sarkar[4], Jun Xu[1,2], and Zhangjie Huang[1,2]

[1] State Key Laboratory of Information Security, Institute of Information Engineering, Chinese Academy of Sciences, Beijing 100 093, China
pengliqiang@iie.ac.cn, hu@is.ac.cn
[2] Data Assurance and Communication Security Research Center, Chinese Academy of Sciences, Beijing 100 093, China
[3] The University of Tokyo, Tokyo, Japan
lywhhit@gmail.com
[4] Indian Institute of Technology Madras, Sardar Patel Road, Chennai 600 036, India
sarkar.santanu.bir@gmail.com

Abstract. In this paper, we analyze the security of two variants of the RSA public key cryptosystem where multiple encryption and decryption exponents are used with a common modulus. For the most well known variant, CRT-RSA, assume that n encryption and decryption exponents (e_l, d_{p_l}, d_{q_l}), where $l = 1, \cdots, n$, are used with a common CRT-RSA modulus N. By utilizing a Minkowski sum based lattice construction and combining several modular equations which share a common variable, we prove that one can factor N when $d_{p_l}, d_{q_l} < N^{\frac{2n-3}{8n+2}}$ for all $l = 1, \cdots, n$. We further improve this bound to d_{p_l} (or d_{q_l}) $< N^{\frac{9n-14}{24n+8}}$ for all $l = 1, \cdots, n$. Moreover, our experiments do better than previous works by Jochemsz-May (Crypto 2007) and Herrmann-May (PKC 2010) when multiple exponents are used. For Takagi's variant of RSA, assume that n key pairs (e_l, d_l) for $l = 1, \cdots, n$ are available for a common modulus $N = p^r q$ where $r \geq 2$. By solving several simultaneous modular univariate linear equations, we show that when $d_l < N^{\left(\frac{r-1}{r+1}\right)^{\frac{n+1}{n}}}$, for all $l = 1, \cdots, n$, one can factor the common modulus N.

Keywords: RSA · Cryptanalysis · Lattice · Coppersmith's method

1 Introduction

Since its invention [16], the RSA public key scheme has been widely used due to its effective encryption and decryption. To obtain high efficiency, some variants of the original RSA were designed. Wiener [24] proposed an algorithm to use the Chinese Remainder Theorem in the decryption phase to accelerate the decryption operation by using smaller exponents d_p and d_q which satisfy $ed_p \equiv 1 \bmod (p - 1)$ and $ed_q \equiv 1 \bmod (q - 1)$ for a modulus $N = pq$ and an

© Springer International Publishing Switzerland 2015
A. Biryukov and V. Goyal (Eds.): INDOCRYPT 2015, LNCS 9462, pp. 105–123, 2015.
DOI: 10.1007/978-3-319-26617-6_6

encryption exponent e. This decryption oriented alternative of RSA scheme is usually called as CRT-RSA. Also for gaining a fast decryption implementation, Takagi [21] proposed another variant of RSA with moduli of the form $N = p^r q$, where $r \geq 2$ is an integer. For Takagi's variant, the encryption exponent e and decryption exponent d satisfy $ed \equiv 1 \pmod{p^{r-1}(p-1)(q-1)}$.

In many applications of the RSA scheme and its variants, either d is chosen to be small or d_p and d_q are chosen to be small for efficient modular exponentiation in the decryption process. However, since Wiener [24] showed that the original RSA scheme is insecure when d is small enough, along this direction many researchers have paid much attention to factoring RSA moduli and its variants under small decryption exponents.

Small Secret Exponent Attacks on RSA and Its Variants. For the original RSA with a modulus $N = pq$, Wiener [24] proved that when $d \leq N^{0.25}$, one can factor the modulus N in polynomial time by a Continued Fraction Method. Later, by utilizing a lattice based method, which is usually called Coppersmith's technique [5] for finding small roots of a modular equation, Boneh and Durfee [2] improved the bound to $N^{0.292}$ under several acceptable assumptions. Then, Herrmann and May [6] used a linearization technique to simplify the construction of the lattice involved and obtained the same bound $N^{0.292}$. Until now, $N^{0.292}$ is still the best result for small secret exponent attacks on the original RSA scheme with full size of e.

For CRT-RSA, Jochemsz and May [10] gave an attack for small d_p and d_q, where p and q are balanced and the encryption exponent e is of full size, i.e. about as large as the modulus $N = pq$. By solving an integer equation, they can factor N provided that the small decryption CRT-exponents d_p and d_q are smaller than $N^{0.073}$. Similarly, Herrmann and May [6] used a linearization technique to obtain the same theoretical bound but better results in experiments.

For Takagi's variant of RSA with modulus $N = p^r q$, May [13] applied Coppersmith's method to prove that one can factor the modulus provided that $d \leq N^{\left(\frac{r-1}{r+1}\right)^2}$. By modifying the collection of polynomials in the construction of the lattice, Lu et al. [12] improved this bound to $d \leq N^{\frac{r(r-1)}{(r+1)^2}}$. Recently, from a new point of view of utilizing the algebraic property $p^r q = N$, Sarkar [17] improved the bound when $r \leq 5$. Especially for the most practical case of $r = 2$, the bound has been significantly improved from $N^{0.222}$ to $N^{0.395}$. The following table lists the existing small decryption exponent attacks on RSA and its variants Table 1.

Multiple Small Secret Exponents RSA. In order to simplify RSA key management, one may be tempted to use a single RSA modulus N for several key pairs (e_i, d_i). Simmons [19] showed that if a massage m is sent to two participants whose public exponents are relatively prime, then m can easily be recovered. However Simmons's attack can not factor N. Hence Howgrave-Graham and Seifert [8] analyzed the case that several available encryption exponents

Table 1. Overview of existing works on small secret exponent attacks on RSA and its variants. The conditions in the last column allow to efficiently factor the modulus N.

Author(s)	Cryptosystem	Bounds
Wiener: 1990 [24]	RSA	$d < N^{0.25}$
Boneh and Durfee: 1999 [2]	RSA	$d < N^{0.292}$
Jochemsz and May: 2007 [10]	CRT-RSA	$d_p, d_q < N^{0.073}$
Herrmann and May: 2010 [6]	CRT-RSA	$d_p, d_q < N^{0.073}$
May: 2004 [13]	Takagi's variant of RSA $N = p^r q$	$d \leq N^{(\frac{r-1}{r+1})^2}$
Lu, Zhang, Peng and Lin: 2014 [12]	Takagi's variant of RSA $N = p^r q$	$d \leq N^{\frac{r(r-1)}{(r+1)^2}}$
Sarkar: 2014 [17]	Takagi's variant of RSA $N = p^2 q$	$d \leq N^{0.395}$

Table 2. Comparison of previous theoretical bounds with respect to the number of decryption exponents.

n	1	2	5	10	20	∞
Howgrave-Graham and Seifert's bound [8]	0.2500	0.3125	0.4677	0.5397	0.6319	1.0000
Sarkar and Maitra's bound [18]		0.2500	0.4167	0.5833	0.6591	0.7024	0.7500
Aono's bound [1]		0.2500	0.4643	0.6250	0.6855	0.7172	0.7500
Takayasu and Kunihiro's bound [23]		0.2929	0.4655	0.6464	0.7460	0.8189	1.0000

$(e_1, \cdots, e_n)$ exist for a common modulus N and the corresponding decryption exponents $(d_1, \cdots, d_n)$ are small. From their result, one can factor N when the n decryption exponents satisfy that $d_l < N^\delta$ for all $l = 1, \cdots, n$, where

$$\delta < \begin{cases} \dfrac{(2n+1)2^n - (2n+1)\binom{n}{\frac{n}{2}}}{(2n-1)2^n + (4n+2)\binom{n}{\frac{n}{2}}}, & \text{if } n \text{ is even, and} \\[4mm] \dfrac{(2n+1)2^n - 4n\binom{n-1}{\frac{n-1}{2}}}{(2n-2)2^n + 8n\binom{n-1}{\frac{n-1}{2}}}, & \text{if } n \text{ is odd.} \end{cases}$$

In [18], Sarkar and Maitra used the strategy of [9] to solve for small roots of an integer equation and improved the bound to $\delta < \frac{3n-1}{4n+4}$. Aono [1] proposed a method to solve several simultaneous modular equations which share a common unknown variable. Aono combined several lattices into one lattice by a Minkowski sum based lattice construction and obtained that when $\delta < \frac{9n-5}{12n+4}$, N can be factored. Shortly afterwards, Takayasu and Kunihiro [23] modified each lattice and collected more helpful polynomials to improve the bound to $1 - \sqrt{\frac{2}{3n+1}}$. In conclusion, an explicit picture of the comparison of previous work is illustrated in Table 2.

Simultaneous Modular Univariate Linear Equations Modulo an Unknown Divisor. In 2001, Howgrave-Graham first considered the problem of solving an univariate linear equation modulo an unknown divisor of a known composite integer,

$$f(x) = x + a \pmod{p},$$

where a is a given integer, and $p \simeq N^\beta$ is an unknown factor of the known N. The size of the root is bounded by $|x| < N^\delta$. Howgrave-Graham proved that one can solve for the root in polynomial time provided that $\delta < \beta^2$.

The generalization of this problem has been considered by Cohn and Heninger [4],

$$\begin{cases} f(x_1) = x_1 + a_1 \pmod{p}, \\ f(x_2) = x_2 + a_2 \pmod{p}, \\ \cdots \\ f(x_n) = x_n + a_n \pmod{p}. \end{cases}$$

In the above simultaneous modular univariate linear equations, $a_1, \cdots, a_n$ are given integers, and $p \simeq N^\beta$ is an unknown factor of N. Based on their result, one can factor N if

$$\frac{\gamma_1 + \cdots + \gamma_n}{n} < \beta^{\frac{n+1}{n}} \text{ and } \beta \gg \frac{1}{\sqrt{\log N}}$$

where $|x_1| < N^{\gamma_1}, \cdots, |x_n| < N^{\gamma_n}$. Then by considering the sizes of unknown variables and collecting more helpful polynomials which are selected to construct the lattice, Takayasu and Kunihiro [22] further improved the bound to

$$\sqrt[n]{\gamma_1 \cdots \gamma_n} < \beta^{\frac{n+1}{n}} \text{ and } \beta \gg \frac{1}{\sqrt{\log N}}.$$

Our Contributions. In this paper, we give an analysis of CRT-RSA and Takagi's variant of RSA with multiple small decryption exponents, respectively. For CRT-RSA, $(e_1, \cdots, e_n)$ are n encryption exponents and $(d_{p_1}, d_{q_1}), \cdots, (d_{p_n}, d_{q_n})$ are the corresponding decryption exponents for a common CRT-RSA modulus N, where $e_1, \cdots, e_n$ are of full size as N. Based on the Minkowski sum based lattice construction proposed by Aono [1], we combine several modular equations which share a common variable and obtain that one can factor N when

$$d_{p_l}, d_{q_l} < N^{\frac{2n-3}{8n+2}}$$

for all $l = 1, \cdots, n$, where n is the number of decryption exponents.

In order to utilize the Minkowski sum based lattice construction to combine the equations, the equations should share a common variable. Hence, we modified each of the equations considered in [10], which results in a worse bound when there is only one pair of encryption and decryption exponents.

However, note that the modular equations

$$k_{p_l}(p-1) + 1 \equiv 0 \pmod{e_l}, \text{ for } l = 1, \cdots, n,$$

share a common root p. Then we can directly combine these n equations by a Minkowski sum based lattice construction, and moreover introduce a new variable q to minimize the determinant of the combined lattice. We can obtain an improved bound that one can factor N when

$$d_{p_l} < N^{\frac{9n-14}{24n+8}}$$

for all $l = 1, \cdots, n$.

Note that, for combining these equations we modified each of the equations considered in [10]. When there are $n = 2$ decryption exponents, our bound is $N^{0.071}$ which is less than the bound $N^{0.073}$ in [10]. Hence, we only improve the previous bound when there are $n \geq 3$ pairs of encryption and decryption exponents for a common CRT-RSA modulus in theory and obtain $N^{0.375}$ asymptotically in n. However, it is nice to see that we successfully factor N when $d_{p_l} < N^{0.035}$ with 3 pairs of exponents in practice and the original bounds are $N^{0.015}$ in [10] and $N^{0.029}$ in [6].

An explicit description of these bounds is illustrated in Fig. 1.

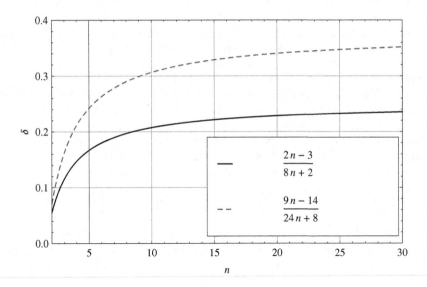

Fig. 1. The recoverable sizes of secret exponents of CRT-RSA. The solid line denotes the range of d_{p_l} and d_{q_l} with respect to n, the dashed line denotes the range of d_{p_l} with respect to n

For Takagi's variant of RSA, assume there exist n encryption and decryption exponents (e_l, d_l), where $l = 1, \cdots, n$ with a common modulus $N = p^r q$, which means there exist l simultaneous modular univariate linear equations. So far, this kind of modular equations is what has been considered in [4, 22].

By an application of their results, we obtain that the modulus can be factored when $d_l \leq N^\delta$ for all $l = 1, \cdots, n$, where

$$\delta < \left(\frac{r-1}{r+1}\right)^{\frac{n+1}{n}}.$$

The rest of this paper is organized as follows. Section 2 is some preliminary knowledge on lattices and the CRT-RSA variant. In Sect. 3, we analyze CRT-RSA with multiple small decryption exponents. Section 4 presents an analysis on Takagi's variant RSA with multiple small decryption exponents. Finally, Sect. 5 is the conclusion.

2   Preliminaries

Let $w_1, w_2, \cdots, w_k$ be k linearly independent vectors in $\mathbb{R}^n$. A lattice $\mathcal{L}$ spanned by $\{w_1, \cdots, w_k\}$ is the set of all integer linear combinations, $c_1 w_1 + \cdots + c_k w_k$, of $w_1, \cdots, w_k$, where $c_1, \cdots, c_k \in \mathbb{Z}$. The k-dimensional lattice $\mathcal{L}$ is a discrete additive subgroup of $\mathbb{R}^n$. The set of vectors $w_1, \cdots, w_k$ is called a basis of the lattice $\mathcal{L}$. The lattice bases are not unique, one can obtain another basis by multiplying any matrix with determinant ± 1, it means that any lattice of dimension larger than 1 has infinitely many bases [15]. Hence, how to find a lattice basis with good properties has been an important problem.

Lenstra et al. [11] introduced the famous L^3 lattice basis reduction algorithm which can output a relatively short and nearly orthogonal lattice basis in polynomial time. Instead of finding the shortest vectors in a lattice, the algorithm finds the L^3 reduced basis with the following useful properties.

Lemma 1 (L^3, [11]). *Let $\mathcal{L}$ be a lattice of dimension k. Applying the L^3 algorithm to $\mathcal{L}$, the outputted reduced basis vectors $v_1, \cdots, v_k$ satisfy that*

$$\|v_i\| \leq 2^{\frac{k(k-i)}{4(k+1-i)}} \det(\mathcal{L})^{\frac{1}{k+1-i}}, \text{ for any } 1 \leq i \leq k.$$

Coppersmith [5] applied the L^3 lattice basis reduction algorithm in order to find small solutions of integer equations and modular equations. Later, Jochemsz and May [9] extended this technique and gave general results to find roots of multivariate polynomials.

For a given polynomial $g(x_1, \cdots, x_k) = \sum_{(i_1, \cdots, i_k)} a_{i_1, \cdots, i_k} x_1^{i_1} \cdots x_k^{i_k}$, we define the norm of g as

$$\|g(x_1, \cdots, x_k)\| = \left(\sum_{(i_1, \cdots, i_k)} a_{i_1, \cdots, i_k}^2\right)^{\frac{1}{2}}.$$

The following lemma due to Howgrave-Graham [7] gives a sufficient condition under which a modular equation can be converted into an integer equation.

Lemma 2 *(Howgrave-Graham, [7]). Let $g(x_1, \cdots, x_k) \in \mathbb{Z}[x_1, \cdots, x_k]$ be an integer polynomial with at most w monomials. Suppose that*

$$g(y_1, \cdots, y_k) \equiv 0 \pmod{p^m} \text{ for } |y_1| \le X_1, \cdots, |y_k| \le X_k, \text{ and}$$

$$\|g(x_1 X_1, \cdots, x_k X_k)\| < \frac{p^m}{\sqrt{w}}.$$

Then $g(y_1, \cdots, y_k) = 0$ holds over the integers.

Suppose we have $w(> k)$ polynomials $b_1, \cdots, b_w$ in the variables $x_1, \ldots, x_k$ such that $b_1(y_1, \ldots, y_k) = \cdots = b_w(y_1, \ldots, y_k) = 0 \bmod p^m$ with $|y_1| \le X_1, \ldots, |y_k| \le X_k$. Now we construct a lattice $\mathcal{L}$ with the coefficient vectors of $b_1(x_1 X_1, \ldots, x_k X_k)$, $\ldots$, $b_w(x_1 X_1, \ldots, x_k X_k)$. After lattice reduction, we get k polynomials $v_1(x_1, \ldots, x_k), \ldots, v_k(x_1, \ldots, x_k)$ such that

$$v_1(y_1, \ldots, y_k) = \cdots = v_k(y_1, \ldots, y_k) = 0 \bmod p^m$$

which correspond to the first k vectors of the reduced basis. Also by the property of the L^3 algorithm, we have

$$\|v_1(x_1 X_1, \ldots, x_k X_k)\| \le \cdots \le \|v_k(x_1 X_1, \ldots, x_k X_k)\| \le 2^{\frac{w(w-1)}{4(w+1-k)}} \det(\mathcal{L})^{\frac{1}{w+1-k}}.$$

Hence by Lemma 2, if

$$2^{\frac{w(w-1)}{4(w+1-k)}} \det(\mathcal{L})^{\frac{1}{w+1-k}} < \frac{p^m}{\sqrt{w}},$$

then we have $v_1(y_1, \ldots, y_k) = \cdots = v_k(y_1, \ldots, k_k) = 0$. Next we want to find $y_1, \ldots, y_k$ from $v_1, \ldots, v_k$.

Once we obtain several polynomial equations over the integers from the L^3 lattice basis reduction algorithm, we can solve for the roots over the integers by calculating the resultants or the Gröbner basis of the polynomials based on the following heuristic assumption. In practical experiments, the following heuristic assumption usually holds.

Assumption 1. *Our lattice-based construction yields algebraically independent polynomials. The common roots of these polynomials can be efficiently computed by using techniques like calculation of the resultants or finding a Gröbner basis.*

Similarly as other lattice reduction works [1, 9, 10, 23], while we present experimental results in support of our attacks, we also like to point out the theoretical results are asymptotic, as we neglect constants in certain cases in the calculations of our attacks.

Minkowski Sum Based Lattice Construction. In [1], Aono proposed a method to construct a lattice for Coppersmith's technique for simultaneous modular equations. In order to make this clear, let us illustrate it by an example. There are two modular equations $f_1 \equiv 0 \, (\mathrm{mod} \, W_1)$ and $f_2 \equiv 0 \, (\mathrm{mod} \, W_2)$.

Based on Coppersmith's technique, to solve for the solutions of f_1 we first select some polynomials $g_1, \cdots, g_n$ which share the same solutions modulo W_1^m. Similarly, we construct polynomials $g_1', \cdots, g_n'$ which share same solutions modulo W_2^m. It is obvious that any polynomial $g_i g_j'$ where $1 \leq i, j \leq n$ has the desired solutions modulo $W_1^m W_2^m$. Then we arrange these polynomials and construct a new lattice with polynomials which have the desired solutions modulo $W_1^m W_2^m$. By an integer linear combination, some of these polynomials which have the same leading monomial can be written as $\sum_{i,j} a_{i,j} g_i g_j'$. To keep the determinant of the lattice small, the integers $a_{i,j}$ are chosen appropriately. This lattice is called a combined lattice obtained from the two lattices, one of which is constructed by the polynomials $g_1, \cdots, g_n$ and another one of which is constructed by the polynomials $g_1', \cdots, g_n'$. Aono proved that the combined lattice is triangular, if each lattice has a triangular basis matrix. The above conclusion could be extend to an arbitrary number of modular equations.

CRT-RSA. Since the RSA public key cryptosystem has been invented [16], this public key scheme has been widely used due to its succinct and effective encryption and decryption. Wiener [24] proposed to use the Chinese Remainder Theorem in the decryption phase. This scheme is usually called CRT-RSA. Based on the work of Sun and Wu [20], one version of this variant can be described as follows:

Algorithm 1. Key generation of CRT-RSA

Input:
   (n, δ_1, δ_2), where $n, \delta_1 n$ and $\delta_2 n$ denote the bitlengths of N, d_p and d_q, respectively.
Output:
   CRT-RSA-instance (N, p, q, e, d_p, d_q).
1: Randomly choose two $\frac{n}{2}$-bit primes $p = 2p_1 + 1$ and $q = 2q_1 + 1$ such that $\gcd(p_1, q_1) = 1$.
2: Randomly generate $(\delta_1 n)$-bit integer d_p and $(\delta_2 n)$-bit integer d_q such that $\gcd(d_p, p - 1) = 1$ and $\gcd(d_q, q - 1) = 1$.
3: Compute $\bar{d} \equiv (d_q - d_p)(p_1^{-1}(\bmod\ q_1))$.
4: Compute $d = d_p + p_1 \cdot \bar{d}$.
5: Compute the encryption exponent e satisfying $ed \equiv (\bmod\ (p - 1)(q - 1))$.
6: The RSA modulus is $N = pq$, the secret key is (d_p, d_q, p, q) and the public key is (N, e).

As described in the key generation algorithm of CRT-RSA, the case that more than one valid encryption and decryption exponents for the same CRT-RSA modulus $N = pq$ may exist, that is, when we are done with Step 1 for choosing a pair (p, q), we generate several different d_p and d_q in the remaining steps. Next, we analyze the weakness in the case that multiple encryption and decryption exponents share a common CRT-RSA modulus.

3   Multiple Encryption and Decryption Exponents Attack of CRT-RSA

In this section, along the idea of [1,8,18,23] we give the following theorems when multiple encryption and decryption exponents are used for a common CRT-RSA modulus. By making a comparison between our results and Jochemsz and May's result [10], we improve the bound when there are 3 or more pairs of encryption and decryption exponents for a common CRT-RSA modulus. And we also improve the experimental results $N^{0.015}$ in [10] and $N^{0.029}$ in [6] to $N^{0.035}$ with 3 pairs of exponents.

Theorem 1. *Let $(e_1, e_2, \cdots, e_n)$ be n CRT-RSA encryption exponents with a common modulus $N = pq$, where $n \geq 3$ and $e_1, e_2, \cdots, e_n$ have roughly the same bitlength as N. Consider that $d_{p_i}, d_{q_i} \leq N^\delta$ for $i = 1, 2, \cdots, n$ are the corresponding decryption exponents. Then under Assumption 1, one can factor N in polynomial time when*

$$\delta < \frac{2n - 3}{8n + 2}.$$

Proof. For one pair of keys (e_l, d_{p_l}, d_{q_l}), we have

$$e_l d_{p_l} - 1 = k_{p_l}(p - 1),$$
$$e_l d_{q_l} - 1 = k_{q_l}(q - 1),$$

where k_{p_l} and k_{q_l} are some integers.

Moreover, by multiplying these two equations, we have that

$$e_l^2 d_{p_l} d_{q_l} - e_l(d_{p_l} + d_{q_l}) + 1 = k_{p_l} k_{q_l}(N - s),$$

where $s = p + q - 1$.

Then $(k_{p_l} k_{q_l}, s, d_{p_l} + d_{q_l})$ is a solution of

$$f_l(x_l, y, z_l) = x_l(N - y) + e_l z_l - 1 \pmod{e_l^2}.$$

Moreover, consider the n modular polynomials

$$f_l(x_l, y, z_l) = x_l(N - y) + e_l z_l - 1 \pmod{e_l^2}, \text{ for } l = 1, \cdots, n. \tag{1}$$

These polynomials have the common root $(x_1, \cdots, x_n, y, z_1, \cdots, z_n) = (k_{p_1} k_{q_1}, \cdots, k_{p_n} k_{q_n}, s, d_{p_1} + d_{q_1}, \cdots, d_{p_n} + d_{q_n})$, and the values of its coefficients can be roughly bounded as $k_{p_l} k_{q_l} \simeq X_l = N^{1+2\delta}$, $s \simeq Y = N^{\frac{1}{2}}$ and $d_{p_l} + d_{q_l} \simeq N^\delta = Z$ for $l = 1, \cdots, n$.

In order to solve for the desired solution of the modular equations $f_l(x_l, y, z_l) = 0 \pmod{e_l^2}$, for $l = 1, \cdots, n$, based on Aono's idea [1], we first selected the following set of polynomials to solve each single equation,

$$S_l = \{x_l^{i_l} z_l^{j_l} f_l^{k_l}(x_l, y, z_l)(e_l^2)^{m-k_l} | 0 \leq k_l \leq m, 0 \leq i_l \leq m - k_l, 0 \leq j_l \leq m - i_l - k_l\},$$

where $l = 1, \cdots, n$ and m is a positive integer.

Each selection for the corresponding equation in (1) generates a triangular basis matrix. Likewise, for each $l = 1, 2, \cdots, n$, we can respectively construct a triangular matrix. Based on the technique of Minkowski sum based lattice construction, these n lattices corresponding to the n triangular matrices can be combined as a new lattice $\mathcal{L}'$ and the basis matrix with polynomials which have the same root as the solutions of the modular equation modulo $(e_1^2 \cdots e_n^2)^m$. Since each basis matrix is triangular, the combined lattice is also triangular. The combined basis matrix has diagonal entries

$$X_1^{i_1} \cdots X_n^{i_n} Y^k Z_1^{j_1} \cdots Z_n^{j_n} (e_1^2)^{m-\min(i_1,k)} \cdots (e_n^2)^{m-\min(i_n,k)},$$

where

$$0 \le i_1, \cdots, i_n \le m, \ 0 \le k \le i_1 + i_2 + \cdots + i_n, \ 0 \le j_1 \le i_1, \cdots, 0 \le j_n \le i_n.$$

Then the determinant of the lattice can be calculated as

$$\det(\mathcal{L}') = \prod_{i_1=0}^{m} \cdots \prod_{i_n=0}^{m} \prod_{k=0}^{i_1+\cdots+i_n} \prod_{j_1=0}^{m-i_1} \cdots \prod_{j_n=0}^{m-i_n} \left(X_1^{i_1} \cdots X_n^{i_n} Y^k Z_1^{j_1} \cdots Z_n^{j_n} \right.$$
$$\left. (e_1^2)^{m-\min(i_1,k)} \cdots (e_n^2)^{m-\min(i_n,k)} \right)$$
$$= X_1^{S_{x_1}} \cdots X_n^{S_{x_n}} Y^{S_y} Z_1^{S_{z_1}} \cdots Z_n^{S_{z_n}} (e_1^2)^{S_{e_1}} \cdots (e_n^2)^{S_{e_n}},$$

where

$$S_{x_1} + S_{x_2} + \cdots + S_{x_n} = \left(\frac{n^2}{18} + \frac{n}{36} \right) \frac{m^{2n+2}}{2^{n-1}} + o(m^{2n+2}),$$

$$S_y = \left(\frac{n^2}{36} + \frac{n}{72} \right) \frac{m^{2n+2}}{2^{n-1}} + o(m^{2n+2}),$$

$$S_{z_1} + S_{z_2} + \cdots + S_{z_n} = \left(\frac{n^2}{18} - \frac{n}{72} \right) \frac{m^{2n+2}}{2^{n-1}} + o(m^{2n+2}),$$

$$S_{e_1} + S_{e_2} + \cdots + S_{e_n} = \left(\frac{n^2}{9} - \frac{n}{72} \right) \frac{m^{2n+2}}{2^{n-1}} + o(m^{2n+2}).$$

On the other hand, the dimension is

$$\dim(\mathcal{L}') = \sum_{i_1=0}^{m} \cdots \sum_{i_n=0}^{m} \sum_{k=0}^{i_1+\cdots+i_n} \sum_{j_1=0}^{m-i_1} \cdots \sum_{j_n=0}^{m-i_n} 1 = \frac{n}{6 \cdot 2^{n-1}} m^{2n+1} + o(m^{2n+1}).$$

Please refer to the appendix to see the detailed calculations.

From Lemmas 1 and 2, we can obtain integer equations when

$$\det(\mathcal{L}')^{\frac{1}{\dim(\mathcal{L}')}} < (e_1^2 \cdots e_n^2)^m. \tag{2}$$

Neglecting the low order terms of m and putting $X_l = N^{1+2\delta}, Y = N^{\frac{1}{2}}, Z_l = N^\delta$ and $e_l^2 \simeq N^2$ into the above inequality (2), the necessary condition can be written as

$$(1 + 2\delta)\left(\frac{n^2}{18} + \frac{n}{36} \right) + \frac{1}{2}\left(\frac{n^2}{36} + \frac{n}{72} \right) + \delta\left(\frac{n^2}{18} - \frac{n}{72} \right) + 2\left(\frac{n^2}{9} + \frac{n}{72} \right) \le \frac{n^2}{3},$$

namely,

$$\delta < \frac{2n-3}{8n+2}.$$

Then we get $2n+1$ polynomials which share the root $(x_1, \ldots, x_n, y, z_1, \ldots, z_n)$. Under Assumption 1, we can find $x_1, \ldots, x_n, y, z_1, \ldots, z_n$ from these polynomials. This concludes the proof of Theorem 1.   □

Moreover, as well as by using Minkowski sum based lattice construction to combine the polynomials $e_l d_{p_l} = k_{p_l}(p-1)+1$, for $l = 1, \cdots, n$, we also introduce an additional variable q to reduce the determinant of our lattice and finally improve our bound of Theorem 1.

More precisely, we firstly construct a lattice which combines the polynomials $f_l(x_l, y) = x_l(y-1) + 1 \pmod{e_l}$, for $l = 1, \cdots, n$ by utilizing Minkowski sum lattice based construction. Then based on an observation of the monomials which appear in the lattice, we found that the desired root p of variable y is a factor of N. Thus, to reduce the determinant of our constructed lattice we can introduce a new variable z which corresponds to q. Since $pq = N$, we can replace yz by N and then by multiplying the inverse of N modulo $e_1 \cdots e_n$. Above all, we can obtain the following theorem.

Theorem 2. *Let $(e_1, e_2, \cdots, e_n)$ be n CRT-RSA encryption exponents with a common modulus $N = pq$, where $n \geq 2$ and $e_1, e_2, \cdots, e_n$ have the roughly same bitlengths as N. Consider that d_{p_l}, d_{q_l} for $l = 1, 2, \cdots, n$ are the corresponding decryption exponents. Assumed that $d_{p_l} < N^\delta$ for $l = 1, 2, \cdots, n$, then under Assumption 1, one can factor N in polynomial time when*

$$\delta < \frac{9n-14}{24n+8}.$$

Proof. For each of the key pairs (e_l, d_{p_l}, d_{q_l}), we have that

$$e_l d_{p_l} = k_{p_l}(p-1) + 1,$$

where k_{p_l} is an integer.

Then (k_{p_l}, p) is a solution of

$$f_l(x_l, y) = x_l(y-1) + 1 \pmod{e_l}.$$

Consider the n modular polynomials

$$f_l(x_l, y) = x_l(y-1) + 1 \pmod{e_l}, \text{ for } l = 1, \cdots, n.$$

Obviously, these polynomials have the common root $(x_1, \cdots, x_n, y) = (k_{p_1}, \cdots, k_{p_n}, p)$, and the sizes of its coefficients can be roughly determined as $k_{p_l} \simeq X_l = N^{\frac{1}{2}+\delta}$, for $l = 1, \cdots, n$ and $p \simeq Y = N^{\frac{1}{2}}$.

In order to solve for the desired solution, similarly we firstly selected the following set of polynomials to solve each single modular equation,

$$S_l = \{x_l^{i_l} f_l^{k_l}(x_l, y)(e_l)^{m-k_l} | 0 \leq k_l \leq m, 0 \leq i_l \leq m - k_l\},$$

where $l = 1, \cdots, n$ and m is a positive integer.

Each selection generates a triangular basis matrix. Then, for $l = 1, \cdots, n$ we construct a triangular matrix respectively. We constructed the basis matrix with polynomials which have the same roots as the solutions of the modular equation modulo $(e_1 \cdots e_n)^m$. By combining these n lattices based on a Minkowski sum based lattice construction, the matrix corresponding to the combined lattice $\mathcal{L}'_1$ is triangular and has diagonal entries

$$X_1^{i_1} \cdots X_n^{i_n} Y^k e_1^{m-\min(i_1,k)} \cdots e_n^{m-\min(i_n,k)},$$

where

$$0 \le i_1, \cdots, i_n \le m, \ 0 \le k \le i_1 + i_2 + \cdots + i_n.$$

Moreover, note that the desired small solution contains the prime factor p, which is a factor of the modulus $N = pq$. Then we introduce a new variable z for another prime factor q, and multiply each polynomial corresponding to each row vector in the $\mathcal{L}'_1$ by a power z^s for some s that will be optimized later. Then, we replace every occurrence of the monomial yz by N because $N = pq$. Therefore, compared to the unchanged polynomials, every monomial $x_1^{i_1} \cdots x_n^{i_n} y^k z^s$ and $k \ge s$ with coefficient $a_{i_1, \cdots, i_n, k}$ is transformed into a monomial $x_1^{i_1} \cdots x_n^{i_n} y^{k-s}$ with coefficient $a_{i_1, \cdots, i_n, k} N^s$. Similarly, when $k < s$, the monomial $x_1^{i_1} \cdots x_n^{i_n} y^k z^s$ with coefficient $a_{i_1, \cdots, i_n, k}$ is transformed into monomial $x_1^{i_1} \cdots x_n^{i_n} z^{s-k}$ with coefficient $a_{i_1, \cdots, i_n, k} N^k$. Let $Z = N^{\frac{1}{2}}$ denote the upper bound of the unknown variable z.

To keep the determinant of the lattice as small as possible, we try to eliminate the factor of N^s and N^k in the coefficients of the diagonal entries. Since $(N, e_1 \cdots e_n) = 1$, we only need to multiply the corresponding polynomial with the inverse of N^s or N^k modulo $(e_1 \cdots e_n)^m$.

Then the determinant of the lattice can be calculated as follows,

$$\det(\mathcal{L}'_1) = X_1^{S_{x_1}} \cdots X_n^{S_{x_n}} Y^{S_y} Z^{S_z} e_1^{S_{e_1}} \cdots e_n^{S_{e_n}},$$

where

$$S_{x_1} + S_{x_2} + \cdots + S_{x_n} = \sum_{i_1=0}^{m} \cdots \sum_{i_n=0}^{m} \sum_{k=0}^{i_1+\cdots+i_n} (i_1 + \cdots i_n),$$

$$S_y = \sum_{i_1=0}^{m} \cdots \sum_{i_n=0}^{m} \sum_{k=s}^{i_1+\cdots+i_n} (k - s),$$

$$S_z = \sum_{i_1=0}^{m} \cdots \sum_{i_n=0}^{m} \sum_{k=0}^{s-1} (s - k),$$

$$S_{e_1} + S_{e_2} + \cdots + S_{e_n} = \sum_{i_1=0}^{m} \cdots \sum_{i_n=0}^{m} \sum_{k=0}^{i_1+\cdots+i_n} (nm - \min(i_1, k) - \cdots - \min(i_n, k)).$$

Since the following formulas hold for any $0 \le a, b \le n$,

$$\sum_{i_1=0}^{m} \cdots \sum_{i_n=0}^{m} i_a i_b = \begin{cases} \frac{1}{3}m^{n+2} + o(m^{n+2}), & (a = b), \\ \frac{1}{4}m^{n+2} + o(m^{n+2}), & (a \ne b), \end{cases}$$

we have that

$$S_{x_1} + S_{x_2} + \cdots + S_{x_n} = (\frac{n^2}{4} + \frac{n}{12})m^{n+2} + o(m^{n+2}),$$

$$S_y = (\frac{\sigma^2 n^2}{2} - \frac{\sigma n^2}{2} + \frac{n^2}{8} + \frac{n}{24})m^{n+2} + o(m^{n+2}),$$

$$S_z = (\frac{\sigma^2 n^2}{2})m^{n+2} + o(m^{n+2}),$$

$$S_{e_1} + S_{e_2} + \cdots + S_{e_n} = (\frac{n^2}{4} + \frac{n}{12})m^{n+2} + o(m^{n+2}).$$

where $s = \sigma n m$ and $0 \le \sigma < 1$.

On the other hand, the dimension of the lattice is

$$\dim(\mathcal{L}_1') = \sum_{i_1=0}^{m} \cdots \sum_{i_n=0}^{m} \sum_{k=0}^{i_1+\cdots+i_n} 1 = \frac{n}{2}m^{n+1} + o(m^{n+1}).$$

From Lemmas 1 and 2, we can obtain integer equations when

$$\det(\mathcal{L}_1')^{\frac{1}{\dim(\mathcal{L}_1')}} < (e_1 \cdots e_n)^m. \tag{3}$$

Neglecting the low order terms of m and putting $X_l = N^{\frac{1}{2}+\delta}, Y = N^{\frac{1}{2}}, Z = N^{\frac{1}{2}}$ and $e_l \simeq N$ into the above inequality (3) for $l = 1, \cdots, n$, the necessary condition can be written as

$$(\frac{1}{2} + \delta)(\frac{n^2}{4} + \frac{n}{12}) + \frac{1}{2}(\frac{\sigma^2 n^2}{2} - \frac{\sigma n^2}{2} + \frac{n^2}{8} + \frac{n}{24}) + \frac{1}{2}(\frac{\sigma^2 n^2}{2}) + (\frac{n^2}{4} + \frac{n}{12}) \le \frac{n^2}{2}.$$

By optimizing $\sigma = \frac{1}{4}$, we finally obtain the following bound on δ

$$\delta < \frac{9n - 14}{24n + 8}.$$

Then under Assumption 1, one can factor N in polynomial time. This concludes the proof of Theorem 2. □

The reason that our result improves over previous work in the literature is based on the following two observations. Firstly, we can combine n polynomials by utilizing the Minkowski sum lattice based construction. Secondly, from the knowledge of $N = pq$, we can optimize the determinant of the lattice by introducing some factor z^s to every polynomials, where z is a new variable corresponding to q and s is an integer which will be optimized during the calculations.

Experimental Results. Note that in the calculations of Theorem 2, we assume that m goes to infinity. Then our result is an asymptotic bound, as we neglect lower order terms of m. If m and n are fixed, the maximum δ satisfying the inequality of condition (3) is easily computed. In Table 3, for each fixed m and n, we list the maximum δ satisfying (3) and the dimension of lattice. The column limit denotes the asymptotic bound.

Table 3. Theoretical bound and lattice dimension for small δ with fixed m.

$n = 2$							
m	5	6	7	8	9	10	∞
s	2	3	3	4	4	5	∞
δ		0.0081	0.0200	0.0244	0.0313	0.0340	0.0385	0.0714
$\dim(\mathcal{L}')$	216	343	512	729	1000	1331	∞
$n = 3$							
m	2	3	4	5	6	7	∞
s	1	2	3	4	4	5	∞
δ		0.0357	0.0746	0.0938	0.1052	0.1127	0.1200	0.1625
$\dim(\mathcal{L}'_1)$	108	352	875	1836	3430	5888	∞

We have implemented the experiment program in Magma 2.11 computer algebra system [3] on a PC with Intel(R) Core(TM) Duo CPU (2.53 GHz, 1.9 GB RAM Windows 7) and carried out the L^3 algorithm [14]. Experimental results are provided in Table 4.

Table 4. Experimental results.

N (bits)	n	theo. of δ	expt. of δ	parameters of lattice	time (in sec.)
1000	3	0.0357	0.0350	$m = 2$, $s = 1$, $\dim(\mathcal{L}'_1) = 108$	3978.213

In the experiments we successfully factored the common modulus N in practice, when there are three decryption exponents and all of them are less than $N^{0.035}$. For this given problem which factor N with small decryption exponent, Jochemsz and May [10] successfully factored N with one small decryption exponent and the bound is $N^{0.015}$, later the bound has been improved to $N^{0.029}$ by utilizing the unraveled linearization technique introduced by Herrmann and May [6]. In other words, we improve both the theoretical and the experimental bound by using more decryption exponents with a common modulus.

Note that in the experiments, we always find many polynomial equations which share the desired solutions over the integers. Moreover we have another equation $yz = N$. Then by calculating the Gröbner basis of these polynomials, we can successfully solve for the desired solutions in less than two hours.

In all experiments we have done for verification of our proposed attack, we indeed successfully collected the roots by using Gröbner basis technique and there was no experimental result to contradict Assumption 1. On the other hand, however, it seems very difficult to prove or demonstrate its validity.

4   Multiple Encryption and Decryption Exponents Attack of Takagi's Variant RSA

Theorem 3. *Let $(e_1, e_2, \cdots, e_n)$ be n encryption exponents of Takagi's variant of RSA with common modulus $N = p^r q$. Consider that $d_1, d_2, \cdots, d_n$ are the corresponding decryption exponents. Then under Assumption 1, one can factor N in polynomial time when*

$$\delta < \left(\frac{r-1}{r+1}\right)^{\frac{n+1}{n}},$$

where $d_l \leq N^\delta$, for $l = 1, \cdots, n$.

Proof. For one modulus $N = p^r q$, there exist n encryption and decryption exponents (e_l, d_l), thus, we have that

$$e_1 d_1 = k_1 p^{r-1}(p-1)(q-1) + 1,$$
$$e_2 d_2 = k_2 p^{r-1}(p-1)(q-1) + 1,$$
$$\cdots$$
$$e_n d_n = k_n p^{r-1}(p-1)(q-1) + 1.$$

Hence, for the unknown $(d_1, \cdots, d_n)$ we have the following modular equations,

$$f(x_1) = e_1 x_1 - 1 \pmod{p^{r-1}},$$
$$f(x_2) = e_2 x_2 - 1 \pmod{p^{r-1}},$$
$$\cdots$$
$$f(x_n) = e_n x_n - 1 \pmod{p^{r-1}}.$$

As it is shown, $(d_1, d_2, \cdots, d_n)$ is a root of simultaneous modular univariate linear equations modulo an unknown divisor, and the size is bounded as $d_l \leq N^\delta$, for $l = 1, \cdots, n$.

Using the technique of [4,22], it can be shown that if

$$\delta < \left(\frac{r-1}{r+1}\right)^{\frac{n+1}{n}},$$

these simultaneous modular univariate linear equations can be solved under Assumption 1, which means $(d_1, \cdots, d_n)$ can be recovered. Then one can easily factor N by calculating the common factor.   □

Table 5. Factoring N with multiple decryption exponents.

r	$\log_2 N$	$\log_2 p$	$n = 2$				$n = 3$			
			theo.	expt.	dim($\mathcal{L}$)	time (in sec.)	theo.	expt.	dim($\mathcal{L}$)	time (in sec.)
2	1500	500	0.272	0.230	66	2022.834	0.291	0.240	84	1537.078

Experimental Results. We have implemented the experiment program in Magma 2.11. In all experiments, we successfully solved for desired solutions $(d_1, d_2, \cdots, d_n)$. Similarly, there was no experimental result to contradict Assumption 1 Table 5.

Notice that, the previous Theorem 3 can be applied for encryption exponents $(e_1, \cdots, e_n)$ of arbitrary sizes. However, if there exist two valid key pairs (e_1, d_1) and (e_2, d_2), where e_1 and e_2 have roughly the same size as the modulus N or some larger values as N^α. Assume that $d_1 \simeq d_2 \simeq N^\delta$, then we can give an analysis as follows.

Given two equations $e_1 d_1 = k_1 p^{r-1}(p-1)(q-1) + 1$ and $e_2 d_2 = k_2 p^{r-1}(p-1)(q-1) + 1$, we eliminate $p^{r-1}(p-1)(q-1)$ and obtain the following equality,

$$k_2(e_1 d_1 - 1) = k_1(e_2 d_2 - 1)$$

which suggests that we look for small solutions of the polynomial

$$f(x, y) = e_2 x + y \pmod{e_1}. \tag{4}$$

Since $(d_2 k_1, k_2 - k_1)$ is a root of $f(x, y) \bmod e_1$. The bound of k_1 can be estimated as $N^{\alpha+\delta-1}$, hence we define the bounds $|d_2 k_1| \simeq X = N^{\alpha+2\delta-1}$ and $|k_2 - k_1| \simeq Y = N^{\alpha+\delta-1}$. For this linear modular equation, we can recover $(d_2 k_1, k_2 - k_1)$ for sufficiently large N provided that $XY < e$, or $\alpha + 2\delta - 1 + \alpha + \delta - 1 < \alpha$.

Thus, to recover $d_2 k_1$ and $k_2 - k_1$ from this lattice-based method, the size of the encryption and decryption exponents should satisfy

$$\alpha + 3\delta < 2,$$

where $\alpha + \delta > 1$.

5   Conclusion

In this paper, we presented some applications of Minkowski sum based lattice construction and gave analyses of the case that multiple pairs of encryption and decryption exponents are used with the common CRT-RSA modulus N. We showed that one can factor N when both $d_{p_i}, d_{q_i} \leq N^{\frac{2l-3}{8l+2}}$ or either d_{p_i} or d_{q_i} is less than $N^{\frac{9l-14}{24l+8}}$, for $i = 1, 2, \cdots, l$. Moreover, we also analyzed the situation when more than one encryption and decryption exponents are used in Takagi's variant of RSA with modulus $N = p^r q$.

Acknowledgements. The authors would like to thank anonymous reviewers for their helpful comments and suggestions. The work of this paper was supported by the National Key Basic Research Program of China (Grants 2013CB834203 and 2011CB302400), the National Natural Science Foundation of China (Grants 61472417, 61402469, 61472416 and 61272478), the Strategic Priority Research Program of Chinese Academy of Sciences under Grant XDA06010702 and XDA06010703, and the State Key Laboratory of Information Security, Chinese Academy of Sciences. Y. Lu is supported by Project CREST, JST.

Appendix

Here we present the detailed calculations of $S_{X_1}, S_Y, S_{Z_1}, S_{e_1}$.

Let $\overset{*}{\sum}$ denotes $\sum\limits_{i_1=0}^{m} \cdots \sum\limits_{i_n=0}^{m} \sum\limits_{j_1=0}^{m-i_1} \cdots \sum\limits_{j_n=0}^{m-i_n}$, for any $0 \leq a, b \leq n$, we have that

$$\overset{*}{\sum} i_a i_b = \begin{cases} \frac{1}{12*2^{n-1}} * m^{2n+2} + o(m^{2n+2}), & (a = b), \\ \frac{1}{18*2^{n-1}} * m^{2n+2} + o(m^{2n+2}), & (a \neq b), \end{cases}$$

and

$$\overset{*}{\sum} i_a j_b = \begin{cases} \frac{1}{24*2^{n-1}} * m^{2n+2} + o(m^{2n+2}), & (a = b), \\ \frac{1}{18*2^{n-1}} * m^{2n+2} + o(m^{2n+2}), & (a \neq b). \end{cases}$$

Then we obtain that

$$\overset{*}{\sum} \sum_{k=0}^{i_1+\cdots+i_n} i_1 + \cdots + i_n = (\frac{n^2}{18} + \frac{n}{36}) * \frac{m^{2n+2}}{2^{n-1}} + o(m^{2n+2}),$$

$$\overset{*}{\sum} \sum_{k=0}^{i_1+\cdots+i_n} j_1 + \cdots + j_n = (\frac{n^2}{18} - \frac{n}{72}) * \frac{m^{2n+2}}{2^{n-1}} + o(m^{2n+2}),$$

$$\overset{*}{\sum} \sum_{k=0}^{i_1+\cdots+i_n} k = \overset{*}{\sum} \frac{(i_1 + \cdots + i_n)^2}{2} + \frac{i_1 + \cdots + i_n}{2}$$

$$= (\frac{n^2}{36} + \frac{n}{72}) * \frac{m^{2n+2}}{2^{n-1}} + o(m^{2n+2}).$$

Moreover,

$$\overset{*}{\sum} \sum_{k=0}^{i_1+\cdots+i_n} \min(i_1, k) = \overset{*}{\sum}(\sum_{k=0}^{i_1} k + \sum_{k=i_1+1}^{i_1+\cdots+i_n} i_1)$$

$$= \overset{*}{\sum}(\frac{i_1(i_1 + 1)}{2} + i_1(i_2 + \cdots + i_n))$$

$$= (\frac{n}{18} - \frac{1}{72}) * \frac{m^{2n+2}}{2^{n-1}} + o(m^{2n+2}).$$

By symmetry, we have

$$\sum^{*} \sum_{k=0}^{i_1+\cdots+i_n} \min(i_1,k) + \cdots + \min(i_n,k) = (\frac{n^2}{18} - \frac{n}{72}) * \frac{m^{2n+2}}{2^{n-1}} + o(m^{2n+2}).$$

The dimension of lattice $\mathcal{L}'$ is

$$\dim(\mathcal{L}') = \sum^{*} \sum_{k=0}^{i_1+\cdots+i_n} 1 = \frac{n}{6*2^{n-1}} * m^{2n+1} + o(m^{2n+1}).$$

References

1. Aono, Y.: Minkowski sum based lattice construction for multivariate simultaneous Coppersmith's technique and applications to RSA. In: Boyd, C., Simpson, L. (eds.) ACISP. LNCS, vol. 7959, pp. 88–103. Springer, Heidelberg (2013)
2. Boneh, D., Durfee, G.: Cryptanalysis of RSA with private key d less than $N^{0.292}$. IEEE IEEE Trans. Inf. Theory **46**(4), 1339–1349 (2000)
3. Bosma, W., Cannon, J.J., Playoust, C.: The MAGMA algebra system I: the user language. J. Symbolic Comput. **24**(3/4), 235–265 (1997)
4. Cohn, H., Heninger, N.: Approximate common divisors via lattices. CoRR abs/1108.2714 (2011)
5. Coppersmith, D.: Small solutions to polynomial equations, and low exponent RSA vulnerabilities. J. Cryptology **10**(4), 233–260 (1997)
6. Herrmann, M., May, A.: Maximizing small root bounds by linearization and applications to small secret exponent RSA. In: Nguyen, P.Q., Pointcheval, D. (eds.) PKC 2010. LNCS, vol. 6056, pp. 53–69. Springer, Heidelberg (2010)
7. Howgrave-Graham, N.: Finding small roots of univariate modular equations revisited. In: Darnell, M.J. (ed.) Cryptography and Coding 1997. LNCS, vol. 1355. Springer, Heidelberg (1997)
8. Howgrave-Graham, N., Seifert, J.-P.: Extending Wiener's attack in the presence of many decrypting exponents. In: Baumgart, R. (ed.) CQRE 1999. LNCS, vol. 1740, pp. 153–166. Springer, Heidelberg (1999)
9. Jochemsz, E., May, A.: A strategy for finding roots of multivariate polynomials with new applications in attacking RSA variants. In: Lai, X., Chen, K. (eds.) ASIACRYPT 2006. LNCS, vol. 4284, pp. 267–282. Springer, Heidelberg (2006)
10. Jochemsz, E., May, A.: A polynomial time attack on RSA with private CRT-exponents smaller than $N^{0.073}$. In: Menezes, A. (ed.) CRYPTO 2007. LNCS, vol. 4622, pp. 395–411. Springer, Heidelberg (2007)
11. Lenstra, A.K., Lenstra, H.W., Lovász, L.: Factoring polynomials with rational coefficients. Mathematische Annalen **261**(4), 515–534 (1982)
12. Lu, Y., Zhang, R., Peng, L., Lin, D.: Solving linear equations modulo unknown divisors: revisited. In: ASIACRYPT 2015 (2015) (to appear). https://eprint.iacr.org/2014/343
13. May, A.: Secret exponent attacks on RSA-type schemes with moduli $N = p^r q$. In: Bao, F., Deng, R., Zhou, J. (eds.) PKC 2004. LNCS, vol. 2947, pp. 218–230. Springer, Heidelberg (2004)
14. Nguên, P.Q., Stehlé, D.: Floating-point LLL revisited. In: Cramer, R. (ed.) EUROCRYPT 2005. LNCS, vol. 3494, pp. 215–233. Springer, Heidelberg (2005)

15. Nguyen, P.Q., Vallée, B. (eds.): The LLL Algorithm - Survey and Applications. Information Security and Cryptography. Springer, Heidelberg (2010)
16. Rivest, R.L., Shamir, A., Adleman, L.M.: A method for obtaining digital signatures and public-key cryptosystems. Commun. ACM **21**(2), 120–126 (1978)
17. Sarkar, S.: Small secret exponent attack on RSA variant with modulus $N = p^r q$. Des. Codes Crypt. **73**(2), 383–392 (2014)
18. Sarkar, S., Maitra, S.: Cryptanalysis of RSA with more than one decryption exponent. Inf. Process. Lett. **110**(8–9), 336–340 (2010)
19. Simmons, G.J.: A weak privacy protocol using the RSA cryptalgorithm. Cryptologia **7**(2), 180–182 (1983)
20. Sun, H., Wu, M.: An approach towards rebalanced RSA-CRT with short public exponent. IACR Cryptology ePrint Archive **2005**, 53 (2005)
21. Takagi, T.: Fast RSA-type cryptosystem modulo $p^k q$. In: CRYPTO 1998. vol. 1462, pp. 318–326 (1998)
22. Takayasu, A., Kunihiro, N.: Better lattice constructions for solving multivariate linear equations modulo unknown divisors. In: Boyd, C., Simpson, L. (eds.) ACISP. LNCS, vol. 7959, pp. 118–135. Springer, Heidelberg (2013)
23. Takayasu, A., Kunihiro, N.: Cryptanalysis of RSA with multiple small secret exponents. In: Susilo, W., Mu, Y. (eds.) ACISP 2014. LNCS, vol. 8544, pp. 176–191. Springer, Heidelberg (2014)
24. Wiener, M.J.: Cryptanalysis of short RSA secret exponents. IEEE Trans. Inf. Theory **36**(3), 553–558 (1990)

Some Results on Sprout

Subhadeep Banik[(✉)]

DTU Compute, Technical University of Denmark, 2800 Kgs. Lyngby, Denmark
subb@dtu.dk

Abstract. Sprout is a lightweight stream cipher proposed by Armknecht and Mikhalev at FSE 2015. It has a Grain-like structure with two state Registers of size 40 bits each, which is exactly half the state size of Grain v1. In spite of this, the cipher does not appear to lose in security against generic Time-Memory-Data Tradeoff attacks due to the novelty of its design. In this paper, we first present improved results on Key Recovery with partial knowledge of the internal state. We show that if 50 of the 80 bits of the internal state are guessed then the remaining bits along with the secret key can be found in a reasonable time using a SAT solver. Thereafter, we show that it is possible to perform a distinguishing attack on the full Sprout stream cipher in the multiple IV setting using around 2^{40} randomly chosen IVs on an average. The attack requires around 2^{48} bits of memory. Thereafter, we will show that for every secret key, there exist around 2^{30} IVs for which the LFSR used in Sprout enters the all zero state during the keystream generating phase. Using this observation, we will first show that it is possible to enumerate Key-IV pairs that produce keystream bits with period as small as 80. We will then outline a simple key recovery attack that takes time equivalent to $2^{66.7}$ encryptions with negligible memory requirement. This although is not the best attack reported against this cipher in terms of the time complexity, it is the best in terms of the memory required to perform the attack.

Keywords: Grain v1 · Sprout · Stream cipher

1 Introduction

Lightweight stream ciphers have become immensely popular in the cryptological research community, since the advent of the eStream project [1]. The three hardware finalists included in the final portfolio of eStream i.e. Grain v1 [11], Trivium [5] and MICKEY 2.0 [3], all use bitwise shift registers to generate keystream bits. After the design of Grain v1 was proposed, two other members Grain-128 [12] and Grain-128a were added to the Grain family mainly with an objective to provide a larger security margin and include the functionality of message authentication respectively. In FSE 2015, Armknecht and Mikhalev proposed the Grain-like stream cipher Sprout [2] with a startling trend: the size of its internal state of Sprout was equal to the size of its Key. After the publication of [4], it is widely accepted that to be secure against generic Time-Memory-Data tradeoff

ⓒ Springer International Publishing Switzerland 2015
A. Biryukov and V. Goyal (Eds.): INDOCRYPT 2015, LNCS 9462, pp. 124–139, 2015.
DOI: 10.1007/978-3-319-26617-6_7

attacks, the internal state of a stream cipher must be atleast twice the size of the secret key. However the novelty of the Sprout design ensured that the cipher remained secure against generic TMD tradeoffs. The smaller internal state makes the cipher particularly attractive for compact lightweight implementations.

1.1 Previous Attacks on Sprout

To the best of our knowledge, four attacks have been reported against Sprout. We present a summary of these attacks:

- In [9], a related Key-chosen IV distinguisher is reported against Sprout. Let K, V denote a Key-IV pair and let K' denote K with the first bit flipped and similarly let V' denote V with the first bit flipped. Then it is easy to see that the probability that the first $80n$ keystream bits produced by K, V and K', V' are equal is given by $\frac{1}{8 \cdot 2^n}$.
- In [13], a fault attack against Sprout is presented. Another attack based on solving a system of non-linear equations by a SAT solver is also presented. The authors guess the values of 54 out of the 80 bits of the internal state. The remaining 106 unknowns, i.e. the remaining 26 internal state bits and the 80 Key bits are found as follows. The authors use the first 450 keystream bits produced by the cipher to populate a bank of non-linear equations in the unknown variables. The resulting system is solved via a SAT solver in around 77 s on average on a system running on a 1.83 GHz processor and 4 GB RAM. The SAT solver on an average returns 6.6 candidate Keys. Thus the authors argue that their findings amount to an attack on Sprout in 2^{54} *attempts*, since 54 bits are initially guessed in this process. However, the authors do not discuss the computational complexity associated with one *attempt* at solution by a SAT solver. If one can perform around 2^e Sprout encryptions in 77 s, then in terms of number of encryptions performed, the attack takes time equivalent to $6.6 \times 2^{54} \times 2^e$ encryptions which is more than 2^{80} if $e > 23$ (which may be achievable with a good implementation of the cipher), and so it is not certain that the work in [13] translates to a feasible attack on Sprout.
- In [10], a list merging technique is employed to determine the internal state and secret key of Sprout that is faster than exhaustive search by 2^{10}. The attack has a memory complexity of 2^{46} bits.
- In [6], a TMD tradeoff attack is outlined using an online time complexity of 2^{33} encryptions and 770 TB of memory. The paper first observes that it is easy to deduce the secret key from the knowledge of the internal state and the keystream. The paper then makes an observation on special states of Sprout that produce keystream without the involvement of the secret key. A method to generate and store such states in tables is first outlined. The online stage consists of inspecting keystream bits, retrieving the corresponding state from the table, assuming of course that the state in question is a special state, and then computing the secret key. The process, if repeated a certain number of times, guarantees that a special state is encountered, from where the correct secret key is found.

1.2   Contribution and Organization of the Paper

We summarize the contributions in this paper as follows:

1. In Sect. 2, we present the mathematical description of the Sprout stream cipher.
2. In Sect. 3, we show that by guessing 50 out of the 80 bits of the internal state, one can determine the remaining bits of the state and the secret key by using a SAT solver. This improves the results presented in [13], but due to reasons mentioned earlier, this does not necessarily amount to cryptanalysis of the cipher.
3. In Sect. 4, we show that it is possible to find two IVs for every secret key that generate 80-bit shifted keystream sequences. Making use of this result we mount a distinguishing attack on Sprout using keystream bits from around 2^{40} randomly chosen IVs, and a memory complexity of around 2^{48} bits. We also show that the time complexity of this attack can be reduced at the cost of more memory.
4. Finally in Sect. 5, we observe that for every secret key there exist around 2^{30} IVs that result in the LFSR landing in the all zero state during the keystream generating phase. Based on this observation, we first show how it is possible to find Key-IV pairs that generate keystream sequences with period as small as 80. Thereafter, we mount a simple key recovery attack that requires time equivalent to $2^{66.7}$ encryptions and negligible memory.
5. In Sect. 6, we conclude the paper by making some wider observations about Sprout and some possible solutions towards making it resistant to the aforementioned cryptanalytic advances.

A summary of the results obtained in this paper with respect to the previous attacks on Sprout is presented in Table 1.

Table 1. Summary of attacks on Sprout

Type of attack	Time complexity	Memory	Reference
Using SAT solver	2^{54} SAT *attempts*	-	[13]
-Do-	2^{50} SAT *attempts*	-	Sect. 3
Guess and determine	2^{70} encryptions	2^{46} bits	[10]
TMD-Tradeoff	2^{33} encryptions	$2^{49.6}$ bits (770 TB)	[6]
Distinguisher	2^{40} encryptions	2^{48} bits	Sect. 4
Guess and determine	$2^{66.7}$ encryptions	Negligible	Sect. 5

2   Description of Sprout

The exact structure of Sprout is explained in Fig. 1. It consists of a 40-bit LFSR and a 40-bit NFSR. Certain bits of both the shift registers are taken as inputs to a

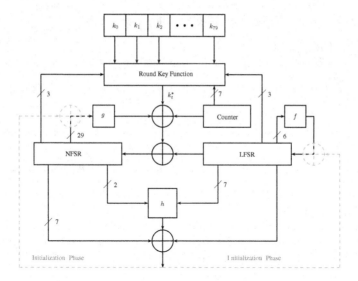

Fig. 1. Block diagram of Sprout

combining Boolean function, whence the keystream is produced. The keystream is produced after performing the following steps:

Initialization Phase: The cipher uses an 80 bit Key and a 70 bit IV. The first 40 most significant bits of the IV is loaded on to the NFSR and the remaining IV bits are loaded on to the first 30 most significant bits of the LFSR. The last 10 bits of the LFSR are initialized with the constant `0x3fe`, i.e. the string of nine 1's followed by a 0. Let $L_t = [l_t, l_{t+1}, \ldots, l_{t+39}]$ and $N_t = [n_t, n_{t+1}, \ldots, n_{t+39}]$ be the 40-bit vectors that denote respectively LFSR and NFSR states at the tth clock interval. During the initialization phase, the registers are updated as follows.

(a) In the first 320 rounds (i.e. $0 \le t \le 319$) of the initialization phase the cipher produces the keystream bit z_t which is not produced as output. This is computed as

$$z_t = l_{t+30} + \sum_{i \in \mathcal{A}} n_{t+i} + h(N_t, L_t).$$

where $\mathcal{A} = \{1, 6, 15, 17, 23, 28, 34\}$ and $h(N_t, L_t) = n_{t+4}l_{t+6} + l_{t+8}l_{t+10} + l_{t+32}l_{t+17} + l_{t+19}l_{t+23} + n_{t+4}l_{t+32}n_{t+38}$.

(b) The LFSR updates as $l_{t+40} = z_t + f(L_t)$, where

$$f(L_t) = l_t + l_{t+5} + l_{t+15} + l_{t+20} + l_{t+25} + l_{t+34}.$$

(c) The NFSR updates as $n_{t+40} = z_t + g(N_t) + c_t^4 + k_t^* + l_0^t$, where c_t^4 denotes the 4th LSB of the modulo 80 up-counter which starts at $t = 0$, k_t^* is the output of the Round Key function defined as:

$$k_t^* = \begin{cases} K_{t \bmod 80}, & \text{if } t < 80, \\ K_{t \bmod 80} \cdot (l_{t+4} + l_{t+21} + l_{t+37} + n_{t+9} + n_{t+20} + n_{t+29}), & \text{otherwise.} \end{cases}$$

Here K_i simply denotes the ith bit of the secret key. The non-linear function $g(N_t)$ is given as:

$$g(N_t) = n_{t+0} + n_{t+13} + n_{t+19} + n_{t+35} + n_{t+39} + n_{t+2}n_{t+25} + n_{t+3}n_{t+5} +$$
$$n_{t+7}n_{t+8} + n_{t+14}n_{t+21} + n_{t+16}n_{t+18} + n_{t+22}n_{t+24} + n_{t+26}n_{t+32} +$$
$$n_{t+33}n_{t+36}n_{t+37}n_{t+38} + n_{t+10}n_{t+11}n_{t+12} + n_{t+27}n_{t+30}n_{t+31}.$$

Keystream Phase: After the initialization phase is completed, the cipher discontinues the feedback of the keystream bit z_t to the update functions of the NFSR and LFSR and makes it available as the output bit. During this phase, the LFSR and NFSR update themselves as $l_{t+40} = f(L_t)$ and $n_{t+40} = g(N_t) + c_t^4 + k_t^* + l_0^t$ respectively.

3   Key Recovery from Partial Knowledge of State

In [13], results were presented pertaining to the recovery of the secret key with partial knowledge of the state. The authors claimed that if all the NFSR bits are known and 14 bits of the LFSR are also known then by using the algebraic equations resulting from the first 450 keystream bits, the keyspace can be reduced to a set of 6.6 candidates on average, by solving the equations through a SAT solver. It was also mentioned that the solver took around 77 s on average to solve the system. Although this does not necessarily lead to an attack, we show in this section that it is possible to propose a better algorithm. Before proceeding we present a brief outline of the algorithm used in [13]:

1. Assume that the entire NFSR state and around m bits of the LFSR are known just after the completion of the key initialization phase. Let us label the time index as $t = 0$ at this instant. The remaining $40 - m$ bits of the LFSR and the 80 bits of the secret key are unknown at this point. The vectors $L_t = [l_t, l_{t+1}, \ldots, l_{t+39}]$ and $N_t = [n_t, n_{t+1}, \ldots, n_{t+39}]$. So initially it is assumed that N_0 is completely known and L_0 is known partially.
2. For $t = 0$ to $N_r - 1$ do
   - Introduce two new unknowns l_{40+t}, n_{40+t} defined as $l_{40+t} = f(L_t)$ and $n_{40+t} = g(N_t) + c_t^4 + k_t^* + l_t$.
   - Form the keystream equation $z_t = l_{t+30} + n_{t+4}l_{t+6} + l_{t+8}l_{t+10} + l_{t+32}l_{t+17} + l_{t+19}l_{t+23} + n_{t+4}l_{t+32}n_{t+38} + \sum_{i \in \mathcal{A}} n_{t+i}$.
3. After forming the above bank of $3N_r$ equations, pass them to a SAT solver.

The authors of [13] claimed that for $m = 14$, $N_r = 450$, the SAT solver was able to narrow down the set of candidate secret keys to 6.6 on average in around 77 s.

3.1   A Few Observations

The ease with which a SAT solver is able to solve a given bank of equations depends on the algebraic degree of the equations so formed [14]. It is clear that the algebraic degree of z_t with respect to the unknowns in L_0 and the secret key increases for increasing t. It is also known that, if the key is known, then the state update during both the keystream phase and the initialization phase are one-to-one and invertible. Indeed, rewriting the functions f, g as $f(L_t) = l_t + f'(L'_t)$ and $g(N_t) = n_t + g'(N'_t)$ (here $L'_t = [l_{t+1}, l_{t+2}, \dots, l_{t+39}]$ and $N'_t = [n_{t+1}, n_{t+2}, \dots, n_{t+39}]$, then if L_t, N_t denote the state at time t, then L_{t-1} is given as $[l_{t-1}, l_t, \dots, l_{t+38}]$ where $l_{t-1} = l_{t+39} + f'(L'_{t-1})$, and since L'_{t-1} is a subset of L_t, we can see that L_{t-1} is completely defined by L_t. Similarly $N_{t-1} = [n_{t-1}, n_t, \dots, n_{t+38}]$ where

$$n_{t-1} = n_{t+39} + l_{t-1} + k^*_{t-1} + c^4_{t-1} + g'(N'_{t-1}).$$

Here too since $N'_{t-1} \subset N_t$, the previous state N_{t-1} is completely defined by L_t, N_t. Keeping this in mind, we formulate the following strategy for key recovery from the partial knowledge of state.

1. We assume that at $t = 320$, all the bits of N_{320} and the first m bits of L_{320} are known. Thereafter, we do the following:
2. For $t = 0$ to 319 do
   - Introduce two new unknowns l_{360+t}, n_{360+t} defined as $l_{360+t} = f(L_{t+320})$ and $n_{360+t} = g(N_{t+320}) + c^4_{t+320} + k^*_{t+320} + l_{t+320}$.

   - Form the keystream equation

$$z_{t+320} = l_{t+350} + \sum_{i \in \mathcal{A}} n_{t+320+i} + h(N_{t+320}, L_{t+320}).$$

3. We now take help of the keystream generated before $t = 320$
4. For $t = 320$ to 1 do
   - Introduce two new unknowns l_{t-1}, n_{t-1} defined as $l_{t-1} = l_{t+39} + f'(L'_{t-1})$ and $n_{t-1} = n_{t+39} + l_{t-1} + k^*_{t-1} + c^4_{t-1} + g'(N'_{t-1})$.
   - Form the keystream equation

$$z_{t-1} = l_{t+29} + \sum_{i \in \mathcal{A}} n_{t-1+i} + h(N_{t-1}, L_{t-1}).$$

5. After preparing this bank of $320 * 3 * 2 = 1920$ equations, we forward it to a SAT Solver.

Since the algebraic degrees of z_{320+t} and z_{320-t} are expected to be the same with respect to the unknowns in L_{320} and the secret key, we achieve the dual purpose of populating our bank of equations with more entries and at the same time control the algebraic degree of the equations to some extent. We performed the experiments with Cryptominisat 2.9.5 [15] solver installed with the SAGE 5.7 [16] computer algebra system on a computer with a 2.1 GHz CPU and 16 GB memory. For $m = 10$, (after guessing 50 bits of the internal state), we were able to find the remaining bits of the state and the correct secret key in around 31 s on average.

4   A Distinguishing Attack

Before we get into details of the distinguisher, let us revisit a few facts about Sprout. We have already shown that if the secret key is known, then the state updates in both the keystream and initialization phases are one-to-one and efficiently invertible. Before proceeding, we give a formal algorithmic description of the state update inversion routines in the keystream and initialization phases, as per the observations in Sect. 3.1. We denote the algorithms by KS^{-1} and Init^{-1} respectively.

Input: L_t, N_t: The LFSR, NFSR state at time t;
Output: L_{t-1}, N_{t-1}: The LFSR, NFSR state at time $t - 1$;

$l_{t-1} \leftarrow l_{t+39} + f'(L'_{t-1});$
$n_{t-1} \leftarrow n_{t+39} + l_{t-1} + k^*_{t-1} + c^4_{t-1} + g'(N'_{t-1});$
$L_{t-1} \leftarrow [l_{t-1}, l_t, l_{t+1}, \ldots, l_{t+38}];$
$N_{t-1} \leftarrow [n_{t-1}, n_t, n_{t+1}, \ldots, n_{t+38}];$
Return L_{t-1}, N_{t-1}

Algorithm 1. Algorithm KS^{-1}

Input: L_t, N_t: The LFSR, NFSR state at time t;
Output: L_{t-1}, N_{t-1}: The LFSR, NFSR state at time $t - 1$;

$z_{t-1} \leftarrow l_{t+29} + \sum_{i \in \mathcal{A}} n_{t-1+i} + h(N_{t-1}, L_{t-1});$
$l_{t-1} \leftarrow l_{t+39} + f'(L'_{t-1}) + z_{t-1};$
$n_{t-1} \leftarrow n_{t+39} + l_{t-1} + k^*_{t-1} + c^4_{t-1} + g'(N'_{t-1}) + z_{t-1};$
$L_{t-1} \leftarrow [l_{t-1}, l_t, l_{t+1}, \ldots, l_{t+38}];$
$N_{t-1} \leftarrow [n_{t-1}, n_t, n_{t+1}, \ldots, n_{t+38}];$
Return L_{t-1}, N_{t-1}

Algorithm 2. Algorithm Init^{-1}

We will use the above subroutines to generate Key-IV pairs that generate 80-bit shifted keystream sequences. To do that we follow the following steps:

1. Fix the secret key K to some constant in $\{0, 1\}^{80}$
2. Fix Success $\leftarrow 0$
3. Do the following till Success $= 1$
   - Select $\mathbf{S} = [s_0, s_1, \ldots, s_{79}] \xleftarrow{R} \{0, 1\}^{80}$ randomly.
   - Assign $N_0 \longleftarrow [s_0, s_1, \ldots, s_{39}], L_0 \longleftarrow [s_{40}, s_{41}, \ldots, s_{79}]$
   - Run Init^{-1} over N_0, L_0 for 320 rounds and store the result as

     $$\mathbf{U} = [u_0, u_1, \ldots, u_{79}].$$

   - Assign $N_{80} \longleftarrow [s_0, s_1, \ldots, s_{39}], L_{80} \longleftarrow [s_{40}, s_{41}, \ldots, s_{79}]$
   - Run KS^{-1} over N_{80}, L_{80} for 80 rounds, followed by Init^{-1} for 320 rounds.
   - Store the result as $\mathbf{V} = [v_0, v_1, \ldots, v_{79}]$.
   - If $u_{70} = u_{71} = \cdots = u_{78} = v_{70} = v_{71} = \cdots = v_{78} = 1$ and $u_{79} = v_{79} = 0$
     then Success $= 1$.

Table 2. Key-IV pairs that produce 80 bit shifted keystream bits. (Note that the first hex character in V_1, V_2 encodes the first 2 IV bits, the remaining 17 hex characters encode bits 3 to 70)

#	K	V_1	V_2
1	8b0b c4c3 781e fe4b 925c	1 03c2cb34d8b8870e5	1 f208a4661d50a1f72
2	be8d d8e2 a818 80c5 eda7	2 d7d0162c62f256ad7	2 5f7c58576e05e3c52

The above algorithm fixes the secret key K, and randomly chooses a state **S** and assumes that for two different IVs V_1, V_2, the state in the 0th round of the keystream phase for (K, V_1) and the 80th round of the keystream phase for (K, V_2) are both equal to **S**. The algorithm then performs the state inversion routines in each case and tries to find V_1 and V_2. A Success occurs when the last 10 bits of both **U**, **V** are equal to the padding 0x3fe used in Sprout. In that case $V_1 = [u_0, u_1, \ldots, u_{69}]$ and $V_2 = [v_0, v_1, \ldots, v_{69}]$ produces exactly 80-bit shifted keystream sequences for the key K. Of course, a Success requires 20 bit conditions to be fulfilled and assuming that **U**, **V** are i.i.d, each iteration of the above algorithm has a success probability of 2^{-20} for any randomly selected **S**. So running the iteration 2^{20} times guarantees one Success on average. By running the above algorithm we were able to obtain several Key-IV pairs that generates 80 bit shifted keystream sequences, which we tabulate in Table 2. Note that the above method can not be used to find Key-IV pairs that generate keystream bits of shift other than multiples of 80. This is because Sprout employs a counter whose 4th LSB is used to update the NFSR. The counter resets after every 80 rounds and so any analysis involving the self-similarity of the initialization phase must be done at intervals of multiples of 80.

Note that it is possible to generate such a Key-IV pair in 2^{10} attempts instead of 2^{20}, if instead of choosing **S**, we first choose K, V_1 randomly, run the forward initialization algorithm to generate **S**, and then assume that **S** is the 80th keystream phase state for some K, V_2 and thereafter run 80 rounds of KS^{-1} and 320 rounds of $Init^{-1}$ to generate **V**. In such a case, Success would be dependent on only the last 10 bits of **V** and hence expected once in 2^{10} attempts. However we present the first algorithm in order to better explain the distinguishing attack.

4.1   The Distinguisher

In the above algorithm for finding Key-IV pairs that generate shifted keystream sequences, once the key is fixed, a Success is expected every 2^{20} attempts and since there are 2^{80} ways of choosing **S**, this implies that for every Key K, there exist $2^{80-20} = 2^{60}$ IV pairs V_1, V_2 such that the key-IV pairs (K, V_1) and (K, V_2) produce exactly 80-bit shifted keystream sequences. So our distinguisher is as follows

1. Generate around 240 keystream bits for the unknown Key K and some randomly generated Initial Vector V.

Table 3. Experimental values of N for smaller versions of Sprout

#	n	N (experimental)	N (theoretical)
1	8	222.4	256
2	9	446.9	512
3	10	911.7	1024
4	11	1865.7	2048

2. Store the keystream bits in some appropriate data structure like a Binary Search Tree.
3. Continue the above steps with more randomly generated IVs V till we obtain two Initial Vectors for K that generate 80-bit shifted keystream.

The only question now remains how many random Initial Vectors do we need to try before we get a match. The answer will become clearer if (for a fixed K) we imagine the space of Initial Vectors as an undirected Graph $G = (W, E)$, where $W = \{0, 1\}^{70}$ is the Vertex set which contains all the possible 70 bit Initial vector values as nodes. An edge $(V_1, V_2) \in E$ if and only if (K, V_1) and (K, V_2) produce 80-bit shifted keystream sequence. From the above discussion, it is clear that the cardinality of E is expected to be 2^{60}. When we run the Distinguisher algorithm for N different Initial Vectors, we effectively add $\binom{N}{2}$ edges to the coverage and a match occurs when one of these edges is actually a member of the Edge-set E. Since there are potentially $\binom{2^{70}}{2}$ edges in the IV space, by the Birthday bound, a match will occur when the product of $\binom{N}{2}$ and the cardinality of E which is around 2^{60} is equal to $\binom{2^{70}}{2}$. From this equation solving for N, we get $N \approx 2^{40}$. This gives a bound for the time and memory complexity of the Distinguisher. The time complexity is around 2^{40} encryptions, and the memory required is of the order of $2^{40} * 240 \approx 2^{48}$ bits.

In general for Sprout like structures that have an n bit LFSR and NFSR with a $2n$-bit secret key and $2n - \Delta$ bit IV (for some $\Delta > 0$), the above equation boils down to

$$\binom{N}{2} * 2^{2n-2\Delta} = \binom{2^{2n-\Delta}}{2}.$$

Solving this equation gives $N \approx 2^n$. In order to verify our theoretical results, we performed experiments on smaller versions of Sprout with $n = 8, 9, 10, 11$ to find the expected value of N in each case. The results have been tabulated in Table 3.

Decreasing the Time Complexity: So far we have been restricting ourselves to 80-bit shifts of keystream sequences. We could easily consider shifts of the form $80 * P$ where P can be any positive integer. The algorithm to find two Initial Vectors V_1, V_2 for any Key K that generates $80 * P$-bit shifted keystream sequence is not very different from the one which finds IVs that generate 80-bit shifted keystream. We present the explicit form of the algorithm for convenience.

1. Fix the secret key K to some constant in $\{0,1\}^{80}$
2. Fix Success $\leftarrow 0$
3. Do the following till Success $=1$
   - Select $\mathbf{S} = [s_0, s_1, \ldots, s_{79}] \xleftarrow{\text{R}} \{0,1\}^{80}$ randomly.
   - Assign $N_0 \longleftarrow [s_0, s_1, \ldots, s_{39}]$, $L_0 \longleftarrow [s_{40}, s_{41}, \ldots, s_{79}]$
   - Run Init^{-1}over N_0, L_0 for 320 rounds and store the result as $\mathbf{U} = [u_0, u_1, \ldots, u_{79}]$.
   - Assign $N_{80*P} \longleftarrow [s_0, s_1, \ldots, s_{39}]$, $L_{80*P} \longleftarrow [s_{40}, s_{41}, \ldots, s_{79}]$
   - Run KS^{-1}over N_{80*P}, L_{80*P} for $80 * P$ rounds, followed by $\mathsf{Init}^{-1}$for 320 rounds.
   - Store the result as $\mathbf{V} = [v_0, v_1, \ldots, v_{79}]$.
   - If $u_{70} = u_{71} = \cdots = u_{78} = v_{70} = v_{71} = \cdots = v_{78} = 1$ and $u_{79} = v_{79} = 0$ then Success $=1$.

The only change is that we assume that $\mathbf{S}$ is the round 0 state for some K, V_1 and the round $80 * P$ state for some K, V_2. We perform the inversion operations accordingly and look for a Success. Arguing just as before, we can say that, for any fixed K and P, there exist 2^{60} IV pairs that generate $80 * P$-bit shifted keystream Sequences. So we redefine our Distinguishing attack as follows:

1. Generate around $80 * P$ keystream bits for the unknown Key K and some randomly generated Initial Vector V.
2. Store the keystream bits in some appropriate data structure like a Binary Search Tree.
3. Continue the above steps with more randomly generated IVs V till we obtain two Initial Vectors for K that generate $80 * i$-bit shifted keystream for some $1 \leq i \leq P$.

We can calculate the expected number of attempts N before we get a match as follows. Redefine the undirected graph $G = (W, E)$, where $W = \{0,1\}^{70}$ is the Vertex set which contains all the possible 70 bit Initial vector values as nodes. An edge $(V_1, V_2) \in E$ if and only if (K, V_1) and (K, V_2) produce $80 * i$-bit shifted keystream sequence for some $0 \leq i \leq P$. The expected cardinality of E is approximately $P * 2^{60}$. Again choosing N Initial Vectors adds $\binom{N}{2}$ edges to the coverage and so the required value of N is given by $\binom{N}{2} * P * 2^{60} = \binom{2^{70}}{2} \Rightarrow N \approx \frac{2^{40}}{\sqrt{P}}$. This implies that the time complexity can be reduced to $\frac{2^{40}}{\sqrt{P}}$ encryptions with the memory complexity at $80 * P * 2^{40}$ bits. For $P = 2^{10}$ say, this results in a time complexity of 2^{35} encryptions and memory of 2^{57} bits.

5   A Key Recovery Attack

We make another observation to begin this section. During the keystream phase, the LFSR pretty much runs autonomously. Which means that if after the initialization phase, the LFSR lands on the all zero state then it remains in this state for the remainder of the keystream phase, i.e. if $L_0 = \mathbf{0}$, then $L_t = \mathbf{0}$ for all $t > 0$. Assuming uniform distribution of L_0, we can argue that for every Key K, this event occurs for 2^{-40} fraction of IVs on average. So for each K, there exists on an average $2^{70-40} = 2^{30}$ IVs which lead to an all zero LFSR after the initialization phase. We shall see two implications of this event.

5.1   Keystream with Period 80

Now once the LFSR enters the all zero state the NFSR runs autonomously. Since the NFSR is a finite state machine of 40 bits only, we can always expect keystream of period less than $80 * 2^{40}$, once the LFSR becomes all zero. Hence for every Key, we expect to find 2^{30} Initial vectors that produce keystream sequences of less than $80 * 2^{40}$. With some effort, we can even find Key-IV pairs that produce keystream with period 80. We will take help of SAT solvers for his. The procedure may be outlined as follows:

1. Select a Key $K \xleftarrow{\text{R}} \{0,1\}^{80}$ randomly.
2. Assume $L_0 = [0,0,0,\ldots,0]$.
3. Assign $N_0 \leftarrow [n_0, n_1, n_2, \ldots, n_{39}]$, where all the n_i are unknowns.
4. For $i = 0$ to 79 do
   - Introduce the unknown n_{40+i}, and add the equation $n_{40+i} = g(N_i) + c_i^4 + k_i^*$ to the equation bank.
5. Add the 40 Equations $n_i = n_{80+i}, \forall\, i \in [0, 39]$ to the equation bank.
6. Pass the equations to the Solver. This effectively asks the solver to solve the vector equation $N_0 = N_{80}$ for the given Key K.
7. If the solver returns the solution $N_0 = [s_0, s_1, \ldots, s_{39}]$ then run the Init^{-1} routine 320 times on $N_0 = [s_0, s_1, \ldots, s_{39}], L_0 = [0,0,\ldots,0]$.
8. Store the result in $\mathbf{B} = [b_0, b_1, \ldots, b_{79}]$.
9. If $b_{70} = b_{71} = \cdots = b_{78} = 1$ and $b_{79} = 0$ then Exit else repeat the above steps with another random secret key.

The steps in the above the above algorithm can be summarized as follows. First select a random secret key K. Then assume that the LFSR is all zero after the initialization phase, and fill the corresponding NFSR state with unknowns. We then populate the equation bank accordingly for the first 80 rounds and ask the solver to solve the vector equation $N_0 = N_{80}$, in the unknowns $n_0, n_1, \ldots, n_{119}$. If the solver returns the solution $N_0 = [s_0, s_1, \ldots, s_{39}]$ then $N_0 = [s_0, s_1, \ldots, s_{39}], L_0 = [0,0,\ldots,0]$ is a valid initial state for the Sprout keystream phase if we can find an IV for the given Key K that results in this state. So we run the Init^{-1} routine 320 times and obtain the resultant vector $\mathbf{B}$. Now if the last ten bits of $\mathbf{B}$ are equal to the 0x3fe pattern used in Sprout, then we can be sure that for the key K and the Initial Vector $V = [b_0, b_1, \ldots, b_{69}]$, the keystream sequence produced is of period exactly 80 since the same state $N_0 = [s_0, s_1, \ldots, s_{39}], L_0 = [0,0,\ldots,0]$ will repeat in the keystream phase every 80 iterations. The above process is expected to produce one such Key-IV pair in 2^{10} attempts. Since the above algorithm can be run for 2^{80} values of the secret key, this implies that there exist around $2^{80-10} = 2^{70}$ Key-IV pairs that produce keystream bits period 80. Table 4 lists a few examples of such Key-IV pairs.

Table 4. Key-IV pairs that produce keystream sequence with period 80. (Note that the first hex character in V encodes the first 2 IV bits, the remaining 17 hex characters encode bits 3 to 70)

#	K	V
1	2819 5612 323c 2357 3518	2 fbfc75bfcb4396485
2	7047 18a0 f88a aff7 7df5	1 4d57f42712b395015

5.2   Application to Key Recovery

It is clear, that for every Key, on average one out of every 2^{40} Initial Vectors lands the LFSR in the all zero state after initialization. In such a situation the algebraic structure of the cipher becomes simpler to analyze. The NFSR update equation becomes

$$n_{t+40} = g(N_t) + c_t^4 + k_t^*,$$

where $k_t^* = K_{t \bmod 80} \cdot (n_{t+9} + n_{t+20} + n_{t+29})$ and the output keystream bit is generated as

$$z_t = n_{t+1} + n_{t+6} + n_{t+15} + n_{t+17} + n_{t+23} + n_{t+28} + n_{t+34}.$$

Given such a situation, this greatly simplifies the guess and determine approach of [10] both in terms of time and memory (although only in the multiple IV mode). To explain the attack better let us define $x_i = n_{i+1}$, for all $i \geq 0$ and so we have $N_1 = [x_0, x_1, x_2, \ldots, x_{39}]$. So for $i = 0$ to 6 we have

$$z_i = x_i + x_{i+5} + x_{i+14} + x_{i+16} + x_{i+22} + x_{i+27} + x_{i+33}.$$

This means that if the attacker knows that $L_0 = \mathbf{0}$, then the first 7 keystream bits $z_0, z_1, z_2, \ldots, z_6$ is dependent on only N_1 and the secret key is not involved directly in the computation. This implies that if the attacker intends to guess N_1 then by observing the first seven keystream bits he can narrow down N_1 to a set of 2^{33} possible candidates in the following way:

1. Guess $x_0, x_1, x_2, \ldots, x_{32}$ first. There are 2^{33} possible candidates.
2. Calculate $x_{i+33} = z_i + x_i + x_{i+5} + x_{i+14} + x_{i+16} + x_{i+22} + x_{i+27}$ for $i = 0$ to 6.

For each of these 2^{33} candidates, the attacker proceeds as follows: he calculates x_{40} from the equation for z_7 as $x_{40} = z_7 + x_7 + x_{12} + x_{21} + x_{23} + x_{24} + x_{31}$ and from x_{40} he calculates k_0^* as $k_0^* = x_{40} + c_0^4 + g(N_1)$. Now we know that $k_0^* = K_0 \cdot (x_8 + x_{19} + x_{28})$. So if $k_0^* = 0$ and $x_8 + x_{19} + x_{28} = 0$ then nothing can be deduced. If $k_0^* = 0$ and $x_8 + x_{19} + x_{28} = 1$ then it can be deduced that $K_0 = 0$. If $k_0^* = 1$ and $x_8 + x_{19} + x_{28} = 1$ then it can be deduced that $K_0 = 1$. If $k_0^* = 1$ and $x_8 + x_{19} + x_{28} = 0$, then a contradiction is reached and it is concluded that the guess for N_1 was incorrect. Thereafter the same procedure with $x_{41}, x_{42} \ldots$ is followed sequentially. We outline the above procedure formally as follows:

1. For Each of the 2^{33} choices of N_1 do the following till a contradiction is arrived at
   A. Assign $i \leftarrow 0$
   B. Do the following:
   - Calculate $x_{i+40} = z_{i+7} + x_{i+7} + x_{i+12} + x_{i+21} + x_{i+23} + x_{i+24} + x_{i+31}$
   - Calculate $k_i^* = x_{i+40} + c_i^4 + g(N_{i+1})$
   - Case 1: $k_i^* = 0$ and $x_{i+8} + x_{i+19} + x_{i+28} = 0 \Rightarrow$ No Deduction
   - Case 2: $k_i^* = 0$ and $x_{i+8} + x_{i+19} + x_{i+28} = 1 \Rightarrow$ If $K_{i \bmod 80}$ is not already assigned then assign $K_{i \bmod 80} = 0$, otherwise if this bit has already been assigned to 0, then we have a contradiction.
   - Case 3: $k_i^* = 1$ and $x_{i+8} + x_{i+19} + x_{i+28} = 1 \Rightarrow$ If $K_{i \bmod 80}$ is not already assigned then assign $K_{i \bmod 80} = 1$, otherwise if this bit has already been assigned to 1, then we have a contradiction.
   - Case 4: $k_i^* = 1$ and $x_{i+8} + x_{i+19} + x_{i+28} = 0 \Rightarrow$ we have a contradiction.
   - If there is a contradiction, then we restart the process with a new guess of N_1.
   - If there is no contradiction then assign $i \leftarrow i+1$ and repeat the process if the entire secret key has not already been found.

Analysis of Time Complexity: The attacker obtains keystream bits for 2^{40} randomly generated IVs and repeats the above routine for every keystream sequence, till the correct Key is found. For the first 80 rounds, the only way we have a contradiction is when Case 4 occurs, i.e. $k_i^* = 1$ and $x_{i+8} + x_{i+19} + x_{i+28} = 0$. So the probability that any guess for N_1 is eliminated in 1 round itself is $\frac{1}{4}$, i.e. assuming that the events $k_i^* = 1$ and $x_{i+8} + x_{i+19} + x_{i+28} = 0$ are independently and uniformly distributed. The probability therefore that it takes 2 rounds to eliminate is $\left(1 - \frac{1}{4}\right) * \frac{1}{4}$. In general, the probability that it takes i steps is roughly $\left(1 - \frac{1}{4}\right)^{i-1} * \frac{1}{4}$. Therefore the average number of rounds θ that a guess takes to eliminate is given by

$$\theta = \sum_{i=1}^{\infty} \frac{i}{4} * \left(1 - \frac{1}{4}\right)^{i-1} = 4.$$

In the analysis we have assumed that the only source of contradiction arises out of Case 4. The actual value of θ is hence slightly smaller than 4. The attacker obtains the keystream for some random IV and then tries all the possible 2^{33} guesses. This takes $\theta \cdot 2^{33} = 2^{35}$ steps for any IV that does not lead to $L_0 = \mathbf{0}$. It has already been pointed out in [6], clocking each Sprout step is equivalent to $2^{-8.34}$ encryptions (a proof is presented in Appendix A). And so for every any IV that does not yield $L_0 = \mathbf{0}$ the total work done is equivalent to $2^{35-8.34} = 2^{26.66}$ encryptions. Now the attacker has to try out around 2^{40} IVs to succeed in getting $L_0 = \mathbf{0}$, and so the total time complexity in this process equals $2^{40+26.66} = 2^{66.66}$ encryptions.

Analysis of Memory Complexity: The memory complexity of the algorithm is surprisingly negligible. Testing each guess of N_1 can be done on the fly and hence the memory complexity is limited to that required to run the loop and store the computed values of the key and the values of the x_i bits. This is in stark contrast to the 2^{46} bits (8 TB) required in [10] or the 770 TB required in [6]. Thus although, the algorithm that we provide is not the best in terms of time complexity, it is certainly best in terms of memory.

6   Discussion and Conclusion

In this paper we outline a Distinguishing attack and a Key Recovery attack on the Sprout stream cipher. We also present some results on Key Recovery from partial knowledge of the state, shifted keystream sequence producing Key-IV pairs and Key-IV pairs producing keystream sequences with period 80. The key recovery attack that we propose is not the best in terms of time complexity but certainly best in terms of the total memory required. It can be pointed out that the attack in [6] was possible due to the non-linear mixing of the secret key during the keystream phase, i.e. $k_t^* = K_{t \bmod 80} \cdot (l_{t+4} + l_{t+21} + l_{t+37} + n_{t+9} + n_{t+20} + n_{t+29})$. This enabled the attacker to identify and generate special internal states that for 40 rounds or so do not involve the secret key bit in the computation of the keystream bit, i.e. those for which $l_{t+4} + l_{t+21} + l_{t+37} + n_{t+9} + n_{t+20} + n_{t+29} = 0$, for 40 consecutive rounds. The attack in [6] would not be directly applicable if the key mixing was linear, for example if $k_t^* = K_{t \bmod 80}$. However even if the key mixing were done linearly, all the attacks presented in this paper would still hold. This reiterates the point that when it comes to designing stream ciphers with shorter internal states, the Sprout architecture needs further tweaks. We briefly summarize what could be the possible solutions to the problems in the Sprout architecture:

State Size: In order to prevent the key recovery attacks of [10] and this paper, one possible solution could be increasing the state size to 100 bits. The cipher could employ two registers of 50 bits each, which would make the attacks of [10] and this paper worse than a brute force search for the secret key.

Use of LFSR: One of the reasons that the 40-bit LFSR (generated by a primitive polynomial) is used in the design, is to guarantee that the resulting keystream has a period which is a multiple $2^{40} - 1$. This is true only when the LFSR is non zero after the initialization phase. However on the rare occasion that the LFSR lands on the all zero state after initialization, it remains in this state for ever, and in the process weakens the algebraic structure of the cipher. One possible solution to this problem is to replace the LFSR with a register that generates a maximal length DeBruijn Sequence [7]. Such an n-bit maximal length producing register cycles through all possible 2^n values, and so will not get stuck at the all zero state. However update functions that produce maximal length sequences are hardware intensive: if $(x_0, x_1, \ldots, x_{n-1})$ represents the n-bit register, then the update function must contain the term $x_1 \cdot x_2 \cdots x_{n-1}$ [8], and so some extra gate area needs to be used.

Reading the IV: Loading the IV on to the register directly is used in ciphers like Grain v1 and Trivium, but for ciphers with shorter internal states this is not a good idea. This helps in **(a)** Finding Key-IV pairs which produce 80 bit shifted keystream bits, **(b)** Mounting the Distinguishing attack in Sect. 4, and **(c)** Finding Key-IV pairs that produce keystream bits with small period. So a different method of reading the IV information into the registers must be found. One possible method could be reading the IV the same way the key is read i.e. bit by bit. For example in rounds $t = 80$ to 149, the IV bit could be included in the update function of the NFSR as

$$n_{t+50} = z_t + g(N_t) + c_t^4 + \nu_t^* + l_0^t,$$

where $\nu_t^* = IV_{t-80}$.

Key Mixing: After the attack of [6] that specifically exploits the non-linear Key mixing used in Sprout, it is quite obvious that the key mixing must be linear to prevent such attacks. Therefore, if the round key bit $k_t^* = K_{t \bmod 80}$, then the attack of [6] can be prevented.

Although, the cipher Sprout may have been cryptanalyzed, the idea of designing a stream cipher with shorter internal states is indeed quite fascinating. This does open up a new research discipline in which the scope to experiment could be boundless.

Appendix A: Cost of Executing One Round of Sprout [6]

To do an exhaustive search, first an initialization phase has to be run for 320 rounds, and then generate 80-bits of keystream to do a unique match. However, since each keystream bit generated matches the correct one with probability $\frac{1}{2}$, 2^{80} keys are tried for 1 clock and roughly half of them are eliminated, 2^{79} for 2 clocks and half of the remaining keys are eliminated, and so on. This means that in the process of brute force search, the probability that for any random key, $(i+1)$ Sprout keystream phase rounds need to be run, is $\frac{1}{2^i}$. Hence, the expected number of Sprout rounds per trial is

$$\sum_{i=0}^{79} \frac{(i+1)2^{80-i}}{2^{80}} = \sum_{i=0}^{79} (i+1)\frac{1}{2^i} \approx 4$$

Add to this the 320 rounds in the initialization phase, the average number of Sprout rounds per trial is 324. As a result, we will assume that clocking the registers once will cost roughly $\frac{1}{320+4} = 2^{-8.34}$ encryptions.

References

1. The ECRYPT Stream Cipher Project. eSTREAM Portfolio of Stream Ciphers. Accessed on 8 September 2008

2. Armknecht, F., Mikhalev, V.: On lightweight stream ciphers with shorter internal states. In: Leander, G. (ed.) FSE 2015. LNCS, vol. 9054, pp. 451–470. Springer, Heidelberg (2015)

3. Babbage, S., Dodd, M.: The stream cipher MICKEY 2.0. ECRYPT Stream Cipher Project Report. http://www.ecrypt.eu.org/stream/p3ciphers/mickey/mickey_p3.pdf

4. Biryukov, A., Shamir, A.: Cryptanalytic time/memory/data tradeoffs for stream ciphers. In: Okamoto, T. (ed.) ASIACRYPT 2000. LNCS, vol. 1976, pp. 1–13. Springer, Heidelberg (2000)

5. De Cannière, C., Preneel, B.: TRIVIUM -Specifications. ECRYPT Stream Cipher Project Report. http://www.ecrypt.eu.org/stream/p3ciphers/trivium/trivium_p3.pdf

6. Esgin, M.F., Kara, O.: Practical Cryptanalysis of Full Sprout with TMD Tradeoff Attacks. To appear in Selected Areas in Cryptography (2015)

7. Fredricksen, H.: A survey of full length nonlinear shift register cycle algorithms. SIAM Rev. **24**(1982), 195–221 (1982)

8. Golomb, S.W.: Shift Register Sequences. Holden-Day Inc., Laguna Hills (1967)

9. Hao, Y.: A Related-Key Chosen-IV Distinguishing Attack on Full Sprout Stream Cipher. http://eprint.iacr.org/2015/231.pdf

10. Lallemand, V., Naya-Plasencia, M.: Cryptanalysis of full Sprout. In: Gennaro, R., Robshaw, M. (eds.) CRYPTO 2015. LNCS, vol. 9215, pp. 663–682. Springer, Heidelberg (2015)

11. Hell, M., Johansson, T., Meier, W.: Grain - A Stream Cipher for Constrained Environments. ECRYPT Stream Cipher Project Report 2005/001 (2005). http://www.ecrypt.eu.org/stream

12. Hell, M., Johansson, T., Meier, W.: A stream cipher proposal: Grain-128. In: IEEE International Symposium on Information Theory (ISIT 2006) (2006)

13. Maitra, S., Sarkar, S., Baksi, A., Dey, P.: Key Recovery from State Information of Sprout: Application to Cryptanalysis and Fault Attack. http://eprint.iacr.org/2015/236.pdf

14. Sarkar, S., Banik, S., Maitra, S.: Differential Fault Attack against Grain family with very few faults and minimal assumptions. http://eprint.iacr.org/2013/494.pdf

15. Soos, M.: CryptoMiniSat-2.9.5. http://www.msoos.org/cryptominisat2/

16. Stein, W.: Sage Mathematics Software. Free Software Foundation Inc. (2009). http://www.sagemath.org. (Open source project initiated by W. Stein and contributed by many)

Linear Cryptanalysis of Reduced-Round SIMECK Variants

Nasour Bagheri[1,2]([✉])

[1] Department of Electrical Engineering,
Shahid Rajaee Teacher Training University, Tehran, Iran
[2] School of Computer Science,
Institute for Research in Fundamental Sciences (IPM), Tehran, Iran
NBagheri@srttu.edu

Abstract. SIMECK is a family of 3 lightweight block ciphers designed by Yang *et al.* They follow the framework used by Beaulieu *et al.* from the United States National Security Agency (NSA) to design SIMON and SPECK. A cipher in this family with K-bit key and N-bit block is called SIMECKN/K. We show that the security of this block cipher against linear cryptanalysis is not as good as its predecessors SIMON. More precisely, while the best known linear attack for SIMON32/64, using Algorithm 1 of Matsui, covers 13 rounds we present a linear attack in this senario which covers 14 rounds of SIMECK32/64. Similarly, using Algorithm 1 of Matsui, we present attacks on 19 and 22 rounds of SIMECK48/96 and SIMECK64/128 respectively, compare them with known attacks on 16 and 19 rounds SIMON48/96 and SIMON64/128 respectively. In addition, we use Algorithm 2 of Matsui to attack 18, 23 and 27 rounds of SIMECK32/64, SIMECK48/96 and SIMECK64/128 respectively, compare them with known attacks on 18, 19 and 21 rounds SIMON32/64, SIMON48/96 and SIMON64/128 respectively.

Keywords: SIMECK · SIMON · SPECK · Linear cryptanalysis

1 Introduction

SIMECK [26] is a new family of lightweight block ciphers designed by Yang *et al.* and inspired by SIMON and SPECK, designed by the NSA [8]. The round function of SIMECK is similar to the round function of SIMON while its key schedule is more similar to the key schedule of SPECK. The aim of SIMECK is to provide optimal hardware and software performance for low-power limited gate devices such as RFID devices by combing good components from both SIMON and SPECK. Variants of this block cipher support plaintext block sizes of 32, 48, 64 and 96 and 128 bits. The key size of those variants are 64, 96 and 128 bits respectively. SIMECKN/K denotes a variant of SIMECK that has a block size of N bits and a key size of K bits.

Although, several works investigated the security of SIMON and SPECK against differential attack [2,3,6,9,22,24], its variants such as impossible

© Springer International Publishing Switzerland 2015
A. Biryukov and V. Goyal (Eds.): INDOCRYPT 2015, LNCS 9462, pp. 140–152, 2015.
DOI: 10.1007/978-3-319-26617-6_8

differential attack [2–4,6,10,12,14,15,21,25] and linear attack [1,4,5,7,11,20]. However, we are not aware of any third party security analysis of SIMECK. In this paper, we present linear cryptanalysis against reduced variants of SIMECK.

Contributions. In this paper, we analyze the security of SIMECK against linear cryptanalytic techniques. In this direction, we present linear characteristics for different variants of SIMECK, that can be used for key recovery attacks on SIMECK reduced to 14, 19 and 22 rounds for the respective block sizes of 32, 48 and 64 bits using Matsui's Algorithm 1. Furthermore, we extend this linear characteristics to attack more rounds using Matsui's Algorithm 2. These attacks covers 18, 23 and 26 rounds for the respective block sizes of 32, 48 and 64. A brief summary of our results on SIMECK and the best known results on the equivalent versions of SIMON are presented in Table 1. It must be noted that designers' security analysis against linear cryptanalysis covers 12, 15 and 19 rounds of SIMECK32/64, SIMECK48/96 and SIMECK64/128 respectively [26, Sect. 5].

Organization. The paper is structured as follows. In Sect. 2 we present a brief description of SIMECK. In Sect. 3 we present the idea of linear attacks on SIMON and apply linear attacks to variants of SIMECK using Matsui's Algorithm 1. In Sect. 3 we extend our attacks on variants of SIMECK using Matsui's Algorithm 2. Finally, we conclude the paper in Sect. 5 and propose possible future directions of research.

Table 1. Linear cryptanalysis of SIMECK, using the Matsui's Algorithms 1 and 2, and comparison with the best known results on the equivalent versions of SIMON, where the information given as the number of the deduced key bits are based on the independent-round-keys assumption.

	Variant	# Attacked rounds	Data	Time	Success probability	# deduced key bits	Reference
Matsui's Algorithm 1	SIMON32/64	13	2^{32}	2^{32}	0.997	1	[4]
	SIMECK32/64	13	2^{30}	2^{30}	0.997	1	Sect. 3
	SIMECK32/64	14	2^{32}	2^{32}	0.841	1	Sect. 3
	SIMON48/96	16	2^{46}	2^{46}	0.997	1	[4]
	SIMECK48/96	18	2^{48}	2^{48}	0.997	1	Sect. 3
	SIMECK48/96	19	2^{46}	2^{46}	0.841	1	Sect. 3
	SIMON64/128	19	2^{58}	2^{58}	0.997	1	[4]
	SIMECK64/128	22	2^{60}	2^{60}	0.997	1	Sect. 3
	SIMECK64/128	23	2^{64}	2^{64}	0.841	1	Sect. 3
Matsui's Algorithm 2	SIMON32/64	17	2^{32}	$2^{61.5}$	< 4 0.477	36	[1]
	SIMECK32/64	18	2^{31}	$2^{63.5}$	0.477	44	Sect. 4
	SIMON48/96	19	2^{47}	2^{82}	0.477	53	[1]
	SIMECK48/96	24	2^{45}	2^{94}	0.477	62	Sect. 4
	SIMON64/128	21	2^{59}	2^{123}	0.477	66	[1]
	SIMECK64/128	27	2^{61}	$2^{120.5}$	0.477	76	Sect. 4

2   Description of the SIMECK Family

SIMECK is a classical Feistel block cipher with the round block size of $2n$ bits and the key size of $4n$, where n is the word size. The number of rounds of cipher is denoted by r and depends on the variant of SIMON which are 32, 36 and 44 rounds for SIMECK32/64, SIMECK48/96 and SIMECK64/128 respectively. For a $2n$-bit string X, we use X_L and X_R to denote the left and right halves of the string respectively. The output of round r is denoted by $X^r = (X_R^r \parallel X_L^r)$ and the subkey used in the round r is denoted by K^r. Given a string X, $(X)_i$ denotes the i-th bit of X. Bitwise circular rotation of string a by b position to the left is denoted by $a \lll b$. Further, $\oplus$ and $\&$ denote bitwise XOR and AND operations respectively. We use P and C to denote a plaintext and a ciphertext respectively.

The function $F : \mathbb{F}_2^n \rightarrow \mathbb{F}_2^n$ used in each round of SIMECK is non-linear and non-invertible, and is applied to the left half of the state, so the state is updated as:

$$X^{r+1} = (F(X_L^r) \oplus X_R^r \oplus K^r \parallel X_L^r). \tag{1}$$

The F function is defined as:

$$F(X) = (X \lll 1) \oplus ((X) \ \& \ (X \lll 5)).$$

The subkeys are derived from a master key. Depending on the size of the master key, the key schedule of SIMECK operates on four n-bit word registers. Detailed description of SIMECK variants structure and key scheduling can be found in [26] but it has no affect on our analysis.

3   Linear Cryptanalysis of SIMECK Using the Matsui's Algorithm 1

Linear cryptanalysis [17] is a classical known-plaintext attack cryptanalytic technique that was employed on several block ciphers such as FEAL-4, DES, Serpent and SAFER [13,16,17,23]. In this section, we present linear characteristics for variants of SIMECK using the Matsui's Algorithm 1 [17].

In the round function of SIMECK, similar to SIMON, the only non-linear operation is the bitwise AND. Note that, given single bits A and B, then $\Pr(A \ \& \ B = 0) = \frac{3}{4}$. Hence, we can extract the following highly biased linear expressions for the F function of SIMECK (there are equivalent linear expressions for the F function of SIMON [4]):

$$
\begin{aligned}
&\text{Approximation 1}: \ \Pr((F(X))_i = (X)_{i-1}) = \tfrac{3}{4}, \\
&\text{Approximation 2}: \ \Pr((F(X))_i = (X)_{i-1} \oplus (X)_i) = \tfrac{3}{4}, \\
&\text{Approximation 3}: \ \Pr((F(X))_i = (X)_{i-1} \oplus (X)_{i-5}) = \tfrac{3}{4}, \\
&\text{Approximation 4}: \ \Pr((F(X))_i = (X)_{i-1} \oplus (X)_i \oplus (X)_{i-5}) = \tfrac{1}{4}.
\end{aligned}
\tag{2}
$$

Given the round function (1) of SIMECK and these linear approximations, we can extract the following linear expressions for the ith round of the SIMECK:

$$(X_L^i)_9 \oplus (X_R^i)_{10} \oplus (K^i)_{10} = (X_L^{i+1})_{10} \qquad (3)$$

$$(X_L^{i+3})_{10} \oplus (X_R^{i+3})_9 \oplus (K^{i+2})_{10} = (X_R^{i+2})_{10} \qquad (4)$$

Each equality in Eq. (3) holds with probability $\frac{3}{4}$. Given that $(X_L^{i+1})_{10} = (X_R^{i+2})_{10}$, as it is shown in Fig. 1, we can use Eq. (3) in a meet in the middle approach to extract a 3-round linear approximation as follows, for which the bias is $\frac{1}{8}$ (the bias of a linear approximation which is hold with the probability of p is defined as $\left| p - \frac{1}{2} \right|$):

$$(X_L^i)_9 \oplus (X_R^i)_{10} \oplus (X_L^{i+3})_{10} \oplus (X_R^{i+3})_9 = (K^i)_{10} \oplus (K^{i+2})_{10}. \qquad (5)$$

Since $(X_R^i)_{10} = (X_L^{i-1})_{10}$ and with the probability of $\frac{3}{4}$, we have $(X_L^i)_9 = (X_L^{i-1})_8 \oplus (X_R^{i-1})_9 \oplus (K^{i-1})_9$, we can add a round to the top of the current

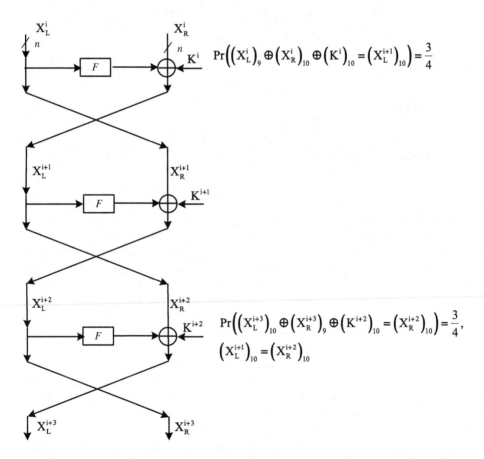

Fig. 1. A 3-round linear approximation for SIMECK32/64.

3-round approximation and produce a 4-round linear expression, with the bias of $\frac{1}{16}$, as follows:

$$(X_L^{i-1})[8, 10] \oplus (X_R^{i-1})_9 \oplus (X_L^{i+3})_{10} \oplus (X_R^{i+3})_9 = (K^{i-1})_9 \oplus (K^i)_{10} \oplus (K^{i+2})_{10}. \tag{6}$$

where $(X)[i_1, ..., i_m] = (X)_{i_1} \oplus ... \oplus (X)_{i_m}$. Similarly, since $(X_L^{i+3})_{10} = (X_R^{i+4})_{10}$ and with the probability of $\frac{3}{4}$ we have $(X_R^{i+3})_9 = (X_R^{i+4})_8 \oplus (X_L^{i+4})_9 \oplus (K^{i+4})_9$, we can add a round to the bottom of the current 4-round approximation and produce a 5-round linear expression, with the bias of $\frac{1}{16}$, as follows:

$$(X_L^{i-1})[8, 10] \oplus (X_R^{i-1})_9 \oplus (X_R^{i+4})[8, 10] \oplus (X_L^{i+4})_9$$
$$= (K^{i-1})_9 \oplus (K^i)_{10} \oplus (K^{i+2})_{10} \oplus (K^{i+4})_9. \tag{7}$$

Following this approach we can extend this linear approximation by adding extra rounds to top and bottom and drive a linear approximation for more rounds of SIMECK. In Tables 2, 3 and 4 sequences of approximation to produce linear characteristics for SIMECK32/64, SIMECK48/96 and SIMECK64/128 are presented. In the last column of each table, number of approximation in each round is presented. Given that for any used approximation in these tables bias is $\frac{1}{4}$, based on the piling-up lemma [17] the bias of a linear characteristic with N approximation would be $2^{N-1} \times (\frac{1}{4})^N = 2^{-(N+1)}$.

It is clear from Table 2 that we can produce a 11-round linear characteristic for SIMECK32/64 with bias 2^{-15} as follows:

$$\begin{pmatrix} (X_R^1)_7 \oplus (X_L^1)[6, 8, 10] \\ \oplus (X_L^{12})_9 \oplus (X_R^{12})[6, 10] \end{pmatrix} = \begin{pmatrix} (K^1)_7 \oplus (K^2)[8, 10] \oplus (K^3)_9 \oplus (K^4)_{10} \\ \oplus (K^6)_{10} \oplus (K^7)_9 \oplus (K^8)[8, 10] \\ \oplus (K^9)_7 \oplus (K^{10})[6, 8, 10] \oplus (K^{11})_9 \end{pmatrix}, \tag{8}$$

Given this 11-round linear characteristic, we can add another round to its top and a round to its bottom to extend the attack up to 13 rounds. The added rounds are related to the plaintext and ciphertext and free of any approximation, because we know the input of F functions for these rounds and key does not affect approximation. In this way we have a 13-round linear characteristic between plaintext and ciphertext of SIMECK32/64 for which the bias is 2^{-15}. Given this linear characteristic, using Matsui's Algorithm 1 with the data complexity of $(2^{-15})^2 = 2^{30}$, an adversary can retrieve 1 bit of the key with the success probability of 0.997 [17, Table 2].

The adversary can use Table 2 to produce a 12-round linear characteristic for SIMECK32/32 with bias of 2^{-17} as follows:

$$\begin{pmatrix} (X_R^1)_7 \oplus (X_L^1)[6, 8, 10] \\ \oplus (X_L^{13})[6, 10] \oplus (X_R^{13})_5 \end{pmatrix} = \begin{pmatrix} (K^1)_7 \oplus (K^2)[8, 10] \oplus (K^3)_9 \oplus (K^4)_{10} \\ \oplus (K^6)_{10} \oplus (K^7)_9 \oplus (K^8)[8, 10] \oplus (K^9)_7 \\ \oplus (K^{10})[6, 8, 10] \oplus (K^{11})_9 \oplus (K^{12})[6, 10] \end{pmatrix}, \tag{9}$$

Given this 12-round linear characteristic, we can add another round to its top and a round to its bottom to extend the attack up to 14 rounds.

Hence, using Matsui's Algorithm 1 with the data complexity of $\frac{1}{4}(2^{-17})^2 = 2^{32}$, the adversary can retrieves 1 bit of the key with the success probability of 0.841 [17, Table 2].

Similarly, it is clear from Table 3 that we can produce a 16-round linear characteristic (Eq. 10) with bias 2^{-24} and a 17-round linear characteristic (Eq. 11) with bias 2^{-25} for SIMECK48/96.

$$\left(\begin{array}{c} (X_R^1)_5 \oplus (X_L^1)[4,6,10] \\ \oplus (X_L^{17})[6,10] \oplus (X_R^{17})_5 \end{array} \right) = \left(\begin{array}{l} (K^1)_5 \oplus (K^2)[6,10] \oplus (K^3)_9 \oplus (K^4)[6,8,10] \\ \oplus (K^5)_7 \oplus (K^6)[8,10] \oplus (K^7)_9 \oplus (K^8)_{10} \\ \oplus (K^{10})_{10} \oplus (K^{11})_9 \oplus (K^{12})[8,10] \oplus (K^{13})_7 \\ \oplus (K^{14})[6,8,10] \oplus (K^{15})_9 \oplus (K^{16})[6,10] \end{array} \right), \tag{10}$$

$$\left(\begin{array}{c} (X_R^1)_5 \oplus (X_L^1)[4,6,10] \\ \oplus (X_L^{18})_5 \oplus (X_R^{18})[4,6,10] \end{array} \right) = \left(\begin{array}{l} (K^1)_5 \oplus (K^2)[6,10] \oplus (K^3)_9 \oplus (K^4)[6,8,10] \\ \oplus (K^5)_7 \oplus (K^6)[8,10] \oplus (K^7)_9 \oplus (K^8)_{10} \\ \oplus (K^{10})_{10} \oplus (K^{11})_9 \oplus (K^{12})[8,10] \oplus (K^{13})_7 \\ \oplus (K^{14})[6,8,10] \oplus (K^{15})_9 \oplus (K^{16})[6,10] \oplus (K^{17})_5 \end{array} \right), \tag{11}$$

Given these linear characteristics, we can add another round to their top and a round to their bottom to extend the attack up to 18 and 19 rounds respectively, free of extra approximation. Hence, using these linear characteristics and Matsui's Algorithm 1 with the data complexity of 2^{48}, the adversary can retrieves 1 bit of the key with the success probability of 0.997 and 0.841 respectively.

Table 4 shows the sequence of approximations to produce a 19-round linear characteristic (Eq. 12) with bias 2^{-30} and a 20-round linear characteristic (Eq. 13) with bias 2^{-33} for SIMECK64/128, which can be extended to attack to 21 and 22 rounds of algorithm respectively. Given those linear characteristics, using Matsui's Algorithm 1, with the data complexity of 2^{60} and 2^{64}, the adversary can retrieve 1 bit of the key with the success probability of 0.997 and 0.841 respectively.

$$\left(\begin{array}{c} (X_R^2)_5 \oplus (X_L^2)[4,6,10] \\ \oplus (X_L^{20})[3,9] \oplus (X_R^{20})[2,6,8,10] \end{array} \right) = \left(\begin{array}{l} (K^2)_5 \oplus (K^3)[6,10] \oplus (K^4)_9 \oplus (K^5)[6,8,10] \\ \oplus (K^6)_7 \oplus (K^7)[8,10] \oplus (K^8)_9 \oplus (K^9)_{10} \\ \oplus (K^{11})_{10} \oplus (K^{12})_9 \oplus (K^{13})[8,10] \oplus (K^{14})_7 \\ \oplus (K^{15})[6,8,10] \oplus (K^{16})_9 \oplus (K^{17})[6,10] \\ \oplus (K^{18})_5 \oplus (K^{19})[4,6,10] \oplus (K^{20})[3,9] \end{array} \right), \tag{12}$$

$$\left(\begin{array}{c} (X_R^1)[4,6,10] \oplus (X_L^1)[3,9] \\ \oplus (X_L^{21})[3,9] \oplus (X_R^{21})[2,6,8,10] \end{array} \right) = \left(\begin{array}{l} (K^1)[4,6,10] \oplus (K^2)_5 \oplus (K^3)[6,10] \oplus (K^4)_9 \oplus \\ (K^5)[6,8,10] \oplus (K^6)_7 \oplus (K^7)[8,10] \oplus (K^8)_9 \\ \oplus (K^9)_{10} \oplus (K^{11})_{10} \oplus (K^{12})_9 \oplus (K^{13})[8,10] \oplus \\ (K^{14})_7 \oplus (K^{15})[6,8,10] \oplus (K^{16})_9 \oplus (K^{17})[6,10] \\ \oplus (K^{18})_5 \oplus (K^{19})[4,6,10] \oplus (K^{20})[3,9] \end{array} \right), \tag{13}$$

Table 2. Sequences of approximation of a 12 round linear characteristic for SIMECK32/64. $\mathcal{A}_L$ and $\mathcal{A}_R$ denote the active bits in the left and right side respectively, Used App. denotes the approximation from Eq. 2 that has been used for the corresponding bit(s) of $\mathcal{A}_R$ and # App denotes that how many approximation has been used in each round.

	$\mathcal{A}_L$	$\mathcal{A}_R$	Used App	# App
1	10, 8, 6	7	1	1
2	9, 9, 7	10, 8	1; 1	2
3	10, 8	9	1	1
4	9	10	1	1
5	10	–	–	0
6	9	10	1	1
7	10,8	9	1	1
8	9, 9, 7	10, 8	1; 1	2
9	10, 8, 6	7	1	1
10	7, 9,5, 7, 5	10, 8, 6	2; 1; 1	3
11	10, 8, 6, 8	9	1	1
12	9, 9, 5	10,6	1; 1	2

Table 3. Sequences of approximation of a 17 round linear characteristic for SIMECK48/96. Notations are similar to the notations used in this table.

	$\mathcal{A}_L$	$\mathcal{A}_R$	Used App	# App
1	10, 6,4	5	1	1
2	9, 9, 5	10,6	1; 1	2
3	10, 8, 6, 8	9	1	1
4	7, 9,5, 7, 5	10, 8, 6	2; 1; 1	3
5	10, 8, 6	7	1	1
6	9, 9, 7	10, 8	1; 1	2
7	10, 8 ·	9	1	1
8	9	10	1	1
9	10	–	–	0
10	9	10	1	1
11	10,8	9	1	1
12	9, 9, 7	10, 8	1; 1	2
13	10, 8, 6	7	1	1
14	7, 9,5, 7, 5	10, 8, 6	2; 1; 1	3
15	10, 8, 6, 8	9	1	1
16	9, 9, 5	10,6	1; 1	2
17	10, 6,4	5	1	1

Table 4. Sequences of approximation of a 20 round linear characteristic for SIMECK64/128. Notations are similar to the notations used in Table 3.

	$\mathcal{A}_L$	$\mathcal{A}_R$	Used App	# App
1	5,9,5,3	10, 6,4	1;1; 1	3
2	10, 6,4	5	1	1
3	9, 9, 5	10,6	1; 1	2
4	10, 8, 6, 8	9	1	1
5	7, 9,5, 7, 5	10, 8, 6	2; 1; 1	3
6	10, 8, 6	7	1	1
7	9, 9, 7	10, 8	1; 1	2
8	10, 8	9	1	1
9	9	10	1	1
10	10	–	–	0
11	9	10	1	1
12	10,8	9	1	1
13	9, 9, 7	10, 8	1; 1	2
14	10, 8, 6	7	1	1
15	7, 9,5, 7, 5	10, 8, 6	2; 1; 1	3
16	10, 8, 6, 8	9	1	1
17	9, 9, 5	10,6	1; 1	2
18	10, 6,4	5	1	1
19	5,9,5,3	10, 6,4	1;1; 1	3
20	10, 6,4,8,4,2	9,3	2,1	2

4   Linear Cryptanalysis of SIMECK Using the Matsui's Algorithm 2

In this section, we use Matsui's Algorithm 2 to recover the key of more rounds of variants of SIMECK. For example, in the case of SIMECK 32/64, given the linear characteristic represented in Eq. 8 with bias 2^{-15}, we guess subkyes of rounds at the beginning and the end of the cipher and determine the correlation of the following linear relation to filter the wrong subkeys:

$$(X_R^i)_7 \oplus (X_L^i)[6, 8, 10] \oplus (X_L^{i+11})_9 \oplus (X_R^{i+11})[6, 10] \qquad (14)$$

With respect to Table 5, we can append a round to the beginning of the cipher to find a new 12-round linear characteristic. Since SIMECK injects the subkey at the end of its round function, then this work does not add any computational complexity. More precisely, for the current 11-round linear characteristic, we evaluate $(X_R^i)_7 \oplus (X_L^i)[6, 8, 10] \oplus (X_L^{i+11})_9 \oplus (X_R^{i+11})[6, 10]$. When we add a round in the backwards direction, i.e. round $i-1$, we can determine $(X_L^i)[6, 8, 10]$ as a function of $F(X_L^{i-1})[6,8,10] \oplus (K^{i-1})[6,8,10] \oplus X_R^{i-1})[6,8,10]$, where we know X_R^{i-1} and X_L^{i-1}. On the other hand, $(X_R^i)_7 = (X_L^{i-1})_7$. Hence, it is possible to use the correlation of the following linear relation to filter the wrong subkeys:

$$(X_L^{i-1})_7 \oplus F(X_L^{i-1})[6,8,10] \oplus X_R^{i-1})[6,8,10] \oplus (X_L^{i+11})_9 \oplus (X_R^{i+11})[6,10].$$

It means that we do not need to know the value of $(K^{i-1})[6,8,10]$ (in Table 5 such bits of key are indicated in red). We can continue our method to add more rounds to the beginning of linear characteristic in the cost of guessing some bits of subkeys. To add more rounds in backward, for example we must guess the bit $(F(X_L^{i-1}))_6 = (X_L^{i-1})_5 \oplus ((X_L^{i-1})_6 \& (X_L^{i-1})_1)$. Given that for any 2-bit AND gate if an input is 0 then the output would be 0, to determine $(F(X_L^{i-1}))_6$ one should guess $(X_L^{i-1})_1$ only if the guessed value for $(X_L^{i-1})_6$ is 1, but it always should guess the value of $(X_L^{i-1})_5$ (this observation originally has been used in [1] to attack SIMON). So, in average we need one bit guess for $(X_L^{i-1})_6$ and $(X_L^{i-1})_1$ (in Table 5 such bits are indicated in blue).

Following this approach, Table 5 shows the bits of subkeys that should be guessed (31 bits of subkey in average) when we add 3 rounds at the top and

Table 5. The keys (in *black*) that should be guessed to attack 18 rounds of SIMECK32/64. The red bits are not required to be guessed and the blue bits cost guessing a half bit on average. Here $i \sim j$ denotes the sequence of numbers $i, i-1, \ldots, j+1, j$, LC is the core linear characteristic, BW is the rounds added at the top and FW is the rounds added at the bottom of the core linear characteristic and AGK denotes average guessed subkey-bits.

		$\mathcal{A}_L$	$\mathcal{A}_R$	active subkeys' bits	AGK.
BW	-2	15~0	14,12,10~0	14,12,10,8,6,3,1,9,7,5,4,2,0	$2^{9.5}$
	-1	14,12,10~0	10~5,3,1	9,7,10,8,5,6,3,1	2^3
	0	10~5, 3,1	10, 8, 6	10, 8, 6	0
LC	1	10, 8, 6	7	-	-
	2	9, 9, 7	10, 8	-	-
	3	10, 8	9	-	-
	4	9	10	-	-
	5	10	–	-	-
	6	9	10	-	-
	7	10,8	9	-	-
	8	9, 9, 7	10, 8	-	-
	9	10, 8, 6	7	-	-
	10	7, 9,5, 7, 5	10, 8, 6	-	-
	11	10, 8, 6, 8	9	-	-
FW	13	10,9,6,5,1	10,6	10,6	0
	14	12,10~8,6~4,1,0	10,9,6,5,1	9,10,6,5,1	2^2
	15	15,12~3,1,0	12,10~8,6~4,1,0	8,12,10,6,1,9,5,4,0	2^6
	16	15,14,12~0	15,12~3,1,0	12,7,15,11~8,6~3,1,0	2^{12}

Table 6. The keys (in *black*) that should be guessed to attack 23 rounds of SIMECK48/96. Notations are similar to the notations used in Table 5.

		$\mathcal{A}_L$	$\mathcal{A}_R$	active subkeys' bits	AGK.
BW	-3	23~12,10~0	23~17,15,13,10~0	7,20,18,15,13,23~21,19,17,10~8,6~0	2^{17}
	-2	23~17,15,13,10~0	23,22,20,18,10~8,6~0	8,2,23,20,18,10,6,1,22,9,5,4,3,0	2^{9}
	-1	23,22,20,18,10~8,6~0	23,10,9,6~3,1	9,3,23,10,6~4,1	2^{3}
	0	23,10,9,6~3,1	10,6,4	10,6,4	0
LC	1	10, 6,4	5	-	-
	2	9, 9, 5	10,6	-	-
	3	10, 8, 6, 8	9	-	-
	4	7, 9,5, 7, 5	10, 8, 6	-	-
	5	10, 8, 6	7	-	-
	6	9, 9, 7	10, 8	-	-
	7	10, 8	9	-	-
	8	9	10	-	-
	9	10	–	-	-
	10	9	10	-	-
	11	10,8	9	-	-
	12	9, 9, 7	10, 8	-	-
	13	10, 8, 6	7	-	-
	14	7, 9,5, 7, 5	10, 8, 6	-	-
	15	10, 8, 6, 8	9	-	-
FW	16	10,9,6,5,1	10,6	10,6	0
	17	20,10~8,6~4,1,0	10,9,6,5,1	9,10,6,5,1	2^{2}
	18	23,20,19,15,10~3,1,0	20,10~8,6~4,1,0	8,20,10,6,1,9,5,4,0	2^{6}
	19	23,22,20~18,15,14,10~0	23,20,19,15,10~3,1,0	7,20,15,23,19,10~8,6~3,1,0	2^{12}

4 rounds at the bottom of the 11-round characteristic of Eq. 8. Hence, we can attack 18 rounds of SIMMECK32/64 using Algorithm 2 of Matsui to recover bits of subkeys. For the data complexity of 2^{31} and the time complexity of $2^{63.5}$ the attack success probability would be 0.477 [19].

Given Eq. 10, as a linear characteristic for SIMECK48/96, is possible to apply the above technique to extend the linear characteristics over more number of rounds. However, the bias of that linear characteristic is 2^{-24}, which means that we can not use it to mount an attack with high success probability [17, 19]. Hence, we use Eq. 15 which covers 15 rounds. Table 6 shows the bits of subkeys that should be guessed (49 bits of subkey in average) when we add 4 rounds at the top and 4 rounds at the bottom of the 15-round characteristic of Eq. 15. Hence, we can attack 23 rounds of SIMECK48/96 using Algorithm 2 of Matsui to recover bits of subkeys. For the data complexity of 2^{45} and the time complexity of 2^{94} the attack success probability would be 0.477 [19].

$$\begin{pmatrix} (X_R^1)_5 \oplus (X_L^1)[4,6,10] \\ \oplus(X_L^{16})_9 \oplus (X_R^{16})[6,10] \end{pmatrix} = \begin{pmatrix} (K^1)_5 \oplus (K^2)[6,10] \oplus (K^3)_9 \oplus (K^4)[6,8,10] \\ \oplus(K^5)_7 \oplus (K^6)[8,10] \oplus (K^7)_9 \oplus (K^8)_{10} \\ \oplus(K^{10})_{10} \oplus (K^{11})_9 \oplus (K^{12})[8,10] \oplus (K^{13})_7 \\ \oplus(K^{14})[6,8,10] \oplus (K^{15})_9 \end{pmatrix},$$
(15)

Similarly, given Eq. 12 with bias 2^{-30}, it is possible to apply this technique to extend the linear characteristics to 27 rounds of SIMECK64/128 (Table 7).

Table 7. The keys (in *black*) that should be guessed to attack 27 rounds of SIMECK64/128. Notations are similar to the notations used in Table 5.

		$\mathcal{A}_L$	$\mathcal{A}_R$	active subkeys' bits	AGK.
BW	-3	31∼20,18,16,10∼0	31∼25,23,21,10∼0	7,28,26,23,21,31∼29,27,25,10∼8,6∼0	2^{17}
	-2	31∼25,23,21,10∼0	31,30,28,26,10∼8,6∼0	8,2,31,28,26,10,6,1,30,9,5,4,3,0	2^{9}
	-1	31,30,28,26,10∼8,6∼0	31,10,9,6∼3,1	9,3,31,10,6∼4,1	2^{3}
	0	31,10,9,6∼3,1	10,6,4	10,6,4	0
LC	1	10, 6,4	5	-	-
	2	9, 9, 5	10,6	-	-
	3	10, 8, 6, 8	9	-	-
	4	7, 9,5, 7, 5	10, 8, 6	-	-
	5	10, 8, 6	7	-	-
	6	9, 9, 7	10, 8	-	-
	7	10, 8	9	-	-
	8	9	10	-	-
	9	10	–	-	-
	10	9	10	-	-
	11	10,8	9	-	-
	12	9, 9, 7	10, 8	-	-
	13	10, 8, 6	7	-	-
	14	7, 9,5, 7, 5	10, 8, 6	-	-
	15	10, 8, 6, 8	9	-	-
	16	9, 9, 5	10,6	-	-
	17	10, 6,4	5	-	-
	18	5,9,5,3	10, 6,4	-	-
	19	10, 6,4,8,4,2	9,3	-	-
FW	20	29,10∼5,3∼1	10,8,6,2	10,8,6,2	0
	21	30∼28,24,10∼0	29,10∼5,3∼1	9,7,3,29,10,8,6,5,2,1	$2^{3.5}$
	22	31∼27,24,23,19,10∼0	30∼28,24,10∼0	8,6,30,29,24,10,3,2,28,9,7,5,4,1,0	2^{10}
	23	31∼22,19,18,14,10∼0	31∼27,24,23,19,10∼0	30,24,19,7,31,29∼27,23,10∼8,6∼0	2^{17}

To attack 27 rounds of SIMECK64/128, the data complexity is 2^{61}, the time complexity is $2^{120.5}$ and the attack success probability would be 0.477 [19].

5   Conclusion and Open Problems

In this paper, we analyzed the security of SIMECK family against linear cryptanalysis techniques. Our results show that each variant of SIMON provides better security against linear cryptanalysis compared to equivalent SIMECK variant. More precisely, the best known attack on SIMON32/64, SIMON48/96 and SIMON64/128 using Mastui's Algorithm 1 covers 13, 16 and 19 rounds respectively while our result on SIMECK32/64, SIMECK48/96, SIMECK64/128 covers 14, 19 and 22 rounds. Moreover, the best known attack on SIMON32/64, SIMON48/96 and SIMON64/128 using Mastui's Algorithm 2 covers 18, 19 and 21 rounds respectively while our result on SIMECK32/64, SIMECK48/96, SIMECK64/128 covers 18, 23 and 27 rounds. Hence, in the perspective of linear cryptanalysis, SIMON provides better security margin compared to SIMECK.

On the other hand, from the point of number of rounds attacked, linear hull [18] shows to be a more promising approach to analyze the security of

SIMON [1,11,20] compared to other attacks. Hence, as a future work, we aim to investigate the security of SIMECK variants against this attack.

References

1. Abdelraheem, M.A., Alizadeh, J., AlKhzaimi, H., Aref, M.R., Bagheri, N., Gauravaram, P., Lauridsen, M.M.: Improved linear cryptanalysis of round reduced SIMON. IACR Cryptology ePrint Archive 2014/681 (2014)
2. Abed, F., List, E., Lucks, S., Wenzel, J.: Differential Cryptanalysis of Reduced-Round Simon. Cryptology ePrint Archive, Report 2013/526 (2013). http://eprint.iacr.org/
3. Abed, F., List, E., Lucks, S., Wenzel, J.: Differential cryptanalysis of round-reduced SIMON and SPECK. In: Cid, C., Rechberger, C. (eds.) FSE 2014. LNCS, vol. 8540, pp. 525–545. Springer, Heidelberg (2015)
4. Alizadeh, J., Alkhzaimi, H.A., Aref, M.R., Bagheri, N., Gauravaram, P., Kumar, A., Lauridsen, M.M., Sanadhya, S.K.: Cryptanalysis of SIMON variants with connections. In: Sadeghi, A.-R., Saxena, N. (eds.) RFIDSec 2014. LNCS, vol. 8651, pp. 90–107. Springer, Heidelberg (2014)
5. Alizadeh, J., Bagheri, N., Gauravaram, P., Kumar, A., Sanadhya, S.K.: Linear Cryptanalysis of Round Reduced SIMON. Cryptology ePrint Archive, Report 2013/663 (2013). http://eprint.iacr.org/
6. AlKhzaimi, H., Lauridsen, M.M.: Cryptanalysis of the SIMON Family of Block Ciphers. IACR Cryptology ePrint Archive 2013/543 (2013)
7. Ashur, T.: Improved linear trails for the block cipher simon. IACR Cryptology ePrint Archive 2015/285 (2015)
8. Beaulieu, R., Shors, D., Smith, J., Treatman-Clark, S., Weeks, B., Wingers, L.: The SIMON and SPECK Families of Lightweight Block Ciphers. Cryptology ePrint Archive, Report 2013/404, 2013. http://eprint.iacr.org/2013/404
9. Biryukov, A., Roy, A., Velichkov, V.: Differential analysis of block ciphers SIMON and SPECK. In: Cid, C., Rechberger, C. (eds.) FSE 2014. LNCS, vol. 8540, pp. 546–570. Springer, Heidelberg (2015)
10. Boura, C., Naya-Plasencia, M., Suder, V.: Scrutinizing and improving impossible differential attacks: applications to CLEFIA, Camellia, LBlock and SIMON. In: Sarkar, P., Iwata, T. (eds.) ASIACRYPT 2014. LNCS, vol. 8873, pp. 179–199. Springer, Heidelberg (2014)
11. Chen, H., Wang, X.: Improved Linear Hull Attack on Round-Reduced Simon with Dynamic Key-guessing Techniques (2015)
12. Chen, Z., Wang, N., Wang, X.: Impossible differential cryptanalysis of reduced round SIMON. IACR Cryptology ePrint Archive 2015/286 (2015)
13. Cho, J.Y., Hermelin, M., Nyberg, K.: A new technique for multidimensional linear cryptanalysis with applications on reduced round serpent. In: Lee, P.J., Cheon, J.H. (eds.) ICISC 2008. LNCS, vol. 5461, pp. 383–398. Springer, Heidelberg (2009)
14. Courtois, N., Mourouzis, T., Song, G., Sepehrdad, P., Susil, P.: Combined algebraic and truncated differential cryptanalysis on reduced-round simon. In: Obaidat, M.S., Holzinger, A., Samarati, P. (eds.) SECRYPT 2014, pp. 399–404. SciTePress (2014)
15. Dinur, I.: Improved differential cryptanalysis of round-reduced Speck. In: Joux, A., Youssef, A. (eds.) SAC 2014. LNCS, vol. 8781, pp. 147–164. Springer, Heidelberg (2014)

16. Nakahara, Jr., J., Preneel, B., Vandewalle, J.: Linear cryptanalysis of reduced-round versions of the SAFER block cipher family. In: Goos, G., Hartmanis, J., van Leeuwen, J., Schneier, B. (eds.) FSE 2000. LNCS, vol. 1978, p. 244–261. Springer, Heidelberg (2001)

17. Matsui, M.: Linear cryptanalysis method for DES cipher. In: Helleseth, T. (ed.) EUROCRYPT 1993. LNCS, vol. 765, pp. 386–397. Springer, Heidelberg (1994)

18. Nyberg, K.: Linear approximation of block ciphers. In: De Santis, A. (ed.) EUROCRYPT 1994. LNCS, vol. 950, pp. 439–444. Springer, Heidelberg (1995)

19. Selçuk, A.A.: On probability of success in linear and differential cryptanalysis. J. Cryptology **21**(1), 131–147 (2008)

20. Shi, D., Hu, L., Sun, S., Song, L., Qiao, K., Ma, X.: Improved Linear (hull) Cryptanalysis of Round-reduced Versions of SIMON. IACR Cryptology ePrint Archive 2014/973 (2014)

21. Sun, S., Hu, L., Wang, M., Wang, P., Qiao, K., Ma, X., Shi, D., Song, L., Fu, K.: Towards Finding the Best Characteristics of Some Bit-oriented Block Ciphers and Automatic Enumeration of (Related-key) Differential and Linear Characteristics with Predefined Properties. IACR Cryptology ePrint Archive 2014/747 (2014)

22. Sun, S., Hu, L., Wang, P., Qiao, K., Ma, X., Song, L.: Automatic security evaluation and (related-key) differential characteristic search: application to SIMON, PRESENT, LBlock, DES(L) and other bit-oriented block ciphers. In: Sarkar, P., Iwata, T. (eds.) ASIACRYPT 2014. LNCS, vol. 8873, pp. 158–178. Springer, Heidelberg (2014)

23. Tardy-Corfdir, A., Gilbert, H.: A known plaintext attack of FEAL-4 and FEAL-6. In: Feigenbaum, J. (ed.) CRYPTO 1991. LNCS, vol. 576, pp. 172–182. Springer, Heidelberg (1992)

24. Wang, N., Wang, X., Jia, K., Zhao, J.: Improved Differential Attacks on Reduced SIMON Versions. IACR Cryptology ePrint Archive 2014/448 (2014)

25. Wang, Q., Liu, Z., Varici, K., Sasaki, Y., Rijmen, V., Todo, Y.: Cryptanalysis of reduced-round SIMON32 and SIMON48. In: Proceedings of Progress in Cryptology - INDOCRYPT 2014–15th International Conference on Cryptology in India, New Delhi, India, 14–17 December 2014, pp. 143–160 (2014)

26. Yang, G., Zhu, B., Suder, V., Aagaard, M.D., Gong, G.: The Simeck family of lightweight block ciphers. In: Güneysu, T., Handschuh, H. (eds.) CHES 2015. LNCS, vol. 9293, pp. 307–329. Springer, Heidelberg (2015)

Improved Linear Cryptanalysis of Reduced-Round SIMON-32 and SIMON-48

Mohamed Ahmed Abdelraheem[1]([✉]), Javad Alizadeh[2], Hoda A. Alkhzaimi[3], Mohammad Reza Aref[2], Nasour Bagheri[4,5], and Praveen Gauravaram[6]

[1] SICS Swedish ICT, Kista, Sweden
mohamed.abdelraheem@sics.se
[2] ISSL, Department of Electrical Engineering,
Sharif University of Technology, Tehran, Iran
alizadja@gmail.com
[3] Section for Cryptology, DTU Compute,
Technical University of Denmark, Lyngby, Denmark
hoalk@dtu.dk
[4] Department of Electrical Engineering,
Shahid Rajaee Teachers Training University, Tehran, Iran
NBagheri@srttu.edu
[5] School of Computer Science,
Institute for Research in Fundamental Sciences (IPM), Tehran, Iran
[6] Queensland University of Technology, Brisbane, Australia
praveen.gauravaram@qut.edu.au

Abstract. In this paper we analyse two variants of SIMON family of light-weight block ciphers against variants of linear cryptanalysis and present the best linear cryptanalytic results on these variants of reduced-round SIMON to date.

We propose a time-memory trade-off method that finds differential/linear trails for any permutation allowing low Hamming weight differential/linear trails. Our method combines low Hamming weight trails found by the correlation matrix representing the target permutation with heavy Hamming weight trails found using a Mixed Integer Programming model representing the target differential/linear trail. Our method enables us to find a 17-round linear approximation for SIMON-48 which is the best current linear approximation for SIMON-48. Using only the correlation matrix method, we are able to find a 14-round linear approximation for SIMON-32 which is also the current best linear approximation for SIMON-32.

The presented linear approximations allow us to mount a 23-round key recovery attack on SIMON-32 and a 24-round Key recovery attack on SIMON-48/96 which are the current best results on SIMON-32 and

---

This work was done while the author was a postdoc at the Technical University of Denmark

Javad Alizadeh, Mohammad Reza Aref and Nasour Bagheri were partially supported by Iran-NSF under grant no. 92.32575.

Praveen Gauravaram is supported by Australian Research Council Discovery Project grant number DP130104304.

© Springer International Publishing Switzerland 2015
A. Biryukov and V. Goyal (Eds.): INDOCRYPT 2015, LNCS 9462, pp. 153–179, 2015.
DOI: 10.1007/978-3-319-26617-6_9

SIMON-48. In addition we have an attack on 24 rounds of SIMON-32 with marginal complexity.

Keywords: SIMON · Linear cryptanalysis · Linear hull · Correlation matrix · Mixed Integer Programming (MIP)

1   Introduction

Over the past few years, the necessity for limited cryptographic capabilities in resource-constraint computing devices such as RFID tags has led to the design of several lightweight cryptosystems [8,12,13,15,17–19,30]. In this direction, Beaulieu *et al.* of the U.S. National Security Agency (NSA) designed SIMON family of lightweight block ciphers that are targeted towards optimal hardware performance [9]. Meeting hardware requirements of low-power and limited gate devices is the main design criteria of SIMON.

SIMON has plaintext block sizes of 32, 48, 64, 96 and 128 bits, each with up to three key sizes. SIMON-N/K denotes a variant of SIMON with block and key sizes of N and K bits respectively. With the proposed block and key lengths, SIMON is a family of ten lightweight block ciphers. Since the publication of SIMON, each cipher in this family has undergone reduced round cryptanalysis against linear [2–6,24], differential [3,4,11,28], impossible differential [14], rectangular [3,4] and integral [29] attacks.

Contributions. In this paper, we analyse the security of SIMON-32 and SIMON-48. First we analyze the security of reduced-round SIMON-32 and SIMON-48 against several variants of linear cryptanalysis and report the best results to date with respect to any form of cryptanalysis in terms of the number of rounds attacked on SIMON-32/64 and 48/96. Our attacks are described below and results are summarised in Table 1.

– We propose a time-memory trade-off method that combines low Hamming weight trails found by the correlation matrix (consumes huge memory) with heavy Hamming weight trails found by the Mixed Integer Programming (MIP) method [26] (consumes time depending on the specified number of trails to be found). The method enables us to find a 17-round linear approximation for SIMON-48 which is the best current approximation.
– We found a 14-round linear hull approximation for SIMON-32 using a squared correlation matrix with input/output masks of Hamming weight ≤ 9.
– Using our approximations, we are able to break 23 and 24 rounds of SIMON-32, 23 rounds of SIMON-48/72 and 24 rounds of SIMON-48/96 with a marginal time complexity $2^{63.9}$.

Previous Results on SIMON Used in Our Paper. The work in [20] provides an explicit formula for computing the probability of a 1-round differential characteristic of the SIMON's non-linear function. It also provides an efficient algorithm for computing the squared correlation of a 1-round linear characteristic of the SIMON nonlinear function which we used in our linear cryptanalysis to SIMON-48.

The work in [24] defines a MIP linear model that finds linear trails for SIMON. The solution of the MIP model sometimes yield a false linear trail but most of the time it yields a valid linear trail. When a solution is found whether valid or invalid, we add a new constraint to the MIP model that prevents the current solution from occurring in the next iteration.

Related Work on SIMON. The most improved results in terms of the number of rounds attacked, data and time complexity presented, up-to-date of this publication, are in the scope of differential, linear and integral attacks as reflected in Table 1. Focusing on the different cryptanalysis results of SIMON-32, SIMON-48/72 and SIMON-48/96, Abed *et al.* [3,4] have presented that classical differential results yield attacks on 18 for the smallest variant and 19 rounds for SIMON-48 with data and time stated in Table 1. This was improved to 21 rounds for SIMON-32 and $22 - 24$ rounds for SIMON-48/72 and SIMON-48/96 by Wang *et al.* [27,28] using dynamic key guessing and automatic enumeration of differential characteristics through imposing conditions on differential paths to reduce the intended key space searched.

Independent to our work, Ashur [7] described a method for finding linear trails that work only against SIMON-like ciphers. This method finds a multivariate polynomial in GF(2) representing the r-round linear approximation under consideration. Each solution of the multivariate polynomial corresponds to a valid trail that is part of the many linear trails that forms the linear approximation. This suggests that the probability that the r-round linear approximation is satisfied is equivalent to the number of solutions for its corresponding multivariate polynomial divided by the size of the solution space. For $r = 2$, the authors mentioned that the space size is 2^{10}. For higher rounds the space gets bigger as many bits will be involved in the corresponding multivariate polynomial. Finding the number of solutions of a multivariate polynomial is a hard problem. To overcome this, the author uses the above method to form what is called a "linear super-trail" which glues two short linear hulls (a short linear hull has a small number of rounds that make it is feasible to find the number of solutions of the corresponding multivariate polynomial) in order to form a super-trail.

In contrast, our time-memory trade-off method which basically combines two different linear trails found using a squared correlation matrix (trails with light Hamming weight) and a mixed integer programming model (trails with heavy Hamming weight) is not SIMON specific, it is very generic and can be used for any permutation allowing low Hamming weight linear/differential trails to find linear/differential trails. As described in Sect. 5.3, we have better attacks on both SIMON-32 (using squared correlation matrix) and SIMON-48 (using time-memory trade-off) compared to the results of [7].

Organization. The paper is structured as follows. In Sect. 2 we describe SIMON. In Sect. 3 concepts and notation required for linear cryptanalysis of SIMON are presented. In Sect. 4 the used Time-Memory Trade-off method is described. In Sect. 5 we used squared correlation matrix to establish a linear hull of SIMON and investigate the data and time complexity for the smallest variant of SIMON. We conclude the paper in Sect. 6.

Table 1. State-of-the-art cryptanalysis of SIMON-(32/64, 48/72, 48/96).

SIMON		Diff.					Imp.Diff.	Z-Corr.	Integ.	Multi.Lin.		Lin.				Lin. Hull	
		[4]	[11]	[28]	[27]	[25]	[14]	[29]	[29]	[7]	[3]	[5]	[2]	[24]	[25]	This work
32/64	#rounds	18	19	21	21	--	19	20	21	24	11	13	17	21	--	**23**
	Time	$2^{46.0}$	$2^{32.0}$	$2^{46.0}$	$2^{55.25}$	--	$2^{62.56}$	$2^{56.96}$	$2^{63.0}$	$2^{63.57}$	--	--	$2^{52.5}$	--	--	$\mathbf{2^{50}}$
	Data	$2^{31.2}$	$2^{31.0}$	$2^{31.0}$	$2^{31.0}$	--	$2^{32.0}$	$2^{32.0}$	$2^{31.0}$	$2^{31.57}$	$2^{23.0}$	$2^{32.0}$	$2^{32.0}$	$2^{30.19}$	--	$\mathbf{2^{30.59}}$
48/72	#rounds	19	20	22	23	16	20	20	--	23	14	16	19	--	--	**23**
	Time	$2^{52.0}$	$2^{52.0}$	$2^{63.0}$	$2^{63.25}$	--	$2^{70.69}$	$2^{59.7}$	--	$2^{68.4}$	--	--	2^{70}	--	--	$\mathbf{2^{62.10}}$
	Data	$2^{46.0}$	$2^{46.0}$	$2^{46.0}$	2^{47}	$2^{44.65}$	2^{48}	2^{48}	--	$2^{44.4}$	$2^{47.0}$	$2^{46.0}$	$2^{46.0}$	--	--	$\mathbf{2^{47.78}}$
48/96	#rounds	19	20	22	24	16	21	21	--	24	14	16	20	21	23	**24**
	Time	$2^{76.0}$	$2^{75.0}$	$2^{71.0}$	$2^{87.25}$	--	$2^{94.73}$	$2^{72.63}$	--	$2^{92.4}$	--	--	$2^{86.5}$	--	--	$\mathbf{2^{83.10}}$
	Data	$2^{46.0}$	$2^{46.0}$	$2^{46.0}$	2^{47}	$2^{44.65}$	$2^{38.0}$	$2^{48.0}$	--	$2^{44.4}$	$2^{47.0}$	$2^{46.0}$	$2^{46.0}$	$2^{42.28}$	$2^{44.92}$	$\mathbf{2^{47.78}}$

2   Description of SIMON

SIMON has a classical Feistel structure with the round block size of $N = 2n$ bits where n is the word size representing the left or right branch of the Feistel scheme at each round. The number of rounds is denoted by r and depends on the variant of SIMON.

We denote the right and left halves of plaintext P and ciphertext C by (P_R, P_L) and (C_R, C_L) respectively. The output of round r is denoted by $X^r = X_L^r \| X_R^r$ and the subkey used in a round r is denoted by K^r. Given a string X, $(X)_i$ denotes the ith bit of X. Bitwise circular left-rotation of string a by b positions to the left is denoted by $a \lll b$. Further, $\oplus$ and &denote bitwise XOR and AND operations respectively.

Each round of SIMON applies a non-linear, non-bijective (and hence non-invertible) function $F : \mathbb{F}_2^n \to \mathbb{F}_2^n$ to the left half of the state. The output of F is added using XOR to the right half along with a round key followed by swapping of two halves. The function F is defined as

$$F(x) = ((x \lll 8)\&(x \lll 1)) \oplus (x \lll 2)$$

The subkeys are derived from a master key. Depending on the size K of the master key, the key schedule of SIMON operates on two, three or four n-bit word registers. We refer to [9] for the detailed description of SIMON structure and key scheduling.

3   Preliminaries

Correlation Matrix. Linear cryptanalysis finds a linear relation between some plaintext bits, ciphertext bits and some secret key bits and then exploits the bias or correlation of this linear relation. In other words, the adversary finds an input mask α and an output mask β which yields a higher absolute *bias* $\epsilon_F(\alpha, \beta) \in [-\frac{1}{2}, \frac{1}{2}]$. In other words

$$Pr[\langle \alpha, X \rangle + \langle \beta, F_K(X) \rangle = \langle \gamma, K \rangle] = \frac{1}{2} + \epsilon_F(\alpha, \beta)$$

deviates from $\frac{1}{2}$ where $\langle \cdot, \cdot \rangle$ denotes an inner product. Let $a = (a_1, \ldots, a_n), b = (b_1, \ldots, b_n) \in \mathbb{F}_2^n$. Then

$$a \cdot b \triangleq a_1 b_1 \oplus \cdots \oplus a_n b_n$$

denotes the *inner product* of a and b. The correlation of a linear approximation is defined as

$$C_F(\alpha, \beta) := 2\epsilon_F(\alpha, \beta)$$

Another definition of the correlation which we will use later is

$$C_F(\alpha, \beta) := \hat{F}(\alpha, \beta)/2^n$$

where n is the block size of F in bits and $\hat{F}(\alpha, \beta)$ is the Walsh transform of F which is defined as follows

$$\hat{F}(\alpha, \beta) := \sum_{x \in \{0,1\}^n} (-1)^{\beta \cdot F(x) \oplus \alpha \cdot x}$$

For a given output mask β, the Fast Walsh Transform algorithm computes the Walsh transforms of an n-bit block size function F for all possible input masks α with output mask β using $n2^n$ arithmetic operations.

In order to find good linear approximations, one can construct a correlation matrix (or a squared correlation matrix). In the following, we explain what is a correlation matrix and show how the average squared correlation over all keys is estimated.

Given a composite function $F : \mathbb{F}_2^n \rightarrow \mathbb{F}_2^n$ such that $F = F_r \circ \cdots \circ F_2 \circ F_1,$, we estimate the correlation of an r-round linear approximation (α_0, α_r) by considering the correlation of each linear characteristic between α_0 and α_r. The correlation of ith linear characteristic $(\alpha_0 = \alpha_{0i}, \alpha_{1i}, \cdots, \alpha_{(r-1)i}, \alpha_r = \alpha_{ri})$ is

$$C_i = \prod_{j=1}^{r} C_{F_j}(\alpha_{(j-1)i}, \alpha_{ji})$$

It is well known [16] that the correlation of a linear approximation is the sum of all correlations of linear trails starting with the same input mask α and ending with the same output mask β, i.e. $C_F(\alpha_0, \alpha_r) = \sum_{i=1}^{N_l} C_i$ where N_l is the number of all possible linear characteristics between (α_0, α_r).

When considering the round keys which affects the sign of the correlation of a linear trail, the correlation of the linear hull (α, β) is

$$C_F(\alpha, \beta) = \sum_{i=1}^{N_l} (-1)^{d_i} C_i,$$

where $d_i \in \mathbb{F}_2$ refers to the sign of the addition of the subkey bits on the ith linear trail. In order to estimate the data complexity of a linear attack, one uses the average squared correlation over all the keys which is equivalent to the sum of the squares of the correlations of all trails, $\sum_i C_i^2$, assuming independent round keys [16].

Let C denotes the correlation matrix of an n-bit key-alternating cipher. C has size $2^n \times 2^n$ and $C_{i,j}$ corresponds to the correlation of an input mask, say α_i, and output mask, say β_j. Now the correlation matrix for the keyed round function is obtained by changing the signs of each row in C according to the round subkey bits or the round constant bits involved. Squaring each entry of the correlation matrix gives us the squared correlation matrix M. Computing M^r gives us the squared correlations after r number of rounds. This can not be used for real block ciphers that have block sizes of at least 32 bits as in the case of SIMON-32/64. Therefore, in order to find linear approximations one can construct a submatrix of the correlation (or the squared correlation) matrix [1,12]. In Sect. 5, we construct a squared correlation submatrix for SIMON in order to find good linear approximations.

3.1   Mixed Integer Programming Method (MIP)

Mouha *et al.*'s [21] presented a mixed integer programming model that minimizes the number of active Sboxes involved in a linear or differential trail. Their work was mainly on byte oriented ciphers. Later, Mouha's framework was extended to accommodate bit oriented ciphers. More recently, at Asiacrypt 2014 [26], the authors described a method for constructing a model that finds the actual linear/differential trail with the specified number of active Sboxes. Of course, there would be many solutions but whenever a solution is found the MIP model is updated by adding a new constraint that discards the current found solution from occurring in the next iteration for finding another solution.

For every input/ouput bit mask or bit difference at some round state, a new binary variable x_i is introduced such that $x_i = 1$ iff the corresponding bit mask or bit difference is non-zero. For every Sbox at each round, a new binary variable a_j is introduced such that $a_j = 1$ if the input mask or difference of the corresponding Sbox is nonzero. Thus, a_j indicates the activity of an Sbox. Now, the natural choice of the *objective function* f of our MIP model is to minimize the number of active Sboxes, i.e., $f = \sum_j a_j$. If our goal from the above integer programming model is to only find the minimum number of active Sboxes existing in a differential/linear trial of a given bit-oriented cipher, then we are only concerned about the binary values which represent the activity of the Sboxes involved in the differential/linear trail a_v. Thus, in order to speed up solving the model, one might consider restricting the activity variables and the dummy variables to be binary and allow the other variables to be any real numbers. This will turn the integer programming model into a Mixed Integer Programming model which is easier to solve than an Integer programming model. However, since we want to find the differential/linear trails which means finding the exact values of all the bit-level inputs and outputs, then all these state variables must be binary which give us an integer programming model rather than a mixed integer programming model.

In order to find the differential/linear trails of a given input/output differential/linear approximation, we set the corresponding binary variables for each input/output to 1 if it is an active bit in the input/output and to 0 otherwise. In this paper, we follow the MIP model for linear cryptanalysis presented in [24] (minimize the number of variables appearing in quadratic terms of the linear approximation of SIMON's non-linear function) and use the algorithm presented in [20] for computing the squared correlation for the SIMON nonlinear function.

In Sect. 4, we propose a hybrid method that combines the matrix method and the MIP method to amplify the differential probability or the squared correlation of a specified input and output differences or masks. Using this method we are able to find a 17-round linear approximation for SIMON-48.

4   Time-Memory Trade-Off Method

Since the matrix method consumes huge memory and the MIP method takes time to enumerate a certain number of trails. It seems reasonable to trade-off the time and memory by combining both methods to get better differential/correlation estimations. Here we combine the correlation matrix method with the recent technique for finding differentials and linear hulls in order to obtain a better estimation for the correlations or differentials of a linear and differential approximations respectively.

The idea is to find good disjoint approximations through the matrix and the mixed integer programming model. Assume that our target is an r-round linear hull (α, β), where α is the input mask and β is the output mask. The matrix method is used to find the resulting correlation from trails that have Hamming weight at most m for each round, from now on we will call them "light trails". The MIP method is used to find the resulting correlation from trails that have Hamming weight at least $m + 1$ at one of their rounds, from now on we will call them "heavy trails".

Now if the target number of rounds is high, then the MIP method might not be effective in finding good estimation for the heavy trails as it will take time to collect all those trails. Therefore, in order to overcome this, we split the cipher into two parts, the first part contains the first r_1 rounds and the second part contains the remaining $r_2 = r - r_1$ rounds. Assume $r_1 > r_2$, where r_2 is selected in such a way that the MIP solution is reachable within a reasonable computation time. Now, we show how to find two disjoint classes that contains heavy trails. The first class contains an r_1-round linear hull (α, γ_i) consisting of light trails found through the matrix method at the first r_1 rounds glued together with an r_2-round linear hulls (γ_i, β) consisting of heavy trails found through the MIP method. We call this class, the lower-round class. The second class basically reverse the previous process, by having an r_1-round linear hull of heavy weight trails found through MIP method glued with an r_2-round linear hull containing light trails found through the matrix method. We call this class the upper-round class. Now, adding the estimations from these two classes (upper-round and lower-round classes) gives us the estimation of the correlation of the heavy trails which will be added to the r-round linear hull of the light trails found through the matrix method. We can also include a middle-round class surrounded by upper lightweight trails and lower lightweight trails found by the matrix method.

Next we describe how to find the heavy trails using MIP with the Big M constraints which is a well known technique in optimization.

4.1　Big M Constraints

Suppose that only one of the following two constraints is to be active in a given MIP model.

$$\text{either} \quad \sum_{i,j} f_i X_{ij} \geq c_1 \tag{1}$$

$$\text{or} \quad \sum_{i,k} g_i X_{ik} \geq c_2 \tag{2}$$

The above situation can be formalized by adding a binary variable y as follows:

$$\sum_{i,j} f_i X_{ij} + My \qquad\qquad \geq c_1 \tag{3}$$

$$\sum_{i,k} g_i X_{ik} + M(1-y) \qquad\qquad \geq c_2 \tag{4}$$

where M is a big positive integer and the value of y indicates which constraint is active. So y can be seen as an indicator variable. One can see that when $y = 0$, the first constraint is active while the second constraint is inactive due to the positive big value of M. Conversely, when $y = 1$, the second constraint is active.

The above formulation can be generalized to the case where we have q constraints under the condition that only p out of q constraints are active. The generalization can be represented as follows:

$$\sum_{i,j} f_i X_{ij} + My_1 \qquad\qquad \geq c_1$$

$$\sum_{i,k} g_i X_{ik} + My_2 \qquad\qquad \geq c_2$$

$$\vdots$$

$$\sum_{i,l} h_i X_{il} + My_q \qquad\qquad \geq c_q$$

$$\sum_{i=1}^{l} y_i = q - p$$

where y_i is binary for all i. Sometimes, we might be interested on the condition where at least p out of the q constraints are active. This can be achieved by simply changing the last equation in the constraints above, $\sum_{i=1}^{l} y_i = q - p$ to $\sum_{i=1}^{l} y_i \leq q - p$. This turns out to be useful in our Hybrid method as it will allow us to find r-round trails which have a heavy Hamming weight on at least one of the r rounds.

5 Linear Hull Effect in SIMON-32 and SIMON-48

In this section we will investigate the linear hull effect on SIMON using the correlation matrix method to compute the average squared correlation.

5.1 Correlation of the SIMON F Function

This section provides an analysis on some linear properties of the SIMON F function regarding the squared correlation. This will assist in providing an intuition around the design rationale when it comes to linear properties of SIMON round Function F. A general linear analysis was applied on the F function of SIMON, with regards to limits around the squared correlations for all possible Hamming weights on input masks α and output masks β, for SIMON-32/64.

5.2 Constructing Correlation Submatrix for SIMON

To construct a correlation submatrix for SIMON, we make use of the following proposition.

Proposition 1. *Correlation of a one-round linear approximation [10]. Let $\alpha = (\alpha_L, \alpha_R)$ and $\beta = (\beta_L, \beta_R)$ be the input and output masks of a one-round linear approximation of SIMON. Let α_F and β_F be the input and output masks of the SIMON F function. Then the correlation of the linear approximation (α, β) is $C(\alpha, \beta) = C_F(\alpha_F, \beta_F)$ where $\alpha_F = \alpha_L \oplus \beta_R$ and $\beta_F = \beta_L = \alpha_R$.*

As our goal is to perform a linear attack on SIMON, we construct a squared correlation matrix in order to compute the average squared correlation (the sum of the squares of the correlations of all trails) in order to estimate the required data complexity. Algorithm 1 constructs a squared correlation submatrix whose input and output masks have Hamming weight less than a certain Hamming weight m, where the correlation matrix is deduced from the algorithm proposed in [20].

The size of the submatrix is $\sum_{i=0}^{m} \binom{2n}{i} \times \sum_{i=0}^{m} \binom{2n}{i}$ where n is the block size of SIMON's F function. One can see that the time complexity is in the order of $2^n \sum_{i=0}^{m} \binom{2n}{i}$ arithmetic operations. The submatrix size is large when $m > 5$, but most of its elements are zero and therefore it can easily fit in memory using a sparse matrix storage format. The table below shows the number of nonzero elements of the squared correlation submatrices of SIMON-32/K when $1 \leq m \leq 9$. These matrices are very sparse. For instance, based on our experimental results when $m \leq 8$, the density of the correlation matrix is very low, namely $\frac{133253381}{15033173 \times 15033173} \approx 2^{-20.7}$.

Algorithm 1. Construction of SIMON's Correlation Submatrix

Require: Hamming weight m, bit size of SIMON's F function n and a *map* function.
Ensure: Squared Correlation Submatrix M
1: **for** all output masks β with Hamming weight $\leq m$ **do**
2:     Extract from β the left/right output masks β_L and β_R.
3:     $\alpha_R \leftarrow \beta_L$.
4:     Compute $C(\alpha_F, \beta_L)$ to SIMON's F function for all possible α_F using the algorithm proposed in [20].
5:     **for** all input masks α_F to SIMON's F function **do**
6:         $c \leftarrow C(\alpha_F, \beta_L)$.
7:         $\alpha_L \leftarrow \alpha_F \oplus \beta_R$.
8:         $\alpha = \alpha_L \| \alpha_R$.
9:         **if** $c \neq 0$ **and** Hamming weight of $\alpha \leq m$ **then**
10:             $i \leftarrow map(\alpha)$. {map α to a row index i in the matrix M}
11:             $j \leftarrow map(\beta)$. {map α to a column index j in the matrix M}
12:             $M(i, j) = c \times c$.
13:         **end if**
14:     **end for**
15: **end for**

5.3 Improved Linear Approximations

One can see that Algorithm 1 is highly parallelizable. This means the dominating factor is the memory complexity instead of time complexity. We constructed a sparse squared correlation matrix of SIMON-32/K with input and output masks that have Hamming weight ≤ 8. Using this matrix, we find a 14-round linear approximations with an average squared correlation $\leq 2^{-32}$ for SIMON-32/K. We also get better estimations for the previously found linear approximations which were estimated before using only a single linear characteristic rather than considering many linear characteristics with the same input and output masks. For example, in [4], the squared correlation of the 9-round single linear characteristic with input mask 0x01110004 and output mask 0x00040111 is 2^{-20}. Using our matrix, we find that this same approximation has a squared correlation $\approx 2^{-18.4}$ with $11455 \approx 2^{13.5}$ trails, which gives us an improvement by a factor of $2^{1.5}$. Note that this approximation can be found using a smaller correlation matrix of Hamming weight ≤ 4 and we get an estimated squared correlation equal to $2^{-18.83}$ and only 9 trails. Therefore, the large number of other trails that cover Hamming weights ≥ 5 is insignificant as they only cause a factor of $2^{0.5}$ improvement.

Also, the 10-round linear characteristic in [6] with input mask 0x01014404 and output mask 0x10004404 has squared correlation 2^{-26}. Using our correlation matrix, we find that this same approximation has an estimated squared correlation $2^{-23.2}$ and the number of trails is $588173 \approx 2^{19.2}$. This gives an improvement by a factor of 2^3. Note also that this approximation can be found using a smaller correlation matrix with Hamming weight ≤ 5 and we get an estimated squared correlation equal to $2^{-23.66}$ and only 83 trails. So the large number of other trails resulting covering Hamming weights ≥ 5 is insignificant as they only cause a factor of $2^{0.4}$ improvement. Both of these approximations give us squared correlations less than 2^{-32} when considering more than 12 rounds.

In the following, we describe our 14-round linear hulls found using a squared correlation matrix with Hamming weight ≤ 8.

Improved 14-round Linear Hulls on SIMON-32 (Squared Correlation Matrix Only). Consider a squared correlation matrix M whose input and output masks have Hamming weight m. When $m \geq 6$, raising the matrix to the rth power, in order to estimate the average squared correlation, will not work as the resulting matrix will not be sparse even when r is small. For example, we are able only to compute M^6 where M is a squared correlation matrix whose masks have Hamming weight ≤ 6. Therefore, we use matrix-vector multiplication or row-vector matrix multiplications in order to estimate the squared correlations for any number of rounds r.

It is obvious that input and output masks with low Hamming weight gives us better estimations for the squared correlation. Hence, we performed row-vector matrix multiplications using row vectors corresponding to Hamming weight one. We found that when the left part of the input mask has Hamming weight one and the right part of input mask is zero, we always get a 14-round squared correlation $\approx 2^{-30.9}$ for four different output masks. Therefore, in total we get 64 linear approximations with an estimated 14-round squared correlation $\approx 2^{-30.9}$.

We also constructed a correlation matrix with masks of Hamming weight ≤ 9 but we have only got a slight improvement for these 14-round approximations by a factor of $2^{0.3}$. We have found no 15-round approximation with squared correlation more than 2^{-32}. Table 2 shows the 14-round approximations with input and output masks written in hexadecimal notation.

Table 2. 14-round linear hulls for SIMON-32/K found, using Hamming weight ≤ 9

α	β				$\log_2 c^2$	$\log_2 N_t$
0x80000000	0x00800020,	0x00800060,	0x00808020,	0x00808060	-30.5815	28.11
0x02000000	0x00028000,	0x00028001,	0x00028200,	0x00028201	-30.5815	28.10
0x00800000	0x80002000,	0x80002080,	0x80006000,	0x80006080	-30.5816	28.06
0x00400000	0x40001000,	0x40001040,	0x40003000,	0x40003040	-30.5815	28.11
0x00040000	0x04000100,	0x04000104,	0x04000300,	0x04000304	-30.5816	28.10
0x00010000	0x01000040,	0x01000041,	0x010000C0,	0x010000C1	-30.5814	28.11

Improved 17-Round Linear Hulls on SIMON-48 (Squared Correlation Matrix + MIP). Using a squared correlation matrix of SIMON-48 having input and output masks with Hamming weight ≤ 6 and size 83278000×83278000, we found that a 17-round linear approximation with input mask 0x404044000001 and output mask $0x000001414044$ ($0x404044000001 \xrightarrow{17-round} 0x000001C04044$) has squared correlation $2^{-49.3611}$. Also the output masks $0x000001414044$ and $0x000001414044$ yield a similar squared correlation $2^{-49.3611}$. Unlike the case for SIMON-32 where we can easily use brute force to compute the squared correlation of a 1-round linear approximation, the squared correlation matrix for SIMON-48 was created using the algorithm proposed in [20]. Again the matrix is sparse and it has $48295112 \approx 2^{25.53}$ nonzero elements.

However, it seems difficult to build matrices beyond Hamming weight 6 for SIMON-48. Therefore we use our time-memory trade-off method to improve the squared correlation of the linear approximation $0x404044000001 \xrightarrow{17-round} 0x000001414044$.

To find the lower class where the heavy trails are on the bottom are glued with the light trails on top. The light trails are found using the matrix method for 11 rounds and the heavy trails are found using the MIP method for 6 rounds. Combining them both we get the 17-round lower class trails. In more detail, we fix the input mask to $0x404044000001$ and we use the matrix method to find the output masks after 11 rounds with the most significant squared correlation. The best output masks are $0x001000004400$, $0x001000004410$ and $0x0010000044C0$, each give an 11-round linear hull with squared correlation $2^{-28.6806}$ coming from 268 light trails. We first create a 6-round MIP model with $0x001000004400$ as an input mask and with the target output mask $0x000001414044$ as the output mask for the 6-round MIP model $0x001000004400 \xrightarrow{6-round} 0x000001414044$. In order to find heavy trails we added the big M constraints described in Sect. 4.1 and set $M = 200$ and all the c_i's to 7 from the end of round 1 to beginning of round 5. So $q = 5$, setting $p = 1$ and using $\sum_{i=1}^{l} y_i \leq q - p = 4$, we guarantee that the trails found will have Hamming weight at least 7 at one of the rounds. The constraints should be set as follows:

$$\sum_{i=0}^{47} s_{48+i} + 200 y_1 \qquad\qquad \geq 7$$

$$\sum_{i=0}^{47} s_{96+i} + 200 y_2 \qquad\qquad \geq 7$$

$$\sum_{i=0}^{47} s_{144+i} + 200 y_3 \qquad\qquad \geq 7$$

$$\sum_{i=0}^{47} s_{192+i} + 200 y_4 \qquad\qquad \geq 7$$

$$\sum_{i=0}^{47} s_{240+i} + 200 y_5 \qquad\qquad \geq 7$$

$$\sum_{i=1}^{5} y_i \leq 4$$

where y_j is a binary variable and $s_{48.j+i}$ is a binary variable representing the intermediate mask value in the jth round at the ith position.

Limiting our MIP program to find 512 trails for the specified approximation, we find that the estimated squared correlation is $2^{-22.3426}$. Combining the light trails with the heavy, we get a 17-round sub approximation whose squared correlation is $2^{-28.6806} \times 2^{-22.3426} = 2^{-51.0232}$. To get a better estimation, we repeated the above procedure for the other output masks $0x001000004410$ and $0x0010000044C0$ and get an estimated squared correlation equivalent to $2^{-28.6806} \times 2^{-24.33967} = 2^{-53.02027}$ and $2^{-28.6806} \times 2^{-24.486272} = 2^{-53.166872}$ respectively. Adding all these three sub linear approximations we get an estimated squared correlation equivalent to $2^{-51.0232} + 2^{-53.02027} + 2^{-53.166872} \approx 2^{-50.4607}$. Moreover, we repeat the same procedure for the 27 next best 11-round linear approximations and we get $2^{-49.3729}$ as a total estimated squared correlation for our 17-round lower class trails ($0x404044000001 \xrightarrow{17-round} 0x000001414044$). All these computations took less than 20 hrs on a standard laptop (See Table 11 in the Appendix).

Similarly to find the upper class where the heavy trails are on the top, are glued with the light trails on bottom. The light trails are found using the matrix method for 11 rounds and the heavy trails are found using the MIP method for 6 rounds under the same big M constraints described above. Combining them both we get the 17-round upper class trails. In more detail, we fix the output mask to $0x000001414044$ and we use the matrix method to find the input masks with the most significant squared correlation after 11 rounds. The best input masks are $0x004400001000$, $0x004410001000$, $0x004C00001000$ and $0x004C10001000$, each give an 11-round linear hull with squared correlation $2^{-28.6806}$ coming from 268 light trails. We first create a 6-round MIP model with $0x004400001000$ as an output mask and the target input mask $0x404044000001$ as the input mask for the 6-round MIP model $0x404044000001 \xrightarrow{6-round} 0x004400001000$. Limiting our MIP program to find 512 trails for the specified approximation, we find that the estimated squared correlation is $2^{-22.3426}$. Combining the light trails with the heavy, we get a 17-round sub approximation whose squared correlation is $2^{-28.6806} \times 2^{-22.3426} = 2^{-51.0232}$. Repeating the above procedure for the other three input masks $0x04410001000$, $0x004C00001000$ and $0x004C10001000$, we get an estimated squared correlation equivalent to $2^{-28.6806} \times 2^{-24.33967} = 2^{-53.02027}$, $2^{-28.6806} \times 2^{-24.486272} = 2^{-53.166872}$ and $2^{-28.6806} \times 2^{-23.979259} = 2^{-52.659859}$ respectively. Adding all these four sub linear approximations we get an estimated squared correlation equivalent to $2^{-51.0232} + 2^{-53.02027} + 2^{-53.166872} + 2^{-52.659859} \approx 2^{-50.1765}$. Repeating the same procedure for the 26 next best input masks and adding them up, we get a total squared correlation equivalent to $2^{-49.3729}$ as a total estimated squared correlation for our 17-round upper class trails ($0x404044000001 \xrightarrow{17-round} 0x000001414044$). All these computations took less than 18 hrs on a standard laptop (See Table 12 in the Appendix).

Adding the contributions of the lower and upper classes found through the above procedure to the contribution of the light trails found through the matrix method, we get $2^{-49.3729} + 2^{-49.3729} + 2^{-49.3611} = 2^{-47.7840} \approx 2^{-47.78}$ as a total estimation for the squared correlation of the 17-round linear hull ($0x404044000001 \xrightarrow{17-round} 0x000001414044$).

5.4   Key Recovery Attack on 24 and 23 Rounds of SIMON-32/K Using 14-Round Linear Hull

We extend the given linear hull for 14 rounds of SIMON-32/K (highlighted masks in the last row of Table 2) by adding some rounds to the beginning and the end of the cipher. The straight-forward approach is to start with the input mask of the 14-round linear hull (e.g. $(\Gamma_0, -)$) and go backwards to add some rounds to the beginning. With respect to Fig. 1, we can append an additional round to the beginning of the cipher. Since SIMON injects the subkey at the end of its round function, this work does not have any computational complexity. More precisely, for the current 14-round linear hull, we evaluate $((X_L^i)_0 \oplus (X_R^{i+14})_6 \oplus (X_L^{i+14})_8)$ to filter wrong guesses. On the other hand, we have $(X_L^i)_0 = (F(X_L^{i-1}))_0 \oplus ((X_R^{i-1})_0 \oplus (K^i)_0$, where $(F(X_L^{i-1}))_0 = (X_L^{i-1})_{14} \oplus ((X_L^{i-1})_{15} \& (X_L^{i-1})_8)$. Hence, if we add a round in the backwards direction, i.e. round $i - 1$, we know X_R^{i-1} and X_L^{i-1} we can determine $F(X_L^{i-1})$. Then it is possible to use the following equation to filter wrong keys, instead of $((X_L^i)_0 \oplus (X_R^{i+14})_6 \oplus (X_L^{i+14})_8)$, where $(K^i)_0$ is an unknown but a constant bit (in Fig. 1 such bits are marked in red):

$$(F(X_L^{i-1}))_0 \oplus (X_R^{i-1})_0 \oplus (K^i)_0 \oplus (X_R^{i+14})_6 \oplus (X_L^{i+14})_8 = (X_L^{i-1})_{14} \oplus ((X_L^{i-1})_{15}$$
$$\& (X_L^{i-1})_8) \oplus (X_R^{i-1})_0 \oplus (K^i)_0 \oplus (X_R^{i+14})_6 \oplus (X_L^{i+14})_8.$$

We can continue our method to add five rounds to the beginning of linear hull at the cost of guessing some bits of subkeys. To add more rounds in the backwards direction, we must guess the bit

$$(F(X_L^{i-1}))_0 = (X_L^{i-1})_{14} \oplus ((X_L^{i-1})_{15} \& (X_L^{i-1})_8).$$

On the other hand, to determine $(F(X_L^{i-1}))_0$ we guess $(X_L^{i-1})_{14}$ and $(X_L^{i-1})_{15}$ only if the guessed value for $(X_L^{i-1})_8$ is 1. Therefore, on average we need one bit guess for $(X_L^{i-1})_{15}$ and $(X_L^{i-1})_8$ (in Fig. 1 such bits are indicated in blue).

The same approach can be used to add five rounds to the end of linear hull at the cost of guessing some bits of subkeys. More details are depicted in Fig. 1.

On the other hand, in [29], Wang *et al.* presented a divide and conquer approach to add extra rounds to their impossible differential trail. We note that it is possible to adapt their approach to extend the key recovery using the exist linear hull over more rounds. Hence, one can use the 14-round linear hull and extend it by adding extra rounds to its beginning and its end. We add five rounds to the beginning and five rounds to the end of the linear hull to attack 24-round variant of SIMON-32/K. This key recovery attack processes as follows:

1. Let T_{max} and T_{cur} be counters (initialized by 0) and SK_{can} be a temporary register to store the possible candidate of the subkey.
2. Collect $2^{30.59}$ known plaintext and corresponding ciphertext pairs (p_i, c_i) for 24-round SIMON-32/64 and store them in a table $\mathcal{T}$.
3. Guess a value for the subkeys involved in the first five rounds of reduced SIMON-32/K, i.e. $(K^{i-4})[0, 2 \ldots 4, 5, 6, 7, 9 \ldots 13, 14] \| (K^{i-3})[4, 5, 6, 8, 11, 12, 13, 14, 15] \| (K^{i-2})[0, 6, 7, 13, 14] \| (K^{i-1}) [8, 15]$ and do as follows (note that the red subkey bits involved in the rounds are the constant bits and do not have to be guessed):
   (a) For any $p_j \in \mathcal{T}$ calculate the partial encryption of the first five rounds of reduced SIMON-32/K and find $\mathcal{V}_j = (X_L^i)[0] \oplus (K^i)[0] \oplus (K^{i-1})[14] \oplus (K^{i-2})[12] \oplus (K^{i-3})[10] \oplus (K^{i-5})[8]$.
   (b) Guess the bits of subkeys $K^{i+19}[0 \ldots 4, 5, 6, 7, 8 \ldots 10, 11, 12, 13, 14, 15]$, $K^{i+18}[1, 2, 3, 4, 5, \ 6, 8, 10, 11, 12, 14, 15]$, $K^{i+17}[0, 3, 4, 6, 7, 12, 13]$, and $K^{i+16}[5, 14]$, step by step.
   (c) For any $c_j \in \mathcal{T}$:
      i. calculate the partial decryption of the last five rounds of reduced SIMON-32/K and find $W_j = (X_L^{i+14})[8] \oplus (X_R^{i+14})[6] \oplus (K^{i+15})[6] \oplus (K^{i+16})[4, 8] \oplus (K^{i+17})[2] \oplus (K^{i+18})[0]$.
      ii. If $\mathcal{V}_j = W_j$ then increase T_{cur}.
   (d) If $T_{max} < T_{cur}$ (or resp. $T_{max} < (2^{32} - T_{cur})$) update T_{max} and SK_{can} by T_{cur} (resp. $2^{32} - T_{cur}$) and the current guessed subkey respectively.
4. Return SK_{can}.

Following the approach presented in [29], guessing the bits of subkeys K^{i+19} $[0 \ldots 4, 5, 6, 7, 8 \ldots 10, 11, 12, 13, 14, 15]$, $K^{i+18}[1, 2, 3, 4, 5, \ 6, 8, 10, 11, 12, 14, 15]$, $K^{i+17}[0, 3, 4, 6, 7, 12, 13]$, and $K^{i+16}[5, 14]$, step by step, to find the amount of $W_j = (X_L^{i+14})[8] \oplus (X_R^{i+14})[6] \oplus (K^{i+15})[6] \oplus (K^{i+16})[4, 8] \oplus (K^{i+17})[2] \oplus (K^{i+18})[0]$, for any c_j, are done as follows:

1. Let T_2 be a vector of 2^{32} counters which correspond to all possible values of $\mathcal{V}_j \| (X_L^{i+19})[0 \ldots 7, \ 10 \ldots 14] \| (X_R^{i+19})[0 \ldots 6, 8 \ldots 15] \| (X_R^{i+18})[8, 9, 15]$ (denoted as S_2^1). Guess the subkey bit $(K^{i+19} \ )[8, 9, 15]$ decrypt partially for each possible value of S_1^1 $(\mathcal{V}_j \| (X_L^{i+19}) \| (X_R^{i+19}))$ to obtain the value of $(X_R^{i+18})[8, 9, 15]$ (and hence S_2^1), then increase the corresponding counter T_{2, S_2^1}.
2. Guess the subkey bits $(K^{i+19})[5, 14]$, $(K^{i+19})[1, 10, 11]$, $(K^{i+19})[12]$, (K^{i+19}) $[13]$, and $(K^{i+19}) \ [0, 2, 3, 4, 6, 7]$ step by step (see Table 3), do similarly to the above and finally get the values of the counters corresponding to the state $\mathcal{V}_j \| (X_L^{i+18})[0 \ldots 6, 8, 10 \ldots 12, 14, 15] \| (X_R^{i+18})$ (denoted as S_0^2).
3. Let X_1 be a vector of 2^{29} counters which correspond to all possible values of $\mathcal{V}_j \| (X_L^{i+18})[0 \ldots 5, \ 8, 10 \ldots 12, 14, 15] \| (X_R^{i+18})[0 \ldots 4, 6 \ldots 15] \| (X_R^{i+17})[6]$ (denoted as S_1^2). Guess the subkey bit $(K^{i+18})[6]$. For each possible value of S_0^2 $(\mathcal{V}_j \| (X_L^{i+18})[0 \ldots 6, 8, 10 \ldots 12, 14, 15] \| (X_R^{i+18}))$, do partial decryption to derive the value of $(X_R^{i+17})[6]$ and add T_{7, S_7^1} to the corresponding counter X_{1, S_1^2} according to the value of S_1^2. After that, guess the subkey bits (K^{i+18})

[15], $(K^{i+18})[1]$, $(K^{i+18})[3,12]$, $(K^{i+18})[2]$, $(K^{i+18})[11]$, $(K^{i+18})[10]$, (K^{i+18}) [14], and $(K^{i+18})[4,5,8]$, step by step (see Table 4). Do similarly to the above and eventually obtain the values of the counters corresponding to the state $\mathcal{V}_j\|(X_L^{i+17})[0',2\ldots4,6,7,12,13]\|(X_R^{i+17})[0\ldots6,8,10\ldots12,14,15]$ (denoted as S_0^3) where $(X_R^{i+17})[0'] = (X_R^{i+17})[0] \oplus (K^{i+18})[0]$.

4. Let Y_1 be a vector of 2^{21} counters which correspond to all possible values of $\mathcal{V}_j\|(X_L^{i+17})[0,2,3,\qquad 6,7,12,13]\|(X_R^{i+17})[0\ldots2,4\ldots6,8,10\ldots12,14,15]\|$ $(X_R^{i+16})[4]$ (denoted as S_1^3). Guess the subkey bit $(K^{i+17})[4]$. For each possible value of S_0^3 $(\mathcal{V}_j\|(X_L^{i+17})[0,2\ldots4,6,7,12,13]\|(X_R^{i+17})$ $[0\ldots6,8,10\ldots 12,14,15])$, do partial decryption to derive the value of $(X_R^{i+16})[4]$ and add X_{9,S_9^2} to the corresponding counter Y_{1,S_1^3} according to the value of S_1^3. After that, guess the subkey bits $(K^{i+17})[3]$, $(K^{i+17})[12]$, $(K^{i+17})[13]$, $(K^{i+17})[7]$, and $(K^{i+17})[0,6]$, step by step (see Table 5). Do similarly to the above and eventually obtain the values of the counters corresponding to the state $\mathcal{V}_j\|(X_L^{i+16})[4,5,8,14]\|(X_R^{i+16})[0,2',3,4,6,7,12,13]$ (denoted as S_0^4) where $\|(X_R^{i+16})[2'] = (X_R^{i+16})[2] \oplus (K^{i+17})[2]$.

5. Let Z_1 be a vector of 2^6 counters which correspond to all possible values of $\mathcal{V}_j\|(X_L^{i+15})[6]\|$ $(X_R^{i+15})[4,5,8,14]$ (denoted as S_1^4) where $(X_R^{i+15})[4'] = (X_R^{i+15})[4] \oplus (K^{i+16})[4]$ and $(X_R^{i+15})[8'] = (X_R^{i+15})[8] \oplus (K^{i+16})[8]$. Guess the subkey bits $(K^{i+16})[5,14]$ and for each possible value of S_0^4 $(\mathcal{V}_j\|(X_L^{i+16})[4,5,8,14]\|(X_R^{i+16})[0,2,3,4,6,7,12,13])$ do partial decryption to derive the value of $(X_R^{i+15})[5,14]$ and add Y_{6,S_6^3} to the corresponding counter Z_{1,S_1^4} according to the value of S_1^4.

6. Let W_{1,S_1^5} be a vector of 2^4 counters which correspond to all possible values of $\mathcal{V}_j\|(X_L^{i+14})[4',8']\|$ $(X_R^{i+14})[6']$ (denoted as S_1^5) where $(X_R^{i+14})[6'] = (X_R^{i+14})[6] \oplus (K^{i+15})[6]$, $(X_L^{i+14})[4'] = (X_L^{i+14})[4] \oplus (K^{i+16})[4] \oplus (K^{i+17})[2] \oplus (K^{i+18})[0]$, and $(X_L^{i+14})[8'] = (X_L^{i+14})[8] \oplus (K^{i+16})[8]$. This state are extracted of S_1^4 and add Z_{1,S_1^4} to the corresponding counter W_{1,S_1^5} according to the value of S_1^5 (See Table 7).

7. Let O be a vector of 2^2 counters which correspond to all possible values of $\mathcal{V}_j\|\mathcal{W}_j$ (Note that $\mathcal{W}_j = (X_L^{i+14})[8] \oplus (X_R^{i+14})[6] \oplus (K^{i+15})[6] \oplus (K^{i+16})[4,8] \oplus (K^{i+17})[2] \oplus (K^{i+18})[0]$ and can be extracted from S_1^5). Each possible value of S_1^5 is converted to $\mathcal{V}_j\|\mathcal{W}_j$ and W_{1,S_1^5} and is added to the relevant counter in O according to the value of $\mathcal{V}_j\|\mathcal{W}_j$. Suppose that O_0 means that $\mathcal{V}_j = 0$ and $\mathcal{W}_j = 0$ and O_3 means that $\mathcal{V}_j = 1$ and $\mathcal{W}_j = 1$. If $O_0 + O_3 \geq T_{max}$ or $2^{32} - (O_0 + O_3) \geq T_{max}$ keep the guessed bits of subkey information as a possible subkey candidate, and discard it otherwise.

Attack Complexity. The time complexity of each sub-step was computed as shown in the Tables 3, 4, 5, 6 and 7. The time complexity of the attack is about $2^{63.9}$. It is clear that, the complexity of this attack is only slightly less than exhaustive search. However, if we reduce the last round and attack 23 round of SIMON-32/K then the attack complexity reduces to 2^{50} which is yet the best key-recovery attack on SIMON-32/K for such number of rounds.

5.5   Key Recovery Attack on SIMON-48/K Using 17-Round Linear Hull

Given the 17-round approximation for SIMON-48, introduced in Sect. 5.3, we apply the approach presented in Sect. 5.4 to extend key recovery over more number of rounds. Our key recovery for SIMON-48/72 and SIMON-48/96 covers 23 and 24 rounds respectively. The data complexity for these attacks is $2^{-47.78}$ and their time complexities are $2^{62.10}$ and $2^{83.10}$ respectively. Since the attack procedure is similar to the approach presented in Sect. 5.4, we do not repeat it. Related tables and complexity of each step of the attack for SIMON-48/96 has been presented in Appendix B (The time complexity of each sub-step was computed as shown in the Tables 8, 9, and 10). To attack SIMON-48/72, we add three rounds in forward direction instead of the current four rounds. Hence, the adversary does not need to guess the average 21 bits of the key in the last round of Fig. 2.

6   Conclusion

In this paper, we propose a time-memory tradeoff that finds better differential/linear approximation. The method benefits from the correlation matrix method and the MIP method to improve the estimated squared correlation or differential probability. Using MIP we can find the trails that are missed by the matrix method. This method enables us to find a 17-round linear hull for SIMON-48. Moreover, we have analyzed the security of some variants of SIMON against different variants of linear cryptanalysis, i.e. classic and linear hull attacks. We have investigated the linear hull effect on SIMON-32/64 and SIMON-48/96 using the correlation matrix of the average squared correlations and presented best linear attack on this variant.

Regarding SIMON-64, the squared correlation matrix which we are able to build and process holds masks with Hamming weight ≤ 6. Using only the matrix and going for more than 20 rounds, the best squared correlation we found has very low squared correlation $< 2^{-70}$ and this is because we are missing good trails with heavy Hamming weights. Applying our time-memory trade-off has not been effective due to the large number of rounds. However, trying to find good trails with heavy Hamming weight in the middle beside the upper and lower classes might yield better results. We note here that we have been looking for fast solutions. It could be that trying to add up many linear trails for some days or weeks can yield better results. Our method seems to be slow due to the slow processing of the huge squared correlation matrix. So it would be very interesting to build a dedicated sparse squared correlation matrix for SIMON-64 in order to speed up the selection of the intermediate masks in our time-memory trade-off method. This will allow us to select many intermediate masks which might yield better results. One interesting target would be also to apply this method to the block cipher PRESENT which also allows low Hamming weight trails and see if we can go beyond the current best 24-round linear approximations [1].

Acknowledgments. The authors would like to thank Lars Knudsen, Stefan Kölbl, Martin M. Lauridsen, Arnab Roy and Tyge Tiessen for many useful discussions about linear and differential cryptanalysis of SIMON. Many thanks go to Anne Canteaut and the anonymous reviewers for their valuable comments and suggestions to improve the quality of the paper.

A   Steps of the Key Recovery Attack on SIMON-32/64

Table 3. Step 1 of key recovery attack on SIMON-32/64

i	Input (S_i^1)	Guessed subkey bit	Output (S_{i+1}^1)	Counter of S_{i+1}^1
0	$(X_L^{i-5})\|(X_R^{i-5})$	$(K^{i-4})[0,2\ldots4,5,6,7,9\ldots13,14]\|(K^{i-3})[4,5,6,8,11,$ $12,13,14,15]\|(K^{i-2})[0,6,7,13,14]\|(K^{i-1})[8,15]$	$\mathcal{V}_j = (X_L^i)[0] \oplus (K^i)[0] \oplus (K^{i-1})[14]$ $\oplus (K^{i-2})[12] \oplus (K^{i-3})[10] \oplus (K^{i-5})[8]$	T_{1,S_1^1}
1	$\mathcal{V}_j\|(X_L^{i+19})$ $\|(X_R^{i+19})$	$(K^{i+19})[8,9,15]$	$\mathcal{V}_j\|(X_L^{i+19})[0\ldots7,10\ldots14]$ $\|(X_R^{i+19})[0\ldots6,8\ldots15]$ $\|(X_R^{i+18})[8,9,15]$	T_{2,S_2^1}
2	$\mathcal{V}_j\|(X_L^{i+19})[0\ldots7,10\ldots14]$ $\|(X_R^{i+19})[0\ldots6,8\ldots15]$ $\|(X_R^{i+18})[8,9,15]$	$(K^{i+19})[5,14]$	$\mathcal{V}_j\|(X_L^{i+19})[0\ldots4,6,7,10\ldots13]$ $\|(X_R^{i+19})[0\ldots6,8\ldots12,14,15]$ $\|(X_R^{i+18})[5,8,9,14,15]$	T_{3,S_3^1}
3	$\mathcal{V}_j\|(X_L^{i+19})[0\ldots4,6,7,10\ldots13]$ $\|(X_R^{i+19})[0\ldots6,8\ldots12,14,15]$ $\|(X_R^{i+18})[5,8,9,14,15]$	$(K^{i+19})[1,10,11]$	$\mathcal{V}_j\|(X_L^{i+19})[0,2\ldots4,6,7,12,13]$ $\|(X_R^{i+19})[0\ldots6,8,10\ldots12,14,15]$ $\|(X_R^{i+18})[1,5,8,9,10,11,14,15]$	T_{4,S_4^1}
4	$\mathcal{V}_j\|(X_L^{i+19})[0,2\ldots4,6,7,12,13]$ $\|(X_R^{i+19})[0\ldots6,8,10\ldots12,14,15]$ $\|(X_R^{i+18})[1,5,8,9,10,11,14,15]$	$(K^{i+19})[12]$	$\mathcal{V}_j\|(X_L^{i+19})[0,2\ldots4,6,7,13]$ $\|(X_R^{i+19})[0\ldots6,8,10\ldots12,14,15]$ $\|(X_R^{i+18})[1,5,8,9,10,11,12,14,15]$	T_{5,S_5^1}
5	$\mathcal{V}_j\|(X_L^{i+19})[0,2\ldots4,6,7,13]$ $\|(X_R^{i+19})[0\ldots6,8,10\ldots12,14,15]$ $\|(X_R^{i+18})[1,5,8,9,10,11,12,14,15]$	$(K^{i+19})[13]$	$\mathcal{V}_j\|(X_L^{i+19})[0,2\ldots4,6,7]$ $\|(X_R^{i+19})[0\ldots6,8,10\ldots12,14,15]$ $\|(X_R^{i+18})[1,5,8,9,10,11,12,13,14,15]$	T_{6,S_6^1}
6	$\mathcal{V}_j\|(X_L^{i+19})[0,2\ldots4,6,7,12,13]$ $\|(X_R^{i+19})[0\ldots6,8,10\ldots12,14,15]$ $\|(X_R^{i+18})[1,5,8,9,10,11,14,15]$	$(K^{i+19})[0,2,3,4,6,7]$	$\mathcal{V}_j\|(X_L^{i+19})[0\ldots6,8,10\ldots12,14,15]$ $\|(X_R^{i+18})$	T_{7,S_7^1}

substep 0: $2^{23} \times 2^{30.59} \times 5/24 = 2^{51.33}$
substep 1: $2^{23} \times 2^{33} \times 2^3 \times 3 \times 1/(16 \times 24) = 2^{52}$
substep 2: $2^{23} \times 2^{32} \times 2^4 \times 2 \times 1/(16 \times 24) = 2^{51.42}$
substep 3: $2^{23} \times 2^{31} \times 2^{6.5} \times 3 \times 1/(16 \times 24) = 2^{53.5}$
substep 4: $2^{23} \times 2^{30} \times 2^{7.5} \times 1/(16 \times 24) = 2^{51.92}$
substep 5: $2^{23} \times 2^{30} \times 2^{8.5} \times 1/(16 \times 24) = 2^{52.92}$
substep 6: $2^{23} \times 2^{30} \times 2^{14} \times 6 \times 1/(16 \times 24) = 2^{61}$

Table 4. Step 2 of key recovery attack on SIMON-32/64

i	Input (S_i^2)	Guessed subkey bit	Output (S_{i+1}^2)	Counter of S_{i+1}^2
0	$\mathcal{V}_j\|(X_L^{i+18})[0\dots6,8,10\dots12,14,15]$ $\|(X_R^{i+18})$	$(K^{i+18})[6]$	$\mathcal{V}_j\|(X_L^{i+18})[0\dots5,8,10\dots12,14,15]$ $\|(X_R^{i+18})[0\dots4,6\dots15]$ $\|(X_R^{i+17})[6]$	X_{1,S_1^2}
1	$\mathcal{V}_j\|(X_L^{i+18})[0\dots5,8,10\dots12,14,15]$ $\|(X_R^{i+18})[0\dots4,6\dots15]$ $\|(X_R^{i+17})[6]$	$(K^{i+18})[15]$	$\mathcal{V}_j\|(X_L^{i+18})[0\dots5,8,10\dots12,14]$ $\|(X_R^{i+18})[0\dots4,6\dots13,15]$ $\|(X_R^{i+17})[6,15]$	X_{2,S_2^2}
2	$\mathcal{V}_j\|(X_L^{i+18})[0\dots5,8,10\dots12,14]$ $\|(X_R^{i+18})[0\dots4,6\dots13,15]$ $\|(X_R^{i+17})[6,15]$	$(K^{i+18})[1]$	$\mathcal{V}_j\|(X_L^{i+18})[0,2\dots5,8,10\dots12,14]$ $\|(X_R^{i+18})[0\dots4,6\dots13]$ $\|(X_R^{i+17})[1,6,15]$	X_{3,S_3^2}
3	$\mathcal{V}_j\|(X_L^{i+18})[0,2\dots5,8,10\dots12,14]$ $\|(X_R^{i+18})[0\dots4,6\dots13]$ $\|(X_R^{i+17})[1,6,15]$	$(K^{i+18})[3,12]$	$\mathcal{V}_j\|(X_L^{i+18})[0,2,4,5,8,10,11,14]$ $\|(X_R^{i+18})[0\dots4,6\dots10,12,13]$ $\|(X_R^{i+17})[1,3,6,12,15]$	X_{4,S_4^2}
4	$\mathcal{V}_j\|(X_L^{i+18})[0,2,4,5,8,10,11,14]$ $\|(X_R^{i+18})[0\dots4,6\dots10,12,13]$ $\|(X_R^{i+17})[1,3,6,12,15]$	$(K^{i+18})[2]$	$\mathcal{V}_j\|(X_L^{i+18})[0,4,5,8,10,11,14]$ $\|(X_R^{i+18})[0,2\dots4,6\dots10,12,13]$ $\|(X_R^{i+17})[1,2,3,6,12,15]$	X_{5,S_5^2}
5	$\mathcal{V}_j\|(X_L^{i+18})[0,4,5,8,10,11,14]$ $\|(X_R^{i+18})[0,2\dots4,6\dots10,12,13]$ $\|(X_R^{i+17})[1,2,3,6,12,15]$	$(K^{i+18})[11]$	$\mathcal{V}_j\|(X_L^{i+18})[0,4,5,8,10,14]$ $\|(X_R^{i+18})[0,2\dots4,6\dots9,12,13]$ $\|(X_R^{i+17})[1,2,3,6,11,12,15]$	X_{6,S_6^2}
6	$\mathcal{V}_j\|(X_L^{i+18})[0,4,5,8,10,14]$ $\|(X_R^{i+18})[0,2\dots4,6\dots9,12,13]$ $\|(X_R^{i+17})[1,2,3,6,11,12,15]$	$(K^{i+18})[10]$	$\mathcal{V}_j\|(X_L^{i+18})[0,4,5,8,14]$ $\|(X_R^{i+18})[0,2\dots4,6\dots8,12,13]$ $\|(X_R^{i+17})[1,2,3,6,10,11,12,15]$	X_{7,S_7^2}
7	$\mathcal{V}_j\|(X_L^{i+18})[0,4,5,8,14]$ $\|(X_R^{i+18})[0,2\dots4,6\dots8,12,13]$ $\|(X_R^{i+17})[1,2,3,6,10,11,12,15]$	$(K^{i+18})[14]$	$\mathcal{V}_j\|(X_L^{i+18})[0,4,5,8]$ $\|(X_R^{i+18})[0,2\dots4,6\dots8,12,13]$ $\|(X_R^{i+17})[1,2,3,6,10,11,12,14,15]$	X_{8,S_8^2}
8	$\mathcal{V}_j\|(X_L^{i+18})[0,4,5,8]$ $\|(X_R^{i+18})[0,2\dots4,6\dots8,12,13]$ $\|(X_R^{i+17})[1,2,3,6,10,11,12,14,15]$	$(K^{i+18})[4,5,8]$	$\mathcal{V}_j\|(X_L^{i+17})[0,2\dots4,6,7,12,13]$ $\|(X_R^{i+17})[0'\dots6,8,10\dots12,14,15]$ where $(X_R^{i+17})[0']=(X_R^{i+17})[0]\oplus(K^{i+18})[0]$	X_{9,S_9^2}

substep 0: $2^{23}\times2^{14}\times2^{30}\times2^{0.5}\times1/16\times24=2^{58.92}$

substep 1: $2^{23}\times2^{14}\times2^{29}\times2\times1/16\times24=2^{58.42}$

substep 2: $2^{23}\times2^{14}\times2^{28}\times2^2\times1/16\times24=2^{58.42}$

substep 3: $2^{23}\times2^{14}\times2^{27}\times2^3\times2\times1/16\times24=2^{58.42}$

substep 4: $2^{23}\times2^{14}\times2^{26}\times2^4\times1/16\times24=2^{58.42}$

substep 5: $2^{23}\times2^{14}\times2^{25}\times2^5\times1/16\times24=2^{58.42}$

substep 6: $2^{23}\times2^{14}\times2^{24}\times2^6\times1/16\times24=2^{58.42}$

substep 7: $2^{23}\times2^{14}\times2^{23}\times2^7\times1/16\times24=2^{58.42}$

substep 8: $2^{23}\times2^{14}\times2^{23}\times2^{9.5}\times3\times1/16\times24=2^{62.5}$

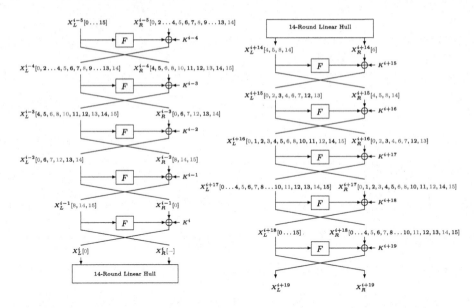

Fig. 1. Adding some rounds to the 14-round linear hull for SIMON-32/K (Color figure online).

Table 5. Step 3 of key recovery attack on SIMON-32/64

i	Input (S_i^3)	Guessed subkey bit	Output (S_{i+1}^3)	Counter of S_{i+1}^3
0	$\mathcal{V}_j\|(X_L^{i+17})[0,2\ldots4,6,7,12,13]$	$(K^{i+17})[4]$	$\mathcal{V}_j\|(X_L^{i+17})[0,2,3,6,7,12,13]$	Y_{1,S_1^3}
	$\|(X_R^{i+17})[0\ldots6,8,10\ldots12,14,15]$		$\|(X_R^{i+17})[0\ldots2,4\ldots6,8,10\ldots12,14,15]$	
	$\|(X_R^{i+17})[0\ldots2,4\ldots6,8,10\ldots12,14,15]$		$\|(X_R^{i+16})[4]$	
1	$\mathcal{V}_j\|(X_L^{i+17})[0,2,3,6,7,12,13]$	$(K^{i+17})[3]$	$\mathcal{V}_j\|(X_L^{i+17})[0,2,6,7,12,13]$	Y_{2,S_2^3}
	$\|(X_R^{i+17})[0\ldots2,4\ldots6,8,10\ldots12,14,15]$		$\|(X_R^{i+17})[0,4\ldots6,8,10\ldots12,14,15]$	
	$\|(X_R^{i+16})[4]$		$\|(X_R^{i+16})[3,4]$	
2	$\mathcal{V}_j\|(X_L^{i+17})[0,2,6,7,12,13]$	$(K^{i+17})[12]$	$\mathcal{V}_j\|(X_L^{i+17})[0,2,6,7,13]$	Y_{3,S_3^3}
	$\|(X_R^{i+17})[0,4\ldots6,8,10\ldots12,14,15]$		$\|(X_R^{i+17})[0,4\ldots6,8,11,12,14,15]$	
	$\|(X_R^{i+16})[3,4]$		$\|(X_R^{i+16})[3,4,12]$	
3	$\mathcal{V}_j\|(X_L^{i+17})[0,2,6,7,13]$	$(K^{i+17})[13]$	$\mathcal{V}_j\|(X_L^{i+17})[0,2,6,7]$	Y_{4,S_4^3}
	$\|(X_R^{i+17})[0,4\ldots6,8,11,12,14,15]$		$\|(X_R^{i+17})[0,4\ldots6,8,14,15]$	
	$\|(X_R^{i+16})[3,4,12]$		$\|(X_R^{i+16})[3,4,12,13]$	
4	$\mathcal{V}_j\|(X_L^{i+17})[0,2,6,7]$	$(K^{i+17})[7]$	$\mathcal{V}_j\|(X_L^{i+17})[0,2,6]$	Y_{5,S_5^3}
	$\|(X_R^{i+17})[0,4\ldots6,8,14,15]$		$\|(X_R^{i+17})[0,4,5,8,14,15]$	
	$\|(X_R^{i+16})[3,4,12,13]$		$\|(X_R^{i+16})[3,4,7,12,13]$	
5	$\mathcal{V}_j\|(X_L^{i+17})[0,2,6]$	$(K^{i+17})[0,6]$	$\mathcal{V}_j\|(X_L^{i+16})[4,5,8,14]$	Y_{6,S_6^3}
	$\|(X_R^{i+17})[0,4,5,8,14,15]$		$\|(X_R^{i+16})[0,2',3,4,6,7,12,13]$	
	$\|(X_R^{i+16})[3,4,7,12,13]$		where $(X_R^{i+16})[2'] = (X_R^{i+16})[2] \oplus (K^{i+17})[2]$	

substep 0: $2^{23} \times 2^{14} \times 2^{9.5} \times 2^{22} \times 2^{0.5} \times 1/(16 \times 24) = 2^{60.42}$
substep 1: $2^{23} \times 2^{14} \times 2^{9.5} \times 2^{21} \times 2^{1.5} \times 1/(16 \times 24) = 2^{60.42}$
substep 2: $2^{23} \times 2^{14} \times 2^{9.5} \times 2^{19} \times 2^{2.5} \times 1/(16 \times 24) = 2^{59.42}$
substep 3: $2^{23} \times 2^{14} \times 2^{9.5} \times 2^{18} \times 2^{3} \times 1/(16 \times 24) = 2^{58.92}$
substep 4: $2^{23} \times 2^{14} \times 2^{9.5} \times 2^{16} \times 2^{3.5} \times 1/(16 \times 24) = 2^{57.42}$
substep 5: $2^{23} \times 2^{14} \times 2^{9.5} \times 2^{15} \times 2^{4.5} \times 2 \times 1/(16 \times 24) = 2^{58.42}$

Table 6. Step 4 of key recovery attack on SIMON-32/64

i	Input (S_i^4)	Guessed subkey bit	Output (S_{i+1}^4)	Counter of S_{i+1}^4
0	$\mathcal{V}_j\|(X_L^{i+16})[4,5,8,14]\|(X_R^{i+16})[0,2,3,4,6,7,12,13]$	$(K^{i+16})[5,14]$	$\mathcal{V}_j\|(X_L^{i+15})[6]\|(X_R^{i+15})[4',5,8',14]$	Z_{1,S_i^4}

where $(X_R^{i+15})[4'] = (X_R^{i+15})[4] \oplus (K^{i+16})[4]$
and $(X_R^{i+15})[8'] = (X_R^{i+15})[8] \oplus (K^{i+16})[8]$

substep 0: $2^{23} \times 2^{14} \times 2^{9.5} \times 2^{4.5} \times 2^{13} \times 2 \times 2 \times 1/(16 \times 24) = 2^{57.42}$

Table 7. Step 5 of key recovery attack on SIMON-32/64

i	Input (S_i^5)	Guessed subkey bit	Output (S_{i+1}^5)	Counter of S_{i+1}^5
0	$\mathcal{V}_j\|(X_L^{i+15})[6]\|(X_R^{i+15})[4,5,8,14]$		$\mathcal{V}_j\|(X_L^{i+14})[4',8']\|(X_R^{i+14})[6']$	W_{1,S_i^5}

where $(X_R^{i+14})[6'] = (X_R^{i+14})[6] \oplus (K^{i+15})[6]$,
$(X_L^{i+14})[4'] = (X_L^{i+14})[4] \oplus (K^{i+16})[4] \oplus (K^{i+17})[2] \oplus (K^{i+18})[0]$, and
$(X_L^{i+14})[8'] = (X_L^{i+14})[8] \oplus (K^{i+16})[8]$.

B   Steps of the Key Recovery Attack on SIMON-48/96

$$\mathcal{V}_j = (X_L^i)[2,6,14,22] \oplus (X_R^i)[0] \oplus (K^i)[2,6,14,22] \oplus (K^{i-1})[0,4,12,20] \oplus (K^{i-2})[2,18]$$
$$\mathcal{W}_j = (X_L^{i+18})[0] \oplus (X_R^{i+18})[2,6,14,16,22] \oplus (K^{i+19})[2,6,14,16,22]$$
$$\oplus (K^{i+20})[0,4,12,20] \oplus (K^{i+21})[2,18] \oplus (K^{i+22})[0]$$

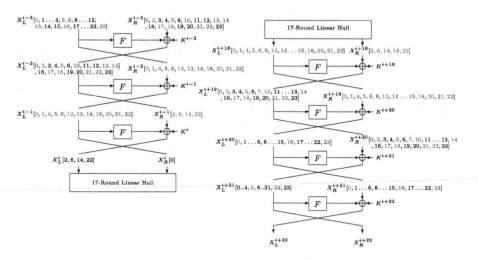

Fig. 2. Adding some rounds to the 17-round linear hull for SIMON-48/96 (Color figure online).

Table 8. Step 1 of key recovery attack on SIMON-48/96

i	Input (S_i^1)	Guessed subkey bit	Output (S_{i+1}^1)	Counter of S_{i+1}^1
0	$(X_L^{i-3})[0\dots6,8\dots23]$ $\|(X_R^{i-3})[0,2\dots6,10\dots14,16\dots23]$	$(K^{i-2})[0,3,4,5,6,10,11,12,13,14,16,17,19,$ $20,21,22,23]\|(K^{i-1})[1,5,6,13,14,18,21,22]$	$\mathcal{V}_j\|(X_L^{i+22})[0\dots6,8\dots23]$ $\|(X_R^{i+22})$	T_{1,S_i^1}
1	$\mathcal{V}_j\|(X_L^{i+22})[0\dots6,8\dots23]$ $\|(X_R^{i+22})$	$(K^{i+22})[10,11,17]$	$\mathcal{V}_j\|(X_L^{i+22})[0\dots6,8,9,12\dots16,18\dots23]$ $\|(X_R^{i+22})[0\dots8,10\dots23]$ $\|(X_R^{i+21})[10,11,17]$	T_{2,S_i^1}
2	$\mathcal{V}_j\|(X_L^{i+22})[0\dots6,8,9,12\dots16,18\dots23]$ $\|(X_R^{i+22})[0\dots8,10\dots23]$ $\|(X_R^{i+21})[10,11,17]$	$(K^{i+22})[16,23]$	$\mathcal{V}_j\|(X_L^{i+22})[0\dots6,8,9,12\dots15,18\dots22]$ $\|(X_R^{i+22})[0\dots8,10\dots14,16\dots23]$ $\|(X_R^{i+21})[10,11,16,17,23]$	T_{3,S_i^1}
3	$\mathcal{V}_j\|(X_L^{i+22})[0\dots6,8,9,12\dots15,18\dots22]$ $\|(X_R^{i+22})[0\dots8,10\dots14,16\dots23]$ $\|(X_R^{i+21})[10,11,16,17,23]$	$(K^{i+22})[9]$	$\mathcal{V}_j\|(X_L^{i+22})[0\dots6,8,12\dots15,18\dots22]$ $\|(X_R^{i+22})[0\dots7,10\dots14,16\dots23]$ $\|(X_R^{i+21})[8,10,11,16,17,23]$	T_{4,S_i^1}
4	$\mathcal{V}_j\|(X_L^{i+22})[0\dots6,8,12\dots15,18\dots22]$ $\|(X_R^{i+22})[0\dots7,10\dots14,16\dots23]$ $\|(X_R^{i+21})[8,10,11,16,17,23]$	$(K^{i+22})[2,3]$	$\mathcal{V}_j\|(X_L^{i+22})[0,1,4\dots6,8,12\dots15,18\dots22]$ $\|(X_R^{i+22})[0,2\dots7,10\dots14,16\dots23]$ $\|(X_R^{i+21})[2,3,8,10,11,16,17,23]$	T_{5,S_i^1}
5	$\mathcal{V}_j\|(X_L^{i+22})[0,1,4\dots6,8,12\dots15,18\dots22]$ $\|(X_R^{i+22})[0,2\dots7,10\dots14,16\dots23]$ $\|(X_R^{i+21})[2,3,8,10,11,16,17,23]$	$(K^{i+22})[1,4\dots6,8,12\dots15,18\dots22]$	$\mathcal{V}_j\|(X_L^{i+21})[0,2\dots7,10\dots14,16\dots23]$ $\|(X_R^{i+21})[0\dots6,8\dots23]$	T_{6,S_i^1}

substep 0: $2^{17} \times 2^{47.78} \times 3/24 = 2^{61.78}$
substep 1: $2^{17} \times 2^{48} \times 2^3 \times 3 \times 1/(24 \times 24) = 2^{60.41}$
substep 2: $2^{17} \times 2^{47} \times 2^4 \times 2 \times 1/(24 \times 24) = 2^{59.83}$
substep 3: $2^{17} \times 2^{46} \times 2^5 \times 1/(24 \times 24) = 2^{58.83}$
substep 4: $2^{17} \times 2^{45} \times 2^7 \times 2 \times 1/(24 \times 24) = 2^{60.83}$
substep 5: $2^{17} \times 2^{44} \times 2^{21} \times 14 \times 1/(24 \times 24) = 2^{76.64}$

Table 9. Step 2 of key recovery attack on SIMON-48/96

i	Input (S_i^2)	Guessed subkey bit	Output (S_{i+1}^2)	Counter of S_{i+1}^2
0	$\mathcal{V}_j\|(X_L^{i+21})[0,2\dots7,10\dots14,16\dots23]$ $\|(X_R^{i+21})[0\dots6,8\dots23]$	$(K^{i+21})[19]$	$\mathcal{V}_j\|(X_L^{i+21})[0,2\dots7,10\dots14,16\dots18,20\dots23]$ $\|(X_R^{i+21})[0\dots6,8\dots16,18\dots23]$ $\|(X_R^{i+20})[19]$	X_{1,S_i^2}
1	$\mathcal{V}_j\|(X_L^{i+21})[0,2\dots7,10\dots14,16\dots18,20\dots23]$ $\|(X_R^{i+21})[0\dots6,8\dots16,18\dots23]$ $\|(X_R^{i+20})[19]$	$(K^{i+21})[12,13]$	$\mathcal{V}_j\|(X_L^{i+21})[0,2\dots7,10,11,14,16\dots18,20\dots23]$ $\|(X_R^{i+21})[0\dots6,8\dots10,12\dots16,18\dots23]$ $\|(X_R^{i+20})[12,13,19]$	X_{2,S_i^2}
2	$\mathcal{V}_j\|(X_L^{i+21})[0,2\dots7,10,11,14,16\dots18,20\dots23]$ $\|(X_R^{i+21})[0\dots6,8\dots10,12\dots16,18\dots23]$ $\|(X_R^{i+20})[12,13,19]$	$(K^{i+21})[11]$	$\mathcal{V}_j\|(X_L^{i+21})[0,2\dots7,10,14,16\dots18,20\dots23]$ $\|(X_R^{i+21})[0\dots6,8,9,12\dots16,18\dots23]$ $\|(X_R^{i+20})[11\dots13,19]$	X_{3,S_i^2}
3	$\mathcal{V}_j\|(X_L^{i+21})[0,2\dots7,10,14,16\dots18,20\dots23]$ $\|(X_R^{i+21})[0\dots6,8,9,12\dots16,18\dots23]$ $\|(X_R^{i+20})[11\dots13,19]$	$(K^{i+21})[0,17]$	$\mathcal{V}_j\|(X_L^{i+21})[2\dots7,10,14,16,18,20\dots23]$ $\|(X_R^{i+21})[0\dots6,8,9,12\dots15,18\dots23]$ $\|(X_R^{i+20})[0,11\dots13,17,19]$	X_{4,S_i^2}
4	$\mathcal{V}_j\|(X_L^{i+21})[2\dots7,10,14,16,18,20\dots23]$ $\|(X_R^{i+21})[0\dots6,8,9,12\dots15,18\dots23]$ $\|(X_R^{i+20})[0,11\dots13,17,19]$	$(K^{i+21})[7]$	$\mathcal{V}_j\|(X_L^{i+21})[2\dots6,10,14,16,18,20\dots23]$ $\|(X_R^{i+21})[0\dots6,8,9,12\dots15,18\dots22]$ $\|(X_R^{i+20})[0,7,11\dots13,17,19]$	X_{5,S_i^2}
5	$\mathcal{V}_j\|(X_L^{i+21})[2\dots6,10,14,16,18,20\dots23]$ $\|(X_R^{i+21})[0\dots6,8,9,12\dots15,18\dots22]$ $\|(X_R^{i+20})[0,7,11\dots13,17,19]$	$(K^{i+21})[10]$	$\mathcal{V}_j\|(X_L^{i+21})[2\dots6,14,16,18,20\dots23]$ $\|(X_R^{i+21})[0\dots6,8,12\dots15,18\dots22]$ $\|(X_R^{i+20})[0,7,10,11\dots13,17,19]$	X_{6,S_i^2}
6	$\mathcal{V}_j\|(X_L^{i+21})[2\dots6,14,16,18,20\dots23]$ $\|(X_R^{i+21})[0\dots6,8,12\dots15,18\dots22]$ $\|(X_R^{i+20})[0,7,10,11\dots13,17,19]$	$(K^{i+21})[4,5]$	$\mathcal{V}_j\|(X_L^{i+21})[2,3,6,14,16,18,20\dots23]$ $\|(X_R^{i+21})[0\dots2,4\dots6,8,12\dots15,18\dots22]$ $\|(X_R^{i+20})[0,4,5,7,10,11\dots13,17,19]$	X_{7,S_i^2}
7	$\mathcal{V}_j\|(X_L^{i+21})[2,3,6,14,16,18,20\dots23]$ $\|(X_R^{i+21})[0\dots2,4\dots6,8,12\dots15,18\dots22]$ $\|(X_R^{i+20})[0,4,5,7,10,11\dots13,17,19]$	$(K^{i+21})[3,6,14,16,20,21,22,23]$	$\mathcal{V}_j\|(X_L^{i+20})[0,1,4\dots6,8,12\dots15,18,20\dots22]$ $\|(X_R^{i+20})[0,2\dots7,10\dots14,16\dots23]$	X_{8,S_i^2}

substep 0: $2^{17} \times 2^{21} \times 2^{44} \times 2 \times 1/24.24 = 2^{73.83}$
substep 1: $2^{17} \times 2^{21} \times 2^{43} \times 2^3 \times 2 \times 1/24.24 = 2^{75.83}$
substep 2: $2^{17} \times 2^{21} \times 2^{42} \times 2^4 \times 1/24.24 = 2^{74.83}$
substep 3: $2^{17} \times 2^{21} \times 2^{41} \times 2^5 \times 2 \times 1/24.24 = 2^{75.83}$
substep 4: $2^{17} \times 2^{21} \times 2^{40} \times 2^{5.5} \times 1/24.24 = 2^{74.33}$
substep 5: $2^{17} \times 2^{21} \times 2^{39} \times 2^6 \times 1/24.24 = 2^{73.83}$
substep 6: $2^{17} \times 2^{21} \times 2^{38} \times 2^{7.5} \times 2 \times 1/24.24 = 2^{75.33}$
substep 7: $2^{17} \times 2^{21} \times 2^{37} \times 2^{14} \times 8 \times 1/24.24 = 2^{82.83}$

Table 10. Step 3 of key recovery attack on SIMON-48/96

i	Input (S_i^3)	Guessed subkey bit	Output (S_{i+1}^3)	Counter of S_{i+1}^3
0	$\mathcal{V}_j \| (X_L^{i+20})[0,1,4\ldots6,8,12\ldots15,18,20\ldots22]$ $\| (X_R^{i+20})[0,2\ldots7,10\ldots14,16\ldots23]$	$(K^{i+20})[1]$	$\mathcal{V}_j \| (X_L^{i+20})[0,4\ldots6,8,12\ldots15,18,20\ldots22]$ $\| (X_R^{i+20})[0,2\ldots7,10\ldots14,16\ldots22]$ $\| (X_R^{i+19})[1]$	Y_{1,S_1^3}
1	$\mathcal{V}_j \| (X_L^{i+20})[0,4\ldots6,8,12\ldots15,18,20\ldots22]$ $\| (X_R^{i+20})[0,2\ldots7,10\ldots14,16\ldots22]$ $\| (X_R^{i+19})[1]$	$(K^{i+20})[5]$	$\mathcal{V}_j \| (X_L^{i+20})[0,4,6,8,12\ldots15,18,20\ldots22]$ $\| (X_R^{i+20})[0,2,4\ldots7,10\ldots14,16\ldots22]$ $\| (X_R^{i+19})[1,5]$	Y_{2,S_2^3}
2	$\mathcal{V}_j \| (X_L^{i+20})[0,4,6,8,12\ldots15,18,20\ldots22]$ $\| (X_R^{i+20})[0,2,4\ldots7,10\ldots14,16\ldots22]$ $\| (X_R^{i+19})[1,5]$	$(K^{i+20})[13]$	$\mathcal{V}_j \| (X_L^{i+20})[0,4,6,8,12,14,15,18,20\ldots22]$ $\| (X_R^{i+20})[0,2,4\ldots7,10,12\ldots14,16\ldots22]$ $\| (X_R^{i+19})[1,5,13]$	Y_{3,S_3^3}
3	$\mathcal{V}_j \| (X_L^{i+20})[0,4,6,8,12,14,15,18,20\ldots22]$ $\| (X_R^{i+20})[0,2,4\ldots7,10,12\ldots14,16\ldots22]$ $\| (X_R^{i+19})[1,5,13]$	$(K^{i+20})[14]$	$\mathcal{V}_j \| (X_L^{i+20})[0,4,6,8,12,15,18,20\ldots22]$ $\| (X_R^{i+20})[0,2,4\ldots7,10,13,14,16\ldots22]$ $\| (X_R^{i+19})[1,5,13,14]$	Y_{4,S_4^3}
4	$\mathcal{V}_j \| (X_L^{i+20})[0,4,6,8,12,15,18,20,22]$ $\| (X_R^{i+20})[0,2,4\ldots7,10,13,14,16\ldots22]$ $\| (X_R^{i+19})[1,5,13,14]$	$(K^{i+20})[21]$	$\mathcal{V}_j \| (X_L^{i+20})[0,4,6,8,12,15,18,20,22]$ $\| (X_R^{i+20})[0,2,4\ldots7,10,13,14,16\ldots18,20\ldots22]$ $\| (X_R^{i+19})[1,5,13,14,21]$	Y_{5,S_5^3}
5	$\mathcal{V}_j \| (X_L^{i+20})[0,4,6,8,12,15,18,20,22]$ $\| (X_R^{i+20})[0,2,4\ldots7,10,13,14,16\ldots18,20\ldots22]$ $\| (X_R^{i+19})[1,5,13,14,21]$	$(K^{i+20})[22]$	$\mathcal{V}_j \| (X_L^{i+20})[0,4,6,8,12,15,18,20]$ $\| (X_R^{i+20})[0,2,4\ldots7,10,13,14,16\ldots18,20,22]$ $\| (X_R^{i+19})[1,5,13,14,21,22]$	Y_{6,S_6^3}
6	$\mathcal{V}_j \| (X_L^{i+20})[0,4,6,8,12,15,18,20]$ $\| (X_R^{i+20})[0,2,4\ldots7,10,13,14,16\ldots18,20,22]$ $\| (X_R^{i+19})[1,5,13,14,21,22]$	$(K^{i+20})[6,8,15,18]$	$\mathcal{V}_j \| (X_L^{i+19})[2,6,14,16,22]$ $\| (X_R^{i+19})[0,1,4\ldots6,8,12\ldots15,18,20\ldots22]$	Y_{7,S_7^3}
7	$\mathcal{V}_j \| (X_L^{i+19})[2,6,14,16,22]$ $\| (X_R^{i+19})[0,1,4\ldots6,8,12\ldots15,18,20\ldots22]$		$\mathcal{V}_j \| (X_L^{i+18})[0]$ $\| (X_R^{i+18})[2,6,14,16,22]$	Y_{7,S_7^3}

substep 0: $2^{17} \times 2^{21} \times 2^{14} \times 2^{35} \times 2^{0.5} \times 1/(24 \times 24) = 2^{78.33}$
substep 1: $2^{17} \times 2^{21} \times 2^{14} \times 2^{34} \times 2^1 \times 1/(24 \times 24) = 2^{77.83}$
substep 2: $2^{17} \times 2^{21} \times 2^{14} \times 2^{33} \times 2^{1.5} \times 1/(24 \times 24) = 2^{77.33}$
substep 3: $2^{17} \times 2^{21} \times 2^{14} \times 2^{32} \times 2^2 \times 1/(24 \times 24) = 2^{76.83}$
substep 4: $2^{17} \times 2^{21} \times 2^{14} \times 2^{31} \times 2^{2.5} \times 1/(24 \times 24) = 2^{76.33}$
substep 5: $2^{17} \times 2^{21} \times 2^{14} \times 2^{30} \times 2^3 \times 1/(24 \times 24) = 2^{75.83}$
substep 6: $2^{17} \times 2^{21} \times 2^{14} \times 2^{29} \times 2^5 \times 4 \times 1/(24 \times 24) = 2^{76.83}$

C MIP Experiments

Table 11 shows the 30 sub approximations that have been used to estimate the squared correlations of the lower class trails. The experiments where the MIP solutions are limited to 512 trails per approximation took exactly 70125.382718 seconds which is less than 20 hrs using a standard laptop.

Table 12 shows the 30 sub approximations that have been used to estimate the squared correlations of the upper class trails. The experiments where the MIP solutions are limited to 512 trails per approximation took exactly 62520.033249 seconds which is less than 18 hrs using a standard laptop.

Table 11. Lower Class Trails found through our time-memory trade-off method, $c_{i1}^2 \equiv$ the squared correlation of the ith 11-round linear approximation with light trails found through the correlation matrix, $c_{i2}^2 \equiv$ the squared correlation of the ith 6-round linear approximation with heavy trails found through the MIP method, $c_{i1}^2 c_{i2}^2 \equiv$ is the squared correlation of the ith 17-round linear approximation and $\sum c_{i1}^2 c_{i2}^2$ is the total estimated squared correlation of the lower class trails of our 17-round linear hull after including $i \leq 30$ linear approximations

i	Matrix trails	$\log_2 c_{i1}^2$	MIP trails	$\log_2 c_{i2}^2$	$\log_2 \sum c_{i1}^2 c_{i2}^2$
1	$404044000001 \xrightarrow{11-round} 001000004400$	-28.6806	$001000004400 \xrightarrow{6-round} 000001414044)$,	-22.342570	-51.023180
2	$404044000001 \xrightarrow{11-round} 001000004410$	-28.6806	$001000004410 \xrightarrow{6-round} 000001414044$	-24.339670	-50.700671
3	$404044000001 \xrightarrow{11-round} 001000004C00$	-28.6806	$001000004C00 \xrightarrow{6-round} 000001414044$	-24.486365	-50.460718
4	$404044000001 \xrightarrow{11-round} 001000004C10$	-28.6806	$001000004C10 \xrightarrow{6-round} 000001414044$	-23.979129	-50.176458
5	$404044000001 \xrightarrow{11-round} 003000004400$	-30.6806	$003000004400 \xrightarrow{6-round} 000001414044$	-22.342570	-49.988669
6	$404044000001 \xrightarrow{11-round} 003000004410$	-30.6806	$003000004410 \xrightarrow{6-round} 000001414044$	-24.339586	-49.945219
7	$404044000001 \xrightarrow{11-round} 003000004420$	-30.6806	$003000004420 \xrightarrow{6-round} 000001414044$	-27.953899	-49.941728
8	$404044000001 \xrightarrow{11-round} 003000004430$	-30.6806	$003000004430 \xrightarrow{6-round} 000001414044$	-26.956545	-49.934784
9	$404044000001 \xrightarrow{11-round} 003000004C00$	-30.6806	$003000004C00 \xrightarrow{6-round} 000001414044$	-24.486642	-49.896909
10	$404044000001 \xrightarrow{11-round} 003000004C10$	-30.6806	$003000004C00 \xrightarrow{6-round} 000001414044$	-24.486642	-49.844727
11	$404044000001 \xrightarrow{11-round} 003000004C20$	-30.6806	$003000004C20 \xrightarrow{6-round} 000001414044$	26.880410	-49.837883
12	$404044000001 \xrightarrow{11-round} 003000005400$	-30.6806	$003000005400 \xrightarrow{6-round} 000001414044$	-31.046525	-49.837503
13	$404044000001 \xrightarrow{11-round} 003000005410$	-30.6806	$003000005410 \xrightarrow{6-round} 000001414044$	-32.568502	-49.837371
14	$404044000001 \xrightarrow{11-round} 003000005420$	-30.6806	$003000005420 \xrightarrow{6-round} 000001414044$	-31.189830	-49.837026
15	$404044000001 \xrightarrow{11-round} 003000005C00$	-30.6806	$003000005C00 \xrightarrow{6-round} 000001414044$	-27.773381	-49.833356
16	$404044000001 \xrightarrow{11-round} 001040004400$	-30.6806	$001040004400 \xrightarrow{6-round} 000001414044$	-22.342570	-49.683331
17	$404044000001 \xrightarrow{11-round} 001040004410$	-30.6806	$001040004410 \xrightarrow{6-round} 000001414044$	-24.339586	-49.648069
18	$404044000001 \xrightarrow{11-round} 001040004420$	-30.6806	$001040004420 \xrightarrow{6-round} 000001414044$	-27.954667	-49.645229
19	$404044000001 \xrightarrow{11-round} 001040004430$	-30.6806	$001040004430 \xrightarrow{6-round} 000001414044$	-26.957186	-49.639576
20	$404044000001 \xrightarrow{11-round} 001040004C00$	-30.6806	$001040004C00 \xrightarrow{6-round} 000001414044$	-24.486272	-49.608628
21	$404044000001 \xrightarrow{11-round} 001040004C10$	-30.6806	$001040004C10 \xrightarrow{6-round} 000001414044$	-23.979129	-49.565757
22	$404044000001 \xrightarrow{11-round} 001040004C20$	-30.6806	$001040004C20 \xrightarrow{6-round} 000001414044$,	-26.879560	-49.560110
23	$404044000001 \xrightarrow{11-round} 001040404400$	-30.6806	$001040404400 \xrightarrow{6-round} 000001414044$	-30.596588	-49.559682
24	$404044000001 \xrightarrow{11-round} 001040404410$	-30.6806	$001040404410 \xrightarrow{6-round} 000001414044$	-27.765884	-49.556637
25	$404044000001 \xrightarrow{11-round} 001040404420$	-30.6806	$001040404420 \xrightarrow{6-round} 000001414044$	-30.819304	-49.556271
26	$404044000001 \xrightarrow{11-round} 001040404C00$	-30.6806	$001040404C00 \xrightarrow{6-round} 000001414044$	-32.191224	-49.556130
27	$404044000001 \xrightarrow{11-round} 003040004400$	-30.6806	$003040004400 \xrightarrow{6-round} 000001414044$	-22.342570	-49.431232
28	$404044000001 \xrightarrow{11-round} 003040004410$	-30.6806	$003040004410 \xrightarrow{6-round} 000001414044$	-24.339753	-49.401570
29	$404044000001 \xrightarrow{11-round} 003040004420$	-30.6806	$003040004420 \xrightarrow{6-round} 000001414044$	-27.954411	-49.399175
30	$404044000001 \xrightarrow{11-round} 003040004C00$	-30.6806	$003040004C00 \xrightarrow{6-round} 000001414044$	-24.486457	-49.372938

Table 12. Upper Class Trails found through our time-memory trade-off method, $c_{i1}^2 \equiv$ the squared correlation of the ith 6-round linear approximation with heavy trails found through the MIP method, $c_{i2}^2 \equiv$ the squared correlation of the ith 6-round linear approximation with light trails found through the correlation matrix, $c_{i1}^2 c_{i2}^2 \equiv$ is the squared correlation of the ith 17-round linear approximation and $\sum c_{i1}^2 c_{i2}^2$ is the total estimated squared correlation of the upper class trails of our 17-round linear hull after including $i \le 30$ linear approximations

i	MIP trails	$\log_2 c_{i1}^2$	Matrix trails	$\log_2 c_{i2}^2$	$\log_2 \sum c_{i1}^2 c_{i2}^2$
1	$404044000001 \xrightarrow{6-round} 004400001000$	-22.342570	$004400001000 \xrightarrow{11-round} 000001414044$	-28.6806	-51.023180
2	$404044000001 \xrightarrow{6-round} 004410001000$	-24.339670	$004410001000 \xrightarrow{11-round} 000001414044$	28.6806	-50.700671
3	$404044000001 \xrightarrow{6-round} 004C00001000$	-24.486272	$004C00001000 \xrightarrow{11-round} 000001414044$	-28.6806	-50.460704
4	$404044000001 \xrightarrow{6-round} 004C10001000$	-23.979129	$004C10001000 \xrightarrow{11-round} 000001414044$	-28.6806	-50.176447
5	$404044000001 \xrightarrow{6-round} 004400003000$	-22.342570	$004400003000 \xrightarrow{11-round} 000001414044$	-30.6806	-49.988659
6	$404044000001 \xrightarrow{6-round} 004410003000$	-24.339753	$004410003000 \xrightarrow{11-round} 000001414044$	-30.6806	-49.945214
7	$404044000001 \xrightarrow{6-round} 004420003000$	-27.955435	$004420003000 \xrightarrow{11-round} 000001414044$	-30.6806	-49.941726
8	$404044000001 \xrightarrow{6-round} 004430003000$	26.956674	$004430003000 \xrightarrow{11-round} 000001414044$	-30.6806	-49.934783
9	$404044000001 \xrightarrow{6-round} 004C00003000$	-24.486272	$004C00003000 \xrightarrow{11-round} 000001414044$	-30.6806	-49.896899
10	$404044000001 \xrightarrow{6-round} 004C10003000$	-23.979129	$004C10003000 \xrightarrow{11-round} 000001414044$	-30.6806	-49.844713
11	$404044000001 \xrightarrow{6-round} 004C20003000$	-26.879317	$004C20003000 \xrightarrow{11-round} 000001414044$	-30.6806	-49.837864
12	$404044000001 \xrightarrow{6-round} 004C20003000$	-31.046525	$005400003000 \xrightarrow{11-round} 000001414044$	-30.6806	-49.837483
13	$404044000001 \xrightarrow{6-round} 005410003000$	-32.568502	$005410003000 \xrightarrow{11-round} 000001414044$	-30.6806	-49.837483
14	$404044000001 \xrightarrow{6-round} 005420003000$	-31.189830	$005420003000 \xrightarrow{11-round} 000001414044$	-30.6806	-49.837007
15	$404044000001 \xrightarrow{6-round} 005C00003000$	-27.77338	$005C00003000 \xrightarrow{11-round} 000001414044$	-30.6806	-49.833337
16	$404044000001 \xrightarrow{6-round} 004400001040$	-22.342570	$004400001040 \xrightarrow{11-round} 000001414044$	-30.6806	-49.683313
17	$404044000001 \xrightarrow{6-round} 004400003040$	-22.342570	$004400003040 \xrightarrow{11-round} 000001414044$	-30.6806	-49.547431
18	$404044000001 \xrightarrow{6-round} 004410001040$	-24.339670	$004410001040 \xrightarrow{11-round} 000001414044$	-30.6806	-49.515307
19	$404044000001 \xrightarrow{6-round} 004410003040$	-24.339670	$004410003040 \xrightarrow{11-round} 000001414044$	-30.6806	-49.483882
20	$404044000001 \xrightarrow{6-round} 004420001040$	-27.955691	$004420001040 \xrightarrow{11-round} 000001414044$	30.6806	-49.481349
21	$404044000001 \xrightarrow{6-round} 004420003040$	-27.954155	$004420003040 \xrightarrow{11-round} 000001414044$	-30.6806	-49.478817
22	$404044000001 \xrightarrow{6-round} 004430001040$	-26.956417	$004430001040 \xrightarrow{11-round} 000001414044$	-30.6806	-49.473776
23	$404044000001 \xrightarrow{6-round} 004C00001040$	-24.486457	$004C00001040 \xrightarrow{11-round} 000001414044$	-30.6806	-49.446160
24	$404044000001 \xrightarrow{6-round} 004C00003040$	-24.486550	$004C00003040 \xrightarrow{11-round} 000001414044$	-30.6806	-49.419065
25	$404044000001 \xrightarrow{6-round} 004C10001040$	-23.979259	$004C10001040 \xrightarrow{11-round} 000001414044$	-30.6806	-49.381407
26	$404044000001 \xrightarrow{6-round} 004C20001040$	-26.879195	$004C20001040 \xrightarrow{11-round} 000001414044$	-30.6806	49.376435
27	$404044000001 \xrightarrow{6-round} 404400001040$	-30.596588	$404400001040 \xrightarrow{11-round} 000001414044$	-30.6806	-49.376058
28	$404044000001 \xrightarrow{6-round} 404410001040$	-27.765898	$404410001040 \xrightarrow{11-round} 000001414044$	-30.6806	-49.373377
29	$404044000001 \xrightarrow{6-round} 404420001040$	-30.819304	$404420001040 \xrightarrow{11-round} 000001414044$	-30.6806	-49.373054
30	$04044000001 \xrightarrow{6-round} 404C00001040$	-32.191224	$404C00001040 \xrightarrow{11-round} 000001414044$	-30.6806	49.372930

References

1. Abdelraheem, M.A.: Estimating the probabilities of low-weight differential and linear approximations on PRESENT-like ciphers. In: Kwon, T., Lee, M.-K., Kwon, D. (eds.) ICISC 2012. LNCS, vol. 7839, pp. 368–382. Springer, Heidelberg (2013)
2. Abdelraheem, M.A., Alizadeh, J., AlKhzaimi, H., Aref, M.R., Bagheri, N., Gauravaram, P., Lauridsen, M.M.: Improved Linear Cryptanalysis of Round Reduced SIMON. IACR Cryptology ePrint Archive, 2014:681 (2014)
3. Abed, F., List, E., Lucks, S., Wenzel, J.: Differential Cryptanalysis of Reduced-Round Simon. IACR Cryptology ePrint Archive 2013:526 (2013)

4. Abed, F., List, E., Lucks, S., Wenzel, J.: Differential cryptanalysis of round-reduced simon and speck. In: Cid, C., Rechberger, C. (eds.) FSE 2014. LNCS, vol. 8540, pp. 525–545. Springer, Heidelberg (2015)

5. Alizadeh, J., Alkhzaimi, H.A., Aref, M.R., Bagheri, N., Gauravaram, P., Kumar, A., Lauridsen, M.M., Sanadhya, S.K.: Cryptanalysis of SIMON variants with connections. In: Sadeghi, A.-R., Saxena, N. (eds.) RFIDSec 2014. LNCS, vol. 8651, pp. 90–107. Springer, Heidelberg (2014)

6. Alizadeh, J., Bagheri, N., Gauravaram, P., Kumar, A., Sanadhya, S.K.: Linear Cryptanalysis of Round Reduced SIMON. IACR Cryptology ePrint Archive 2013:663 (2013)

7. Ashur, T.: Improved linear trails for the block cipher simon. Cryptology ePrint Archive, Report 2015/285 (2015). http://eprint.iacr.org/

8. Aumasson, J.-P., Henzen, L., Meier, W., Naya-Plasencia, M.: QUARK: a lightweight hash. In: Mangard, S., Standaert, F.-X. (eds.) CHES 2010. LNCS, vol. 6225, pp. 1–15. Springer, Heidelberg (2010)

9. Beaulieu, R., Shors, D., Smith, J., Treatman-Clark, S., Weeks, B., Wingers, L.: The SIMON and SPECK Families of lightweight block ciphers. IACR Cryptology ePrint Archive **2013**, 404 (2013)

10. Biham, E.: On matsui's linear cryptanalysis. In: De Santis, A. (ed.) EUROCRYPT 1994. LNCS, vol. 950, pp. 341–355. Springer, Heidelberg (1995)

11. Biryukov, Alex, Roy, Arnab, Velichkov, Vesselin: Differential analysis of block ciphers SIMON and SPECK. **8540**, 546–570 (2015)

12. Bogdanov, A., Knezevic, M., Leander, G., Toz, D., Varici, K., Verbauwhede, I.: SPONGENT: A Lightweight Hash Function. In: Preneel and Takagi [22], pp. 312–325

13. Bogdanov, A.A., Knudsen, L.R., Leander, G., Paar, C., Poschmann, A., Robshaw, M., Seurin, Y., Vikkelsoe, C.: PRESENT: an ultra-lightweight block cipher. In: Paillier, P., Verbauwhede, I. (eds.) CHES 2007. LNCS, vol. 4727, pp. 450–466. Springer, Heidelberg (2007)

14. Boura, C., Naya-Plasencia, M., Suder, V.: Scrutinizing and improving impossible differential attacks: applications to CLEFIA, Camellia, LBlock and SIMON. In: Sarkar, P., Iwata, T. (eds.) ASIACRYPT 2014. LNCS, vol. 8873, pp. 179–199. Springer, Heidelberg (2014)

15. De Cannière, C., Preneel, B.: Trivium. In: Robshaw and Billet [23], pp. 244–266

16. Daemen, J., Rijmen, V.: The Design of Rijndael: AES - The Advanced Encryption Standard. Information Security and Cryptography. Springer, Heidelberg (2002)

17. Guo, J., Peyrin, T., Poschmann, A.: The PHOTON Family of Lightweight Hash Functions. In: Rogaway, P. (ed.) CRYPTO 2011. Lecture Notes in Computer Science, vol. 6841, pp. 222–239. Springer, Heidelberg (2011)

18. Guo, J., Peyrin, T., Poschmann, A., Robshaw, M.J.B.: The LED Block Cipher. In: Preneel and Takagi [22], pp. 326–341

19. Hell, M., Johansson, T., Maximov, A., Meier, W.: The grain family of stream ciphers. In: Robshaw and Billet [23], pp. 179–190

20. Kölbl, S., Leander, G., Tiessen, T.: Observations on the SIMON Block Cipher Family. In: Gennaro, R., Robshaw, M. (eds.) CRYPTO 2015. LNCS, vol. 9215, pp. 161–185. Springer, Heidelberg (2015)

21. Mouha, N., Wang, Q., Gu, D., Preneel, B.: Differential and Linear Cryptanalysis Using Mixed-Integer Linear Programming. In: Wu, C.-K., Yung, M., Lin, D. (eds.) Inscrypt 2011. LNCS, vol. 7537, pp. 57–76. Springer, Heidelberg (2012)

22. Preneel, B., Takagi, T. (eds.): CHES

23. Robshaw, M.J.B., Billet, O. (eds.): New Stream Cipher Designs - The eSTREAM Finalists. LNCS, vol. 4986. Springer, Heidelberg (2008)

24. Shi, D., Lei, H., Sun, S. Song, L., Qiao, K., Ma, X.: Improved Linear (hull) Cryptanalysis of Round-reduced Versions of SIMON. IACR Cryptology ePrint Archive 2014: 973 (2014)

25. Sun, S., Lei, H., Wang, M., Wang, P., Qiao, K., Ma, X., Shi, D., Song, L., Kai, F.: Towards Finding the Best Characteristics of Some Bit-oriented Block Ciphers and Automatic Enumeration of (Related-key) Differential and Linear Characteristics with Predefined Properties. IACR Cryptology ePrint Archive 2014: 747 (2014)

26. Sun, S., Hu, L., Wang, P., Qiao, K., Ma, X., Song, L.: Automatic security evaluation and (Related-key) differential characteristic search: application to SIMON, PRESENT, LBlock, DES(L) and other bit-oriented block ciphers. In: Sarkar, P., Iwata, T. (eds.) ASIACRYPT 2014. LNCS, vol. 8873, pp. 158–178. Springer, Heidelberg (2014)

27. Wang, N., Wang, X., Jia, K., Zhao, J.: Differential Attacks on Reduced SIMON Versions with Dynamic Key-guessing Techniques. IACR Cryptology ePrint Archive 2014: 448 (2014)

28. Wang, N., Wang, X., Jia, K., Zhao, J.: Improved Differential Attacks on Reduced SIMON Versions. IACR Cryptology ePrint Archive 2014: 448 (2014)

29. Wang, Qingju, Liu, Zhiqiang, Kerem Varici, Yu., Sasaki, Vincent Rijmen, Todo, Yosuke: Cryptanalysis of Reduced-Round SIMON32 and SIMON48. In: Meier, Willi, Mukhopadhyay, Debdeep (eds.) INDOCRYPT 2014. Lecture Notes in Computer Science, vol. 8885, pp. 143–160. Springer, Heidelberg (2014)

30. Yang, G., Zhu, B., Suder, V., Aagaard, M.D., Gong, G.: The simeck family of lightweight block ciphers. In: Güneysu, T., Handschuh, H. (eds.) CHES 2015. LNCS, vol. 9293, pp. 307–329. Springer, Heidelberg (2015)

Some Results Using the Matrix Methods on Impossible, Integral and Zero-Correlation Distinguishers for Feistel-Like Ciphers

Thierry P. Berger[1] and Marine Minier[2(✉)]

[1] XLIM (UMR CNRS 7252), Université de Limoges,
123 Avenue A. Thomas, 87060 Limoges Cedex, France
`thierry.berger@xlim.fr`
[2] INRIA, INSA-Lyon, CITI, Université de Lyon, Lyon, France
`marine.minier@insa-lyon.fr`

Abstract. While many recent publications have shown strong relations between impossible differential, integral and zero-correlation distinguishers for SPNs and Feistel-like ciphers, this paper tries to bring grist to the mill to this research direction by first, studying the Type-III, the Source-Heavy (SH) and the Target-Heavy (TH) Feistel-like ciphers regarding those three kinds of distinguishers. Second, this paper tries to make a link between the matrix methods used to find such distinguishers and the adjacency matrix of the graph of a Feistel-like cipher.

Keywords: Block ciphers · Feistel-like ciphers · Impossible differential · Zero-correlation · Integral · Matrix · $\mathcal{U}$-method and UID-method

1 Introduction

Impossible differential (ID) [3,11], integral (INT) [13] and multidimensional zero-correlation (ZC) [8] distinguishers are efficient ways to mount attacks against SPNs or Feistel-like block ciphers. Contrary to classical linear or differential distinguishers, the involved property holds with probability 1.

More precisely, an *ID distinguisher* looks for differentials with probability 0, i.e. which are impossible, a *ZC distinguisher* takes advantage of linear approximations that have probability 1/2 to happen, i.e. which have a correlation equal to 0 and finally an *INT distinguisher* predicts with probability 1 the sum taken on some ciphertexts computed from particular plaintexts (for example some parts of the plaintext set could be equal to constant whereas the other parts take all possible values).

In the literature, two ways have been conducted in parallel: the first one tries using a matrix representation of the round function to describe automatic methods to find elementary IDs with the so called $\mathcal{U}$-method [10], more sophisticated IDs on Feistel-like ciphers with the so-called UID-method [14] or on SPNs [19]; or ZC distinguishers [15]; or INT distinguishers [21]. The second one tries to exploit the matrix representation to find strong links between those different

© Springer International Publishing Switzerland 2015
A. Biryukov and V. Goyal (Eds.): INDOCRYPT 2015, LNCS 9462, pp. 180–197, 2015.
DOI: 10.1007/978-3-319-26617-6_10

attacks: between ZC and ID in [4], between ZC and INT in [7], and between the three in [5,6] and in [16,17].

So, many general schemes have been studied regarding the different attacks (for example, in [14], many Feistel-like ciphers have been studied regarding ID distinguishers) but only [5,6] and [16,17] consider the three attacks at the same time but limit their studies to the case of Type-I and Type-II Feistel-like ciphers and of SPN schemes.

In this paper, we first sum up in Table 1 the best existing distinguishers in those three attack models for most of existing Feistel-like ciphers and complete this Table (the bold results are the new ones). Note that the notation used in Table 1 is detailed in Sect. 2. More precisely, we study in details Type-III, Source-Heavy (SH) and Targer-Heavy (TH) especially regarding respectively ZC and INT distinguishers. We also derived a link between the $\mathcal{U}$-method and the adjacency matrix of the graph of a Feistel-like cipher under specific conditions.

This paper is organized as follows: some preliminary notation and all the existing works are summed up in Sect. 2. In Sect. 3, we give some insights on the distinguishers we found in the ID, ZC and/or INT contexts for the Type-III, the SH and the TH Feistel-like ciphers. In Sect. 4 we show how the adjacency matrix of the graph of a Feistel-like cipher could be directly used to compute the best distinguishers obtained through the $\mathcal{U}$-method or the methods derived from this one.

Table 1. The best ID, ZC and INT distinguishers for different Feistel-like ciphers with $b = 4$ branches found using the UID-method. The bold results are the new ones. In most cases, the notation l_1 and l_2 means that l_1 and l_2 are independent.

Block Cipher	Best ID dist.	Best ZC dist.	Best INT dist.
Gen-Skipjack	16 [14]	16 [4]	19 [12]
(rule-A)	$(0, l_1, 0, 0) \nrightarrow (l_2, l_2, 0, 0)$	$(l_1, 0, 0, 0) \rightarrow (l_2, l_2, 0, 0)$	$(m_3, m_3, m_1, m_2) \rightarrow (0, ?, ?, ?)$
Gen-Skipjack	16 [4]	16 [4]	19 [12]
(rule-B)	$(l_1, l_1, 0, 0) \nrightarrow (l_2, 0, 0, 0)$	$(l_1, l_1, 0, 0) \rightarrow (0, l_2, 0, 0)$	$(m_3, m_3, m_1, m_2) \rightarrow (0, ?, ?, ?)$
Type-I	19 [14]	**19** [4]	16 [16]
(Gen-CAST256)	$(0, 0, 0, l_1) \nrightarrow (l_1, 0, 0, 0)$	$(l_1, 0, 0, 0) \rightarrow (0, l_1, 0, 0)$	$(0, m_1, m_2, m_3) \rightarrow (?, 0, ?, ?)$
Type-II	9 [6]	9 [6]	8 [6]
(RC6)	$(0, 0, 0, l_1) \nrightarrow (0, 0, l_1, 0)$	$(l_1, 0, 0, 0) \nrightarrow (0, 0, 0, l_1)$	$(0, m_1, m_2, m_3) \rightarrow (?, ?, 0, ?)$
Type-II	7 [6]	7 [6]	8 [6]
(Nyberg)	$(0, 0, l_1, 0) \nrightarrow (l_2, 0, 0, 0)$	$(l_1, 0, 0, 0) \rightarrow (0, 0, 0, l_2)$	$(0, m_1, m_2, m_3) \nrightarrow (?, ?, ?, 0)$
Type-III	6 [20]	**6**	6 [20]
	$(0, 0, 0, l_1) \nrightarrow (0, 0, l_2, 0)$	$(0, l_1, 0, 0) \rightarrow (0, 0, 0, l_1)$	$(0, l_1, m_1, m_2) \rightarrow (?, ?, ?, 0)$
Type-III [20]	5 [20]	**5**	5 [20]
(1)	$(0, 0, 0, l_1) \nrightarrow (0, l_2, 0, 0)$	$(0, l_1, 0, 0) \rightarrow (0, 0, 0, l_1)$	$(0, l_1, m_1, m_2) \rightarrow (?, 0, 0, 0)$
Type-III [20]	5 [20]	**5**	5 [20]
(2)	$(0, 0, 0, l_1) \nrightarrow (0, 0, l_2, 0)$	$(l_1, 0, 0, 0) \rightarrow (0, 0, l_1, 0)$	$(0, 0, l_1, m_1) \rightarrow (?, ?, 0, 0)$
TH	11 [14]	11 [4]	**10**
(Gen-MARS)	$(0, 0, 0, l_1) \nrightarrow (l_1, 0, 0, 0)$	$(l_1, l_1, l_1, 0) \rightarrow (0, l_1, l_1, l_1)$	$(m_1, m_2, m_3, m_1) \rightarrow \sum = 0$
SH	11 [14]	11 [16]	**11**
(Gen-SMS4)	$(l_1, l_1, l_1, 0) \nrightarrow (l_1, l_1, l_1, 0)$	$(0, 0, 0, l_1) \nrightarrow (l_1, 0, 0, 0)$	$(m_2, m_1, l_1, l_1) \rightarrow (0, ?, ?, ?)$
Four-Cell	18 [14]	12 [4]	18 [21]
	$(l_1, 0, 0, 0) \nrightarrow (l_2, l_2, 0, 0)$	$(0, 0, 0, l_1) \nrightarrow (l_2, l_2, l_2, l_2)$	$(m_1, m_2, m_3, 0) \rightarrow \sum = 0$

2   Preliminaries

2.1   General Matrix Representation

From the notation and definitions introduced in [4,5,17], we have the following definition that covers the matrix representation of all the Feistel-like ciphers with bijective components studied in this paper:

Definition 1. *Omitting key and constant addition, the round function of a GFN with b branches of c-bit blocks, b even, can be matricially represented as a combination of four $b \times b$ matrices $\mathcal{F}$, $\mathcal{X}$, $\mathcal{Y}$ and $\mathcal{P}$ with coefficients $\{0, 1, F_i\}$ where the $\{F_i\}_{i \leq b-1}$ denote the internal non-linear bijective and balanced functions.*

- *Representing the non-linear layer (F-layer), the non-zero coefficients of the matrix $\mathcal{F}$ are equal to 1 in the diagonal and have coefficient F_i in row j and column ℓ if the input of the function F_i is given by the ℓ-th branch and the output is Xor-ed to the j-th branch.*
- *Representing the permutation of the branches (P-layer), the matrix $\mathcal{P}$ is a permutation matrix with only one non-zero coefficient per line and column.*
- *Representing the eventual linear layers (L-layer), the matrices $\mathcal{X}$ and $\mathcal{Y}$ contain only binary coefficients: the diagonal contains 1 and there are at maximum two 1 per row and column.*

From these four matrices, a Feistel-like round function can be represented by a $b \times b$ matrix $\mathcal{R}$ as $\mathcal{R} = \mathcal{P} \cdot \mathcal{X} \cdot \mathcal{F} \cdot \mathcal{Y}$, the inverse of the round function is $\mathcal{R}^{-1} = \mathcal{Y}^{-1} \cdot \mathcal{F} \cdot \mathcal{X}^{-1} \cdot \mathcal{P}^{-1}$.

Remark 1. As noticed in [4], in fact, only one linear layer is required if we omit the first $\mathcal{Y}$ linear transform. Indeed, in this case, the round function could be rewritten as $\mathcal{R}_* = \mathcal{Y} \cdot \mathcal{P} \cdot \mathcal{X} \cdot \mathcal{F} = \mathcal{P} \cdot (\mathcal{P}^{-1} \cdot \mathcal{Y} \cdot \mathcal{P} \cdot \mathcal{X}) \cdot \mathcal{F} = \mathcal{P} \cdot \mathcal{X}'_* \cdot \mathcal{F}$ with $\mathcal{X}'_* = \mathcal{P}^{-1} \cdot \mathcal{Y} \cdot \mathcal{P} \cdot \mathcal{X}$. We decide to maintain two L-layers for the sake of clarity. In many cases, such as Feistel Type-I or Feistel Type-II, we have: $\mathcal{Y} = \mathcal{X} = Id$. When not required and for the sake of simplicity, we sometimes identify the F_i to a single F function.

Example 1. Figure 1 represents one round of a Source-Heavy (SH) scheme (MARS-like cipher) and its equivalent representation according to Definition 1, one round of a Type-III and one round of a Type-II cipher.

For the SH scheme, the four matrices are:

$$\mathcal{Y} = \mathcal{X} = \begin{pmatrix} 1 & 0 & 0 & 0 \\ 0 & 1 & 0 & 0 \\ 0 & 1 & 1 & 0 \\ 0 & 1 & 0 & 1 \end{pmatrix}, \mathcal{F} = \begin{pmatrix} 1 & 0 & 0 & 0 \\ F & 1 & 0 & 0 \\ 0 & 0 & 1 & 0 \\ 0 & 0 & 0 & 1 \end{pmatrix}, \mathcal{P} = \begin{pmatrix} 0 & 1 & 0 & 0 \\ 0 & 0 & 1 & 0 \\ 0 & 0 & 0 & 1 \\ 1 & 0 & 0 & 0 \end{pmatrix} \text{ thus } \mathcal{R} = \begin{pmatrix} F & 1 & 0 & 0 \\ F & 0 & 1 & 0 \\ F & 0 & 0 & 0 \\ 1 & 0 & 0 & 0 \end{pmatrix}$$

For the Type-III, the four matrices are:

$$\mathcal{Y} = \mathcal{X} = Id, \mathcal{F} = \begin{pmatrix} 1 & 0 & 0 & 0 \\ F & 1 & 0 & 0 \\ 0 & F & 1 & 0 \\ 0 & 0 & F & 1 \end{pmatrix}, \mathcal{P} = \begin{pmatrix} 0 & 1 & 0 & 0 \\ 0 & 0 & 1 & 0 \\ 0 & 0 & 0 & 1 \\ 1 & 0 & 0 & 0 \end{pmatrix} \text{ thus } \mathcal{R} = \begin{pmatrix} F & 1 & 0 & 0 \\ 0 & F & 1 & 0 \\ 0 & 0 & F & 1 \\ 1 & 0 & 0 & 0 \end{pmatrix}$$

For the Type-II, we have:

$$\mathcal{Y} = \mathcal{X} = Id, \ \mathcal{F} = \begin{pmatrix} 1 & 0 & 0 & 0 \\ F & 1 & 0 & 0 \\ 0 & 0 & 1 & 0 \\ 0 & 0 & F & 1 \end{pmatrix}, \ \mathcal{P} = \begin{pmatrix} 0 & 1 & 0 & 0 \\ 0 & 0 & 1 & 0 \\ 0 & 0 & 0 & 1 \\ 1 & 0 & 0 & 0 \end{pmatrix} \ \text{thus} \ \mathcal{R} = \begin{pmatrix} F & 1 & 0 & 0 \\ 0 & 0 & 1 & 0 \\ 0 & 0 & F & 1 \\ 1 & 0 & 0 & 0 \end{pmatrix}$$

From [4, 15], we need another matrix representation to compute the ZC distinguishers on Feistel-like schemes using the matrix method. Indeed, from those results the so-called $\mathcal{U}$ and UID methods allow to find ZC distinguishers when applying those methods on the mirror representation of the round function.

Definition 2. *For a GFN, given the matrix representation of the round function $\mathcal{R} = \mathcal{P} \cdot \mathcal{X} \cdot \mathcal{F} \cdot \mathcal{Y}$, we call mirror function the round function described by the matrix $\mathcal{M} = \mathcal{P} \cdot \mathcal{X}^T \cdot \mathcal{F}^T \cdot \mathcal{Y}^T$, where the notation $\mathcal{A}^T$ denotes the transposition of the matrix $\mathcal{A}$. In the same way, we have: $\mathcal{R}_* = \mathcal{P} \cdot \mathcal{X}'_* \cdot \mathcal{F}$ and thus $\mathcal{M}_* = \mathcal{P} \cdot \mathcal{X}'^T_* \cdot \mathcal{F}^T$.*

2.2 Matrix Methods on GFNs for Finding ID, ZC and INT Distinguishers

We do not give here all the details concerning the matrix methods existing in the literature [10, 14, 15, 21], we only recall the general results given in [5] omitting to formally give the inconsistency rules and the passing rules through a XOR or an F function.

General Passing Rules and Inconsistency Rules. In the matrix methods, the input/output of the application of s rounds are represented by two column vectors V_0/W_0 of size b. Each vector coefficient belongs to the set $\{0, l_i, m_j, r_k\}$ where i, j and k are local counters. 0 means zero difference in the ID context, zero mask in the ZC context and constant value in the INT context. l_i means a known

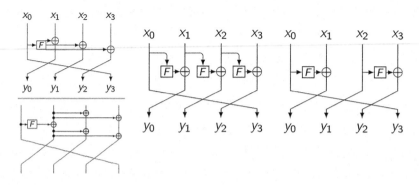

Fig. 1. One round of a Source-Heavy (SH) (MARS-like scheme) on the left, one round of a Type-III in the middle and one round of a Type-II on the right.

non-zero difference, a known non-zero mask and a known permutation property (i.e. a word takes all possible values of the set exactly once) according to each context. m_j means a non-zero difference, a non-zero mask and a permutation property according to each context. And finally, r_k means an unknown difference, an unknown mask and an unknown property according to each context. From V_0 and W_0, the matrix method consists in computing $V_i = \mathcal{A}^i \cdot V_0$ and $W_j = \mathcal{B}^j \cdot W_0$ where $\mathcal{A}$ and $\mathcal{B}$ are matrices among $\mathcal{R}, \mathcal{R}^{-1}, \mathcal{M}, \mathcal{M}^{-1}$. We denote by $V_i[p]$ or $W_j[p]$ the coordinate p of V_i or W_j. Of course, the matrix multiplication is "symbolic" and some rules must be defined to perform the operations (essentially XOR operations and F functions) linked with the matrix multiplication.

As demonstrated in [5], the rules to pass a XOR or an F function during the matrix multiplication when the matrix method is used in the different ID, ZC and INT contexts are the same. More precisely, passing an F function transforms a 0 into a 0, an l_i into an m_j, an m_j into an m_{j+1} and an r_k into an r_{k+1}. The same kind of rules holds for a XOR. For example a $l_i \oplus l_i$ becomes a 0 whereas a $m_i \oplus m_j$ becomes an r_k[1].

From the previous notation, we could easily derive inconsistencies for the ID and ZC contexts whereas the required properties for an INT distinguisher are given in Proposition 3. In the ID and ZC contexts, [5] gives 3 possible kinds of inconsistency:

– ID and ZC of type 1: $\exists\ p \in \{0, \cdots, (b-1)\}\ |\ (V_i[p] = 0$ and $W_j[p] = m_k)$ or $(V_i[p] = m_k$ and $W_j[p] = 0)$.
– ID and ZC of type 2: $\exists\ p \in \{0, \cdots, (b-1)\}\ |\ V_i[p] = l_k$ and $W_j[p] = l_{k'}$.
– ID and ZC of type 3: $\exists\ p \in \{0, \cdots, (b-1)\}\ |\ (V_i[p] = l_k \oplus m_{k'}$ and $W_j[p] = l_k)$ or $(V_i[p] = l_k$ and $W_j[p] = l_k \oplus m_{k'})$

ID Distinguisher. Using the matrix representation of the round function and the matrix methods described in [10,14], we can derive an ID distinguisher on a Feistel-like cipher as

Proposition 1. *Given V_0^{ID} and W_0^{ID} a representation of the input and output differences, we have an ID distinguisher (V_0^{ID}, W_0^{ID}) on $s_0 + s_1$ rounds if we have an inconsistency (as previously defined) between $\mathcal{R}^{s_0} \cdot V_0^{ID}$ and $\mathcal{R}^{-s_1} \cdot W_0^{ID}$.*

ZC Distinguisher. In [15], the matrix method to find ZC distinguishers is presented. It uses mirror matrix $\mathcal{M}$ instead of $\mathcal{R}$.

Proposition 2. *Given V_0^{ZC} and W_0^{ZC} a representation of the input and output linear masks, we have a ZC distinguisher (V_0^{ZC}, W_0^{ZC}) on $s_0 + s_1$ rounds if we have an inconsistency (as previously defined) between $\mathcal{M}^{s_0} \cdot V_0^{ZC}$ and $\mathcal{M}^{-s_1} \cdot W_0^{ZC}$.*

INT Distinguisher. In [21], the authors are interested in structural integral attacks (i.e. attacks that produce particular properties on input/output c-bit words) and proposed an algorithm to automatically find INT distinguishers.

---

[1] We refer to [5] for the complete table describing the XOR effect in the matrix method.

The algorithm uses the same rules and the same matrix representation $\mathcal{R}$ as in the ID context, only the termination rules differ.

Proposition 3 ([5,21]). *Given Z^{INT} a representation of the state in the INT context, we have an INT distinguisher on $s_0 + s_1$ rounds if the following rules are fulfilled:*

- *Termination at the end of an INT distinguisher: If $W_0^{INT} = \mathcal{R}^{s_1} \cdot Z^{INT}$ is such that there exists two set $I, J \subset \{0, \cdots, (b-1)\}$ with $\oplus_{p \in I} W_0^{INT}[p] = \oplus_{j \in J} m_j$ and s_1 is the greatest integer for this property to hold.*
- *Termination at the beginning of an INT distinguisher: If $V_0^{INT} = \mathcal{R}^{-s_0} \cdot Z^{INT}$ is such that $\exists\, i \in \{0, \cdots, (b-1)\}$ with $V_0^{INT}[i] = 0$ (with classical notation $= C$) and s_0 is the greatest integer for this property to hold.*

2.3   Equivalence Between ID, ZC and INT Distinguishers

Many papers [4,5,7,17] show the equivalence between the three kinds of distinguishers. We sum up in this Subsection all those results. First and from [4,17], we have the following Theorem to characterize the link between ID and ZC distinguishers.

Theorem 1 ([4,17]). *Let $\mathcal{R}$ be the matrix representation of the round function of a GFN and $\mathcal{M}$ be the matrix representation of its mirror function as given in Definition 2. If there exists a $b \times b$ permutation matrix $\mathcal{Q}$ such that $\mathcal{R} = \mathcal{Q} \cdot \mathcal{M} \cdot \mathcal{Q}^{-1}$ or $\mathcal{R} = \mathcal{Q} \cdot \mathcal{M}^{-1} \cdot \mathcal{Q}^{-1}$ we deduce that: an impossible differential distinguisher on s rounds involving a number of differentials equal to M exists if and only if a zero-correlation linear distinguisher on s rounds involving M linear masks exists.*

Remark 2. As shown in Table 1 and in [4], the following schemes verify this Theorem and thus their ID and ZC distinguishers reach the same number of rounds: Gen-Skipjack (rule-A alone), Gen-Skipjack (rule-B alone), Type-I (Gen-CAST256), all the Type-II (RC6, Nyberg and the ones proposed in [18]), all the Type-III, Target Heavy (TH, Gen-MARS), Source Heavy (SH, Gen-SMS4). Finally, only Four-Cell does not verify this property.

The following Theorem from [17] gives a simple equivalence between a ZC distinguisher and some particular INT distinguishers. Moreover, in [17], some results are given concerning the link between an ID distinguisher and an INT distinguisher in the case where the block cipher is a Feistel with two branches and with internal SP functions or the block cipher is an SPN.

Theorem 2 ([17]). *A non trivial ZC distinguisher of a block cipher always implies the existence of an INT distinguisher.*

More precisely, suppose that A and B are the two subspaces linked with the vectors V_0^{ZC} and W_0^{ZC}. It means that if we take $a \in A^*$, $b \in B$, $\{0, a\} \to b$ is a ZC distinguisher. Thus, an INT distinguisher could be deduced from this ZC distinguisher as if $V = \{0, a\}$, then $b \cdot E(x)$ where E is the block cipher under study, is balanced on $V^{\perp} = \{x \in (\mathbb{F}^c)^b | v \cdot x = 0, v \in V\}$.

However, as in this paper we focus on structural INT distinguishers, this kind of distinguishers do not interest us because Theorem 2 does not apply when the input/output masks are dependent, i.e. $l_1 \neq l_2$, $l_1 = l_2$ or $m_1 = F(l_1)$. We finally called INT1 distinguishers the distinguishers constructed using Theorem 2 and INT2 distinguishers, structural distinguishers built using the matrix method described in Proposition 3. Moreover, from [17], an INT1 distinguisher could always been deduced from an ID or a ZC distinguisher. So, the best INT1 distinguisher depends on the best ID or ZC distinguisher which is not the case for an INT2 distinguisher because when using the matrix method, the matrices used for building an INT2 distinguisher are different from the ones used to build ID or ZC distinguishers.

3   Distinguishers

In this Section, we will describe the way we have completed Table 1 with several discussions considering always that $b = 4$. We first concentrate our efforts on the Type-III schemes because up to our knowledge this is the less studied schemes in the three attack contexts. More precisely, we only found [20] that studies how to improve diffusion of Type-III and gives some results on best ID and INT distinguishers for Type-III and the modified Type-III they proposed.

3.1   Results on Type-III

In this Subsection, we will study the classical Type-III with 4 branches regarding ID, ZC and INT distinguishers and also the two variants proposed in [20] that improve the diffusion delay by changing the $\mathcal{P}$ matrix. The two new schemes that we will call Type-III(1) and Type-III(2) have the following permutation matrices: $\mathcal{P}_1 = \begin{pmatrix} 0\,1\,0\,0 \\ 0\,0\,0\,1 \\ 0\,0\,1\,0 \\ 1\,0\,0\,0 \end{pmatrix}$ and $\mathcal{P}_2 = \begin{pmatrix} 0\,0\,1\,0 \\ 0\,1\,0\,0 \\ 0\,0\,0\,1 \\ 1\,0\,0\,0 \end{pmatrix}$.

ID Distinguishers. In Table 3 of [20], the best ID distinguishers found for Type-III, Type-III(1) and Type-III(2) are on 6, 5 and 5. We retrieve those results. More precisely, the ID distinguishers of the three schemes are summed up in Table 2 using the notation of Subsect. 2.2. We will define later an ID of type 4.

Among the six ID distinguishers presented in Table 2, one is of type 1, three are of type 3 (i.e. the difference values in input/output are both equal to l_1) as defined in Sect. 2 and the two others (called type 4) work only if an inconsistency of type $l_1 \nrightarrow m_5$ or of type $m_3 \nrightarrow l_2$ (remember that l_i is a fixed known non-zero difference and that m_i is an unknown non-zero difference) exists with "$m_5 = F(F(F(l_2))) \neq l_1$" or "$m_3 = F(l_1) \neq l_2$".

These kind of inconsistencies have been studied in [9]. More precisely, the authors defined the following set:

Table 2. The best ID distinguishers for Type-III, Type-III(1) and Type-III(2). The notation means $(x_0, x_1, x_2, x_3) = (0, 0, 0, l_1)$ for example.

Round Nb.	Type-III ID type 4	Type-III ID type 3	Type-III(1) ID type 4	Type-III(1) ID type 1	Type-III(2) ID type 4	Type-III(2) ID type 3
↓ 0	$(0,0,0,l_1)$	$(0,0,0,l_1)$	$(0,0,0,l_1)$	$(0,0,0,l_1)$	$(0,0,0,l_1)$	$(0,0,0,l_1)$
1	$(0,0,l_1,0)$	$(0,0,l_1,0)$	$(0,l_1,0,0)$	$(0,l_1,0,0)$	$(0,0,l_1,0)$	$(0,0,l_1,0)$
2	$(0,l_1,m_1,0)$	$(0,l_1,m_1,0)$	$(l_1,0,m_1,0)$	$(l_1,0,m_1,0)$	$(l_1,0,m_1,0)$	$(l_1,0,m_1,0)$
3	$(l_1,r_4,m_2,0)$	$(l_1,r_4,m_2,0)$	(m_3,m_2,m_1,l_1)	(m_3,m_2,m_1,l_1)	(m_3,m_2,m_1,l_1)	(m_3,m_2,m_1,l_1)
3	(m_5,r_1,r_2,r_3)	$(l_1 \oplus m_4,r_1,r_2,r_3)$	(l_2,m_4,m_5,m_6)	$(0,0,m_4,l_1 \oplus m_5)$	(l_2,m_4,m_5,m_6)	$(m_4,m_5,m_6,l_1 \oplus m_5)$
4	(l_2,m_3,m_4,m_5)	$(m_1,m_2,m_3,l_1 \oplus m_4)$	$(0,0,0,l_2)$	$(0,l_1,m_4,0)$	$(0,0,0,l_2)$	$(0,0,l_1,m_4)$
5	$(0,0,0,l_2)$	$(0,0,l_1,m_1)$	$(0,l_2,0,0)$	$(l_1,0,0,0)$	$(0,0,l_2,0)$	$(l_1,0,0,0)$
↑ 6	$(0,0,l_2,0)$	$(0,l_1,0,0)$	-	-	-	-
condition	$m_5 = \Delta F(\Delta F(\Delta F(l_2))) \neq l_1$	-	$m_3 = \Delta F(l_1) \neq l_2$	-	$m_3 = \Delta F(l_1) \neq l_2$	-

Definition 3 ([9]). *For a function $F(\cdot)$ and a set ΔS of input differences, we define the output difference set $\Delta F(\Delta S)$ to be the set containing all the output differences that are feasible by an input difference in ΔS. The quantity $\max_{|\Delta S| > 0} \frac{|\Delta F(\Delta S)|}{|\Delta S|}$ is called the differential expansion rate of $F(\cdot)$ and define as the maximal increase in the size of a difference set through the round function.*

We could map this definition by saying that to hold the three ID distinguishers of type 4 given in Table 2 must verify on their input/output the two following rules according the studied scheme $l_1 \notin \Delta F(\Delta F(\Delta F(\{l_2\})))$ and $l_2 \notin \Delta F(\{l_1\})$.

As noticed in [9], if F is a $n = 4$ or 8-bit bijective Sbox, the differential expansion rate of one value, i.e. $\Delta F(\{\alpha\})$, is equal to half of the values, i.e. 8 or 128.

Note also that an equivalent property could be defined in the case of a ZC distinguisher (this property will be used latter in this article).

Definition 4. *For a function $F(\cdot)$ and a set ΓL of input linear masks, we define the output linear mask set $\Gamma F(\Gamma L)$ to be the set containing all the output linear masks that are feasible by an input linear mask in ΓL. The quantity $\max_{|\Gamma L| > 0} \frac{|\Gamma F(\Gamma L)|}{|\Gamma L|}$ is called the linear expansion rate of $F(\cdot)$ and define as the maximal increase in the size of a linear mask set through the round function.*

ZC Distinguishers. We study in this Subsection the best ZC distinguishers we found for Type-III, Type-III(1) and Type-III(2). They have respectively 6, 5 and 5 rounds and are given in Table 3.

INT Distinguishers. We focus in this paragraph on the INT distinguishers of Type-III schemes that are of type INT2 as defined in Subsect. 2.2. The results we obtain are summed up in Table 4.

More precisely, for a Type-III, we obtain the following 6-round INT2 distinguisher: for all the $2^{(b-1)c} = 2^{3c}$ possible values of $(0, l_1, m_4, m_5)$ where l_1, m_4 and m_5 take all possible 2^c values, then after 6 rounds, the sum taken over all the $2^{(b-1)c}$ input values on the first position is equal to 0. Only one distinguisher is given here but many such distinguishers could be generated. The 5 rounds INT2 distinguishers of Type-III(1) and of Type-III(2) are derived in the same way but with only $2^{(b-2)c}$ input values for Type-III(2).

Table 3. The best type 3 ZC distinguishers for Type-III, Type-III(1) and Type-III(2).

Round Number	Type-III	Type-III(1)	Type-III(2)
↓ 0	$(0, l_1, 0, 0)$	$(0, l_1, 0, 0)$	$(l_1, 0, 0, 0)$
1	$(l_1, 0, 0, m_1)$	$(l_1, 0, 0, m_1)$	$(0, 0, 0, l_1)$
2	$(m_3, m_2, m_1, m_4 \oplus l_1)$	$(m_3, m_1, m_2, m_4 \oplus l_1)$	(m_1, m_2, l_1, m_3)
3	$(r_2, r_1, m_4 \oplus l_1, r_3)$	$(r_2, m_4 \oplus l_1, r_2, r_3)$	$(m_4 \oplus l_1, r_1, m_3, r_2)$
3	$(r_6, r_5, l_1, 0)$	$(m_5, l_1, 0, 0)$	$(l_1, 0, m_2, m_1)$
4	$(m_5, l_1, 0, 0)$	$(l_1, 0, 0, 0)$	$(0, 0, m_1, l_1)$
5	$(l_1, 0, 0, 0)$	$(0, 0, 0, l_1)$	$(0, 0, l_1, 0)$
↑ 6	$(0, 0, 0, l_1)$	-	-

3.2 Other Distinguishers

TH and SH Distinguishers When F Has an SP Structure. Linking together the results of [17] and of [4], the TH and SH schemes are mirror representations of one another according to Theorem 1 and the following ZC distinguisher on 12 rounds of an instantiated SH scheme (SMS4) has been introduced in Appendix B of [17] $(0, 0, 0, l_1) \rightarrow_{ENC_6^{SH}} (m_1 \oplus m_2 \oplus r_1, d \oplus m_2 \oplus r_1, m_1 \oplus r_1, m_1 \oplus m_2)$ is incompatible with $(l_1, 0, 0, 0) \rightarrow_{DEC_6^{SH}} (m_3 \oplus m_4, m_3 \oplus r_2, d \oplus m_4 \oplus r_2, m_3 \oplus m_4 \oplus r_2)$ if and only if $m_1 \notin \Gamma F(\{l_1\})$ and $l_1 \oplus m_1 \notin \Gamma F(\{l_1\})$ and $m_1 \notin \Gamma F(\{m_1\})$ and $m_1 \oplus l_1 \notin \Gamma F(\{m_1 \oplus l_1\})$.

From those results we could directly derive the following 12-round ZC and ID distinguishers for the TH and SH constructions. The previous 12-round ZC distinguisher on a SH scheme is directly equivalent to a 12-round ID distinguisher on a TH scheme due to the mirror property. The conditions stay the same except that the ΔF function is used instead of ΓF.

This distinguisher could be directly derived into a 12-round ID distinguisher for a SH scheme: $(l_1, l_1, l_1, 0) \rightarrow_{ENC_6^{TH}} (l_1, m_1, l_1 \oplus m_2, l_1 \oplus r_1)$ is incompatible with $(0, l_1, l_1, l_1) \rightarrow_{DEC_6^{TH}} (l_1 \oplus r_2, l_1 \oplus m_5, m_4, l_1)$ if and only if the condition $r_1 = r_2 = 0$ could not be fulfilled with $r_1 = F(m_1 \oplus m_2)$ and $r_2 = F(F(m_4) \oplus m_4)$ which happens if $m_1 \oplus F(m_4) \neq l_1$ and $m_2 \oplus m_4 \neq l_1$. Those two conditions could be fulfilled if l_1 is chosen such that $m_1 \notin \Delta F(\{l_1\})$ and $l_1 \oplus m_1 \notin \Delta F(\{l_1\})$

Table 4. The best INT2 distinguishers for Type-III, Type-III(1) and Type-III(2).

Round Number	Type-III	Type-III(1)	Type-III(2)
↑ 0	$(0, l_1, m_4, m_5)$	$(0, l_1, m_4, m_5)$	$(0, 0, l_1, m_1)$
↓ 1	$(l_1, 0, 0, 0)$	$(l_1, 0, 0, 0)$	$(l_1, 0, 0, 0)$
2	$(m_1, 0, 0, l_1)$	$(m_1, 0, 0, l_1)$	$(0, m_1, 0, l_1)$
3	$(m_2, 0, l_1, m_1)$	$(m_2, l_1, 0, m_1)$	$(m_2, m_1, l_1, 0)$
4	(m_3, l_1, r_1, m_2)	(r_1, m_1, m_2, m_3)	(r_1, r_2, m_3, m_2)
5	$(l_1 \oplus F(m_3), r_2, r_3, r_4)$	$(r_2, \sum F(m_3) \oplus m_2 = 0,$ $\sum F(m_2) \oplus r_1 = 0, \sum r_1 = 0)$	$(r_3, r_4, \sum F(m_3) \oplus m_1 = 0,$ $\sum r_1 = 0)$
6	$(r_5, r_6, r_7, \sum l_1 \oplus F(m_3) = 0)$	-	-

Table 5. The best INT2 distinguishers for SH and TH.

Round Nb.	SH	TH
0	$(l_1 \oplus m_2, m_1, l_1, l_1)$	$(m_3, m_3 \oplus m_5, m_4 \oplus m_5, m_3 \oplus m_5)$
1	(m_1, l_1, l_1, l_1)	(m_3, m_4, m_3, m_3)
2	$(l_1, l_1, l_1, 0)$	(l_1, m_3, m_3, m_3)
3	$(l_1, l_1, 0, l_1)$	$(0, 0, 0, l_1)$
4	$(l_1, 0, l_1, l_1)$	$(0, 0, l_1, 0)$
↑ 5	$(0, l_1, l_1, l_1)$	$(0, l_1, 0, 0)$
↓ 6	(l_1, l_1, l_1, m_3)	$(l_1, 0, 0, 0)$
7	$(l_1, l_1, m_3, l_1 \oplus m_4)$	(m_1, m_1, m_1, l_1)
8	$(l_1, m_3, l_1 \oplus m_4, l_1 \oplus r_1)$	$(m_1 \oplus m_2, m_1 \oplus m_2, l_1 \oplus m_2, m_1)$
9	$(m_3, l_1 \oplus m_4, l_1 \oplus r_1, l_1 \oplus r_2)$	$(r_1 \oplus m_1 \oplus m_2, r_1 \oplus l_1 \oplus m_2, r_1 \oplus m_1, m_1 \oplus m_2)$
10	$(l_1 \oplus m_4, l_1 \oplus r_1, l_1 \oplus r_2, m_3 \oplus r_3)$	$\sum$ all words $= 0$
11	$(\sum l_1 \oplus r_1 = 0, l_1 \oplus r_2, m_3 \oplus r_3, r_5)$	-

and $m_1 \notin \Delta F(\{m_1\})$ and $m_1 \oplus l_1 \notin \Delta F(\{m_1 \oplus l_1\})$. This distinguisher also corresponds to the 12-round ZC distinguisher for a TH scheme.

Note that the conditions used for those distinguishers to work mainly depend on the F function. Indeed, if F is a single S-box, the two conditions $m_1 \notin \Delta F(\{m_1\})$ and $m_1 \oplus l_1 \notin \Delta F(\{m_1 \oplus l_1\})$ could not be fulfilled. But if F has an SP structure as in SMS4 there could exist some values that verify those two conditions depending on the P-layer properties. Thus, in some particular instantiated cases, those 12 rounds distinguishers could exist.

However, as those distinguishers reduce the number of inputs/outputs by adding some particular constraints, an important issue (not treated in this paper) concerns the possible extensions into attacks of those distinguishers: do they allow to attack the same number of rounds when adding some rounds to guess the subkeys at the beginning and at the end?

INT2 Distinguishers. In Table 5, we present the results we obtain in terms of INT2 distinguishers on SH and TH schemes. Note that for the instantiated SH called SMS4, an INT1 distinguisher on 12 rounds is given in [17].

For the SH construction with $b = 4$ branches, the 11-round INT2 distinguisher works as follows: for the $2^{(b-1)c} = 2^{3c}$ possible values $(l_1 \oplus m_2, m_1, l_1, l_1)$ where $l_1 \oplus m_2$, m_1 and l_1 take all possible 2^c values, then after 11 rounds, the sum taken over all the $2^{(b-1)c}$ input values on the first position is equal to 0. Note also that only one distinguisher is given here but many such distinguishers could be generated as soon as the F function is bijective.

Surprisingly, for the TH construction, we are only able to exhibit a 10-round INT1 distinguisher that works as follows: for the $2^{(b-1)c} = 2^{3c}$ possible values $(m_3, m_3 \oplus m_5, m_4 \oplus m_5, m_3 \oplus m_5)$ where m_3, $m_3 \oplus m_5$ and $m_4 \oplus m_5$ take all possible 2^c values, then after 10 rounds, the sum of all the 4 output words taken over all the $2^{(b-1)c}$ input values is equal to 0. Many other such distinguishers could be generated according the initial position of the first l_1 as soon as the F function is bijective.

ZC Distinguishers on 19 Rounds of Type-I and on 12 Rounds of Four-Cell. Those two distinguishers are given in Appendix A. Note that the existence of these distinguishers is known from [4] but not the detailed paths given here for the first time.

Concluding Remarks. Whereas the link between ID and ZC distinguishers has been clearly established in [4], the link between ID or ZC and INT2 distinguishers seems less evident. Indeed, the number of rounds on which the best INT2 distinguisher applies could be smaller (Type-I, Type-II, TH) than both ID and ZC best distinguishers, equal to the number of rounds of the best ID distinguisher (all Type-III, SH, Four-Cell) or greater (Gen-Skipjack rule A and B, Type-II Nyberg as shown in [6]). Of course, it is a direct consequence of the matrix method that differently finds those three kinds of distinguishers.

4   Matrix Method: Another Point of View

Under the same conditions than the ones given in Subsect. 2.2, we could directly read the three types of distinguishers on the successive powers of the matrix representation.

4.1   Links Between the Matrix Representation and the Adjacency Matrix of the Graph

To any round function of an iterative block cipher acting on b blocks, we can associate a connection graph between the input blocks and the output blocks of a round function.

Taking as example the Type-II round function described in the middle of Fig. 1, the matrix of the round function $\mathcal{R} = \begin{pmatrix} F & 1 & 0 & 0 \\ 0 & 0 & 1 & 0 \\ 0 & 0 & F & 1 \\ 1 & 0 & 0 & 0 \end{pmatrix}$ could also be seen as the adjacency matrix $\mathcal{A}$ with entries in $\mathbb{Z}[F]$ associated with the following graph.

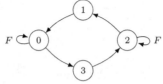

Suppose now that we consider s_0 iterations of a round function. The coefficient $A_{i,j}^{s_0}$ gives the number of paths between the i-th block in input and the j-th block in output. With the Type-II example, we have: $\mathcal{A}^3 = \begin{pmatrix} F^3 & F^2 & 2F & 1 \\ 1 & 0 & F^2 & F \\ 2F & 1 & F^3 & F^2 \\ F^2 & F & 1 & 0 \end{pmatrix}$.

It means that there exists one path from the input x_0 to the output y_1, two paths from the input x_0 to the output y_2, and no path from the input x_1 to the output y_1. Clearly, we derive directly an ID distinguisher from this last property.

The diameter of such graph, i.e. the least value of s_0 such that A^{s_0} does not have any zero coefficient, is then a meaningful measure called diffusion delay in [2] and in [1]. The minimum diffusion delay corresponds also to the maximum diffusion rounds as defined in [18]. We denote this value as d_0.

Since $\mathcal{A}^4 = \begin{pmatrix} F^4+1 & F^3 & 3F^2 & 2F \\ 2F & 1 & F^3 & F^2 \\ 3F^2 & 2F & F^4+1 & F^3 \\ F^3 & F^2 & 2F & 1 \end{pmatrix}$, the diffusion delay of Type-II is $d_0 = 4$.

However, if we look in more details to the output of the 4-th round, we can observe that the dependency between x_1 and y_1 is in fact linear due to the coefficient $\mathcal{A}^4_{1,1} = 1$, more precisely, $y_1 = x_1 \oplus c$ where c is a constant term depending only on x_0, x_2 and x_3. With the matrix representation, we are able to distinguish the XOR-linear operations (1 in the matrix $\mathcal{A}^{s_0}$) and the action of the F functions (with a polynomial form $\sum a \cdot F^b$ in the matrix $\mathcal{A}^{s_0}$).

Computing $\mathcal{A}^5$, we obtain $\mathcal{A}^5 = \begin{pmatrix} F^5+F^3 & F^4+1 & 4F^3 & 3F^2 \\ 3F^2 & 2F & F^4+1 & F^3 \\ 4F^3 & 3F^2 & F^5+3F & F^4+1 \\ F^4+1 & F^3 & 3F^2 & 2F \end{pmatrix}$.

As all the coefficients of the matrix $\mathcal{A}^5$ are polynomials of degree at least 1, it means that for any input x_i and any output y_j after 5 rounds there exists at least one path from x_i to y_j that has crossed at least one F function. We call the smallest integer satisfying this condition the depth of the GFN and we denote it by d_1.

4.2 Link Between Powers of the Adjacency Matrix and the $\mathcal{U}$-Method

In this Subsection, we limit our study to quasi-involutive Generalized Feistel Networks (GFNs) and their matrix representation as defined in [2] with the additional condition that the F functions must be bijective.

Definition 5. *A matrix $\mathcal{R}$ with coefficients in $\{0, 1, F\} \subset \mathbb{Z}[F]$ is a quasi-involutive GFN matrix if it can be written as $\mathcal{R} = \mathcal{P} \cdot \mathcal{F}$ (i.e. $\mathcal{X} = \mathcal{Y} = Id$ if Definition 1 is used) such that $\mathcal{P}$ is a permutation matrix and the matrix $\mathcal{F}$ satisfies the following conditions:*

1. *the main diagonal is filled with 1,*
2. *the off-diagonal coefficients are either 0 or F,*
3. *for each index i, row i and column i cannot both have an F coefficient.*
4. *for each index i, there is at most one F per row and per column.*

As shown in [2], the condition on $\mathcal{F}$ is equivalent to $\mathcal{F}^{-1} = 2\mathcal{I} - \mathcal{F}$ in $\mathbb{Z}[F]$, which simply means that the matrix representation of the decryption round function is $\mathcal{R}^{-1} = \mathcal{F} \cdot \mathcal{P}^T$ which can be rewritten $\mathcal{R}^{-1} = \mathcal{P}' \cdot \mathcal{F}'$ with $\mathcal{P}' = \mathcal{P}^T = \mathcal{P}^{-1}$ and $\mathcal{F}' = \mathcal{P} \cdot \mathcal{F} \cdot \mathcal{P}'$, which is also a quasi-involutive GFN matrix.

This definition covers Type-I, Type-II and its generalizations introduced in [18] and also Type-III and its generalizations introduced in [20] but does not cover Gen-Skipjack Rule-A and Rule-B, Four-Cell, TH and SH Feistel-like ciphers. The restricted conditions added to this new definition compared to Definition 1 allow to match the order of evaluation of an F function and of a XOR operation in a GFN to products of adjacency matrices for which the addition is a concatenation of paths.

We are now interested in using this restricted definition to study, for a given GFN with its associated matrix $\mathcal{A} = \mathcal{FP}$, the coefficients of $\mathcal{A}^{s_0}_{i,j}$. There are only the 5 following possibilities[2]:

1. If $\mathcal{A}^{s_0}_{i,j} = 0$, it means that, after s_0 rounds, the output block y_j does not depend on the input value x_i.
2. If $\mathcal{A}^{s_0}_{i,j} = 1$, it means that, after s_0 rounds, for any fixed input value x_k, $k \neq i$, the output block y_j is of the form $y_j = x_i \oplus c$, where c is a constant depending only on the x_k values, $k \neq i$.
3. If $\mathcal{A}^{s_0}_{i,j} = F^k$, i.e. is a monic monomial of degree at least 1, it means that, after s_0 rounds, there is only one path from x_i to y_j, and this path has crossed at least one F function. As the F function is bijective, it means that, after s_0 rounds, the output block y_j is of the form $y_j = p(x_i)$, where p is a random permutation depending on the x_k with $k \neq i$.
4. If $\mathcal{A}^{s_0}_{i,j} = F^k + 1$, i.e. a monic monomial of degree at least 1 added with the unit, it means that, after s_0 rounds, there is exactly two paths from x_i to y_j, and one of those paths has crossed at least one F function. As the F function is bijective, it means that, after s_0 rounds, the output block y_j is of the form $y_j = a(x_i) = p(x_i) + x_i$, where p is a random permutation. Thus it means that the dependency between y_j and x_i is the composition of two known paths.
5. If $\mathcal{A}^{s_0}_{i,j}$, is a polynomial such that the evaluation of this polynomial in $F = 1$ is at least 2 and not of the previous particular form, i.e. there are at least two paths from x_i to y_j, and at least two paths have crossed at least one F function. It means that, after s_0 rounds, the output block y_j is of the form $y_j = f(x_i)$, where f is an unknown function.

Note that the distinction between the two last cases makes sense only if the F functions are permutations, which is the case in our definitions.

We denote by d_2 the minimum value of s_0 such that all the entries of $\mathcal{A}^{s_0}$ correspond to the 5-th case. With the Type-II example, we substitute respectively p, a and f to the coefficients of $\mathcal{A}^4$ corresponding to cases 3, 4 and 5. This

leads to $\mathcal{A}^4 = \begin{pmatrix} a & p & f & f \\ f & 1 & p & p \\ f & f & a & p \\ p & p & f & 1 \end{pmatrix}$.

Suppose now that we want to construct an ID distinguisher using this matrix representation. For example, if we consider a vector of the form $(0, 0, 0, \delta)^T$,

---

[2] In particular a coefficient 2 could not appear due to the restricted previous definition where a receiver could not receive twice.

multiplying it by the matrix $\mathcal{A}^4$ leads to $(f(\delta), p(\delta), p(\delta), \delta)^T$. This result can be directly interpreted as $(r_1, m_1, m_2, l_1)^T$ considering the input vector $(0, 0, 0, l_1)^T$ using the previous notation. As an ID distinguisher is composed of a search in the encryption direction and a search in the decryption direction, we also introduce the corresponding adjacency matrix $\mathcal{B}$ for the decryption round as

$$\mathcal{B} = \begin{pmatrix} 0 & 0 & 0 & 1 \\ 1 & 0 & 0 & F \\ 0 & 1 & 0 & 0 \\ 0 & F & 1 & 0 \end{pmatrix}, \text{ and then } \mathcal{B}^5 = \begin{pmatrix} p & f & f & a \\ a & f & f & f \\ f & a & p & f \\ f & f & a & f \end{pmatrix}. \text{ As done in the encryption direction, if}$$

an input vector of difference in the decryption direction is of the form $(0, 0, \delta, 0)^T$, we obtain after 5 decryption rounds, the vector $(f(\delta), f(\delta), p(\delta), a(\delta))^T$, which could be written as $(r_1, r_2, m_3, m_4 \oplus l_1)$ with the previous notation creating an inconsistency of type 3 between the input $(0, 0, 0, \delta)^T$ and the output $(0, 0, \delta, 0)^T$.

Thus using the notation p, a and f, we could reinterpret the 3 kinds of inconsistencies: an inconsistency of type 1 means that there exists three indices i, j and k such that $(\mathcal{A}^{s_0}_{i,j} = 0$ and $\mathcal{B}^{s_1}_{i,k} = p)$ or $(\mathcal{A}^{s_0}_{i,j} = p$ and $\mathcal{B}^{s_1}_{i,k} = 0)$; an inconsistency of type 2 means that there exists three indices i, j and k such that $\mathcal{A}^{s_0}_{i,j} = \mathcal{B}^{s_1}_{i,k} = 1$; an inconsistency of type 3 means that there exists three indices i, j and k such that $(\mathcal{A}^{s_0}_{i,j} = 1$ and $\mathcal{B}^{s_1}_{i,k} = a)$ or $(\mathcal{A}^{s_0}_{i,j} = a$ and $\mathcal{B}^{s_1}_{i,k} = 1)$.

This method is in fact equivalent to the $\mathcal{U}$-method and gives the same results. Clearly, it cannot be extended to the case of differences located on several blocks in input and in output. Indeed, first the sequential order of the operations is not included in our adjacency matrix method and second, it seems that when several coordinates of the input vectors are not equal to 0, it becomes very difficult to read the matrix and thus to find the corresponding ID distinguisher. Thus our adjacency matrix method fully captures the $\mathcal{U}$-method but not the UID-method. In this last case, the adjacency matrix method could be combined with the resolution matrix method proposed in [19] to simplify the search step of the UID-method. More precisely, the aim of the algorithm will be to compute all the possible column combinations of the successive powers of the adjacency matrix.

Moreover, as the UID-method and the $\mathcal{U}$-method give the same results when considering quasi-involutive GFNs, this is why we limit our study to this case where the best ID distinguishers have only one active coordinate in the output/input vectors. This fact is mainly linked with the form of the adjacency matrices of quasi-involutive GFNs and especially to the fact that the matrix $\mathcal{F}$ contains at most one F on each row and each column. It means that a difference on a block cannot cross distinct F functions at a given round and thus avoid correlations between differences on distinct blocks in input.

Note that we could also find the corresponding ZC and INT2 distinguishers induced by the $\mathcal{U}$-method always using the adjacency matrix method. Indeed in the case of an ID distinguisher, we look at $\mathcal{A}^{s_0}$ and at $\mathcal{B}^{s_1}$ computed from the matrices $\mathcal{R}$ and $\mathcal{R}^{-1}$. In the case of a ZC distinguisher, we do the same study but on the adjacency matrices computed from the matrices $\mathcal{M}$ and $\mathcal{M}^{-1}$ corresponding with the mirror graph. For an INT2 distinguisher, we are looking for the best possible composition between $\mathcal{A}^{s_0}$ and $\mathcal{B}^{s_1}$ on a given coordinate, i.e. we look at the same column in $\mathcal{A}^{s_0}$ and $\mathcal{B}^{s_1}$. However, in this case, we need to specify if the permutations p of $\mathcal{A}^{s_0}$ and $\mathcal{B}^{s_1}$ are the same or not and thus to denote them by $p_1, \cdots p_t$.

Thus, we have the following conjecture concerning the ID distinguishers that could be directly mapped in the case of ZC and INT2 distinguishers. First, Let B be the least number of rounds for which a structural ID distinguisher does not exist (here a structural distinguisher is independent of the properties of the F functions). Let d_i^+ and d_i^-, $i \in \{0, 1, 2\}$ respectively denote the diffusion delay and the two kinds of depth for encryption and decryption directions ($+$ for encryption, $-$ for decryption) previously defined.

Following our experimental results and our previous remarks, we propose the following conjectures:

Conjecture 1. We consider the family of quasi-involutive GFNs defined in Definition 5 corresponding to the adjacency matrix $\mathcal{A} = \mathcal{P} \cdot \mathcal{F}$.

- If there exists an ID distinguisher on s rounds, the $\mathcal{U}$-method provides an ID distinguisher on s rounds (*i.e.* the UID-method is not more efficient than the $\mathcal{U}$-method when considering quasi-involutive GFNs).
- The bound B on the minimal number of rounds for which a structural ID distinguisher exists (*i.e.* an ID distinguisher which is independent of the properties of the F functions) satisfies the following inequalities: $B \leq d_1^+ + d_1^-$, $B \leq d_0^+ + d_2^-$ and $B \leq d_2^+ + d_0^-$.

Note that this conjecture is false for Gen-Skipjack rule-A and rule-B, Source Heavy, Target Heavy GFNs and Four-Cell where the UID-method gives better results than the $\mathcal{U}$-method because $\mathcal{X} \neq Id$ or $\mathcal{Y} \neq Id$.

We sum up in Table 8 given in Appendix B the value of d_0, d_1 and d_2 for the Feistel-like ciphers we have studied in this paper, with $d_i = \max(d_i+, d_i-)$ for i from 0 to 2 regarding the matrices $\mathcal{R}$ and $\mathcal{R}^{-1}$. Note that those results could be slightly modified when computing the same bounds for $\mathcal{M}$ and $\mathcal{M}^{-1}$.

5   Conclusion

In this paper, we have first completed the missing parts of Table 1 in terms of ID, ZC and INT2 distinguishers using the classical matrix methods. We have essentially focused our work on Type-III, SH and TH Feistel-like ciphers. We have also proposed another view of how to use the matrix representation to find such distinguishers. We finally conjecture that this new point of view leads to upper bounds on the best ID, ZC and INT2 distinguishers that could be found using the matrix methods especially the $\mathcal{U}$-method.

In future works, we want to see if those bounds could be also applied in the case of the UID-method and in the context of classical security proofs. Another research direction concerns the extension of the matrix methods to other attacks such as meet-in-the middle attacks against Feistel-like and SPN ciphers.

Acknowledgment. The authors would like to thank Céline Blondeau for our fruitful discussions and the anonymous referees for their valuable comments. This work was partially supported by the French National Agency of Research: ANR-11-INS-011.

A ZC Distinguishers on 19 Rounds of Type-I and on 12 Rounds of Four-Cell

A.1 ZC Distinguishers on 19 Rounds of Type-I

If the round function of a Type-I is bijective, then the 19-round ZC linear hull $(l_1, 0, 0, 0) \rightarrow (0, l_1, 0, 0)$ has zero correlation. The details of this ZC distinguisher is given in Table 6.

Table 6. 19-Round ZC dinstinguisher for Type-I

Round Nb.	x_0	x_1	x_2	x_3
0 ↓	l_1	0	0	0
1	0	0	0	l_1
2	0	0	l_1	0
3	0	l_1	0	0
4	l_1	0	0	m_1
5	0	0	l_1	m_1
6	0	m_1	l_1	0
7	m_1	l_1	0	m_2
8	l_1	0	m_2	r_1
9	0	m_2	r_1	l_1
10	m_2	r_1	l_1	m_3
11	r_1	l_1	m_3	r_2
12	l_1	m_3	r_2	r_3
12	$m_7 \oplus l_1$	m_6	m_5	m_4
13	m_6	m_5	m_4	l_1
14	m_5	m_4	l_1	0
15	m_4	l_1	0	0
16	l_1	0	0	0
17	0	0	0	l_1
18	0	0	l_1	0
19 ↑	0	l_1	0	0

A.2 ZC Distinguishers on 12 Rounds of Four-Cell

If the round function of Four-Cell is bijective, then the 12-round ZC linear hull $(0, 0, 0, l_1) \rightarrow (l_2, l_2, l_2, l_2)$ has zero correlation. The details of this ZC distinguisher is given in Table 7.

B Table of the Values of d_0, d_1 and d_2

It is easy to see that the bounds given in Conjecture 1 are false for SH, TH and Gen-Four-Cell ciphers as for example the best ID distinguisher given by the UID-method on Gen-Four-Cell is on 18 rounds whereas the best value of B is upper bounded by 14.

Table 7. 12-Round ZC dinstinguisher for Four-Cell

Round Nb.	x_0	x_1	x_2	x_3
$0 \downarrow$	0	0	0	l_1
1	0	0	l_1	0
2	0	l_1	0	0
3	l_1	0	0	0
4	m_1	m_1	m_1	m_1
5	$m_1 \oplus m_2$	$m_1 \oplus m_2$	$m_1 \oplus m_2$	m_2
7	0	m_5	$m_3 \oplus m_4$	0
6	m_5	$m_3 \oplus m_4$	0	0
5	m_4	m_3	m_3	m_3
4	0	0	0	m_3
3	0	0	m_3	0
2	0	m_3	0	0
1	m_3	0	0	0
$0 \uparrow$	l_2	l_2	l_2	l_2

Table 8. Value of d_0, d_1 and d_2 for different Feistel-like schemes with k branches.

Feistel Type	d_0	d_1	d_2
Gen-Skipjack (rule-A)	$k^2 - 2k + 2$	$k^2 - k + 1$	k^2
Gen-Skipjack (rule-B)	$k^2 - 2k + 2$	$k^2 - k + 1$	k^2
Type-1 (Gen-CAST256)	$(k-1)^2 + 1$	$k(k-1) + 1$	$k^2 + 1$
Type-2 (RC6)	k	$k + 1$	$k + 3$
Type-2 Nyberg	k	$k + 1$	$k + 3$
Type-2 [18]	$2\log_2 k$	$2\log_2 k + 1$	$2\log_2 k + 1$
Type-3	k	$k + 1$	$k + 2$
Type-3(1) and (2) [20] ($k = 4$)	4	4	6
Type-3(1) and (2) [20] ($k = 8$)	4	5	6
SH (Gen-SMS4)	k	$k + 1$	$k + 2$
TH (Gen-MARS)	k	$k + 1$	$k + 2$
Gen-Four-Cell	k	$2k - 1$	$2k - 1$

References

1. Arnault, F., Berger, T.P., Minier, M., Pousse, B.: Revisiting LFSRs for cryptographic applications. IEEE Trans. Inf. Theory **57**(12), 8095–8113 (2011)
2. Berger, T.P., Minier, M., Thomas, G.: Extended generalized feistel networks using matrix representation. In: Lange, T., Lauter, K., Lisoněk, P. (eds.) SAC 2013. LNCS, vol. 8282, pp. 289–305. Springer, Heidelberg (2014)
3. Biham, E., Biryukov, A., Shamir, A.: Cryptanalysis of Skipjack reduced to 31 rounds using impossible differentials. In: Stern, J. (ed.) EUROCRYPT 1999. LNCS, vol. 1592, pp. 12–23. Springer, Heidelberg (1999)
4. Blondeau, C., Bogdanov, A., Wang, M.: On the (In)equivalence of impossible differential and zero-correlation distinguishers for Feistel- and Skipjack-type ciphers. In: Boureanu, I., Owesarski, P., Vaudenay, S. (eds.) ACNS 2014. LNCS, vol. 8479, pp. 271–288. Springer, Heidelberg (2014)

5. Blondeau, C., Minier, M.: Analysis of impossible, integral and zero-correlation attacks on type-ii generalized Feistel networks using the matrix method. In: Leander, G. (ed.) FSE 2015. LNCS, vol. 9054, pp. 92–113. Springer, Heidelberg (2015)

6. Blondeau, C., Minier, M.: Relations between Impossible, Integral and Zero-correlation Key-Recovery Attacks (extended version). Cryptology ePrint Archive, Report 2015/141 (2015). http://eprint.iacr.org/

7. Bogdanov, A., Leander, G., Nyberg, K., Wang, M.: Integral and multidimensional linear distinguishers with correlation zero. In: Wang, X., Sako, K. (eds.) ASIACRYPT 2012. LNCS, vol. 7658, pp. 244–261. Springer, Heidelberg (2012)

8. Bogdanov, A., Rijmen, V.: Zero-correlation linear cryptanalysis of block ciphers. IACR Cryptology ePrint Arch. **2011**, 123 (2011)

9. Bouillaguet, C., Dunkelman, O., Fouque, P.-A., Leurent, G.: New insights on impossible differential cryptanalysis. In: Miri, A., Vaudenay, S. (eds.) SAC 2011. LNCS, vol. 7118, pp. 243–259. Springer, Heidelberg (2012)

10. Kim, J.-S., Hong, S.H., Sung, J., Lee, S.-J., Lim, J.-I., Sung, S.H.: Impossible differential cryptanalysis for block cipher structures. In: Johansson, T., Maitra, S. (eds.) INDOCRYPT 2003. LNCS, vol. 2904, pp. 82–96. Springer, Heidelberg (2003)

11. Knudsen, L.: DEAL-a 128-bit block cipher. Complexity **258**(2), 216 (1998)

12. Knudsen, L., Wagner, D.: Integral cryptanalysis nes/doc/uib/wp5/015. NESSIE Report (2001). http://www.cosic.esat.kuleuven.be/nessie/reports/phase2/uibwp5-015-1.pdf

13. Knudsen, L.R., Wagner, D.: Integral cryptanalysis. In: Daemen, J., Rijmen, V. (eds.) FSE 2002. LNCS, vol. 2365, pp. 112–127. Springer, Heidelberg (2002)

14. Luo, Y., Lai, X., Wu, Z., Gong, G.: A unified method for finding impossible differentials of block cipher structures. Inf. Sci. **263**, 211–220 (2014)

15. Soleimany, H., Nyberg, K.: Zero-correlation linear cryptanalysis of reduced-round LBlock. Des. Codes Crypt. **73**(2), 683–698 (2014)

16. Sun, B., Liu, Z., Rijmen, V., Li, R., Cheng, L., Wang, Q., Alkhzaimi, H., Li, C.: Links among impossible differential, integral and zero correlation linear cryptanalysis. Cryptology ePrint Archive, Report 2015/181 (2015). http://eprint.iacr.org/

17. Sun, B., Liu, Z., Rijmen, V., Li, R., Cheng, L., Wang, Q., Alkhzaimi, H., Li, C.: Links among impossible differential, integral and zero correlation linear cryptanalysis. In: Gennaro, R., Robshaw, M. (eds.) CRYPTO 2015, Part I. LNCS, vol. 9215, pp. 95–115. Springer, Heidelberg (2015)

18. Suzaki, T., Minematsu, K.: Improving the generalized Feistel. In: Hong, S., Iwata, T. (eds.) FSE 2010. LNCS, vol. 6147, pp. 19–39. Springer, Heidelberg (2010)

19. Wu, S., Wang, M.: Automatic search of truncated impossible differentials for word-oriented block ciphers. In: Galbraith, S., Nandi, M. (eds.) INDOCRYPT 2012. LNCS, vol. 7668, pp. 283–302. Springer, Heidelberg (2012)

20. Yanagihara, S., Iwata, T.: Improving the permutation layer of type 1, type 3, source-heavy, and target-heavy generalized Feistel structures. IEICE Trans. **96–A**(1), 2–14 (2013)

21. Zhang, W., Su, B., Wu, W., Feng, D., Wu, C.: Extending higher-order integral: an efficient unified algorithm of constructing integral distinguishers for block ciphers. In: Bao, F., Samarati, P., Zhou, J. (eds.) ACNS 2012. LNCS, vol. 7341, pp. 117–134. Springer, Heidelberg (2012)

Improved Meet-in-the-Middle Attacks on 7 and 8-Round ARIA-192 and ARIA-256

Akshima, Donghoon Chang, Mohona Ghosh[✉], Aarushi Goel,
and Somitra Kumar Sanadhya

Indraprastha Institute of Information Technology, Delhi (IIIT-D), New Delhi, India
{akshima12014,donghoon,mohonag,aarushi12003,somitra}@iiitd.ac.in

Abstract. The ARIA block cipher has been established as a Korean encryption standard by Korean government since 2004. In this work, we re-evaluate the security bound of reduced round ARIA-192 and ARIA-256 against meet-in-the-middle (MITM) key recovery attacks in the single key model. We present a new 4-round distinguisher to demonstrate the best 7 & 8 round MITM attacks on ARIA-192/256. Our 7-round attack on ARIA-192 has data, time and memory complexity of 2^{113}, $2^{135.1}$ and 2^{130} respectively. For our 7-round attack on ARIA-256, the data/time/memory complexities are 2^{115}, $2^{136.1}$ and 2^{130} respectively. These attacks improve upon the previous best MITM attack on the same in all the three dimensions. Our 8-round attack on ARIA-256 requires 2^{113} cipher calls and has time and memory complexity of $2^{245.9}$ and 2^{138} respectively. This improves upon the previous best MITM attack on ARIA-256 in terms of time as well as memory complexity. Further, in our attacks, we are able to recover the actual secret key unlike the previous cryptanalytic attacks existing on ARIA-192/256. To the best of our knowledge, this is the first actual key recovery attack on ARIA so far. We apply multiset attack - a variant of meet-in-the-middle attack to achieve these results.

Keywords: Block cipher · ARIA · Key Recovery · Differential characteristic · Multiset attack

1 Introduction

The block cipher ARIA, proposed by Kwon et al. in ICISC 2003 [12], is a 128-bit block cipher that adopts substitution-permutation network (SPN) structure, similar to AES [3], and supports three key sizes -128-bit, 192-bit and 256-bit. The first version of ARIA (version 0.8) had 10/12/14 rounds for key sizes of 128/192/256 respectively and only two kinds of S-boxes were employed in its substitution layer [2,19]. Later ARIA version 0.9 was announced at ICISC 2003 in which four kinds of S-boxes were used. This was later upgraded to ARIA version 1.0 [9], the current version, which was standardized by Korean Agency for Technology and Standards (KATS) - the government standards organization of South Korea as the 128-bit block encryption algorithm (KS X 1213)

© Springer International Publishing Switzerland 2015
A. Biryukov and V. Goyal (Eds.): INDOCRYPT 2015, LNCS 9462, pp. 198–217, 2015.
DOI: 10.1007/978-3-319-26617-6_11

in December 2004. In this version, the number of rounds was increased to 12/14/16 and some modifications in the key scheduling algorithm were introduced. ARIA has also been adopted by several standard protocols such as IETF (RFC 5794 [11]), SSL/TLS (RFC 6209 [10]) and PKCS #11 [13].

ARIA block cipher has been subjected to reasonable cryptanalysis in the past 12 years since its advent. In [1], Biryukov et al. analyzed the first version (version 0.8) of ARIA and presented several attacks such as truncated differential cryptanalysis, dedicated linear attack, square attack etc. against reduced round variants of ARIA. In the official specification document of the standardized ARIA (version 1.0) [12], the ARIA developers analyzed the security of ARIA against many classical cryptanalyses such as differential and linear cryptanalysis, impossible and higher order differential cryptanalysis, slide attack, interpolation attack etc. and claimed that ARIA has a better resistance against these attacks as compared to AES. In [18], Wu et al. presented a 6-round impossible differential attack against ARIA which was improved in terms of attack complexities by Li et al. in [15]. In [16], Li et al. presented a 6-round integral attack on ARIA followed by Fleischmann et al. [8] who demonstrated boomerang attacks on 5 and 6 rounds of ARIA. Du et al. in [6], extended the number of rounds by one and demonstrated a 7-round impossible differential attack on ARIA-256. In [17], Tang et al., applied meet-in-the-middle (MITM) attack to break 7 and 8-rounds of ARIA-192/256. In Table 1, we summarize all the existing attacks on ARIA version 1.0.

In this work, we improve the attack complexities of the 7 and 8-round MITM attack on ARIA-192/256. Our work is inspired from the multiset attack demonstrated by Dunkelman et al. on AES in [7]. Multiset attack is a variant of meet-in-the-middle attack presented by Demirci et al. on AES in [4]. Demirci et al.'s attack involves constructing a set of functions which map one active byte in the first round to another active byte after 4-rounds of AES. This set of functions depend on 'P' parameters and can be described using a table of 2^P ordered 256-byte sequence of entries. This table is precomputed and stored, thus allowing building a 4-round distinguisher and attacking upto 8 rounds of AES. Due to structural similarities between ARIA and AES, a similar attack was applied to 7 & 8-rounds of ARIA by Tang et al. in [17]. The bottleneck of this attack is a very high memory complexity which is evident in the attacks on ARIA as well as shown in Table 1. To reduce the memory complexity of Demirci's attacks on AES, Dunkelman et al. in [7], proposed multiset attack which replaces the idea of storing 256 ordered byte sequences with 256 unordered byte sequences (with multiplicity). This reduced both memory and time complexity of MITM attack on AES by reducing the parameters to 'Q' (where, Q<P). They also introduced the novel idea of differential enumeration technique to significantly lower the number of parameters required to construct the multiset from 'Q' to 'R' (where, R<Q<P), thus further decreasing the attack complexities on AES. Derbez et al. in [5] improved Dunkelman et al.'s attack by refining the differential enumeration technique. By using rebound-like techniques [14], they showed that the number of reachable multisets are much lower than those counted in Dunkelman et al.'s

Table 1. Comparison of cryptanalytic attacks on ARIA version 1.0. The entries are arranged in terms of decreasing time complexities for each category of attacked rounds.

Rounds attacked	Attack type	Time complexity	Data complexity	Memory complexity	Reference
5	Boomerang Attack	2^{110}	2^{109}	2^{57}	[8]
	Integral Attack	$2^{76.7}$	$2^{27.5}$	$2^{27.5}$	[16]
	Impossible Differential	$2^{71.6}$	$2^{71.3}$	2^{72}	[15]
	Meet-in-the-middle	$2^{65.4}$	2^{25}	$2^{122.5}$	[17]
6	Integral Attack	$2^{172.4}$	$2^{124.4}$	$2^{124.4}$	[16]
	Meet-in-the-middle	$2^{121.5}$	2^{56}	$2^{122.5}$	[17]
	Impossible Differential	2^{112}	2^{121}	2^{121}	[18]
	Boomerang Attack	2^{108}	2^{128}	2^{56}	[8]
	Impossible Differential	$2^{104.5}$	$2^{120.5}$	2^{121}	[15]
7	Impossible Differential	2^{238}	2^{125}	2^{125}	[6]
	Boomerang Attack	2^{236}	2^{128}	2^{184}	[8]
	Meet-in-the-middle	$2^{185.3}$	2^{120}	2^{187}	[17]
	Meet-in-the-middle (ARIA-192)	$2^{135.1}$	2^{113}	2^{130}	This work, Sect. 4
	Meet-in-the-middle (ARIA-256)	$2^{136.1}$	2^{115}	2^{130}	This work, Sect. 4
8	Meet-in-the-middle (ARIA-256)	$2^{251.6}$	2^{56}	2^{252}	[17]
	Meet-in-the-middle (ARIA-256)	$2^{245.9}$	2^{113}	2^{138}	This work, Sect. 5

attack. This improvement allowed mounting of comparatively efficient attacks on AES and also enabled extension of number of rounds attacked. Though the results of this line of work are quite interesting, yet they have not been explored further. Coupled with the fact that the security of ARIA has not been analyzed much after Fleischmann et al.'s attack in Indocrypt 2010 [8], motivated us to investigate the effectiveness of multiset attack on ARIA.

In our attacks, we construct a new 4-round distinguisher for ARIA. As a result, our attacks significantly reduce the data/time/memory complexities of the previous 7-round MITM attack on ARIA-192/256 shown in [17]. Our 8-round attack also improves upon the time and memory complexities of the previous best 8-round MITM attack on ARIA-256 [17] but at the expense of increase in the data complexity. The key schedule algorithm of ARIA does not allow recovery of master key from a subkey unlike AES [3]. This is likely the reason why none of the previous attacks have shown the actual key retrieval on any ARIA variant. However, depending upon the key expansion of ARIA, recovery of specific subkeys allows extracting the actual secret key. In our 7 and 8-round attack on ARIA-192/256, we exploit this key scheduling property to demonstrate the actual secret key recovery in ARIA. To the best of our knowledge, we are the first to demonstrate actual key recovery on ARIA.

Our Contribution. The main contributions of this work are as follows:

- We present the best 7-round MITM based key recovery attack on ARIA 192/256 and 8-round attack on ARIA-256.
- We apply multiset attack to construct a new 4-round distinguisher on ARIA-192 and ARIA-256.
- Our 7-round attack on ARIA-192 has data/time/memory complexity of 2^{113}, $2^{135.1}$ and 2^{130} respectively.
- Our 7-round attack on ARIA-256 has data/time/memory complexity of 2^{115}, $2^{136.1}$ and 2^{130} respectively.
- Our 8-round attack on ARIA-192/256 has data/time/memory complexity of 2^{113}, $2^{245.6}$ and 2^{138} respectively.
- We present the first actual master key recovery on our attacks on ARIA-192/256.

Our results are summarized in Table 1.

Organization. The paper is organized as follows. In Sect. 2, we provide a brief description of ARIA followed by important notations adopted throughout the work. In Sect. 3, we give details of our distinguisher so constructed on 4-rounds of ARIA. In Sect. 4, we present our 7-round attack followed by Sect. 5, where we demonstrate our 8-round attack on ARIA and show actual key recovery. Finally in Sect. 6, we summarize and conclude our paper.

2   Preliminaries

In this section, we first describe ARIA and then mention the key notations and definitions used in our cryptanalysis technique to facilitate better understanding.

2.1   Description of ARIA

The block cipher ARIA adopts substitution-permutation network in its design and is structurally similar to Advanced Encryption Standard (AES). The ARIA specification defines 3 key sizes - 128-bit, 192-bit and 256-bit with block size limited to a fixed 128-bit size for all the three alternatives. Each ARIA variant has different number of rounds per full encryption, i.e., 12, 14 and 16 rounds for ARIA-128, ARIA-192 and ARIA-256 respectively. The 128-bit internal state and key state are treated as a byte matrix of 4×4 size, where the bytes are numbered from 0 to 15 column wise (as shown in Fig. 1). Each round consists of 3 basic operations (as shown in Fig. 2):

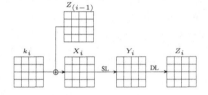

Fig. 1. Byte numbering in a state of ARIA   **Fig. 2.** i^{th} round of ARIA.

1. *Add Round Key (ARK)* - This step involves an exclusive-or operation with the round subkey. The key schedule of ARIA consists of two phases:
   - A nonlinear expansion phase, in which the 128-bit, 192-bit or 256-bit master key is expanded into four 128-bit words W_0, W_1, W_2, W_3 by using a 3-round 256-bit Feistel cipher.
   - A linear key schedule phase in which the subkeys are generated via simple XORs and rotation of W_0, W_1, W_2, W_3 each.
2. *Substitution Layer (SL)* - It uses four types of 8-bit S-boxes S_1, S_2 and their inverses S_1^{-1} and S_2^{-1}. Each S-Box is defined to be an affine transformation of the inversion function over $GF(2^8)$. The S_1 S-box is the same as used in AES. ARIA has two types of substitution layers for even and odd rounds respectively. In each odd round, the substitution layer is (LS, LS, LS, LS) where $LS = (S_1, S_2, S_1^{-1}, S_2^{-1})$ operates one column and in each even round, the substitution layer is $(LS^{-1}, LS^{-1}, LS^{-1}, LS^{-1})$ where $LS^{-1} = (S_1^{-1}, S_2^{-1}, S_1, S_2)$ operates on one column as well.
3. *Diffusion Layer (DL)* - This layer consists of a 16×16 involutional binary matrix with branch number 8. Given an input state y and output state z, the diffusion layer is defined as:

$$z[0] = y[3] \oplus y[4] \oplus y[6] \oplus y[8] \oplus y[9] \oplus y[13] \oplus y[14]$$
$$z[1] = y[2] \oplus y[5] \oplus y[7] \oplus y[8] \oplus y[9] \oplus y[12] \oplus y[15]$$
$$z[2] = y[1] \oplus y[4] \oplus y[6] \oplus y[10] \oplus y[11] \oplus y[12] \oplus y[15]$$
$$z[3] = y[0] \oplus y[5] \oplus y[7] \oplus y[10] \oplus y[11] \oplus y[13] \oplus y[14]$$
$$z[4] = y[0] \oplus y[2] \oplus y[5] \oplus y[8] \oplus y[11] \oplus y[14] \oplus y[15]$$
$$z[5] = y[1] \oplus y[3] \oplus y[4] \oplus y[9] \oplus y[10] \oplus y[14] \oplus y[15]$$
$$z[6] = y[0] \oplus y[2] \oplus y[7] \oplus y[9] \oplus y[10] \oplus y[12] \oplus y[13]$$

$$z[7] = y[1] \oplus y[3] \oplus y[6] \oplus y[8] \oplus y[11] \oplus y[12] \oplus y[13]$$
$$z[8] = y[0] \oplus y[1] \oplus y[4] \oplus y[7] \oplus y[10] \oplus y[13] \oplus y[15]$$
$$z[9] = y[0] \oplus y[1] \oplus y[5] \oplus y[6] \oplus y[11] \oplus y[12] \oplus y[14]$$
$$z[10] = y[2] \oplus y[3] \oplus y[5] \oplus y[6] \oplus y[8] \oplus y[13] \oplus y[15]$$
$$z[11] = y[2] \oplus y[3] \oplus y[4] \oplus y[7] \oplus y[9] \oplus y[12] \oplus y[14]$$
$$z[12] = y[1] \oplus y[2] \oplus y[6] \oplus y[7] \oplus y[9] \oplus y[11] \oplus y[12]$$
$$z[13] = y[0] \oplus y[3] \oplus y[6] \oplus y[7] \oplus y[8] \oplus y[10] \oplus y[13]$$
$$z[14] = y[0] \oplus y[3] \oplus y[4] \oplus y[5] \oplus y[9] \oplus y[11] \oplus y[14]$$
$$z[15] = y[1] \oplus y[2] \oplus y[4] \oplus y[5] \oplus y[8] \oplus y[10] \oplus y[15]$$

In the last round, diffusion layer is replaced by key xoring to generate the ciphertext. The key schedule algorithm of ARIA [11] is divided into two phases - *Initialization phase* and *Round Key Generation phase*. In the initialization phase, for ARIA-256, first we compute KL and KR for the master key K as follows:

$$KL \parallel KR = K \parallel 0...0$$

where, $\mid KL \mid = \mid KR \mid = 128$-bits and number of zeroes padded to K equals 128, 64 and 0 for $\mid K \mid$ equal to 128, 192 and 256 respectively.

Then, four 128-bit values W_0, W_1, W_2 and W_3 are set as:

$$W_0 = KL \tag{1}$$
$$W_1 = F_o(W_0, CK_1) \oplus KR \tag{2}$$
$$W_2 = F_e(W_1, CK_2) \oplus W_0 \tag{3}$$
$$W_3 = F_o(W_2, CK_3) \oplus W_1 \tag{4}$$

where, F_o and F_e are ARIA odd and even round functions and CK_1, CK_2 and CK_3 are pre-defined constants. In the round key generation phase, the following round subkeys are generated as follows:

$$K_1 = W_0 \oplus (W_1 >>> 19) \tag{5}$$
$$K_2 = W_1 \oplus (W_2 >>> 19) \tag{6}$$
$$K_3 = W_2 \oplus (W_3 >>> 31) \tag{7}$$
$$K_4 = (W_0 >>> 19) \oplus W_3 \tag{8}$$
$$K_5 = W_0 \oplus (W_1 >>> 31) \tag{9}$$
$$K_6 = W_1 \oplus (W_2 >>> 31) \tag{10}$$
$$K_7 = W_2 \oplus (W_3 >>> 31) \tag{11}$$
$$K_8 = (W_0 >>> 31) \oplus W_3 \tag{12}$$
$$K_9 = W_0 \oplus (W1 <<< 61) \tag{13}$$

For further details, one can refer [11].

2.2   Notations and Definitions

The following notations are followed throughout the rest of the paper.

P :	Plaintext
C :	Ciphertext
k_i :	Subkey of round i
k_i^* :	$DL^{-1}(k_i)$, where, DL^{-1} is the inverse diffusion layer
X_i :	State obtained after ARK in round i
Y_i :	State obtained after SL in round i
Z_i :	State obtained after DL in round i
Δs :	Difference in a state s
$s_i[m]$:	m^{th} byte of a state s in round i, where $0 \le m \le 15$
$s_i[p, \dots, r]$:	p^{th} byte, $\dots$, r^{th} byte of state s in round i, where $0 \le p, r \le 15$

In our attacks, rounds are numbered from 1 to R, where R = 7 or 8. A *full* round consists of all the three round operations, i.e., ARK, SL and DL whereas a *half* round denotes a round in which the DL operation is omitted.

We utilize the following definitions for our attacks.

Definition 1 (δ-list). We define the δ-list as an ordered list of 256 16-byte distinct elements that are equal in 15 bytes. Each of the 15 equal bytes is called as passive byte whereas the one byte that takes all possible 256 values is called the active byte [3]. We denote the δ-list as $(x^0, x^1, x^2, \dots, x^{255})$ where x^j indicates the jth 128-bit member of the δ-list. As mentioned in the notations section, x_i^j [m] represents the mth byte of x^j in round i.

Definition 2 (Multiset). A multiset is a set of elements in which multiple instances of the same element can appear. A multiset of 256 bytes, where each byte can take any one of the 256 possible values, can have $\binom{2^8+2^8-1}{2^8} \approx 2^{506.17}$ different values.

Two crucial properties that will be used in our attacks are as follows:

Property 1. For a given input-output difference (denoted as $(\Delta Y, \Delta Z)$) state over a diffusion layer operation (as shown in Fig. 3), if the 7-bytes of ΔY [3, 4, 6, 8, 9, 13, 14] have equal differences, say y, then it will lead to non-zero difference only at byte 0 of ΔZ (instead of full state diffusion) after the diffusion layer operation. Rest all bytes of ΔZ will be passive. Thus, under the given constraints, probability of the differential trail $\Delta Y \rightarrow \Delta Z$ is 1.

Proof. As per the diffusion layer specification of ARIA, each output byte of state Z is a xored sum of 7 input bytes of state Y. The same property is preserved in case of differences as well, i.e., each output byte difference of Z is a xored sum of 7 input byte difference of Y. In lieu of this, for each output byte, if even number

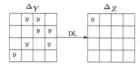

Fig. 3. Differential property of diffusion layer

of corresponding input bytes (i.e., 2, 4 or 6) have equal differences, then they cancel out each other. In the above trail, 7 bytes of Y, i.e., Y [3, 4, 6, 8, 9, 13, 14] have equal differences 'y', whereas the rest of the bytes have zero differences. Hence, all output bytes except ΔZ [0] have zero differences since their xored sum have either 2 or 4 equal input byte differences. E.g.,

$$\Delta Z[0] = \Delta Y[3] \oplus \Delta Y[4] \oplus \Delta Y[6] \oplus \Delta Y[8] \oplus \Delta Y[9] \oplus \Delta Y[13] \oplus \Delta Y[14]$$
$$= y \oplus y \oplus y \oplus y \oplus y \oplus y \oplus y = y$$
$$\Delta Z[1] = \Delta Y[2] \oplus \Delta Y[5] \oplus \Delta Y[7] \oplus \Delta Y[8] \oplus \Delta Y[9] \oplus \Delta Y[12] \oplus \Delta Y[15]$$
$$= 0 \oplus 0 \oplus 0 \oplus y \oplus y \oplus 0 \oplus 0 = 0$$
$$\Delta Z[11] = \Delta Y[2] \oplus \Delta Y[3] \oplus \Delta Y[4] \oplus \Delta Y[7] \oplus \Delta Y[9] \oplus \Delta Y[12] \oplus \Delta Y[14]$$
$$= 0 \oplus y \oplus y \oplus 0 \oplus y \oplus 0 \oplus y = 0$$

Similar equations can be constructed for other output bytes of Z as well. Thus, Property 1 holds true.

Property 2. For a given ARIA S-box, say S_1 and any non-zero input - output difference pair, say $(\Delta_i$ - $\Delta_o)$ in F_{256}, there exists one solution in average, say y, for which the equation, $S_1(y) \oplus S_1(y \oplus \Delta_i) = \Delta_o$, holds true (since ARIA uses AES S-box as S_1 [5]). This property is also applicable to other ARIA S-boxes, i.e., S_2, S_1^{-1} and S_2^{-1}.

The time complexity of the attack is measured in terms of number of full round (7 or 8) ARIA encryptions required. The memory complexity is measured in units of 128-bit ARIA blocks required.

3   Distinguishing Property of 4-round ARIA

Given a list of 256 distinct bytes $(M^0, M^1, \ldots, M^{255})$, a function $f : \{0,1\}^{128} \mapsto \{0,1\}^{128}$ and a 120-bit constant U, we define a multiset v as follows:

$$C^i = f(M^i \parallel U), \text{where } (0 \leq i \leq 255)$$
$$v = \{C^0[0] \oplus C^0[0], C^1[0] \oplus C^0[0], \ldots, C^{255}[0] \oplus C^0[0]\}$$

Note that, $(M^0 \parallel U, M^1 \parallel U, \ldots, M^{255} \parallel U)$ forms a δ-list and atleast one element of the multiset is always zero.

Distinguishing Property. Let us consider $\mathcal{F}$ to be a family of permutations on 128-bit. Then, given any list of 256 distinct bytes $(M^0, M^1, \ldots, M^{255})$, the aim is to find how many multisets v are possible when, $f \xleftarrow{\$} \mathcal{F}$ and $U \xleftarrow{\$} \{0,1\}^{120}$.

In case, when $\mathcal{F}$ = family of all permutations on 128-bit and $f \xleftarrow{\$} \mathcal{F}$. Under such setting, since in the multiset v, we have 255 values that are chosen uniformly and independently from the set $\{0, 1, \ldots, 255\}$ (as one element, say $C^0[0] \oplus C^0[0]$, is always 0), the total possible multisets v are $\binom{2^8 - 1 + 2^8 - 1}{2^8 - 1} \approx 2^{505.17}$.

In case, when $\mathcal{F}$ = 4-full rounds of ARIA and $f \xleftarrow{\$} \mathcal{F}$. Here, $f \xleftarrow{\$} \mathcal{F}$ $\Leftrightarrow K \xleftarrow{\$} \{0,1\}^k$ and $f = E_K$, where, k = 128 (for ARIA-128), 192 (for ARIA-192) or 256 (for ARIA-256). Let us consider, 4-full rounds of ARIA as shown in Fig. 4 where, multiset v is defined as $v = \{Z_4^0[0] \oplus Z_4^0[0], Z_4^1[0] \oplus Z_4^0[0], \ldots, Z_4^{255}[0] \oplus Z_4^0[0]\}$. Then, we state the following *Observation 1*.

<u>Observation 1.</u> The multiset v is determined by the following 30 single byte parameters only:

- $X_2^0[3, 4, 6, 8, 9, 13, 14]$ (7-bytes)
- $X_3^0[0, 1, 2, 3, 4, 5, 6, 7, 8, 9, 10, 11, 12, 13, 14, 15]$ (full 16-byte state)
- $X_4^0[3, 4, 6, 8, 9, 13, 14]$ (7-bytes)

Thus, the total number of multisets possible is $2^{30 \times 8} = \mathbf{2^{240}}$ since, each 30-bytes defines one multiset.

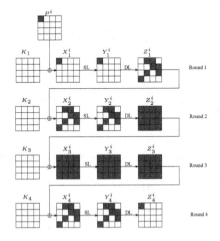

Fig. 4. 4-Round distinguisher in ARIA. Here, P^i denotes $(M^i \parallel U)$ and X_j^i, Y_j^i, Z_j^i denote intermediate states corresponding to P^i in round j. The round subkeys K_i, where, $1 \leq i \leq 4$ are generated from the master key K.

Proof. In round 1, the set of differences $\{X_1^0[0] \oplus X_1^0[0], X_1^1[0] \oplus X_1^0[0], \ldots,$ $X_1^{255}[0] \oplus X_1^0[0]\}$ (or, equivalently, set of differences at $X_1[0]$) are known as there are exactly 256 differences possible. Since S-box S_1 is injective, exactly 256 values exist in the set $\{Y_1^0[0] \oplus Y_1^0[0], Y_1^1[0] \oplus Y_1^0[0], \ldots\ldots, Y_1^{255}[0] \oplus Y_1^0[0]\}$ as well. Due to DL and ARK operations being linear, the set of differences at $X_2[3, 4, 6, 8, 9, 13, 14]$ are known (according to diffusion layer (DL) definition discussed in Sect. 2). Owing to the non-linearity of the substitution layer, the set of differences at $Y_2[3, 4, 6, 8, 9, 13, 14]$ cannot be known and one cannot move forward. To alleviate this problem, it is sufficient to know $X_2^0[3, 4, 6, 8, 9, 13, 14]$, i.e., values of the active bytes of the first state (out of 256 states) at X_2 as it enables calculating the active bytes of the other X_2^i states (where, $1 \leq i \leq 255$) and cross SL in round 2. Again, since DL and ARK operations are linear, the set of differences $\{X_3^0 \oplus X_3^0,$ $X_3^1 \oplus X_3^0, \ldots, X_3^{255} \oplus X_3^0\}$ is known. In order to know the set of values $\{X_3^0, X_3^1,$ $\ldots, X_3^{255}\}$ for crossing the SL in round 3, it is sufficient to know the value of the full state X_3^0 which is given as a parameter.

By similar logic, as explained above, the set of differences $\{X_4^0 \oplus X_4^0, X_4^1 \oplus X_4^0,$ $\ldots, X_4^{255} \oplus X_4^0\}$ are known. Now, at this stage, if only $X_4^0[3, 4, 6, 8, 9, 13, 14]$ bytes are known, the SL layer in round 4 can be crossed and the set of 256 values $\{Z_4^0[0], Z_4^1[0], \ldots, Z_4^{255}[0]\}$ at Z_4 can be computed. Then the value of multiset v $= \{Z_4^0[0] \oplus Z_4^0[0], Z_4^1[0] \oplus Z_4^0[0], \ldots, Z_4^{255}[0] \oplus Z_4^0[0]\}$ can be determined easily as well. This shows that the multiset v depends on 30 parameters and can take 2^{240} possible values. □

Since, there are 2^{240} possible multisets at $Z_4[0]$, if we precompute and store these values in a hash table, then the precomputation complexity goes higher than brute force for ARIA-192. In order to reduce the number of multisets, we apply the Differential Enumeration technique suggested by Dunkelman et al. in [7] and improved by Derbez et al. in [5]. We call the improved version proposed in [5] as *Refined Differential Enumeration*.

Refined Differential Enumeration. The basic idea behind this technique is to choose a list of 256 distinct bytes $(M^0, M^1, \ldots, M^{255})$ such that several of the parameters that are required to construct the multiset equal some predetermined constants.

To achieve so, let us construct a truncated differential for four full rounds of ARIA, in which the input and output differences are non-zero at byte 0 only (as shown in Fig. 5).

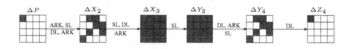

Fig. 5. 4-Round truncated differential in ARIA

The probability of this trail is 2^{-120} as follows: the one byte difference at $\Delta P[0]$ propagates to 7-byte difference in ΔX_2 and 16-byte difference in ΔY_3 with probability 1. Next, the probability that full state difference in ΔY_3 leads to 7-byte difference in ΔY_4 is 2^{-72} (since 9 bytes of ΔY_4, i.e., $\Delta Y_4[0, 1, 2, 5, 7, 10, 11, 12, 15]$) have zero difference). Further, the probability that random differences in ΔY_3 yield equal differences in the active bytes of ΔY_4 i.e., $\Delta Y_4[3, 4, 6, 8, 9, 12, 13]$ is 2^{-48}.[1] Therefore, the total probability of $\Delta Y_3 \rightarrow \Delta Y_4$ is $2^{-(72+48)} = 2^{-120}$. Then, by the virtue of *Property 1* (mentioned in Sect. 2), 7-byte difference in ΔY_4 yields a single byte difference in $\Delta Z_4[0]$ with probability 1. Thus, the overall probability of the differential from $\Delta P \rightarrow \Delta Z_4$ is 2^{-120}.

In other words, we require 2^{120} plaintext pairs to get a right pair. Once, we get a right pair, say (P^0, P^1), we state the following *Observation 2*:

Observation 2. Given a right pair (P^0, P^1) that follows the truncated differential trail shown in Fig. 5, then the 30 parameters corresponding to P^0 mentioned in *Observation 1* can take one of at most 2^{128} fixed 30-byte values (out of the total 2^{240} possible values) where, each of these 2^{128} 30-byte values are defined by each of the 2^{128} values of the 16 following parameters:

- $\Delta Y_1[0]$
- $X_2^0[3, 4, 6, 8, 9, 13, 14]$
- $Y_4^0[3, 4, 6, 8, 9, 13, 14]$
- $\Delta Z_4[0]$

Proof. Given a right pair (P^0, P^1), the knowledge of these 16 new parameters allows us to compute all the differences shown in Fig. 4. This is so because, knowledge of $\Delta Y_1[0]$ allows computation of $\Delta Z_1[3, 4, 6, 8, 9, 13, 14]$ and $\Delta X_2[3, 4, 6, 8, 9, 13, 14]$. Then, if the values of $X_2^0[3, 4, 6, 8, 9, 13, 14]$ are known, one can compute the corresponding $X_2^1[3, 4, 6, 8, 9, 13, 14]$, cross the SL layer in round 2 and calculate the full state difference ΔX_3. Similarly, from the bottom side, knowledge of $\Delta Z_4[0]$ allows computation of $\Delta Y_4[3, 4, 6, 8, 9, 13, 14]$. Then, if the values of $Y_4^0[3, 4, 6, 8, 9, 13, 14]$ are known, one can easily determine $Y_4^1[3, 4, 6, 8, 9, 13, 14]$, compute the corresponding $X_4^0[3, 4, 6, 8, 9, 13, 14]$ and $X_4^1[3, 4, 6, 8, 9, 13, 14]$ respectively and subsequently full state ΔY_3. Then, using the differential property of ARIA S-boxes (*property 2* mentioned in Sect. 2), the possible values of X_3^0 and X_3^1 can be computed.   □

Thus, the knowledge of these 16 bytes given in *Observation 2* allows computation of the corresponding 30 parameters described in *Observation 1*. Hence, total possible values of these 30 single byte parameters are at most $2^{16 \times 8} = 2^{128}$. Moreover, since these computations do not require the knowledge of key bytes, they can be easily precomputed.

Using *Observations 1* and *2*, we state the following third *Observation 3*:

Observation 3. Given $(M^0, M^1, \ldots, M^{255})$ and $f \xleftarrow{\$} \mathcal{F}$ and $U \xleftarrow{\$} \{0,1\}^{120}$, such that $M^0 \parallel U$ and $M^j \parallel U$, (where, $j \in \{0, 1, \ldots, 255\}$) is a right pair

---

[1] Random differences in 16-bytes of ΔY_3 yield random differences in the 7 active bytes of ΔX_4 which in turn lead to random differences in the active bytes of ΔY_4. The probability that these random differences in the 7-bytes of ΔY_4 are equal is 2^{-48}.

that follows differential trail shown in Fig. 5, then at most 2^{128} multisets v are possible at $Z_4[0]$.

Proof. From *Observation 1*, we know that each 30-byte parameter defines one multiset and *Observation 2* restricts the possible values of these 30-byte parameters to 2^{128}. Thus, at most 2^{128} multisets are only possible for ARIA.   □

As the number of multisets in case of 128-bit random permutation ($= 2^{505.17}$) is much higher than 4-round ARIA ($= 2^{128}$), a valid distinguisher is constructed.

4   Key Recovery Attack on 7-Round ARIA-192/256

In this section, we use our *Observation 3* to launch a meet-in-the-middle attack on 7-round ARIA-192/256 to recover the key. The distinguisher is placed from round 2 to round 5, i.e., δ-list is constructed in state X_2 with byte 0 being the active byte and multiset is checked in $Z_5[0]$ (as shown in Fig. 6). One round at the top and two rounds at the bottom are added to the 4-round distinguisher. The attack consists of the following two phases:

Precomputation Phase. Compute and store the 2^{128} possible multisets at $\Delta Z_5[0]$ in a hash table based on *Observation 2*.

Online Phase. If we extend the differential trail (shown in Fig. 5) by one round backwards, such that 7-bytes (3, 4, 6, 8, 9, 13 and 14) are active in the plaintext, then with a probability of 2^{-48}, these 7 active bytes will induce a non-zero difference of one byte in $X_2[0]$. Thus, we require $2^{120+48} = 2^{168}$ plaintext pairs to start our online phase. For each of these pairs, we will guess the subkey candidates for which the pair becomes a right pair and construct the corresponding δ-list. The steps of the online phase are:

1. Encrypt 2^{57} structures of 2^{56} plaintexts each, where bytes 3, 4, 6, 8, 9, 13 and 14 take all possible values and rest of the bytes are constants.[2].
2. For each structure, store the ciphertexts in a hash table and look for pairs in which the difference in bytes 0, 1, 2, 5, 7, 10, 11, 12, 15 of the ciphertext is zero. Out of the total 2^{168} pairs, only 2^{96} pairs are expected to remain.
3. For each of the remaining 2^{96} plaintext pairs do the following:
   (a) Guess 7 bytes of $K_8[3, 4, 6, 8, 9, 13, 14]$ and check whether ΔY_6 has non zero difference only in byte 0 or not. Out of the 2^{56} possible values for K_8, only 2^8 key guesses are expected to remain (since with probability 2^{-48}, each will yield equal differences in the active bytes of ΔZ_6). Since we are only interested in checking the difference at $\Delta Y_6[0]$, $K_7[0]$ is not required to be guessed at this stage.
   (b) Guess 7 bytes of $K_1[3, 4, 6, 8, 9, 13, 14]$ and check whether ΔZ_1 has non zero difference only in byte 0 or not. Out of the 2^{56} possible values for K_1, only 2^8 key guesses are expected to remain.
   (c) For each of the $2^8 \times 2^8 = 2^{16}$ remaining guesses of 14 active bytes of K_1 and K_8:

---

[2] One structure has $2^{56} \times 2^{55} = 2^{111}$ plaintext pairs. Therefore, 2^{57} structures have $2^{57+111} = 2^{168}$ plaintext pairs.

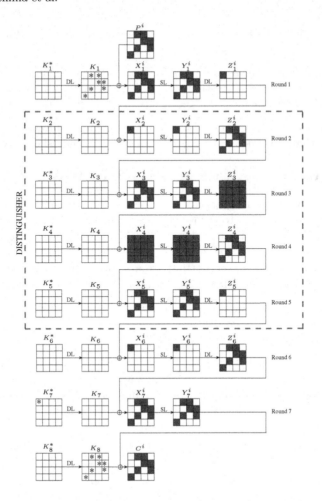

Fig. 6. 7-round attack on ARIA-192/256. The subkey bytes derived are star marked.

- Take one of the members of the pair and find its δ-list at $Z_1[0]$ using the knowledge of 7 active bytes of K_1.[3]
- Get the corresponding ciphertexts of the resulting plaintext set of the δ-list from the hash table. Guess byte $K_7^*[0] = DL^{-1}(K_7[0]) = K_7[3] \oplus K_7[4] \oplus K_7[6] \oplus K_7[8] \oplus K_7[9] \oplus K_7[13] \oplus K_7[14]$ and using the knowledge of $K_8[3, 4, 6, 8, 9, 13, 14]$, partially decrypt the ciphertexts of the δ-list to obtain the multiset at $\Delta Z_5[0]$ (which is same as that constructed in $\Delta X_6[0]$).
- Check whether this multiset exists in the precomputed table or not. If not, then discard the corresponding key guess.

---

[3] Encrypt the chosen right pair message to one full round using $k_1[3, 4, 6, 8, 9, 13, 14]$ and compute $Z_1[0]$. Xor other $Z_1[0]$ byte with 255 other values and decrypt them back to obtain the other plaintexts.

The probability for a wrong guess to pass the test is $2^{128} \times 2^{-467.6} = 2^{-339.6}$.[4] Since we try only $2^{96+16+8} = 2^{120}$ multisets, only the right subkey should verify the test with a probability close to 1.

Complexities. The time complexity of the precomputation phase is $2^{128} \times 2^8 \times 2^{-1.9} = 2^{134.1}$.[5] ARIA encryptions. The time complexity of the online phase is dominated by step 3(c) which is $2^{96} \times 2^{16} \times 2^8 \times 2^8 \times 2^{-2.9} = 2^{125.1}$ ARIA encryptions. Clearly the time complexity of this attack is dominated by the precomputation phase. It was shown in [5] that each 256-byte multiset requires 512-bit space. Hence, the memory complexity of the attack is $2^{128} \times 2^2 = 2^{130}$ 128-bit ARIA Blocks. The data complexity of the attack is 2^{113} plaintexts.

4.1   Recovering the Actual Master Key for 7-Round ARIA-192

In the above attack, 7-bytes of subkeys K_1 and 7-bytes of K_8 as well as 1 byte of K_7^* were recovered. In order to recover the master key do the following:

1. Guess 16-bytes of W_0.
   (a) Using the guessed value of W_0 and 7-bytes of K_1 recovered in the attack, we can deduce 56-bit of W_1 from Eq. 5. It is observed that 16-bit of this 56-bit of W_1 deduced, are part of 11th, 12th and 13th bytes and rest 40-bits are part of first 8 bytes.
   (b) Calculate $F_o(W_0, CK_1)$. We already know that for ARIA-192, $KR\,[8, 9,...,15] = 0$. Thus, $W_1[8, 9, ..., 15]$ equals corresponding bytes of $F_o(W_0, CK_1)$ following from Eq. 2.
   (c) Discard the guesses of W_0 for which the common 16-bit of W_1 computed in (a) and (b) do not match. 2^{112} guesses of W_0 are expected to remain.
2. For each of the remaining guesses of W_0, guess 24-bits of $W_1[0, 1, ...7]$ other than the 40-bits deduced in 1(a) to know the 2^{24} possible values of W_1 corresponding to each of W_0.
3. For each remaining guesses of W_0 and corresponding guesses of W_1, deduce W_2 and W_3 from Eqs. 3 and 4.
   (a) Following Eq. 12, deduce K_8 and compare its bytes 3, 4, 6, 8, 9, 13 and 14 with the values of the same 7-bytes of K_8 recovered from the attack. Discard the guesses of W_0 and W_1 in case of mismatch of these 7-bytes of K_8. Repeat the same process for 1-byte of K_7^*. This is a 8-byte and 64-bit filtering. Out of 2^{136}, 2^{72} guesses of W_0 and W_1 are expected to remain which can be tested by brute force to obtain the correct master key.

The time complexity of the recovering process of step 3 is maximum. It is equal to $2^{136} \times (2/7) = 2^{134.2}$ 7-round ARIA encryptions as we need to compute

---

[4] Note that the probability of randomly having a match is $2^{-467.6}$ and not $2^{-505.17}$ since the number of ordered sequences associated with a multiset is not constant [7].

[5] The normalization factor of $2^{-1.9}$ is calculated by calculating the ratio of number of S-Box operations required in the precomputation phase to the total number of S-Box operations performed in 7-Round ARIA encryption. Similarly all other normalization factors have been calculated.

2 rounds of ARIA to deduce W_2 and W_3 and all other operations have negligible complexity as they are simple linear operations.

Therefore, the final time complexity of the attack is $2^{134.2} + 2^{134} = 2^{135.1}$. Other complexities remain the same.

4.2   Recovering the Actual Master Key for 7-Round ARIA-256

In the above attack, 7-byte of subkey K_1 and 7-byte of subkey K_8 as well as 1 byte of K_7^* were recovered. As shown in Fig. 6, we have obtained a trail such that 1st byte is active at X_2. In order to recover all 16-bytes of subkey K_1, we can repeat the attack 4 times by modifying the trail such that we get a different byte active at X_2:

- bytes 3,4,6,8,9,13,14 to obtain byte 0 active at X_2
- bytes 2,5,7,8,9,12,15 to obtain byte 1 active at X_2
- bytes 1,4,6,10,11,12,15 to obtain byte 2 active at X_2
- bytes 0,5,7,10,11,13,14 to obtain byte 3 active at X_2

The time and data complexity of the attack will become 4 times of the time and data complexties mentioned in the 7-round attack in Sect. 4 respectively. Then we do the following to recover the master key:

1. Guess 16-bytes of W_0
2. For each guess of W_0, using the value of K_1 recovered from the attack, we obtain W_1 from Eq. 2. Then we follow the step 3 as mentioned in Sect. 4.1.

The time complexity of recovering the master key is $2^{128} \times (2/7) = 2^{126.2}$ 7-round ARIA encryptions.

Therefore, the final time complexity of the attack is $(4 \times 2^{134}) + 2^{126.2} = 2^{136}$. The data complexity of the attack becomes 2^{115} while the memory complexity remains same.

5   Key Recovery Attack on 8-Round ARIA-256

In this section, we describe our meet-in-the-middle attack on 8-round ARIA-256.

5.1   Construction of 4.5-Round Distinguisher

For the 8-round attack, the distinguisher constructed in Fig. 4 is extended by half round forwards upto Y_5 (DL operation is omitted). The distinguisher for 8-round attack is shown in Appendix A. Similar to *Observation 1*, we state the following *Observation 4*:

Observation 4. Given $(M^0, M^1, \ldots, M^{255})$ and $f \xleftarrow{\$} \mathcal{F}$ and $U \xleftarrow{\$} \{0,1\}^{120}$, where, f represents 4.5 rounds of ARIA, the multiset $v = \{Y_5^0[0] \oplus Y_5^0[0], Y_5^0[0] \oplus Y_5^1[0], \ldots, Y_5^0[0] \oplus Y_5^{255}[0]\}$ is determined by the following 31 1-byte parameters:

- $X_2^0[3, 4, 6, 8, 9, 13, 14]$
- $X_3^0[0, 1, 2, 3, 4, 5, 6, 7, 8, 9, 10, 11, 12, 13, 14, 15]$ (full 16-byte state)
- $X_4^0[3, 4, 6, 8, 9, 13, 14]$
- $X_5^0[0]$

The number of possible multisets is $2^{31 \times 8} = 2^{248}$. The proof for this is similar to that described for *Observation 1* in Sect. 3.

Number of Admissible Multisets. The differential trail shown in Fig. 5 can be extended 0.5 round forwards to ΔY_5 in which only byte 0 is active with probability 1, i.e., the probability of differential trail: $\Delta P \rightarrow \Delta Y_5$ remains 2^{-120}. Then, similar to *Observation 2*, we state the following *Observation 5*.

Observation 5. Given a right pair (P^0, P^1) that follows the truncated differential trail $(\Delta P \rightarrow \Delta Y_5)$, then the 31 parameters corresponding to P^0 mentioned in *Observation 4* can take one of at most 2^{136} fixed 31-byte values (out of the total 2^{248} possible values) where, each of these 2^{136} 31-byte values are defined by each of the 2^{136} values of the 17 following parameters:

- $\Delta Y_1[0]$
- $X_2^0[3, 4, 6, 8, 9, 13, 14]$
- $Y_4^0[3, 4, 6, 8, 9, 13, 14]$
- $\Delta Z_4[0]$
- $X_5^0[0]$

The proof of this *Observation* is similar to the proof of *Observation 2* described in Sect. 3. From, *Observations 4 and 5*, we can say that the total number of admissible multisets is $2^{17 \times 8} = \mathbf{2^{136}}$.

5.2   Key Recovery Attack

In this section, we discuss our 8-round attack. The distinguisher is placed from round 2 to round 5.5, i.e., δ-list is constructed in state X_2 with byte 0 being the active byte and multiset is checked in $Y_6[0]$ (as shown in Fig. 7). One round at the top and three rounds at the bottom are added to the 4.5-round distinguisher. The attack consists of the following two phases:

Precomputation Phase. Compute and store the 2^{136} possible multisets at $\Delta Y_6[0]$ in a hash table based on *Observation 5*.

Online Phase. The steps of the online phase are:

1. Encrypt 2^{57} structures of 2^{56} plaintexts each, where bytes 3, 4, 6, 8, 9, 13 and 14 take all possible values and rest of the bytes are constants. Store the ciphertexts in a hash table.
2. For each of the 2^{168} plaintext pairs do the following:
   (a) For each 2^8 guesses of $\Delta Z_1[0]$, resolve input-output differences at SL layer of round 1 (using *Property 2*) and deduce the corresponding value of $K_1[3, 4, 6, 8, 9, 13, 14]$.

(b) For each $2^8 \times 2^{56} = 2^{64}$ guesses of $\Delta Y_6[0]$ and ΔY_7 [3, 4, 6, 8, 9, 13, 14], resolve input-output differences at SL layers in round 7 and round 8 respectively and deduce corresponding K_8^* [3, 4, 6, 8, 9, 13, 14] and full subkey K_9.

(c) For each of the $2^{64+8} = 2^{72}$ guesses of 30 bytes of K_1, K_8^* and K_9:
  - Take one of the members of the pair and find its δ-list using the knowledge of 7 active bytes of K_1.
  - Get the corresponding ciphertexts of the resulting plaintext set of the δ-list from the hash table. Using the knowledge of K_9 and K_8^* [3, 4, 6,

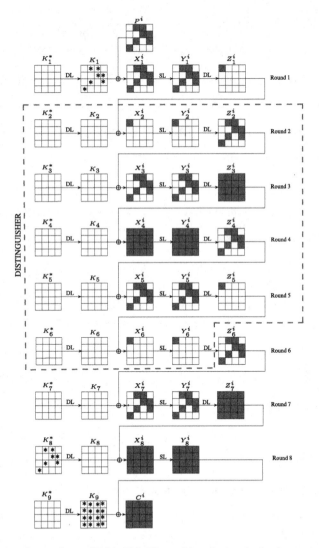

Fig. 7. 8-round attack on ARIA-256. The subkey bytes derived are star marked.

8, 9, 13, 14], partially decrypt the ciphertexts of the δ-list to compute the multiset at $\Delta Y_6[0]$.

- Check whether this multiset exists in the precomputed table or not. If not, then discard the corresponding key guess.

The probability for a wrong guess to pass the test is $2^{136} \times 2^{-467.6} = 2^{-331.6}$. Since, we try only $2^{168+72} = 2^{240}$ multisets, only the right subkey should verify the test with a probability close to 1.

Complexities. The time complexity of the precomputation phase is $2^{136} \times 2^8 \times 2^{-2} = 2^{142}$ ARIA encryptions. The time complexity of the online phase is dominated by step 2(c) which is $2^{168} \times 2^{72} \times 2^8 \times 2^{-2.1} = 2^{245.9}$ ARIA encryptions. Clearly the time complexity of this attack is dominated by the online phase. The memory complexity of the attack is $2^{136} \times 2^2 = 2^{138}$ 128-bit ARIA Blocks. The data complexity of the attack is 2^{113} plaintexts.

5.3   Recovering the Actual Master Key

In the above attack, 7-bytes of subkeys k_1 and k_8 as well as full subkey k_9 were recovered. Once these bytes are known, the remaining bytes in k_1 and k_8 can be found by exhaustive search without affecting the overall complexity of the 8-round attack. When full subkeys k_1 and k_9 are known then the master key K can be recovered as follows. Since, Eqs. 5 and 6 are two equations in two variables, they can be solved through standard matrix method by constructing a (256 × 256) binary matrix. We found the rank of this matrix to be 240 suggesting 2^{16} solutions for the tuple (W_0 and W_1). Once, values of W_0 and W_1 are known, KL and KR can be obtained through Eqs. 1 and 2 respectively. Thus, we get 2^{16} solutions for the master key K. Then through brute-force, the original key can be easily recovered.

6   Conclusions

In this work, we explore the space of multiset attacks as applied to key recovery attack on ARIA-192 and ARIA-256. We improve the previous 7-round and 8-round attacks on these structures and show the best attacks on them. We achieve these results by constructing a new 4-round distinguisher on ARIA and applying MITM attacks on the rest of the rounds. We also show recovery of the actual master key through our 8-round attack on ARIA-256. To our best knowledge, this is the first attempt in this direction. Currently, the number of attacked rounds remains 8 and it would be an interesting problem to try applying multiset attacks to break more rounds of ARIA.

References

1. Biryukov, A., De Canniere, C., Lano, J., Ors, S.B., Preneel, B.: Security and performance analysis of ARIA, version 1.2. Technical report, Katholieke Universiteit Leuven, Belgium (2004). http://www.cosic.esat.kuleuven.be/publications/article-500.pdf
2. De Cannière, C.: Analysis and Design of Symmetric Encryption Algorithms. PhD thesis, Katholieke Universiteit Leuven, Belgium, May 2007
3. Daemen, J., Rijmen, V.: The Design of Rijndael: AES - The Advanced Encryption Standard. Information Security and Cryptography. Springer, Heidelberg (2002)
4. Demirci, H., Selçuk, A.A.: A meet-in-the-middle attack on 8-round AES. In: Nyberg, K. (ed.) FSE 2008. LNCS, vol. 5086, pp. 116–126. Springer, Heidelberg (2008)
5. Derbez, P., Fouque, P.-A., Jean, J.: Improved key recovery attacks on reduced-round AES in the single-key setting. In: Johansson, T., Nguyen, P.Q. (eds.) EUROCRYPT 2013. LNCS, vol. 7881, pp. 371–387. Springer, Heidelberg (2013)
6. Du, C., Chen, J.: Impossible differential cryptanalysis of ARIA reduced to 7 Rounds. In: Heng, S.-H., Wright, R.N., Goi, B.-M. (eds.) CANS 2010. LNCS, vol. 6467, pp. 20–30. Springer, Heidelberg (2010)
7. Dunkelman, O., Keller, N., Shamir, A.: Improved single-key attacks on 8-round AES-192 and AES-256. J. Cryptology **28**(3), 397–422 (2015)
8. Fleischmann, E., Forler, C., Gorski, M., Lucks, S.: New boomerang attacks on ARIA. In: Gong, G., Gupta, K.C. (eds.) INDOCRYPT 2010. LNCS, vol. 6498, pp. 163–175. Springer, Heidelberg (2010)
9. Korean Agency for Technology and Standards. 128 bit block encryption algorithm ARIA - Part 1: General (in Korean). KS X 1213-1:2009, December 2009
10. Kim, W.-H., Lee, J., Park, J.-H., Kwon, D.: Addition of the ARIA Cipher Suites to Transport Layer Security (TLS). RFC 6209, April 2011. https://tools.ietf.org/html/rfc6209
11. Kwon, D., Kim, J., Lee, J., Lee, J., Kim, C.: A Description of the ARIA Encryption Algorithm. RFC 5794, March 2010. https://tools.ietf.org/html/rfc5794
12. Kwon, D., et al.: New block cipher: ARIA. In: Lim, J.-I., Lee, D.-H. (eds.) ICISC 2003. LNCS, vol. 2971. Springer, Heidelberg (2004)
13. RSA Laboratories. Additional PKCS #11 Mechanisms. PKCS #11 v2.20 Amendment 3 Revision 1, January 2007
14. Lamberger, M., Mendel, F., Rechberger, C., Rijmen, V., Schläffer, M.: Rebound distinguishers: results on the full whirlpool compression function. In: Matsui, M. (ed.) ASIACRYPT 2009. LNCS, vol. 5912, pp. 126–143. Springer, Heidelberg (2009)
15. Li, R., Sun, B., Zhang, P., Li, C.: New impossible differential cryptanalysis of ARIA. IACR Cryptology ePrint Archive, 2008:227 (2008). http://eprint.iacr.org/2008/227
16. Li, Y., Wu, W., Zhang, L.: Integral attacks on reduced-round ARIA block cipher. In: Kwak, J., Deng, R.H., Won, Y., Wang, G. (eds.) ISPEC 2010. LNCS, vol. 6047, pp. 19–29. Springer, Heidelberg (2010)
17. Tang, X., Sun, B., Li, R., Li, C., Yin, J.: A meet-in-the-middle attack on reduced-round ARIA. J. Syst. Softw. **84**(10), 1685–1692 (2011)
18. Wenling, W., Zhang, W., Feng, D.: Impossible differential cryptanalysis of reduced-round ARIA and camellia. J. Comput. Sci. Technol. **22**(3), 449–456 (2007)
19. Z'aba, M.R.: Analysis of linear relationships in block ciphers. Master's thesis, Queensland University of Technology, May 2010

A   4.5 Round Distinguisher on ARIA-256

In Fig. 8, we show the 4.5 round distinguisher require for the 8-round attack on ARIA-256 demonstrated in Sect. 5.

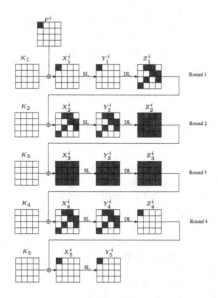

Fig. 8. 4.5-Round distinguisher in ARIA

Structural Evaluation for Generalized Feistel Structures and Applications to LBlock and TWINE

Huiling Zhang[1,2,3]($\boxtimes$) and Wenling Wu[1,2,3]

[1] TCA Laboratory, SKLCS, Institute of Software,
Chinese Academy of Sciences, Beijing, China
[2] State Key Laboratory of Cryptology, P.O. Box 5159, Beijing 100878, China
[3] University of Chinese Academy of Sciences, Beijing, China
{zhanghuiling,wwl}@tca.iscas.ac.cn

Abstract. The generalized Feistel structure (GFS) is the variant of Feistel structure with $m > 2$ branches. While the GFS is widely used, the security is not well studied. In this paper, we propose a generic algorithm for searching integral distinguishers. By applying the algorithm, we prove that the low bound for the length of integral distinguishers is $m^2 + m - 1$ and $2m+1$ for Type-1 GFS and Type-2 GFS, respectively. Meanwhile, we evaluate the security of the improved Type-1 and Type-2 GFSs when the size of each branch and the algebraic degree of F-functions are specified. Our results show that the distinguishers are affected by the parameters to various levels, which will provide valuable reference for designing GFS ciphers. Although our search algorithm is generic, it can improve integral distinguishers for specific ciphers. For instance, it constructs several 16-round integral distinguishers for LBlock and TWINE, which directly extends the numbers of attacked rounds.

Keywords: Generalized Feistel structure · Integral distinguisher · Algebraic degree · Division property · LBlock · TWINE

1 Introduction

Feistel structure is a basic symmetric cryptographic primitive, which provides many superior design features, for example, both the encryption and decryption algorithms can be achieved with a single scheme and the round function can be non-bijective. The generalized Feistel structure (GFS) introduced by Nyberg [4] is the variant of Feistel structure with $m > 2$ branches. Many GFSs exist in the literature so far. The most popular versions are Type-1 as in CAST-256 [1] and Type-2 as in CLEFIA [6]. They inherit the superior features from Feistel structure, moreover, have advantages of high parallelism, simple design and suitability for low cost implementation. Recently, lightweight cryptography has become a hot topic. Thus the GFS is an attractive structure for a lightweight symmetric key primitive such as a block cipher or a hash function.

© Springer International Publishing Switzerland 2015
A. Biryukov and V. Goyal (Eds.): INDOCRYPT 2015, LNCS 9462, pp. 218–237, 2015.
DOI: 10.1007/978-3-319-26617-6_12

In 2010, Suzaki et al. introduced the improved Type-2 GFS by replacing the cycle shift in Type-2 GFS with the optimal permutation [7]. More precisely, they proposed the maximum diffusion round (DR) which is the minimum number of rounds such that every output nibble depends on every input nibble. And then they found that the cycle shift does not provide optimum DR when $m \geq 6$. Hence, they exhaustively searched all the optimum permutations for $m \leq 16$, and gave a generic construction whose DR is close to the lower bound when m is a power of 2. In [12], Yanagihara and Iwata did the similar work for Type-1 GFS. They showed that better DR can also be obtained if one uses other permutations, moreover, an generic construction of optimum permutations for arbitrary m was devised. As shown in [5, 7, 12], the improved GFS has more secure than the standard GFS.

Integral attack was firstly proposed by Daemen et al. to evaluate the security of Square cipher [2], and then it was unified by Knudsen and Wagner in FSE 2002 [3]. The crucial part is the construction of the integral distinguisher, i.e., choosing a set of plaintexts such that the states after several rounds have a certain property, e.g., the XOR-sum of all states equals to 0 with probability 1. This property is called balanced in the following. The integral attack tends to be one of bottlenecks for the security of the GFS as shown in [5]. [7, 12] evaluated the security of Type-1, Type-2 and their improved versions against integral attack. Their results show there exist m^2- and $2m$- round integral distinguishers for Type-1 GFS and Type-2 GFS, respectively. 2DR- or $(2DR-1)$-round integral distinguishers exist for the improved Type-2 GFS. However, the specific properties of the F-functions are not utilized in the evaluations, and F-functions are restricted to be bijective.

In EUROCRYPT 2015 [10], Todo proposed a new notion, named division property, which is a generalized integral property. Based on the division property, he introduced a path search algorithm to derive the integral distinguishers for Feistel or SPN ciphers. The algorithm has several desirable features, such as, it can take advantage of the algebraic degree of the F-functions, and it can effectively construct the integral distinguisher even though the F-functions are non-bijective. Therefore, generalizing and applying the algorithm to the GFS will be very meaningful.

Our Contributions. In this paper, we evaluate the security of the GFS against integral attack. We first study the propagation characteristic of the division property for the GFS. Due to the rapid expansion of the vectors in the division property, it is difficult to directly trace the propagation when $m \geq 14$ even if we perform it by computer. Therefore, a technique named early reduce technique is devised to simplify the procedure, which works by detecting and discarding "useless" vectors. Then we propose a generic algorithm of constructing integral distinguishers. By using our algorithm, we prove that integral distinguishers for Type-1 GFS and Type-2 GFS with bijective F-functions can be extended by $m - 1$ and 1 rounds, respectively. And we show $(m^2 + m - 2)$- and $2m$- round distinguishers exist even though the F-functions are non-bijective.

For the improved GFS, our results indicate that distinguishers vary with several parameters, such as the number of branches, size of each branch, algebraic degree of F-functions, permutation layer and whether F-functions are bijective or not, which is not reflected from previous analysis. Hence, our results can provide valuable reference for designing GFS ciphers.

Although our search algorithm is generic, it can improve integral distinguishers for specific ciphers. We construct for the first time several 16-round integral distinguishers for LBlock and TWINE. The integral attacks can thus be applied to 23-round LBlock, 23-round TWINE-80 and 24-round TWINE-128.

Paper Outline. Section 2 describes the GFS we focus in this paper and gives a brief review on the division property. The path search algorithm and the improved integral distinguishers for the GFS are shown in Sect. 3. In Sect. 4, we apply the improvements to the integral attacks against LBlock and TWINE. Finally, we conclude this paper in Sect. 5.

2   Preliminaries

In this section, we introduce the generalized Feistel structure and review the definition of the division property.

2.1   Generalized Feistel Structure

A GFS divides its input into $m > 2$ *branches* of n bits each, where n is defined as the *branch size*. The round function can be separated into two successive layers, as done in [7], a F-function layer and a permutation layer. The F-function layer is made of F-functions whose inputs are some of the branches and whose outputs are added to the other branches. The permutation layer is a shuffle of m branches. In this paper, we focus on the generalized Type-1 GFS and the generalized Type-2 GFS as shown in Fig. 1. Note that they are Type-1 and Type-2 GFS, respectively, when the permutation is the left cycle shift.

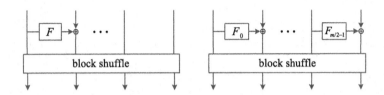

Fig. 1. Generalized Type-1 (left) and Type-2 (right) GFS

For convenience, we assume all F-functions are with algebraic degree d and $d < n$, which is reasonable for applicable ciphers. Let $P = \{p_0, p_1, \cdots, p_{m-1}\}$ denote the permutation layer moving i-th branch of the input to p_i-th branch (we number the branches from left to right, starting with 0 for the left-most branch), for example $P = \{3, 0, 1, 2\}$ for Type-1 GFS with 4 branches. Then, a GFS with parameters n, m, d and P can be defined as $[n, m, d, P]$-GFS.

2.2 Division Property

Some notations are first described for clarity. We use $\oplus$ and $+$ to distinct the XOR of $\mathbb{F}_2^n$ and the addition of $\mathbb{Z}$, and accordingly, $\bigoplus$ and $\sum$ represents XOR sum and addition sum, respectively. Denote the hamming weight of $u \in \mathbb{F}_2^n$ by $w(u)$ which is calculated as

$$w(u) = \sum_{0 \le i \le n-1} u[i],$$

where $u[i]$ is the i-th bit. Furthermore, denote the vectorial hamming weight of $U = (u_0, \cdots, u_{m-1}) \in (\mathbb{F}_2^n)^m$, $(w(u_0), \cdots, w(u_{m-1}))$, by $W(U)$. Let $K = (k_0, \cdots, k_{m-1})$ and $K' = (k'_0, \cdots, k'_{m-1})$ be the vectors in $\mathbb{Z}^m$. We define $K \succeq K'$ if $k_i \ge k'_i$ for $0 \le i \le m - 1$, otherwise, $K \nsucceq K'$.

Subset $\mathbb{S}_K^{n,m}$. Let $\mathbb{S}_K^{n,m}$ be a subset of $(\mathbb{F}_2^n)^m$ for any vector $K = (k_0, \cdots, k_{m-1})$, where $0 \le k_i \le n$. The subset $\mathbb{S}_K^{n,m}$ is composed of all $U \in (\mathbb{F}_2^n)^m$ satisfying $W(U) \succeq K$, i.e.,

$$\mathbb{S}_K^{n,m} = \{U \in (\mathbb{F}_2^n)^m | W(U) \succeq K\}.$$

Bit Product Functions π_u and π_U. Let $\pi_u : \mathbb{F}_2^n \to \mathbb{F}_2$ be a function for $u \in \mathbb{F}_2^n$. For any $x \in \mathbb{F}_2^n$, $\pi_u(x)$ is the AND of $x[i]$ for i satisfying $u[i] = 1$. Namely, the bit product function π_u is defined as

$$\pi_u(x) = \prod_{u[i]=1} x[i].$$

Let $\pi_U : (\mathbb{F}_2^n)^m \to \mathbb{F}_2$ be a function for $U = (u_0, u_1, \cdots, u_{m-1}) \in (\mathbb{F}_2^n)^m$. For any $X = (x_0, x_1, \cdots, x_{m-1}) \in (\mathbb{F}_2^n)^m$, $\pi_U(X)$ is calculated as

$$\pi_U(X) = \prod_{i=0}^{m-1} \pi_{u_i}(x_i).$$

Definition 1 (Division Property). *[10] Let Λ be a multi-set whose elements take values in $(\mathbb{F}_2^n)^m$, and $K^{(j)}$ $(0 \le j \le q - 1)$ are m-dimensional vectors whose elements take a value between 0 and n. When the multi-set Λ has the division property $\mathcal{D}_{K^{(0)}, K^{(1)}, \cdots, K^{(q-1)}}^{n,m}$, it fulfils the following conditions: the check-sum, $\bigoplus_{X \in \Lambda} \pi_U(X)$, equals to 0 if $U \in \{V \in (\mathbb{F}_2^n)^m | W(V) \nsucceq K^{(0)}, \cdots, W(V) \nsucceq K^{(q-1)}\}$. Moreover, the checksum becomes unknown if there exist i satisfying $W(U) \succeq K^{(i)}$.*

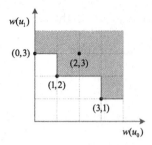

Fig. 2. Division Property $\mathcal{D}^{4,2}_{(0,3),(1,2),(2,3),(3,1)}$

We call U in the bit product function π_U as the *mask*, then view the division property from a vivid perspective as: it divides the set of all masks (i.e., $(\mathbb{F}_2^n)^m$) into two subsets, $\Gamma_?$ and Γ_0, where $\Gamma_?$ is the subset whose element results in an unknown checksum and Γ_0 is the subset whose element results in the zero-sum. Specifically, it has $\Gamma_? = \mathbb{S}^{n,m}_{K^{(0)}} \cup \cdots \cup \mathbb{S}^{n,m}_{K^{(q-1)}}$ and $\Gamma_0 = (\mathbb{F}_2^n)^m \setminus \Gamma_?$. Taking $\mathcal{D}^{4,2}_{(0,3),(1,2),(2,3),(3,1)}$ for an example, $\Gamma_?$ consists of all (u_0, u_1) located in the shadow area of Fig. 2. Note that this division property equals to $\mathcal{D}^{4,2}_{(0,3),(1,2),(3,1)}$, because they lead to the same division.

The division property is useful to construct integral distinguishers. The basic idea is that we choose a set of plaintexts satisfying certain division property and trace its propagation through $r+1$ encryption rounds until it has $\Gamma_0 \setminus \{(0^m)^n\} = \phi$. Since the cipher reduced to r rounds can be distinguished from a random permutation according to the checksum, a r-round integral distinguisher is thus constructed.

3   Improved Integral Distinguishers for GFS

In this section, we first study the propagation characteristic of the division property and construct an algorithm of searching integral distinguishers for the GFS. Meanwhile, a technique is proposed to optimize the program for wider applications. Finally, we apply the algorithm to evaluate the security of the GFS against integral attack.

3.1   Propagation Characteristic of the Division Property

The F-function layer of Type-2 GFS can be divided into three successive operations: "Type-2 copy", "Type-2 substitution" and "Type-2 compression" as depicted on Fig. 3, that is similar to [10] done for Feistel structure. We describe the propagation characteristics for these operations in Proposition 2–4 whose proofs are shown in Appendix A.

Proposition 1 (Type-2 Copy). *Let* $G : (\mathbb{F}_2^n)^m \rightarrow (\mathbb{F}_2^n)^{3m/2}$ *be the Type-2 copy, which accepts* $(x_0, \cdots, x_{m-1})$ *and produces* $(y_0, \cdots, y_{3m/2-1})$ *as* $(x_0, x_0, x_1,$

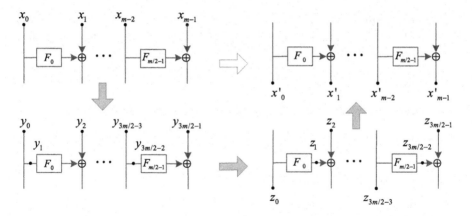

Fig. 3. Equivalent operations for GFS

$\cdots, x_{m-2}, x_{m-2}, x_{m-1})$. *If a multi-set of the inputs has division property* $\mathcal{D}_K^{n,m}$, *then the multi-set of the outputs has division property* $\mathcal{D}_{K^{(0)}, \cdots, K^{(q-1)}}^{n,3m/2}$, *where* $\{K^{(0)}, \cdots, K^{(q-1)}\}$ *is calculated as*

$$\{(i_0, (k_0 - i_0), k_1, \cdots, i_{m/2-1}, (k_{m-2} - i_{m/2-1}), k_{m-1}) | 0 \le i_j \le k_{2j}, 0 \le j < m/2\}.$$

Proposition 2 (Type-2 Substitution). *Let G be the Type-2 substitution, which accepts $(y_0, \cdots, y_{3m/2-1}) \in (\mathbb{F}_2^n)^m$ and produces $(z_0, \cdots, z_{3m/2-1}) \in (\mathbb{F}_2^n)^m$ as $(y_0, F_0(y_1), y_2, \cdots, y_{3m/2-3}, F_{m/2-1}(y_{3m/2-2}), y_{3m/2-1})$. If a multi-set of the inputs has division property $\mathcal{D}_{K^{(0)}, \cdots, K^{(q-1)}}^{n,3m/2}$, then the multi-set of the outputs has division property $\mathcal{D}_{K'^{(0)}, \cdots, K'^{(q-1)}}^{n,3m/2}$, where*

$$K'^{(j)} = \left(k_0^{(j)}, \left\lceil k_1^{(j)}/d \right\rceil, k_2^{(j)}, \cdots, k_{3m/2-3}^{(j)}, \left\lceil k_{3m/2-2}^{(j)}/d \right\rceil, k_{3m/2-1}^{(j)} \right).$$

Moreover, when the F-functions are bijective, we view $\lceil n/d \rceil$ as n.

Proposition 3 (Type-2 Compression). *Let $G : (\mathbb{F}_2^n)^{3m/2} \to (\mathbb{F}_2^n)^m$ be the Type-2 compression, which accepts $(z_0, \cdots, z_{3m/2-1})$ and produces $(x'_0, \cdots, x'_{m-1})$ as $(z_0, (z_1 \oplus z_2), \cdots, z_{3m/2-3}, (z_{3m/2-2} \oplus z_{3m/2-1}))$. If a multi-set of the inputs has division property $\mathcal{D}_{K^{(0)}, \cdots, K^{(q-1)}}^{n,3m/2}$, then the multi-set of the outputs has division property $\mathcal{D}_{K'^{(0)}, \cdots, K'^{(q-1)}}^{n,m}$, where*

$$K'^{(j)} = (k_0^{(j)}, (k_1^{(j)} + k_2^{(j)}), \cdots, k_{3m/2-3}^{(j)}, (k_{3m/2-2}^{(j)} + k_{3m/2-1}^{(j)})).$$

Following Proposition 2–4, the propagation characteristic for Type-2 GFS is easily achieved. For simplicity, we write the division property $\mathcal{D}_{K^{(0)}, \cdots, K^{(q-1)}}^{n,m}$ as a set of vectors, $\{K^{(0)}, \cdots, K^{(q-1)}\}$.

Theorem 1. *For $[n, m, d, P]$-Type-2 GFS, if a multi-set of its inputs has division property $\mathcal{D}_K^{n,m}$, then the multi-set of the outputs from one encryption round has division property $\{\sigma(i_0, \lceil(k_0 - i_0)/d\rceil + k_1, \cdots, i_{m/2-1}, \lceil(k_{m-2} - i_{m/2-1})/d\rceil + k_{m-1})|0 \leq i_j \leq k_{2j}, 0 \leq j < m/2\}$, where σ moves i-th component of the input to p_i-th component.*

In the similar manner, we get the propagation characteristic for Type-1 GFS.

Theorem 2. *For $[n, m, d, P]$-Type-1 GFS, if a multi-set of its inputs has division property $\mathcal{D}_K^{n,m}$, then the multi-set of the outputs from one encryption round has division property $\{\sigma(i, \lceil(k_0 - i)/d\rceil + k_1, k_2, k_3, \cdots, k_{m-1}) \mid 0 \leq i \leq k_0\}$, where σ moves i-th component of the input to p_i-th component.*

3.2   Path Search Algorithm for GFS

The most troublesome problem for the propagation of division property is the rapid expansion of the vectors, which makes the procedure time-consuming and costing mass memory. Therefore, we devise a technique to discard "useless" vectors early. After that, we propose the path search algorithm.

Early Reduce Technique. This technique is based on the following observation:

Observation 1. *Let K and K' be two vectors which respectively propagates to $\{K^{(0)}, \cdots K^{(q-1)}\}$ and $\{K'^{(0)}, \cdots K'^{(q'-1)}\}$ through the round function. If there exists a vector, $K'^{(j)} \in \{K'^{(0)}, \cdots K'^{(q'-1)}\}$, such that $K^{(i)} \succcurlyeq K'^{(j)}$ for each $K^{(i)} \in \{K^{(0)}, \cdots K^{(q-1)}\}$, $\Omega \cup \{K, K'\}$ and $\Omega \cup \{K'\}$ propagate to the same division property for any vector set Ω.*

Note that $K \succcurlyeq K'$ certainly satisfies with the condition, however, not limitation to it. An example is $K = (1,3)$, $K' = (4,0)$ for $[4,2,3,\{1,0\}]$-Type-2 GFS with the bijective F-function, which is actually Feistel structure. K propagates to $\{(4,0),(3,1)\}$, while K' propagates to $\{(4,0),(1,1),(0,4)\}$. It has $(4,0) \succcurlyeq (4,0)$ and $(3,1) \succcurlyeq (1,1)$, therefore, $\{(1,3),(0,4)\}$ and $\{(0,4)\}$ propagate to the same division property as depicted on Fig. 4.

The early reduce technique discards vector K if there exist $K' \in \Omega$ satisfying Observation 1. It can amazingly reduce the division property, meanwhile, it does not change the division property achieved through one round function. To show the effectiveness, we compare the numbers of vectors before and after applying the technique to TWINE in Table 1.

Table 1. Comparison of the numbers of vectors

Round		1	2	3	4	5	6	7	8	9	10	11	12	13	14	15	16
Num.	Before	1	2	3	5	11	31	184	1967	22731	113440	124827	42756	7072	952	164	44
	After	1	2	3	5	10	30	110	841	3709	10560	8976	2139	415	71	36	8

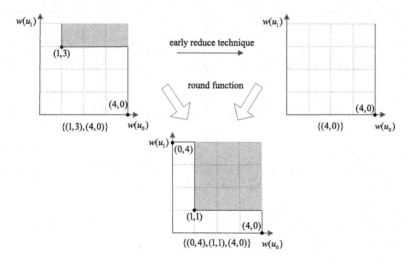

Fig. 4. An Example for early reduce technique

Path Search Algorithm. We then devise a generic algorithm of constructing integral distinguishers for the GFS, which is described in Algorithm 1. *Round-Prop* propagates the division property through the round function by Theorem 1 (or 2). *TermCondition* judges whether the division property satisfies the terminal condition: if

$$1 < \max_{0 \leq j \leq q-1} \left\{ \sum_{0 \leq i \leq m-1} k_i^j \right\},$$

it returns true, otherwise, returns false. *AddDivision* adds vectors, $K_0^i, \cdots K_{q_i-1}^i$, to the vector set, Ω, and updates p which denotes the number of vectors in Ω. *SizeReduce* discards the vector $K \in \Omega$ if there exists vector $K' \in \Omega$ satisfying $K \succcurlyeq K'$, meanwhile, updates the value of p. *EarlyReduce* further reduces Ω by using the early reduce technique.

For Type-2 GFS, *EarlyReduce* is implemented as follows: we first create a list saving all vectors $K = (k_0, k_1)$ for each $K' = (k_0', k_1')$, which satisfies that there exists a vector $K'^{(j')} \in \{(i, \lceil (k_0' - i)/d \rceil + k_1') | 0 \leq i \leq k_0'\}$ such that $K^{(j)} \succcurlyeq K'^{(j')}$ for each $K^{(j)} \in \{(i, \lceil (k_0 - i)/d \rceil + k_1) | 0 \leq i \leq k_0\}$. Then, if (k_{2i}, k_{2i+1}) locals in the list of (k_{2i}', k_{2i+1}') for $0 \leq i < m/2$, we discard $K = (k_0 \cdots k_{m-1})$ when $K' = (k_0' \cdots k_{m-1}')$ is in the vector set Ω. Notice that, the function may change the result of *TermCondition*, therefore, we need to set a threshold to decide whether it will be performed. We suggest the threshold to be 20000 for $n = 4, m = 16$.

3.3   Improved Integral Distinguishers for GFS

We evaluate the security of $[n,m,d,P]$-Type-2 GFS and $[n,m,d,P]$-Type-1 GFS against integral attack by Algorithm 1. A low bound of the length of distinguishers

Algorithm 1. Path search algorithm for the GFS

Input: Parameters n, m, d, P of the GFS and division property of the plaintext set $K = (k_0, k_1, \cdots, k_{m-1})$.

Output: The number of rounds of the integral distinguisher.

$0 \Rightarrow r$

$RoundProp(n, m, d, P, K) \Rightarrow \{K_0, K_1, \cdots K_{q-1}\}$

while $TermCondition(\{K_0, K_1, \cdots K_{q-1}\})$ **do**

    $r + 1 \Rightarrow r$

    $\emptyset \Rightarrow \Omega, 0 \Rightarrow p$

    for $i = 0$ to $q - 1$ **do**

        $RoundProp(n, m, d, P, K_i) \Rightarrow \{K_0^i, K_1^i, \cdots K_{q_i-1}^i\}$

        $AddDivision(\{K_0^i, K_1^i, \cdots K_{q_i-1}^i\}) \Rightarrow (\Omega, p)$

        $SizeReduce(\Omega) \Rightarrow (\Omega, p)$

        if threshold $\leq p$ **then**

            $EarlyReduce(\Omega) \Rightarrow (\Omega, p)$

        end if

    end for

    $p \Rightarrow q, \Omega \Rightarrow \{K_0, K_1, \cdots K_{q-1}\}$

end while

return r

---

for Type-2 (or Type-1) GFS is first given, and then the lengths of the distinguishers for improved Type-2 (or Type-1) GFSs are specified.

Results on $[n, m, d, P]$-Type-2 GFS. We prove the following conclusion for Type-2 GFS.

Theorem 3. *For Type-2 GFS with $m \leq 16$ branches of size n, there always exist the integral distinguishers with at least $2m + 1$ rounds when F-functions are bijective, moreover, there exist the integral distinguishers with at least $2m$ rounds when F-functions are non-bijective.*

Proof. For simplicity, we prove the case when Type-2 GFS with $m = 4$ branches and bijective F-functions. The general case follows by a similar manner. Firstly, assume the degree of F-functions to be $n-1$ and $n \geq 4$. We start with the division property $K = \{(n-1, n, n, n)\}$ and trace its propagation by Algorithm 1, as shown in Table 2. Since the division property after 10 rounds will be $\{(0,0,0,1),(0,0,1,0), (0,1,0,0),(1,0,0,0)\}$, which reaches the terminal condition, integral distinguishers with 9 rounds are proved to be existed. Then, in the same way, we can prove the existence of 10-round distinguishers for $n = 3$. Due to the fact that the lower degree of F-functions, the longer distinguishers achieved by our path search algorithm, the results are actually low bounds.

For improved Type-2 GFSs, the shuffles do not show the regularity as the cycle shift, which leads to the absence of a similar conclusion. However, we search the integral distinguishers for each most common parameter. The results are summarized in Table 3 ($m \leq 8$) and Table 6 ($8 < m \leq 16$, in Appendix C).

Table 2. Propagation of division property for Type-2 GFS.

Round	Division property
0	$\{(n\text{-}1,n,n,n)\}$
1	$\{(n,n,n,n\text{-}1)\}$
2	$\{(n,1,n,n),(n,n,n\text{-}1,n)\}$
3	$\{(2,n,n,1),(1,n,n,n),(n,n\text{-}1,n,n)\}$
4	$\{(n,1,2,2),(n,n,1,2),(n,n,n,1),(n\text{-}1,n,n,n)\}$
5	$\{(2,0,3,1),(2,2,2,1),(1,0,3,n),(1,2,2,n),(n,1,2,n),(n,n,1,n)\}$
6	$\{(1,0,2,0),(1,3,1,0),(0,0,2,2),(0,3,1,2),(3,2,1,0),(2,2,1,2),(0,3,n,1)\}$
7	$\{(1,0,1,0),(1,2,0,0),(0,0,1,1),(0,2,0,1),(0,0,3,0),(0,2,2,0),(3,1,0,0),(2,1,0,3)\}$
8	$\{(1,1,0,0),(0,1,0,1),(0,0,1,0),(0,3,0,0),(2,0,0,0),(1,0,0,3)\}$
9	$\{(0,0,1,0),(0,1,0,0),(1,0,0,0),(0,0,0,2)\}$

Table 3. Integral distinguishers for improved type-2 GFS

m	Type	P	DR	IND [7]	IND for $[n,m,d,P]$-Type-2 GFS							
					$[n,d]$=[4,3]		$[n,d]$=[8,7]		$[n,d]$=[16,3]		$[n,d]$=[32,7]	
				bij	bij	nbij	bij	nbij	bij	nbij	bij	nbij
6	No. 1	$\{3,0,1,4,5,2\}$	5	10	11	10	11	10	12	12	11	10
8	No. 1	$\{3,0,1,4,7,2,5,6\}$	6	11	13	13	12	12	15	15	13	13
	No. 2	$\{3,0,7,4,5,6,1,2\}$	6	11	13	12	12	12	13	13	12	12

Compared with the integral distinguishers in [7], our results extend the length of distinguishers by at least one round when the F-functions are bijective. Moreover, the distinguishers on Type-2 GFSs with non-bijective F-functions are constructed for the first time. Our results also indicate that the security of structures is sensitive to the parameters for different degrees. For example, No. 1 and No. 2 structures with $m = 8$ have the same length of distinguishers when n=4, d=3 and F-functions are bijective, however, No. 1 has longer distinguishers than No. 2 when n=16, d=3 and F-functions are bijective. This difference may help designers choosing the suitable structure to gain more security.

Results on $[n, m, d, P]$-Type-1 GFS. Similar to the proof of Theorem 3, we get the conclusion for Type-1 GFS.

Theorem 4. *For Type-1 GFS with $m \leq 16$ branches of size n, there always exist the integral distinguishers with at least $m^2 + m - 1$ rounds when F-functions are bijective, moreover, there exist the integral distinguishers with at least $m^2 + m - 2$ rounds when F-functions are non-bijective.*

We also search the integral distinguishers for improved Type-1 GFSs. The results are summarized in Table 4 ($m \leq 8$) and Table 5 ($8 < m \leq 16$, in Appendix B). An interesting observation is that our integral distinguishers have the same length with impossible differential distinguishers in [12] for all improved Type-1 GFSs when $[n, d] = [4, 3]$ or $[8, 7]$. Besides, the value of DR for the same m does not affect the length of integral distinguishers when the F-function is bijective.

Table 4. Integral distinguishers for improved Type-1 GFS

m	P	DR	ID	IND [12] bij	IND for $[n,m,d,P]$-Type-1 GFS					
					$[n,d]=[4,3]$		$[n,d]=[8,7]$		$[n,d]=[16,3]$	
					bij	nbij	bij	nbij	bij	nbij.
4	$\{2,0,3,1\}$	10	19	16	19	18	19	18	23	23
5	$\{2,0,3,4,1\}$	17	29	25	29	28	29	28	34	34
	$\{2,3,1,4,0\}$	14	29	21	29	27	29	27	32	32
6	$\{2,0,3,4,5,1\}$	26	41	36	41	40	41	40	47	47
7	$\{2,0,3,4,5,6,1\}$	37	55	49	55	54	55	54	62	62
	$\{2,3,4,5,1,6,0\}$	27	55	37	55	52	55	52	59	59
8	$\{2,0,3,4,5,6,7,1\}$	50	71	64	71	70	71	70	79	79
	$\{2,3,4,5,1,6,7,0\}$	38	71	50	71	68	71	68	77	77

4   Applications to LBlock and TWINE

Although our search algorithm is generic, it can improve integral distinguishers for specific ciphers. We construct several 16-round integral distinguishers for LBlock and TWINE, which directly leads to the extension of the numbers of attacked rounds for integral attack.

4.1   Integral Attack on LBlock

LBlock is a 32-round lightweight block cipher with 64-bit block and 80-bit master key. It adopts a Feistel structure with a twist: an 8-bit rotation is performed on the branch being XOR with the output of the Feistel function. The Feistel function is made of a key addition, a S-box layer and a nibble permutation. We denote $X_L^i \| X_R^i$ the internal state which is the input to the i-th round (or the output from $(i-1)$-th round), and further describe 8 nibbles inside of X_L^i and X_R^i as $X_L^i = X_L^i[0] \| X_L^i[1] \cdots \| X_L^i[7]$ and $X_R^i = X_R^i[0] \| X_R^i[1] \cdots \| X_R^i[7]$, respectively. A plaintext is load into the state $X_L^0 \| X_R^0$ which is then processed as Fig. 5 (left), and finally, $X_L^{32} \| X_R^{32}$ is produced as the ciphertext.

Keyschedule. The keyschedule generates 32 round keys from the master key. Firstly, the master key is loaded to a key register, denoted by $K = k_{79} k_{78} \cdots k_1 k_0$. After that, extract leftmost 32 bits of current content of the register as round key K^0. And then update the key register as follow:

1. $K <<< 29$
2. $[k_{79}k_{78}k_{77}k_{76}] = S_9[k_{79}k_{78}k_{77}k_{76}]$, $[k_{75}k_{74}k_{73}k_{72}] = S_8[k_{75}k_{74}k_{73}k_{72}]$
3. $[k_{50}k_{49}k_{48}k_{47}k_{46}] \oplus [i]_2$
4. Output the left most 32 bits of the register as round key K^i

where S_8 and S_9 are 4-bit S-boxes, and $[i]_2$ is the binary form of i for $1 \leq i \leq 31$.

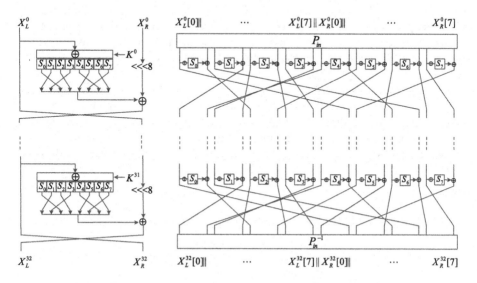

Fig. 5. LBlock (left) and the equivalent representation (right)

Improved Integral Distinguishers. As shown in Fig. 5, LBlock is equal to a $[4, 16, 3, \{9, 4, 13, 0, 3, 6, 7, 2, 1, 12, 5, 8, 11, 14, 15, 10\}]$-Type-2 GFS cipher, except a shuffle and its inverse is applied to the plaintext and the ciphertext, respectively, where the shuffle P_{in} is $\{0, 2, 4, 6, 8, 10, 12, 14, 9, 13, 3, 7, 1, 5, 11, 15\}$. Therefore, we construct several 16-round integral distinguishers for LBlock by Algorithm 1, which improves the 15-round distinguisher proposed by designers in [11]. For example, choosing a set of 2^{63} plaintexts which are constant at one bit and are active at other 63 bits, then the state X_R^{16} is balanced.

Key Recovery. Appending 7 rounds to the integral distinguisher, we can attack 23-round LBlock with 2^{76} encryption, 2^{63} chosen plaintexts and 2^{67} bytes memory, which improved the previous best integral attack by one more round. We first give a high-level description of the key recovery.

1. Query 2^{63} plaintexts which are constant at one bit and are active at other bits.
2. Compute $\bigoplus(S(X_L^{16}[0] \oplus K^{16}[0]))$ by guessing 60-bit key.

3. Compute $\bigoplus(X_L^{17}[2])$ by guessing 40-bit key independently.
4. Find matches between two results, and get corresponding 74-bit key as key candidates.
5. For 2^{70} key candidates, we exhaustively search remaining 6-bit key to recover the master key.

Details of Step 2 is given in Appendix D. We obtain a list with 2^{60} entries which contains 64-bit information: $\bigoplus(S(X_L^{16}[0] \oplus K^{16}[0]))$ and corresponding 60-bit guessed key. This procedure costs $2^{69.8}$ 23-round encryptions. Due to the Feistel structure, Step 3 costs much less time to produce a list with 2^{40} entries, which contains 44-bit information: $\bigoplus X_L^{17}[2]$ and corresponding 40-bit guessed key $K^{22}[0,1,4,5,7]\|K^{21}[0,2]\|K^{20}[4,5]\|K^{21}[7]_{(0)}\|K^{20}[6]_{(2,3)}\|K^{20}[7]_{(0)}$. In total, 78 bits key are guessed in key recovery as shown in Fig. 8, however, there exist only 74-bit significant key information, because 4-bit guessed in K^{18} can be deduced from remaining 74-bit key. Therefore, we obtain 2^{70} key candidates after Step 4. For 6-bit key is remained unknown as shown in Fig. 6, we guess it and exhaustively search the right master key combining with the key candidates.

K^{21}

State in key register state when $i = 21$

▨ : Guessed key   ▢ : Unknown key

Fig. 6. Key state after key recovery

The time complexity of the attack is determined by exhaustively searching, that is $2^6 \times 2^{70} = 2^{76}$ 23-round encryptions. The data complexity is 2^{63} chosen plaintexts. Besides, we need $2^{60} \times 15 \times 4 \times 2^{-3} = 2^{61}$ bytes of memory to save 15 nibbles of ciphertext involved in the computation of $X_L^{16}[0]$.

4.2   Integral Attack on TWINE

TWINE is a Type-2 GFS block cipher with 16 branches of 4 bits each. It supports two key lengths, 80-bit and 128-bit, which we write as TWINE-80 and TWINE-128, respectively. They only differ by the key-schedule and both have 36 rounds. The i-round of TWINE is depicted in Fig. 7, where X^i is the input which is also expressed by $X^i = X^i[0]\|X^i[1]\cdots\|X^i[15]$ and the S-box S is a 4-bit permutation with algebraic degree 3. We denote the j-th nibble of i-th round key K^i by $K^i[j]$ for $0 \le j \le 7$.

Keyschedule. The keyschedule produces 36 round keys from the master key. Firstly, the key register is initialized to the master key, and then the key register are updated by a sparse GFS using only 2 S-box per updating procedure for

$$X^i = X^i[0] \| X^i[1] \cdots X^i[15]$$

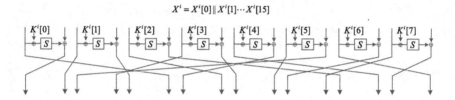

Fig. 7. Round function of TWINE

TWINE-80 and 3 for TWINE-128. Finally fixed 8 nibbles are extracted from the key register as the round key. For more details, please refer to [8].

In [8], the designers gave several 15-round integral distinguishers for TWINE. For example, considering a set of 2^{60} plaintexts which are constant for the left-most nibble (indexed by 0) and are active for other nibbles, the state after 15 rounds has 4 balanced nibbles indexed by 1, 3, 13, 15. Then, they launched the integral attack on 22-round TWINE-80 with the time, data and memory complexity being 2^{77} encryption, 2^{60} chosen plaintexts and 2^{70} bytes. The time complexity can be further reduced to $2^{68.4}$ encryption with the data complexity increased by a factor of 4. In a similar manner, 23-round TWINE-128 can be attacked with $2^{106.14}$ encryptions, $2^{62.81}$ chosen plaintexts and 2^{106} bytes memory. These results are both the best integral attacks up to now.

Improved Integral Distinguishers. We discover several 16-round integral distinguishers for TWINE by applying our path search algorithm. If we choose 2^{63} plaintexts which are constant at any one bit and are active at other 63 bits, the state after 16 encryption rounds is balanced for any nibble with odd index.

Key Recovery. Due to the keyschedule, 1-th nibble is the optimal choice for the attack considering the time complexity, therefore, we can attack 23-round TWINE-80 by following the key recovery procedure in [8] directly. Note that a structure for our distinguisher contains 2^{63} plaintexts instead of 2^{60}. Using one structure, the time, data and memory complexities of the attack are thus 2^{77} encryption, 2^{63} chosen plaintexts and 2^{70} bytes, respectively. Similarly, we can attack 24-round TWINE-128 with 2^{124} encryption, 2^{63} chosen plaintexts and 2^{106} bytes.

5 Conclusion

In this paper, we first studied the propagation characteristic of the division property for the GFS, and then proposed a generic algorithm of searching the integral distinguishers. Meanwhile, we devised the early reduce technique, which is useful to optimize the time and memory complexities. By using our algorithm, we evaluated the security of the GFS. The results show that the length of integral distinguishers can be extended by at least $m - 1$ and 1 rounds for Type-1 and

Type-2 GFS with m branches, respectively. For improved Type-1 and Type-2 GFSs, distinguishers depend on the specific parameters of the structure, such as m, branch size, algebraic degree of F-functions, permutation layer and whether F-functions are bijective or not. Finally, the algorithm was applied to LBlock and TWINE. We constructed several 16-round integral distinguishers, which lead to the integral attacks on 23-round LBlock, 23-round TWINE-80 and 24-round TWINE-128.

Acknowledgments. We would like to thank the anonymous reviewers for their useful comments and suggestions. The research presented in this paper is supported by the National Basic Research Program of China (No. 2013CB338002) and National Natural Science Foundation of China (No. 61272476, No. 61232009 and No. 61202420).

A   Proofs for Proposition 2-4

Proposition 2

Proof. We give the proof for the case of $m = 4$, which can then simply be transferred to the general case. Denote Λ and Λ' as the multi-set of inputs and outputs, respectively. The checksum of Λ' for $U = (u_0, \cdots, u_5)$ has

$$
\begin{aligned}
&\bigoplus_{Y \in \Lambda'} \pi_U(Y) \\
&= \bigoplus_{X \in \Lambda} \pi_{(u_0, \ldots, u_5)}(x_0, x_0, x_1, x_2, x_2, x_3) \\
&= \bigoplus_{X \in \Lambda} \pi_{u_0}(x_0) \times \pi_{u_1}(x_0) \times \pi_{u_2}(x_1) \times \pi_{u_3}(x_2) \times \pi_{u_4}(x_2) \times \pi_{u_5}(x_3) \\
&= \bigoplus_{X \in \Lambda} \pi_{u_0 \vee u_1}(x_0) \times \pi_{u_2}(x_1) \times \pi_{u_3 \vee u_4}(x_2) \times \pi_{u_5}(x_3) \\
&= \bigoplus_{X \in \Lambda} \pi_{(u_0 \vee u_1, u_2, u_3 \vee u_4, u_5)}(X)
\end{aligned}
$$

where $\vee$ is OR. When $(w(u_0 \vee u_1), w(u_2), w(u_3 \vee u_4), w(u_5)) \not\succeq K$, the result is always 0. Its sufficient condition is $(w(u_0) + w(u_1), w(u_2), w(u_3) + w(u_4), w(u_5)) \not\succeq K$. Therefore, the division property of Λ' is $\{(i_0, k_0 - i_0, k_1, i_1, k_2 - i_1, k_3) | 0 \le i_0 \le k_0, 0 \le i_1 \le k_2\}$.

Proposition 3

Proof. The proposition describes a special case of Rule 1 in [9]. Readers can refer to [9] for the details of the proof.

Proposition 4

Proof. We prove the case when $m = 4$. The general case follows by a similar manner. Denote Λ and Λ' as the multi-set of inputs and the multi-set of outputs,

Table 5. Integral distinguishers for improved Type-1 GFS with $8 < m \leq 16$

m	P	DR	ID	IND [12]	Our IND			
					$[n,d]$=[4, 3]		$[n,d]$=[8, 7]	
				bij	bij	nbij	bij	nbij.
9	$\{2,3,1,4,5,6,7,8,0\}$	58	89	73	89	87	89	87
	$\{2,3,4,5,6,7,1,8,0\}$	44	89	57	89	85	89	85
10	$\{2,3,4,5,1,6,7,8,9,0\}$	66	109	82	109	106	109	106
11	$\{2,3,1,4,5,6,7,8,9,10,0\}$	92	131	111	131	129	131	129
	$\{2,3,4,5,1,6,7,8,9,10,0\}$	83	131	101	131	128	131	128
	$\{2,3,4,5,6,7,1,8,9,10,0\}$	74	131	91	131	127	131	127
	$\{2,3,4,5,6,7,8,9,1,10,0\}$	65	131	81	131	126	131	126
12	$\{2,3,4,5,6,7,8,9,1,10,11,0\}$	82	155	100	155	150	155	150
13	$\{2,3,1,4,5,6,7,8,9,10,11,12,0\}$	134	181	157	181	179	181	179
	$\{2,3,4,5,1,6,7,8,9,10,11,12,0\}$	123	181	145	181	178	181	178
	$\{2,3,4,5,6,7,1,8,9,10,11,12,0\}$	112	181	133	181	177	181	177
	$\{2,3,4,5,6,7,8,9,1,10,11,12,0\}$	101	181	121	181	176	181	176
	$\{2,3,4,5,6,7,8,9,10,11,1,12,0\}$	90	181	109	181	175	181	175
14	$\{2,3,4,5,1,6,7,8,9,10,11,12,13,0\}$	146	209	170	209	206	209	206
	$\{2,3,4,5,6,7,8,9,1,10,11,12,13,0\}$	122	209	144	209	204	209	204
15	$\{2,3,1,4,5,6,7,8,9,10,11,12,13,14,0\}$	184	239	211	239	236	239	236
	$\{2,3,4,5,6,7,1,8,9,10,11,12,13,14,0\}$	158	239	183	239	234	239	234
	$\{2,3,4,5,6,7,8,9,10,11,12,13,1,14,0\}$	119	239	141	239	232	239	232
16	$\{2,3,4,5,1,6,7,8,9,10,11,12,13,14,15,0\}$	198	271	211	271	268	271	268
	$\{2,3,4,5,6,7,8,9,1,10,11,12,13,14,15,0\}$	170	271	183	271	266	271	266
	$\{2,3,4,5,6,7,8,9,10,11,12,13,1,14,15,0\}$	142	271	141	271	264	271	264

respectively. The checksum of Λ' for $U = (u_0, \cdots, u_3)$ has

$$
\bigoplus_{X' \in \Lambda'} \pi_{(u_0,u_1,u_2,u_3)}(X')
$$
$$
= \bigoplus_{Z \in \Lambda} \pi_U((z_0, z_1 \oplus z_2, z_3, z_4 \oplus z_5))
$$
$$
= \bigoplus_{Z \in \Lambda} \pi_{u_0}(z_0) \times \pi_{u_1}(z_1 \oplus z_2) \times \pi_{u_2}(z_3) \times \pi_{u_3}(z_4 \oplus z_5)
$$
$$
= \bigoplus_{Z \in \Lambda} \left(\pi_{u_0}(z_0) \times (\bigoplus_{c_1 \prec u_1} \pi_{u_1}(z_1) \times \pi_{u_1 \oplus c_1}(z_2)) \times \pi_{u_2}(z_3) \times (\bigoplus_{c_2 \prec u_3} \pi_{u_3}(z_4) \times \pi_{u_3 \oplus c_2}(z_5)) \right)
$$
$$
= \bigoplus_{Z \in \Lambda} \bigoplus_{c_1 \prec u_1} \bigoplus_{c_2 \prec u_3} \left(\pi_{(u_0,u_1,u_1 \oplus c_1,u_2,u_3,u_3 \oplus c_2)}(Z) \right)
$$

where $c \prec u$ denotes the elements of F_2^n satisfying c AND u equals to c. Obviously, it has $w(c) + w(u \oplus c) = w(u)$ if $c \prec u$. When $(w(u_0), w(u_1), w(u_1) - w(c_1), w(u_2), w(u_3), w(u_3) - w(c_2)) \not\succeq K$ for any $c_1 \prec u_1$ and any $c_2 \prec u_3$, the result is always 0. Thereafter, the division property of Λ' is $\{K'^{(j)} = (k_0^j, (k_1^j + k_2^j), k_3^j, (k_4^j + k_5^j))|0 \leq j \leq q - 1\}$.

B   Results on Improved Type-1 GFS for $8 < m \leq 16$

Table 5 shows integral distinguishers for improved Type-1 GFS when the number of branches is more than 8.

C   Results on Improved Type-2 GFS for $8 < m \leq 16$

Table 6 shows integral distinguishers for improved Type-2 GFS when the number of branches is more than 8.

Table 6. Integral distinguishers for improved Type-2 GFS with $8 < m \leq 16$

m	Type	P	DR	IND [7]	Our IND			
					$[n,d]{=}[4,3]$		$[n,d]{=}[8,7]$	
				bij	bij	nbij	bij	nbij.
	No. 1	$\{5,0,7,2,9,6,3,8,1,4\}$	7	13	14	14	14	13
10	No. 2	$\{3,0,1,4,7,2,5,8,9,6\}$	7	13	15	14	14	14
	No. 3	$\{3,0,7,4,1,6,5,8,9,2\}$	7	13	14	13	13	13
	No. 1	$\{3,0,7,2,9,4,11,8,5,10,1,6\}$	8	15	17	16	16	16
12	No. 2	$\{3,0,7,2,11,4,1,8,5,10,9,6\}$	8	16	17	17	17	16
	No. 3	$\{7,0,9,2,11,4,1,8,5,10,3,6\}$	8	15	16	15	16	15
	No. 4	$\{5,0,9,2,1,6,11,4,3,10,7,8\}$	8	15	17	17	16	16
	No. 1	$\{1,2,9,4,3,6,13,8,7,10,11,12,5,0\}$	8	15	17	16	16	16
	No. 2	$\{1,2,9,4,13,6,7,8,5,10,3,12,11,0\}$	8	15	16	15	16	15
14	No. 14	$\{1,2,11,4,13,6,7,8,5,12,9,10,3,0\}$	8	15	16	16	16	16
	No. 16	$\{5,2,9,4,1,6,13,10,11,8,7,0,3,12\}$	8	15/16	17	17	17	16
	No. 20	$\{7,2,1,4,9,6,5,10,3,12,13,0,11,8\}$	8	15	16	16	16	15
	No. 1	$\{1,2,9,4,15,6,5,8,13,10,7,14,11,12,3,0\}$	8	16	17	16	17	16
16	No. 7	$\{1,2,11,4,3,6,7,8,15,12,5,14,9,0,13,10\}$	8	15	17	16	16	16
	No. 10	$\{7,2,13,4,11,8,3,6,15,0,9,10,1,14,5,12\}$	8	15	16	15	16	15

D   Details of the Attack on LBlock

We need to guess 60-bit key to compute the value of $\bigoplus(S(X_L^{16}[0] \oplus K^{16}[0]))$ according to the keyschedule. These guessed keys are marked by gray cubes in Fig. 8, and the procedure is as follows:

1. Query 2^{63} plaintexts which are constant at one bit and are active at other bits.
2. Count whether each 15-nibble value $X_L^{23}[0,1,2,3,4,6,7]\|X_R^{23}[0,1,2,3,4,5,6,7]$ appears even or odd times, and pick the values which appear odd times.
3. Guess $K^{22}[3]$, and then compute $X_R^{22}[3]$. Compress the data into 2^{56} texts of the value of $X_L^{23}[0,2,3,4,6,7]\|X_R^{23}[0,1,2,4,5,6,7]\|X_R^{22}[3]$ appearing odd times.

4. Guess $K^{22}[5]$, and then compute $X_R^{22}[6]$. Compress the data into 2^{52} texts of the value of $X_L^{23}[0,2,3,6,7]||X_R^{23}[0,1,2,4,6,7]||X_R^{22}[3,6]$ appearing odd times.

5. Guess $K^{22}[1]$, and then compute $X_R^{22}[2]$. Compress the data into 2^{48} texts of the value of $X_L^{23}[2,3,6,7]||X_R^{23}[0,2,4,6,7]||X_R^{22}[3,6,2]$ appearing odd times.

6. Guess $K^{21}[6]$, and then compute $X_R^{21}[1]$. Compress the data into 2^{44} texts of the value of $X_L^{23}[2,3,6,7]||X_R^{23}[0,2,4,6]||X_R^{22}[3,2]||X_R^{21}[1]$ appearing odd times.

7. Guess $K^{22}[4]$, and then compute $X_R^{22}[0]$. Compress the data into 2^{44} texts of the value of $X_L^{23}[2,3,7]||X_R^{23}[0,2,4,6]||X_R^{22}[0,2,3]||X_R^{21}[1]$ appearing odd times.

8. Guess $K^{22}[6]$, and then compute $X_R^{22}[1]$. Compress the data into 2^{44} texts of the value of $X_L^{23}[2,3]||X_R^{23}[0,2,4,6]||X_R^{22}[0,1,2,3]||X_R^{21}[1]$ appearing odd times.

9. Guess $K^{22}[0]$, and then compute $X_R^{22}[4]$. Compress the data into 2^{44} texts of the value of $X_L^{23}[3]||X_R^{23}[0,2,4,6]||X_R^{22}[0,1,2,3,4]||X_R^{21}[1]$ appearing odd times.

10. Guess $K^{21}[4]$, and then compute $X_R^{21}[0]$. Compress the data into 2^{40} texts of the value of $X_L^{23}[3]||X_R^{23}[0,2,4]||X_R^{22}[0,1,2,3]||X_R^{21}[0,1]$ appearing odd times.

11. Due to the keyschedule, $K^{20}[0]$ is determined by rightmost two bits in $K^{22}[5]$ and leftmost two bits in $K^{22}[6]$, which are all guessed. We can directly compute $X_R^{20}[4]$ and compress the data into 2^{36} texts of the value of $X_L^{23}[3]$ $||X_R^{23}[0,2,4]||X_R^{22}[0,1,3]||X_R^{21}[1]||X_R^{20}[4]$ appearing odd times.

12. Due to the keyschedule, $K^{20}[1]$ is determined by rightmost two bits in $K^{22}[6]$ and leftmost two bits in $K^{22}[7]$. We only need guess the leftmost two bits in $K^{22}[7]$. Compute $X_R^{20}[2]$ and compress the data into 2^{36} texts of the value of $X_L^{23}[3]||X_R^{23}[0,2,4] ||X_R^{22}[0,1,3]||X_R^{20}[2,4]$ appearing odd times.

13. Guess $K^{21}[1]$ and compute $X_R^{21}[2]$ and compress the data into 2^{32} texts of the value of $X_L^{23}[3]||X_R^{23}[2,4]||X_R^{22}[0,3]||X_R^{21}[2]||X_R^{20}[2,4]$ appearing odd times.

14. Guess $K^{20}[2]$ and compute $X_R^{20}[5]$ and compress the data into 2^{28} texts of the value of $X_L^{23}[3]||X_R^{23}[2,4]||X_R^{22}[0]||X_R^{20}[2,4,5]$ appearing odd times.

15. Guess $K^{21}[0]$ and compute $X_R^{21}[4]$ and compress the data into 2^{28} texts of the value of $X_L^{23}[3]||X_R^{23}[2,4]||X_R^{21}[4]||X_R^{20}[2,4,5]$ appearing odd times.

16. Due to the keyschedule, $K^{19}[5]$ is determined by $K^{22}[3]$ and $K^{22}[4]$. Compute $X_R^{19}[6]$ and compress the data into 2^{24} texts of the value of $X_L^{23}[3]||X_R^{23}[2,4]|| X_R^{20}[2,4] ||X_R^{19}[6]$ appearing odd times.

17. Guess $K^{22}[2]$ and then compute $X_R^{22}[5]$ and compress the data into 2^{20} texts of the value of $X_R^{22}[5]||X_R^{23}[4]||X_R^{20}[2,4]||X_R^{19}[6]$ appearing odd times.

18. Guess $K^{21}[5]$ and then compute $X_R^{21}[6]$ and compress the data into 2^{16} texts of the value of $X_R^{21}[6]||X_R^{20}[2,4]||X_R^{19}[6]$ appearing odd times.

19. Due to the keyschedule, $K^{19}[4]$ is determined by $K^{22}[2]$ and $K^{22}[3]$. Compute $X_R^{19}[0]$ and compress the data into 2^{12} texts of the value of $X_R^{20}[2]||X_R^{19}[0,6]$ appearing odd times.

20. Guess $K^{18}[0]$ and compute $X_R^{18}[4]$ and compress the data into 2^8 texts of the value of $X_R^{19}[6]\|X_R^{18}[4]$ appearing odd times.
21. Due to the keyschedule, $K^{17}[4]$ is determined by the rightmost three bits of $K^{20}[2]$ and the leftmost bit of $K^{20}[3]$. Guess the leftmost bit of $K^{20}[3]$, compute $X_R^{17}[0]$ and compress the data into 2^4 texts of the value of $X_R^{17}[0]$ appearing odd times.
22. Due to the keyschedule, $K^{16}[0]$ is determined by the rightmost bit of $K^{21}[3]$ and the leftmost three bits of $K^{21}[4]$. Guessing the rightmost bit of $K^{21}[3]$, and then we compute the sum $\bigoplus(S(X_L^{16}[0] \oplus K^{16}[0]))$.

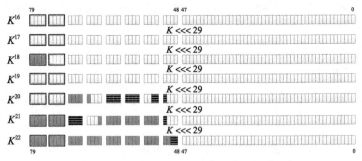

▮ : Guessed key to compute $\oplus\big(S(X_L^{16}[0]\oplus K^{16}[0])\big)$

▮ and ▤ : Guessed key in key recovery

Fig. 8. Key state for key recovery

Complexity for Computing $\bigoplus(S(X_L^{16}[0] \oplus K^{16}[0]))$. The complexity for each step is estimated as a product of the previous date size and the total number of guessed bits. In total,

$$
\begin{aligned}
&2^4 \times 2^{60} + 2^8 \times 2^{56} + 2^{12} \times 2^{52} + 2^{16} \times 2^{48} + 2^{20} \times 2^{44} \\
&+2^{24} \times 2^{44} + 2^{28} \times 2^{44} + 2^{32} \times 2^{44} + 2^{32} \times 2^{40} + 2^{34} \times 2^{36} \\
&+2^{38} \times 2^{36} + 2^{42} \times 2^{32} + 2^{46} \times 2^{28} + 2^{46} \times 2^{28} + 2^{50} \times 2^{24} \\
&+2^{54} \times 2^{20} + 2^{54} \times 2^{16} + 2^{58} \times 2^{12} + 2^{59} \times 2^8 + 2^{40} \times 2^4 \\
&= 2^{77.3}.
\end{aligned}
$$

That is $2^{77.3} \times \frac{1}{8} \times \frac{1}{23} \approx 2^{69.8}$ 23-round encryptions. After step 22, we obtain a list with 2^{60} entries which contains 64-bit information: $\bigoplus(S(X_L^{16}[0] \oplus K^{16}[0]))$ and corresponding 60-bit guessed key.

References

1. Adams, C.: The CAST-256 encryption algorithm. In: AES proposal (1998)
2. Daemen, J., Knudsen, L.R., Rijmen, V.: The block cipher SQUARE. In: Biham, E. (ed.) FSE 1997. LNCS, vol. 1267, pp. 149–165. Springer, Heidelberg (1997)

3. Knudsen, L.R., Wagner, D.: Integral cryptanalysis. In: Daemen, J., Rijmen, V. (eds.) FSE 2002. LNCS, vol. 2365, p. 112. Springer, Heidelberg (2002)
4. Nyberg, K.: Generalized Feistel networks. In: Kim, K., Matsumoto, T. (eds.) ASIACRYPT 1996. LNCS, vol. 1163, pp. 91–104. Springer, Heidelberg (1996)
5. Shibutani, K.: On the diffusion of generalized Feistel structures regarding differential and linear cryptanalysis. In: Biryukov, A., Gong, G., Stinson, D.R. (eds.) SAC 2010. LNCS, vol. 6544, pp. 211–228. Springer, Heidelberg (2011)
6. Shirai, T., Shibutani, K., Akishita, T., Moriai, S., Iwata, T.: The 128-bit blockcipher CLEFIA (extended abstract). In: Biryukov, A. (ed.) FSE 2007. LNCS, vol. 4593, pp. 181–195. Springer, Heidelberg (2007)
7. Suzaki, T., Minematsu, K.: Improving the generalized Feistel. In: Hong, S., Iwata, T. (eds.) FSE 2010. LNCS, vol. 6147, pp. 19–39. Springer, Heidelberg (2010)
8. Suzaki, T., Minematsu, K., Morioka, S., Kobayashi, E.: TWINE: a lightweight block cipher for multiple platforms. In: Knudsen, L.R., Wu, H. (eds.) SAC 2012. LNCS, vol. 7707, pp. 339–354. Springer, Heidelberg (2013)
9. Todo, Y.: Integral cryptanalysis on full MISTY1. In: Gennaro, R., Robshaw, M. (eds.) CRYPTO 2015. LNCS, vol. 9215, pp. 413–432. Springer, Heidelberg (2015)
10. Todo, Y.: Structural evaluation by generalized integral property. In: Oswald, E., Fischlin, M. (eds.) EUROCRYPT 2015. LNCS, vol. 9056, pp. 287–314. Springer, Heidelberg (2015)
11. Wu, W., Zhang, L.: LBlock: a lightweight block cipher. In: Lopez, J., Tsudik, G. (eds.) ACNS 2011. LNCS, vol. 6715, pp. 327–344. Springer, Heidelberg (2011)
12. Yanagihara, S., Iwata, T.: On permutation layer of Type 1, Source-Heavy, and Target-Heavy generalized Feistel structures. In: Lin, D., Tsudik, G., Wang, X. (eds.) CANS 2011. LNCS, vol. 7092, pp. 98–117. Springer, Heidelberg (2011)

Side Channel Attacks

Cryptanalysis of Two Fault Countermeasure Schemes

Subhadeep Banik$^{(\boxtimes)}$ and Andrey Bogdanov

DTU Compute, Technical University of Denmark, 2800 Kgs. Lyngby, Denmark
{subb,anbog}@dtu.dk

Abstract. In this paper, we look at two fault countermeasure schemes proposed very recently in literature. The first proposed in ACISP 2015 constructs a transformation function using a cellular automata based linear diffusion, and a non-linear layer using a series of bent functions. This countermeasure is meant for the protection of block ciphers like AES. The second countermeasure was proposed in IEEE-HOST 2015 and protects the Grain-128 stream cipher. The design divides the output function used in Grain-128 into two components. The first called the masking function, masks the input bits to the output function with some additional randomness and computes the value of the function. The second called the unmasking function, is computed securely using a different register and undoes the effect of the masking with random bits. We will show that there exists a weakness in the way in which both these schemes use the internally generated random bits which make these designs vulnerable. We will outline attacks that cryptanalyze the above schemes using 66 and 512 faults respectively.

Keywords: AES · Fault analysis · Grain-128 · Infective countermeasures

1 Introduction

There has been a lot of effort to design hardware based countermeasures to prevent fault attacks on AES-128 circuits. Most of these countermeasures can be classified into two broad categories: **(a)** Detection based and **(b)** Infection based. As the name suggests, detection based measures aim to detect the injection of fault by performing various intermediate checks during the course of the encryption operation [7,12,15]. The functionality is achieved by comparing two or more data blocks output by the encryption circuit. Since the comparison operation is itself prone to faults, Infection based countermeasures have also become popular [8,14,16]. In this approach, the circuit is designed in such a fashion that even if an attacker is able to inject a fault in the circuit, he can not utilize the corrupted output to find the secret key. Most of these countermeasures work by introducing additional operations in between or after the encryption algorithm that make it difficult for an adversary to deduce simple enough algebraic relations to deduce the secret key.

© Springer International Publishing Switzerland 2015
A. Biryukov and V. Goyal (Eds.): INDOCRYPT 2015, LNCS 9462, pp. 241–252, 2015.
DOI: 10.1007/978-3-319-26617-6_13

As already mentioned, the main philosophy behind infective countermeasures is to ensure that a faulty ciphertext produced by the system can not be exploited by the attacker to obtain any non-trivial information about the secret key. There have been several infective countermeasures proposed in literature but most of them have a hardware overhead of over 100 %. In [8], a countermeasure using redundant and dummy round functions was proposed. A countermeasure using random masks was proposed in [16]. Both these fault protection schemes were cryptanalyzed in [3]. In response, an improved countermeasure was proposed in [19], that replaced the output of the cipher with a random 128 bit string whenever the system detected a fault. This method was again cryptanalyzed in [4]. In this paper, we will look at two different infective countermeasures that have been proposed very recently. The first [9], was designed mainly to protect block ciphers like AES [6]. The design includes two identical AES modules, the outputs of which are xored and fed into a transformation function composed of the following functions

1. A linear diffusion function based on the principles of cellular automata,
2. A non-linear mixing function that additionally utilizes some internally generated randomness.

The output of the transformation function is xored back to the outputs of one of the AES modules, and produced as ciphertext. Assuming that the adversary does not have the capability to reproduce the same fault in both the AES modules, any difference introduced by a fault in one of the modules is transformed by the non-linear function so that any simple algebraic relation can not be derived between the faulty ciphertext and the roundkey. The second countermeasure proposed in [10] has been designed protect the Grain-128 [11] stream cipher. The work is significant because not many architectures have been proposed to protect stream ciphers from fault attacks. The design decomposes the output boolean function of Grain-128 into two component functions. The first called the masking function, masks the inputs to the output function with certain random bits generated internally. The second called the unmasking function, (which is computed securely using a different register) undoes the effect of the masking, so that the GF(2) sum of the masking and unmasking function equals the output function of Grain-128. We will show that method of utilizing the internal randomness in both these schemes has some weakness which can be used to cryptanalyze them.

1.1 Organization of the Paper

In Sect. 2, we will first provide a complete architectural and mathematical description of the infective countermeasure proposed in ACISP 2015 [9]. In Sect. 3, we will begin by outlining a fault attack on an unprotected AES implementation. We will then go on to reveal a weakness in the non-linear mixing function used in this scheme that makes this function easy to invert. Using this observation we will propose a method that allows the attacker to deduce the secret key using around 66 faults on average. In Sect. 4, we will provide a preliminary mathematical description of Grain-128 and the countermeasure as proposed in [10]. Thereafter in Sect. 5 we will then point out two weaknesses in the scheme. First we show, that due to a flaw in the masking function, any fault localized on a specific LFSR location reveals non-trivial information about the internal state of the cipher. This can be used to mount a state recovery attack, that reveals the entire internal state in less than 512 faults. The second weakness comes from the fact that although the design tries to protect the output function of the cipher, the NFSR update function is left completely unprotected. Using this result a fault attack using as less as 4 randomly applied faults can be mounted. Section 6 concludes the paper.

2 Countermeasure Proposed in ACISP 2015 [9]

The scheme proposed in [9] can be described as follows. The design takes the xor of the outputs of two identical AES modules and passes it through a transformation function $\mathcal{T}$. This function is composed of a sequence of two functions. The first is a cellular automata based linear diffusion function. The output of the linear diffusion function is then input to a non-linear mixing function. The mixing function additionally uses some random bits which are generated internally by a cellular automata based random number generator. The output of the mixing function is then xored back with the output of one of the AES modules and produced as ciphertext. The architecture is described pictorially in Fig. 1.

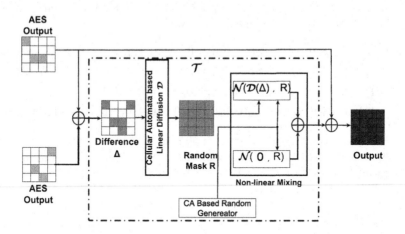

Fig. 1. Infective countermeasure of ACISP 2015 [9]

The architecture described above makes two assumptions. The first is that the attacker does not have the capability to inject the same fault in both the AES modules. Otherwise the output of both the AES modules is identical and so the input to $\mathcal{T}$ is zero, and since $\mathcal{T}$ maps the zero input to zero, the attacker gets back the original faulty ciphertext. The second assumption is that since the attacker does not know the random bits used to compute the output of the mixing function, it is not possible for him to deduce algebraic relations between the ciphertext and the roundkey.

2.1 Linear Diffusion Function

The linear diffusion function $\mathcal{D} : \{0,1\}^{128} \rightarrow \{0,1\}^{128}$ is based on the principles of a 3-neighborhood cellular automata. The state update operation in such a system can be expressed equivalently as pre-multiplication with a 128×128 binary matrix. In this specific case, the function $\mathcal{D}$ is constructed as follows. One iteration of the automata is first designed using the following primitive polynomial over GF(2):

$$p(x) = x^{128} + x^{29} + x^{27} + x^2 + 1$$

Thus, if X_t and X_{t+1} denote the 128 bit vectors that are input and output respectively of a single iteration of the automata, then these vectors are related as $X_{t+1} = A \cdot X_t$, where A is a 128×128 binary tridiagonal matrix whose ij^{th} element a_{ij} is given as follows:

$$a_{ij} = \begin{cases} m_i, & \text{if } i = j, \\ 1, & \text{if } |i - j| = 1, \\ 0, & \text{otherwise.} \end{cases}$$

The element a_{ii} in the principal diagonal of the matrix A is taken as the ith element m_i of the vector M defined as follows:

$$
\begin{aligned}
M = [\, & 0,1,0,0,1,0,0,0,1,0,0,0,1,0,0,0,0,0,1,0,1,1,1,1,1,0,1,1,1,1,0,1,0, \\
& 1,1,0,0,1,1,1,0,0,0,0,0,0,1,1,0,0,0,1,1,0,1,0,0,1,1,1,1,0,1,0,0,1, \\
& 1,1,1,0,1,0,0,1,1,1,1,0,0,1,1,1,0,0,0,0,0,0,1,1,1,0,0,1,1,0,1,0,1, \\
& 1,1,1,0,1,1,1,1,1,0,1,0,0,0,0,0,1,0,0,0,1,0,0,0,1,0,0,1,0 \,]
\end{aligned}
$$

The vector M is constructed using the polynomial $p(x)$ by following the methods outlined in [5]. The update function described by the single iteration of this automata has a period of $2^{128} - 1$, and achieves full diffusion of any bit difference in 127 iterations. The function $\mathcal{D}$ is constructed using 255 iterations of the automata, i.e., $\mathcal{D}(X) = A^{255} \cdot X$. However, the matrix A is invertible and hence the function $\mathcal{D}$ is efficiently invertible.

2.2 Non-linear Mixing Function

The non-linear mixing function $\mathcal{N} : \{0,1\}^{128} \times \{0,1\}^{128} \rightarrow \{0,1\}^{128}$ is a constructed using a series of bent functions. The function uses a 128 bit random string R which is generated using a cellular automata based random number generator Given $\mathbf{X} = [x_0, x_1, \ldots, x_{127}]$, $\mathbf{R} = [r_0, r_1, \ldots, r_{127}]$, $\mathcal{N}(\mathbf{X}, \mathbf{R}) = \mathbf{Y} = [y_0, y_1, \ldots, y_{127}]$ is defined as:

$$c_i = \bigoplus_{j=0}^{i} x_j r_j \oplus x_{i-1} x_i \oplus r_{i-1} r_i$$

$$y_i = x_i \oplus r_i \oplus c_{i-1}$$

for $0 \leq i \leq 127$, with $x_{-1} = r_{-1} = 0$ with $c_{-1} = c_{127}$. Each y_i is a bent function of algebraic degree 2 and nonlinearity $2^{2i+1} - 2^{i+2}$. Since the fault protection mechanism must output the original ciphertext if no fault is injected, the transformation function $\mathcal{T}$ must map the zero input to zero. For this reason, the output of the nonlinear layer is taken as $\mathcal{S}(\mathbf{X}) = \mathcal{N}(\mathbf{X}, \mathbf{R}) \oplus \mathcal{N}(\mathbf{0}, \mathbf{R})$.

3   Cryptanalysis of the Fault Countermeasure Scheme

3.1   Basic Fault Attack on Unprotected AES

We outline the basic fault attack on AES which finds the secret key by injecting a random byte fault before the 9th round MixColumn (MC) operation [13, Chap. 4.2]. Assuming that a random byte fault has been injected in the first element of a column, the attacker computes a list $\mathcal{L}$ of possible differences at the output column of the MixColumn operation. The list $\mathcal{L}$ thus contains 4×255 four-byte elements. This is a one time operation. As can be seen in Fig. 2, a fault injected in the first byte of the AES state before the 9th round MixColumn will result in a faulty ciphertext that differs with the original ciphertext in byte positions $1, 8, 11, 14$. So given a pair of fault-free and faulty ciphertexts C, C_f, the attacker guesses 4 bytes of the 10th roundkey K_{10} (i.e. the 1st, 8th, 11th and 14th bytes) and computes the four differences Δ_i ($i = 1, 8, 11, 14$) as follows:

$$\Delta_i = SB^{-1}\left(C[i] \oplus K_{10}[i]\right) \oplus SB^{-1}\left(C_f[i] \oplus K_{10}[i]\right),$$

(note that the ith byte of any block X is represented as $X[i]$). Each tuple $(\Delta_1, \Delta_8, \Delta_{11}, \Delta_{14})$ is then compared with the elements contained in the list $\mathcal{L}$. The candidates $(K_{10}[1], K_{10}[8], K_{10}[11], K_{10}[14])$ for which a match is found are gathered in another list $\mathcal{E}$. With one pair (C, C_f), the list $\mathcal{E}$ contains 1,036 elements on average. By using another pair (C, C_f) with a fault injected into the same column, the corresponding four bytes of the last round key are uniquely

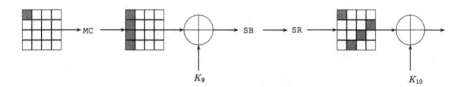

Fig. 2. Basic Attack on unprotected AES

determined with a 98 % probability. Similar analysis holds for faults injected into the 2nd, 3rd and 4th columns. Therefore the last round key can be recovered by using eight faulty ciphertexts with faults induced at chosen locations.

3.2   Cryptanalysis of the Scheme

The fault protection scheme as outlined in [9] can be outlined as follows. The scheme first computes the difference Δ in the two ciphertexts, the first of which is produced by a fault in one of the AES modules, and the second produced by the other AES module which is fault-free. This difference Δ is then passed through the transformation function $\mathcal{T}$ which is basically the composition of the functions $\mathcal{S} \circ \mathcal{D}$. The transformation function is probabilistic since it uses some randomness generated internally by the system. Now instead of $C_f = C \oplus \Delta$, the system outputs $C \oplus \mathcal{T}(\Delta)$. Now intuitively it is clear that $\mathcal{T}$ needs to be a one way function, because if the attacker can obtain the value of Δ from the value of $\mathcal{T}(\Delta)$, he can compute the value of $C_f = C \oplus \Delta$ and perform the attack described in Sect. 3.1. In this section we will prove that $\mathcal{T}$ is not a one way function due to which the security of the scheme collapses. Note that we have already established that $\mathcal{D}$ is efficiently invertible, and hence if we can show that $\mathcal{S}$ is also invertible, we would have proven that $\mathcal{T}$ is not one way.

Lemma 1. *$\mathcal{S}$ is not a one way function.*

Proof. For any fixed input $\mathbf{X} = [x_0, x_1, \ldots, x_{127}]$, $\mathcal{S}(\mathbf{X})$ is queried a few times. Since the function $\mathcal{S}$ uses internally generated randomness, it outputs a different value every time. The task therefore would be to recover $\mathbf{X}$ from $\mathcal{S}(\mathbf{X}) = \mathcal{N}(\mathbf{X}, \mathbf{R}) \oplus \mathcal{N}(\mathbf{0}, \mathbf{R})$ for different values of $\mathbf{R}$. From the description given in Sect. 2.2, the algebraic relation between $\mathcal{S}(\mathbf{X}) = [s_0, s_1, \ldots, s_{127}]$, $\mathbf{X} = [x_0, x_1, \ldots, x_{127}]$, $\mathbf{R} = [r_0, r_1, \ldots, r_{127}]$ is given as:

$$s_0 = x_0 + \bigoplus_{j=0}^{127} r_j x_j + x_{126} x_{127}$$

$$s_1 = x_1 + r_0 x_0$$

$$s_2 = x_2 + r_0 x_0 + r_1 x_1 + x_0 x_1$$

$$s_3 = x_3 + r_0 x_0 + r_1 x_1 + r_2 x_2 + x_1 x_2$$

$$s_4 = x_4 + r_0 x_0 + r_1 x_1 + r_2 x_2 + r_3 x_3 + x_2 x_3$$

$$\vdots$$

$$s_i = x_i + \bigoplus_{j=0}^{i-1} r_j x_j + x_{i-2} x_{i-1}$$

Define the sequence w_i as follows:

$$w_i = \begin{cases} s_1, & \text{if } i = 0, \\ s_i + s_{i+1}, & \text{if } 0 < i < 127, \\ s_{127} + s_0, & \text{if } i = 127, \end{cases}$$

It can be checked that $w_0 = x_1 + r_0 x_0$, $w_1 = x_1 + x_2 + x_0 x_1 + r_1 x_1$ and $w_i = x_i + x_{i+1} + x_{i-1} x_i + x_{i-2} x_{i-1} + r_i x_i$, for all $i > 1$. Note that

$$w_0 = x_1 \text{ if } x_0 = 0, \text{ and } w_0 = x_1 + r_0 \text{ if } x_0 = 1.$$

Now, if we compute the value of w_0 for different outputs of the function $\mathcal{S}(\mathbf{X})$ (which are generated for different values of the internal random string $\mathbf{R}$), then the value of w_0 will be a constant and equal to x_1 if and only if $x_0 = 0$. If $x_0 = 1$, then $w_0 = x_1 + r_0$, and w_0 will evaluate to a different bit value each time, depending on the value of the random bit r_0. This argument can be extended to any i. If and only if $x_i = 0$, $w_i = x_i + x_{i+1} + x_{i-1} x_i + x_{i-2} x_{i-1}$ and will thus evaluate to a constant for every query. If $x_i = 1$, then the value of w_i has linear dependence on the random bit r_i and is more or less uniformly randomly distributed over the set $\{0, 1\}$. So our algorithm to invert $\mathcal{S}$ is as follows. Query the function $\mathcal{S}$ for a fixed $\mathbf{X}$ around N times. Then compute the vector $W = [w_0, w_1, \ldots, w_{127}]$ for each query $\mathcal{S}(\mathbf{X})$. If w_i evaluates to the same value over all the queries, then we conclude that $x_i = 0$, otherwise we conclude that $x_i = 1$. Computer simulations have confirmed that $N = 8.3$ queries on average are required to fully determine the value of $\mathbf{X}$.

Corollary 1. $\mathcal{T}$ *is not a one way function.*

Proof. Since $\mathcal{T}^{-1} = \mathcal{D}^{-1} \circ \mathcal{S}^{-1}$, and we have established that $\mathcal{S}^{-1}$ and $\mathcal{D}^{-1}$ are both efficiently calculable, we can conclude as such.

Fault Attack on the Scheme: We have just established that the function $\mathcal{T}$ is invertible, if one ensure that the same input is fed to the non-linear function around 8 times. So the attacker proceeds as follows:

1. He first obtains the fault-free ciphertext from the device.
2. He resets the device and applies a fault in the 1st column of the AES state before the 9th round MixColumn and obtains the faulty ciphertext $C + \mathcal{T}(\Delta)$.
3. He repeats the process 8 times, and each time he applies the same fault in the device. This ensures that the input to the nonlinear function $\mathcal{S}$ is the same each time. Note that this can be achieved by using optical fault [18] to flip the logic at a particular register location during each fault injection process.
4. The attacker uses the procedure outlined in Lemma 1 and Corollary 1, to obtain the value of Δ.
5. He then uses the attack outlined Sect. 3.1 to deduce 4 bytes of the 10th roundkey. This requires another fault-free and faulty ciphertext pair with fault in the same column and so the Steps 1–4 need to executed once more.
6. The process is repeated for the 2nd, 3rd and 4th columns of the AES state to obtain the full roundkey.

Fault Complexity: Since finding 4 bytes of the 10th roundkey, requires around $2 \times 8.3 = 16.6$ faults on average, the entire roundkey can be deduced in $4 \times 16.6 \approx 66$ faults on average.

4   Countermeasure Proposed in HOST 2015 [10]

Before we proceed to describe the infective countermeasure proposed in [10], we will give a short mathematical description of the Grain-128 stream cipher. Grain-128 consists of a 128 bit LFSR and a 128 bit NFSR. The state is initialized with a 128 bit Key which is loaded on to the NFSR and a 96 bit IV and a 32 bit pad $P = 0x$ `ffff ffff` which is loaded on to the LFSR. The LFSR state is update according to the rule:

$$y_{t+128} \overset{\Delta}{=} f(Y_t) = y_{t+96} + y_{t+81} + y_{t+70} + y_{t+38} + y_{t+7} + y_t.$$

The NFSR state is updated as follows

$$x_{t+128} = y_t + g(X_t), \quad \text{where}$$

$$g(X_t) = x_t + x_{t+26} + x_{t+56} + x_{t+91} + x_{t+96} + x_{t+3}x_{t+67} + x_{t+11}x_{t+13} +$$
$$x_{t+17}x_{t+18} + x_{t+27}x_{t+59} + x_{t+40}x_{t+48} + x_{t+61}x_{t+65} + x_{t+68}x_{t+84}$$

The output keystream bit z_t in some round t is produced as

$$\sum_{j \in A} x_{t+j} + y_{t+93} + h(x_{t+12}, y_{t+8}, y_{t+13}, y_{t+20}, x_{t+95}, y_{t+42}, y_{t+60}, y_{t+79}, y_{t+95})$$

where $A = \{2, 15, 36, 45, 64, 73, 89\}$ and $h(s_0, \ldots, s_8) = s_0 s_1 + s_2 s_3 + s_4 s_5 + s_6 s_7 + s_0 s_4 s_8$. The cipher is executed for 256 rounds without producing any output, during which the output bit z_t is fed back to the update functions of the LFSR and NFSR. Thereafter the feedback is discontinued and the cipher starts producing output.

4.1   Fault Protection Scheme in [10]

The fault protection scheme of [10] can be described as follows. The output function h used in Grain-128 is decomposed into two component functions: a masking function h_{masked} and an unmasking function $\mathcal{M}$. The function h_{masked} is computed as follows: a nine bit random string $\epsilon_0, \epsilon_1, \ldots, \epsilon_8$ is generated by an internal mechanism and then the function is computed as follows:

$$h_{masked} = (s_0 + \epsilon_0)(s_1 + \epsilon_1) + (s_2 + \epsilon_2)(s_3 + \epsilon_3) + (s_4 + \epsilon_4)(s_5 + \epsilon_5) +$$
$$(s_6 + \epsilon_6)(s_7 + \epsilon_7) + s_0 s_4(s_8 + \epsilon_8) + (s_0 + s_4)s_8\epsilon_8$$

The unmasking function $\mathcal{M}$ is computed so that $h = h_{masked} + \mathcal{M}$. The function $\mathcal{M}$ is computed securely via a different 128 bit register which stores the values of $\mathcal{M}$ for 128 consecutive iterations. The process is described pictorially in Fig. 3.

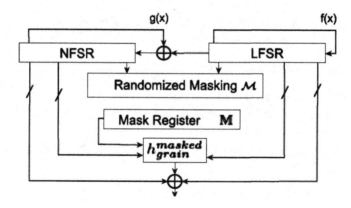

Fig. 3. Fault protection scheme in HOST 2015 [10]

5   Cryptanalysis of the Fault Countermeasure Scheme

5.1   Fault Attack on Unprotected Grain-128

There have been several fault attacks on Grain-128 reported in literature [1,2,17]. The basic philosophy in these attacks is the same. The attacker exploits the low algebraic degree of the output function h. For example if the attacker applies an optical fault which flips the logic at the bit denoted by the input variable s_2, then the difference between the fault-free keystream bit z and the faulty keystream bit z^f is given as

$$z + z^f = h(s_0, s_1, s_2, \ldots, s_8) + h(s_0, s_1, 1 + s_2, \ldots, s_8) = s_3$$

This therefore leaks the value of one state bit of the internal state of Grain-128. By applying several faults an studying the faulty keystream patterns, the attacker can easily determine the entire internal state of Grain-128.

However if one uses the fault protection scheme proposed in [10], then since the value of the unmasking function is computed securely, the difference between the faulty and fault-free keystream bit is given as:

$$z + z^f = h_{masked}(s_0, s_1, s_2, \ldots, s_8) + h_{masked}(s_0, s_1, 1 + s_2, \ldots, s_8) = s_3 + \epsilon_3$$

Since ϵ_3 is an internally generated random bit which the attacker does not know, this prevents the leakage of state information. However we will demonstrate that there exists two weaknesses in this scheme which still allows the attacker to determine the values of the internal state bits.

5.2   First Weakness

If the attacker faults the input bit s_8 (which corresponds to the 95th bit of the LFSR), then the difference between the faulty and fault-free keystream bit is given as

$$z + z^f = h_{masked}(s_0, s_1, s_2, \ldots, s_8) + h_{masked}(s_0, s_1, s_2, \ldots, 1 + s_8)$$
$$= s_0 s_4 + (s_4 + s_0)\epsilon_8$$

So the attacker proceeds as follows:

- He obtains the fault-free keystream bit z_t in some round t.
- He resets the device and injects a fault in the 95th LFSR bit (s_8) in the round t, and obtains the faulty bit z_t^f.
- He repeats the process N times so that he accumulates N different values of the keystream difference $z_t + z_t^f$.

Now if and only if $s_4 + s_0 = 0$, the value of $z_t + z_t^f = s_0 s_4$, and thus the above process will yield the same value of $z_t + z_t^f$ for each new fault injection. However if $s_4 + s_0 = 1$, then the value of $z_t + z_t^f$ is more or less uniformly randomly distributed over the set $\{0, 1\}$. Thus a fault in the bit s_8 leaks additional information about the internal state. If the attacker is able to execute this process for $0 \le t < \tau$ number of keystream rounds, then this leaks the following information about the state

1. It reveals the value of $s_0 + s_4$ in every round t.
2. If $s_0 + s_4 = 0$ in some round, it additionally leaks the value of $s_0 s_4$ in that round.

Armed with this information the attacker can proceed with the fault attack as follows. He creates an equation bank containing the following equations in the internal state variables for every round $0 \le t \le \tau$:

A. He adds an equation for the fault-free keystream bit z_t:

$$z_t = \sum_{j \in A} x_{t+j} + y_{t+93} + h(x_{t+12}, y_{t+8}, y_{t+13}, \ldots, y_{t+95})$$

B. He adds an equation for the value of the bit $s_0 + s_4$ at each round t (denote this bit value by the term a_t).

$$a_t = x_{t+12} + x_{t+95}$$

C. If $s_0 + s_4 = 0$ at any round t (i.e. if $a_t = 0$), he additionally adds an equation for $s_0 s_4$ (denote this bit value by the term b_t).

$$b_t = x_{t+12} \cdot x_{t+95}$$

The above equation bank is fed to a suitable equation solver which tries to determine the value of the internal state bits. In our experiments, we used the Cryptominisat-2.8 SAT Solver, which determined the solution of the above system in around 0.2 s on average on a system running on a 2.5 GHz processor and 16 GB internal memory for $\tau = 256$.

Fault Complexity: For each round t, we require around $N = 2$ faults on average to determine the value of $s_0 + s_4$. Since a total of $\tau = 256$ keystream rounds are used, the total number of faults required is around $\tau \times N = 512$.

5.3 Second Weakness

The second weakness of the fault protection scheme arises from the fact that the designers make no effort to protect either the NFSR update function g or the seven additional bits from the NFSR that are xored to the function h to produce the output keystream bit. Any random fault applied in the NFSR would therefore propagate along the NFSR through the update function g and the since the seven NFSR bits are unprotected, if the differential introduced by the fault appears on one of these bits they may reveal non-trivial information about the internal state bits. In fact in the work presented in [17], the attacker can apply random faults in the NFSR and by constructing an equation bank for every faulty and fault-free keystream bit, the attacker is able to find the entire internal state of Grain-128 in 5–6 faults by using a SAT based solver within 6 min on average. It is clear that the countermeasure scheme does not counteract the attack presented in [17]. Hence, in order for the scheme in [10] to be secure, it not only must design a better masking function h_{masked}, it must also take steps to protect **(a)** the NFSR update function g and **(b)** the seven NFSR bits that are added to the h function to produce the output keystream bit.

6   Conclusion

In this paper, we looked at the security of two fault countermeasure schemes proposed very recently in literature and proposed attacks on them requiring 66 and 512 faults respectively. We conclude that in both the schemes, the manner in which the designs use the internally generated random bits, make them vulnerable to attack. Additionally, the countermeasure used to protect Grain-128 is simply inadequate since no effort is made to protect the NFSR update function g or the seven additional bits that are xored to the output function h. From the discussion it is evident that the transformation function used in the first scheme needs to be a one way function, failing which the scheme would not provide any security. In the second scheme, not only must a better masking scheme be designed, but some additional effort must be expended to protect the other critical components of the circuit.

References

1. Banik, S., Maitra, S., Sarkar, S.: A differential fault attack on the grain family of stream ciphers. In: Prouff, E., Schaumont, P. (eds.) CHES 2012. LNCS, vol. 7428, pp. 122–139. Springer, Heidelberg (2012)
2. Banik, S., Maitra, S., Sarkar, S.: A differential fault attack on the grain family under reasonable assumptions. In: Galbraith, S., Nandi, M. (eds.) INDOCRYPT 2012. LNCS, vol. 7668, pp. 191–208. Springer, Heidelberg (2012)

3. Battistello, A., Giraud, C.: Fault analysis of infective AES computation. In: FDTC 2013, pp. 101–107. IEEE Computer Society (2013)
4. Battistello, A., Giraud, C.: Fault cryptanalysis of CHES 2014 symmetric infective countermeasure. IACR Cryptology ePrint Archive. http://eprint.iacr.org/2015/500.pdf
5. Catell, K., Muzio, J.C.: Synthesis of one-dimensional linear hybrid cellular automata. IEEE Trans. Comput. Aided Des. Integr. Circuits Syst. **15**(3), 325–335 (1996)
6. Daemen, J., Rijmen, V.: The Design of Rijndael: AES - The Advanced Encryption Standard. Springer-Verlag, Berlin (2002)
7. Genelle, L., Giraud, C., Prouff, E.: Securing AES implementation against fault attacks. In: FDTC 2009, pp. 51–62 (2009)
8. Gierlichs, B., Schmidt, J.-M., Tunstall, M.: Infective computation and dummy rounds: fault protection for block ciphers without check-before-output. In: Hevia, A., Neven, G. (eds.) LatinCrypt 2012. LNCS, vol. 7533, pp. 305–321. Springer, Heidelberg (2012)
9. Ghosh, S., Saha, D., Sengupta, A., Roy Chowdhury, D.: Preventing fault attacks using fault randomization with a case study on AES. In: Foo, E., Stebila, D. (eds.) ACISP 2015. LNCS, vol. 9144, pp. 343–355. Springer, Heidelberg (2015)
10. Ghosh, S., Roy Chowdhury, D.: Preventing fault attack on stream cipher using randomization. In: 2015 IEEE International Symposium on Hardware Oriented Security and Trust (HOST), pp. 88–91 (2015)
11. Hell, M., Johansson, T., Meier, W.: A stream cipher proposal: Grain-128. In: IEEE International Symposium on Information Theory (ISIT 2006) (2006)
12. Ishai, Y., Sahai, A., Wagner, D.: Private circuits: securing hardware against probing attacks. In: Boneh, D. (ed.) CRYPTO 2003. LNCS, vol. 2729, pp. 463–481. Springer, Heidelberg (2003)
13. Joye, M., Tunstall, M. (eds.): Fault Analysis in Cryptography. Information Security and Cryptography. Springer, Berlin (2012)
14. Joye, M., Manet, P., Rigaud, J.B.: Strengthening hardware AES implementations against fault attacks. IET Inf. Secur. **1**, 106–110 (2007)
15. Karpovsky, M., Kulikowski, K., Taubin, A.: Robust protection against fault-injection attacks on smart cards implementing the advanced encryption standard. In: International Conference on Dependable Systems and Networks (DSN 2004), pp. 93–101. IEEE Computer Society (2004)
16. Lomné, V., Roche, T., Thillard, A.: On the need of randomness in fault attack countermeasures - application to AES. In: FDTC 2012, pp. 85–95. IEEE Computer Society (2012)
17. Sarkar, S., Banik, S., Maitra, S.: Differential fault attack against grain family with very few faults and minimal assumptions. IEEE Trans. Comput. **64**(6), 1647–1657 (2015)
18. Skorobogatov, S.P., Anderson, R.J.: Optical fault induction attacks. In: Kaliski Jr., B.S., Koç, Ç.K., Paar, C. (eds.) CHES 2002. LNCS, vol. 2523. Springer, Heidelberg (2003)
19. Tupsamudre, H., Bisht, S., Mukhopadhyay, D.: Destroying fault invariant with randomization. In: Batina, L., Robshaw, M. (eds.) CHES 2014. LNCS, vol. 8731, pp. 93–111. Springer, Heidelberg (2014)

Differential Fault Analysis of SHA-3

Nasour Bagheri[1,2]([✉]), Navid Ghaedi[1], and Somitra Kumar Sanadhya[3]

[1] Electrical Engineering Department,
Shahid Rajaee Teacher Training University, Tehran, Iran
na.ghaedi@gmail.com
[2] The School of Computer Science,
Institute for Research in Fundamental Sciences (IPM), Tehran, Iran
Nbagheri@srttu.edu
[3] IIIT-Delhi, Delhi, India
somitra@iiitd.ac.in

Abstract. In this paper we present the first differential fault analysis (DFA) of SHA-3. This attack can recover the internal state of two versions of SHA-3 (namely, SHA3-512 and SHA3-384) and can be used to forge MAC's which are using these versions of SHA-3. Assuming that the attacker can inject a random single bit fault on the intermediate state of the hash computation, and given the output of the SHA-3 version for a correct message and 80 faulty messages, we can extract 1592 out of the 1600 bits of the compression function's internal state. To the best of our knowledge, this is the first public analysis of SHA-3 against DFA. Although our results do not compromise any security claim of SHA-3, it shows the feasibility of DFA on this scheme and possibly other Sponge based MACs and increases our understanding of SHA-3.

Keywords: SHA-3 · Keccak · Differential fault analysis

1 Introduction

The new SHA-3 standard [23] is adapted from the Keccak hash function [9]. Keccak hash function is a family of sponge based hash function [8] with *Keccak-f$[r+c]$* as the primitive. The parameters r and c are the bit rate and the capacity and determine the width of the *Keccak-f* permutation. In the case of SHA-3, the internal state size is $b = 1600$ bits while the output size $n \in \{224, 256, 384, 512\}$ bits. In this paper, we concentrate on the standard Keccak versions submitted to the SHA-3 competition and denote them by SHA3-224, SHA3-256, SHA3-384 and SHA3-512, depending on the output size n.

Keccak has received significant attention of the cryptography community, both during and after the SHA-3 competition. Some of the prominent works analyzing Keccak are [3,7,11–13,15,17–21,29,33–36]. However, these numerous analysis have not compromised any security claim of Keccak and there exists a big gap between the number of rounds practically broken and the number of rounds of Keccak suggested by the designers.

© Springer International Publishing Switzerland 2015
A. Biryukov and V. Goyal (Eds.): INDOCRYPT 2015, LNCS 9462, pp. 253–269, 2015.
DOI: 10.1007/978-3-319-26617-6_14

Differential Fault Analysis (DFA), first introduced by Biham and Shamir in [10], derives information about the secret key from the physical implementation of the cipher by examining the differences between a fault-free encryption and several faulty ones. A fault may be injected by introducing an external impact on the processing device by means of voltage variation, glitch, laser, etc. However, neither the fault location nor the bit-value at the fault location may be known to the attacker. This attack model has been successfully applied to several block ciphers, stream ciphers, hash functions and authenticated encryption functions, where the attacks on DES [27,39], AES [1,22,25,26,31,32,37,41], Grain [5,16], Mickey [4,30], SHA-1 compression function [28], Grøstl [24], Streebog [2] and APE [40], CLOC and SILC authenticated encryption schemes [14] are examples. In the case of the hash functions, DFA make sense when the hash function is used in a message authenticated mode such as secret IV-MAC [38], HMAC or NMAC [6]. In such applications, DFA could be used to recover the secret key or perform forgery against the MAC. Since using a hash function is common in constructing a secure MAC scheme, it is important to investigate the security of a hash function against the DFA attack. However, to the best of our knowledge, there is no public report on DFA on SHA-3 or Keccak.

On the other hand, Dinur et al. [20] recently have analyzed the keyed modes of Keccak sponge function where it is used to generate bit stream for stream cipher or it is used as a building block for message authentication codes(MACs) and authenticated encryption (AE) schemes. Motiviated by that work, where the internal state of Keccak permutation is extracted using cube attack, in this paper the internal state of Keccak permutation is reconstructed with fault injection. Therefore, this attack could be used to recover the secret key or do forgery against MACs based on Keccak. It also maybe applicable to recover the secret key and the initial value of stream cipher based on Keccak if the attack scenario change to known plaintext attack. In addition, in the keyed version of SHA-3, if it is used with nonce and there is a restriction on the repeating the nonce, then the given attack does not work.

1.1   Contribution

We present differential fault analysis on two versions of SHA-3, namely SHA3-512 and SHA3-384. The attack model follows the approach used already in DFA against Grøstl [24], Streebog [2] and CLOC & SILC authenticated encryption schemes [14] where the adversary injects a single bit fault in a random position of the internal state. The presented attack can recover the complete internal state of SHA-3 given the first 320 least significant bits of its output (called a plane in Keccak) for the correct message and enough faulty messages. Since the output of SHA3-384 and SHA3-512 includes a complete plane, we apply our attack to these variants of SHA-3. We then present a theoretical bound on the number of detected bits of the Keccak state after injecting a single bit fault on N messages and compare the same with simulation results. Our theoretical analysis shows that injecting 80 randomly distributed single bit faults on internal state are enough to recover 1592 bits out of 1600 bits of the internal state.

1.2   Paper Organization

The rest of the paper is organized as follows: in Sect. 2, the notations used in the paper are presented, and in Sect. 2.2, the Keccak hash function is briefly described. In Sect. 3, we present our differential fault analysis on SHA3-384 and SHA3-512. In Sect. 4, we present theoretical bounds and simulation results for state bit recovery. Finally, we conclude the paper in Sect. 5.

2   Notations and Preliminaries

2.1   Notations

Throughout the paper, we use the following notations:

- $A^i(x, y, z)$:   denotes a single bit of state at the beginning of the ith round.
- $H(M)$:   denotes the output of SHA-3.
- $plane_{H(m)}(0)$:   denotes 320 consecutive bits of output of SHA-3 that could be used to reconstruct $plane(0)$ of the state.
- $plane_{H(m)}(0)'$:   denotes 320 consecutive bits of faulty output of SHA-3 that could be used to reconstruct $plane(0)$ of the state.
- θ^i: denotes the θ step at i^{th} round.
- χ^i: denotes the χ step at i^{th} round.
- π^i: denotes the π step at i^{th} round.
- ρ^i: denotes the ρ step at i^{th} round.
- ι^i: denotes the ι step at i^{th} round.
- B: denotes the state before χ step in the penultimate round and B_b^l denotes the bth bit of the lth lane of B.
- C: denotes the state after θ step in the last round and C_b^l denotes the bth bit of the lth lane of C.

2.2   Description of Keccak Hash function

Keccak [9] is a family of hash functions and the winner of the SHA-3 competition. Some of its variants were adapted as SHA-3 [23]. It is a sponge based hash function based on Keccak-$f[b, n_r]$ permutation. Figure 1 illustrates the sponge construction, based on permutation $f : \{0,1\}^r \times \{0,1\}^c \rightarrow \{0,1\}^r \times \{0,1\}^c$. In the sponge construction, r is the bitrate and called *rate* and c is the security parameter and called *capacity*. Larger r provides higher speed and larger c provides better security. The state size of the hash function is determined by $b = c + r$. In Keccak $b \in \{25, 50, 100, 200, 400, 800, 1600\}$ and for the case of SHA-3, the state size is 1600 bits.

In Keccak-$f[b, n_r]$, b and n_r denote the state size and number of the rounds of the permutation respectively. The SHA-3 standard uses Keccak-$f[1600,24]$ permutation from [9]. Depending on the output length n, 4 versions of SHA-3 use $c = 2n$ for $n \in \{224, 256, 384, 512\}$, and are called SHA3-224, SHA3-256 etc.

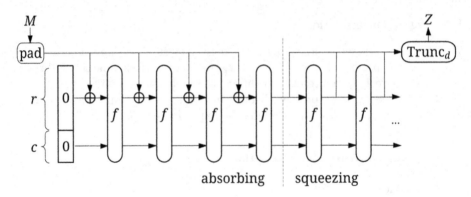

Fig. 1. The sponge construction based on permutation f [23].

Keccak-f[1600,24] has 24 rounds, indexed from 0 to 23, and each round performs 5 consecutive permutations on the state, denoted by θ, ρ, π, χ, and ι. The only non-linear permutation is χ and the only round dependent permutation is ι. The input to these permutations is constructed as a 3-dimensional array (x, y, z) where the dimensions of x, y and z are 5, 5 and 64 respectively. Denoting the Array by A, each bit of the array can be described as $A(x, y, z)$, where $0 \leq x \leq 4, 0 \leq y \leq 4$ and $0 \leq z \leq 63$. The initial state of Keccak-f[1600,24] (i.e. before the application of the 0th round) is denoted by $A^0(.,.,.)$ and the output of Keccak-f[1600,24] after the 23rd round is denoted by $A^{24}(.,.,.)$.

Keccak-f[1600,24] state can be defined in different parts. This naming convention is helpful in describing Keccak-f[1600,24]:

- A row is a set of 5 bits with constant y and z coordinates, i.e. $A(*, y, z)$.
- A column is a set of 5 bits with constant x and z coordinates, i.e. $A(x, *, z)$.
- A lane is a set of 64 bits with constant x and y coordinates, i.e. $A(x, y, *)$.
- A sheet is a set of 320 bits with constant x coordinate, i.e. $A(x, *, *)$.
- A plane is a set of 320 bits with constant y coordinate, i.e. $A(*, y, *)$.
- A slice is a set of 25 bits with constant z coordinate, i.e. $A(*, *, z)$.

In this paper, each lane of the state is specified according to Table 1.

Table 1. Lane numbering, each square represents a lane in the state.

0	1	2	3	4
5	6	7	8	9
10	11	12	13	14
15	16	17	18	19
20	21	22	23	24

Next we briefly describe the 5 permutations used in each round of SHA-3. θ. As the first step in each round, the role of θ is to XOR each bit $A(x, y, z)$ with bits in column $(x - 1, *, z)$ and $(x + 1, *, z - 1)$. Hence, θ can be represented as follows:

$$\theta : A(x, y, z) \leftarrow A(x, y, z) + \sum_{y'=0}^{4} A(x - 1, y', z) + \sum_{y'=0}^{4} A(x + 1, y', z - 1).$$

ρ. In this step, the bits are rotated in their lane by $T(x, y)$ positions, which is a predefined offset value for each lane. Table 2 shows these offset values.

Table 2. Offset values of ρ for Keccak-$p[1600,24]$

	$x = 3$	$x = 4$	$x = 0$	$x = 1$	$x = 2$
$y = 2$	25	39	3	10	43
$y = 1$	55	20	36	46	6
$y = 0$	28	27	0	1	62
$y = 4$	56	14	18	2	61
$y = 3$	21	8	41	45	15

π. Permutation π is used to rearrange the positions of the lanes in the array. Each lane in position $(x, y, *)$ is moved to the new position $(x', y', *)$, where $x' = y$ and $y' = (x + 3y)$ mode 5.

χ. This is the only non-linear permutaion in each round. A row in position $(*, y, z)$ is processed and replaced by a new row by this permutation. Each input bit affects 3 bits at the output of χ as described below.

$$A(x', y', z') = A(x, y, z) \oplus (\overline{A(x + 1, y, z)} \ \& \ A(x + 2, y, z)),$$

where $\bar{x} = x \oplus 1$ and $\&$ is the bitwise AND operator.

ι. The final step in each round is the application of ι, which is the only round dependent step. In the ith round of this step, a round dependent value $RC(i)$ is XOR'ed with $Lane(0)$. The values of $RC(i)$ can be found in [9]. However, they do not impact our attack.

3 DFA Attack on SHA3-384, and SHA3-512

In this section, we show how to obtain the bits of internal state that do not appear in the output of SHA3-384, and SHA3-512. We assume a single bit fault is injected at the beginning of penultimate round of Keccak-$p[1600,24]$. In the rest of the paper, for simplicity we denote Keccak-$p[1600,24]$ permutation by Keccak-p. We use the following observations on χ function in the our DFA attack on SHA-3 variants.

Observation 1.   Suppose a single bit fault is injected in the input of χ and we are given the difference of the correct and the faulty output of χ. In this case, it is easy to extract two bits of input of χ given the differential output for χ. This is due to the fact that bitwise AND operation in χ leaks information of its input, if a single bit of its input was corrupted.

Let a single bit fault be injected in position (f, i, j) in the input to χ then it leads to a single bit difference in the output of χ with probability '1'. Moreover, it also leads to a single bit difference in $(f - 1, i, j)$ if $A(f + 1, i, j) = 1$' and a single bit difference in $(f - 2, i, j)$ if $A(f - 1, i, j) = 0$. It can be seen that if we have differential output of χ, we can extract two bits of its input.

Observation 2.   Given any input bit of the χ function, we receive some linear equations of the input bits at the output of χ. Denoting the input and output of χ by $x_0, \ldots, x_4$ and $y_0, \ldots, y_4$ respectively, given the input bit x_i it appears at the output as $y_{i-2} = x_{i-2} \oplus (\overline{x}_{i-1} \& x_i)$, $y_{i-1} = x_{i-1} \oplus (\overline{x}_i \& x_{i+1})$ and $y_i = x_i \oplus (\overline{x}_{i+1} \& x_{i+2})$. It is clear that given x_i we achieve two linear equations of the inputs of χ at its output, see Table 3. In Table 4, we represent a case where all output bits are a linear function of the unknown inputs or they are constant.

Table 3. The relation between input bits (x_i) and output bits (y_i) of χ assuming that x_4 is known and the rest of the inputs are unknown. Here, NL is used to denote non-linear function. (In the linear function, the output depends only on the XOR of its inputs. Otherwise the function is non- linear.)

x_0	x_1	x_2	x_3	x_4	y_0	y_1	y_2	y_3	y_4
x_0	x_1	x_2	x_3	0	NL	NL	x_2	$x_3 \oplus x_0$	NL
x_0	x_1	x_2	x_3	1	NL	NL	$x_2 \oplus \overline{x}_3$	x_3	NL

Table 4. The relation between input's bits (x_i) and output's bits (y_i) of χ assuming that x_0 and x_2 are unknown and the rest of the inputs are known.

x_0	x_1	x_2	x_3	x_4	y_0	y_1	y_2	y_3	y_4
x_0	0	x_2	0	0	$x_0 \oplus x_2$	0	x_2	x_0	0
x_0	0	x_2	0	1	$x_0 \oplus x_2$	0	$\overline{x}_2$	0	1
x_0	0	x_2	1	0	$x_0 \oplus x_2$	$\overline{x}_2$	x_2	$\overline{x}_0$	0
x_0	0	x_2	1	1	$x_0 \oplus x_2$	$\overline{x}_2$	x_2	1	1
x_0	1	x_2	0	0	x_0	1	x_2	x_0	$\overline{x}_0$
x_0	1	x_2	0	1	x_0	1	$\overline{x}_2$	0	$\overline{x}_0$
x_0	1	x_2	1	0	x_0	$\overline{x}_2$	x_2	$\overline{x}_0$	$\overline{x}_0$
x_0	1	x_2	1	1	x_0	$\overline{x}_2$	x_2	1	$\overline{x}_0$

Observation 3.   Assuming that we know some input bits of the χ function, the output bits will be non-linear if and only if two consecutive input bits of χ are unknown. This observation directly comes from the fact that for any output bit we have $y_i = x_i \oplus (\overline{x}_{i+1} \& x_{i+2})$. In this equation the only no-linear part is the AND operation and, assuming that we know some input bits, it remains non-linear if and only if both x_{i+1} and x_{i+2} are unknown. Table 3 represents an example of the case where some of the output bits are non-linear function of unknown bits and some are linear.

Next, we explain how this single bit propagates in one round.

- **Propagation of Active Bit in One Round.**   Assume that we have injected a single bit fault on $A^{22}(x, y, z)$, which is the input of penultimate round of the Keccak-p. Hence, we have $\Delta A^{22}(x, y, z) = 1$ and $\Delta A^{22}(x', y', z') = 0$, for any $(x', y', z') \neq (x, y, z)$. On the other hand, this active bit affects 11 bits after θ^{22} in the following positions and converts them to active:

$$(x, y, z), (x-1, 0, z+1), (x-1, 1, z+1), (x-1, 2, z+1), (x-1, 3, z+1), (x-1, 4, z+1),$$

$$(x+1, 0, z), (x+1, 1, z), (x+1, 2, z), (x+1, 3, z), (x+1, 4, z),$$

The rest of the bits would remain inactive after the θ^{22} function. It is clear from offset values of ρ in Table 2 that, excluding the case where $x = 1$, ρ^{22} function moves active bits to different slices. Hence, after ρ^{22}, we expect to receive 11 slices with only one active bit in each slice or 9 slices with only one active bit in each slice and $slice(z + 1)$ with two active bits in the positions $(0, 0, z+1)$ and $(1, 0, z+1)$. The π^{22} function changes the location of the active bits in their slices. Hence, the number of active bits after the π^{22} function remains unaffected. On the other hand, if we have active bits in $(0, 0, z + 1)$ and $(1, 0, z + 1)$ after ρ^{22}, the π^{22} function will move these active bits to $(0, 0, z+1)$ and $(0, 2, z+1)$. Hence, up to the end of the π^{22} function, we have 11 rows each having only one active bit and the rest of the rows having no active bits. Next we apply χ^{22} to the internal state. It is clear that if any row has no active bit then it does not generate any active bit after χ^{22}. Hence, the number of active bits for those rows that have no active bits remains zero after χ^{22}. On other hand, if $Row(i, j)$ includes single active bit in position (f, i, j), then after χ^{22} it leads to an active bit in (f, i, j) with probability '1', leads to an active bit in $(f - 1, i, j)$ if $A(f + 1, i, j) = 1$ and leads to an active bit in $(f - 2, i, j)$ if $A(f - 1, i, j) = 0$. Finally, the ι^{22} function keeps the number of active bits in the state unaffected and produces $A^{23}(x, y, z)$.

According to Observation 1, these 11 active bits, that are located in different rows at beginning of χ^{22}, on the differential inputs of χ^{22} leak 22 bits of internal state. It must be noted that these 22 bits of internal state before χ^{22} are distinct.

Thanks to θ^{23}, the θ function of the last round, any of those 22 bits of internal state affects each plane of internal state after θ^{23} and its differential output. Then

```
Differential state at the beginning of round 22 after injection of a single bit fault
|----------------1|----------------|----------------|----------------|----------------|----------------|
|----------------|----------------|----------------|----------------|----------------|----------------|
|----------------|----------------|----------------|----------------|----------------|----------------|
|----------------|----------------|----------------|----------------|----------------|----------------|
|----------------|----------------|----------------|----------------|----------------|----------------|

Differential output after theta in round 22:
|----------------1|----------------1|----------------|----------------|----------------|----------------2|
|----------------|----------------1|----------------|----------------|----------------|----------------2|
|----------------|----------------1|----------------|----------------|----------------|----------------2|
|----------------|----------------1|----------------|----------------|----------------|----------------2|
|----------------|----------------1|----------------|----------------|----------------|----------------2|

Differential output after rho in round 22:
|----------------1|----------------2|----------------|----------------|--------1--------|--------|
|----------------|----1-----------|----------------|----------------|--------2-----|
|----------------|----------4--|----------------|----------------|----1----------|
|----------------|----2-----------|----------------|----------------|----------2--|
|----------------|----------------4|----------------|----------------|--------8---|
Differential output after pi in round 22:
|----------------1|----1-----------|----------------|----------------|--------8---|
|----------------|----------2-----|----------------|----2-----------|----------------|
|----------------2|----------------|----------------|----------2--|----------------|
|--------1-------|----------------|----------------|----------------|----------------|
|----------------|----------------|-----1----------|----------------|----------------4|
Differential output after chi in round 22:
|----x----------1|----1-----------|------------x---|----------x--x|----x-------8--x|
|--------x-----|----x-----2-----|----x----------|----2-----------|--------x-----|
|----------------2|----------------x--|----------x--|----------2-x|--------------x|
|-------1----x--|----------------x--|----4-------|--------x-------|--------x-------|
|-----x----------|-----x----------|-----1--------x|----------------x|--------------4|
```

The state is described as a matrix of 5×5 lanes of 64 bits, ordered from left to right, where each lane is given in hexadecimal using little-endian format.

x represents a bit of χ input (B_b^l) according to observation 1. We stress that the differential characteristic has probability 1.

Fig. 2. Injection a single bit fault and it's differential path

the internal state goes through ρ^{23}, π^{23}, χ^{23} and ι^{23} to produce the final output. In Fig. 2, a differential characteristic is shown under the assumption that a single bit fault is injected on $A^{22}(0, 0, 0)$.

The output of SHA3-384 and SHA3-512 $(H(M))$ include one complete plane. Given that plane of the output of $H(M)$, it is possible to invert ι^{23}, χ^{23}, π^{23} and ρ^{23} for this plane. Hence, it is possible to compute $(\rho^{23})^{-1}((\pi^{23})^{-1}((\chi^{23})^{-1}$ $((\iota^{23})^{-1}(plane_{H(M)}(0)))))$. Following this fact, given the output of SHA3-384 (or SHA3-512) it is possible to invert a fraction of output that are included in $plane(0)$ and determine the internal state of $lane(0)$, $lane(6)$, $lane(12)$, $lane(18)$ and $lane(24)$ after θ^{23}. These lanes comes from $sheet(0)$, $sheet(1)$, $sheet(2)$, $sheet(3)$ and $sheet(4)$ respectively. In addition, any active bit in $sheet(i)$ affects all the lanes in $sheet(i - 1)$ and $sheet(i + 1)$. Hence, we can be sure that any of these 22 target bits appear linearly in $lane(0)$, $lane(6)$, $lane(12)$, $lane(18)$ and $lane(24)$ after θ^{23}. Suppose a single bit fault is injected on $A^{22}(0, 0, 0)$, the equations of these lanes are shown as follow:

$$c^0 = 00000000000000000B_{45}^6 1 C_{44}^0 00 B_{40}^{21} 00000000000000 B_{28}^{19} 000001000 B_{10}^{16} B_9^{11} 00000001 B_1^{14} C_0^0$$

$$c_0^0 = 1 \oplus B_0^4, C_{44}^0 = B_{44}^4 \oplus B_{44}^0$$

$$c^6 = 00000000000000000B_{45}^7 B_{45}^6 B_{44}^0 001 B_{40}^{20} 00000000000001000000 B_{21}^5 0000 B_{15}^2 00001 C_{10}^6 000000 B_2^{22} 011$$

$$c_{10}^6 = B_{10}^{15} \oplus B_9^{12}$$

$$c^{12} = 0000000000000001 B_{45}^6 1000 B_{40}^{21} 00000000000 B_{28}^{18} 000000010000 B_{15}^3 00000 B_{10}^{16} B_9^{11} 00000 B_2^{23} B_1^{13} B_0^3 0$$

$$c^{18} = 00000000000000000C_{45}^{18} 000001000000000 B_{28}^{19} B_{28}^{18} 00000 B_{21}^9 000001 B_{15}^2 00001 B_9^{12} 000001 C_2^{18} B_0^4 0$$

$$c_2^{18} = B_2^{22} \oplus B_1^{14}, C_{45}^{18} = B_{45}^6 \oplus B_{44}^4$$

$$c^{24} = 00000000000000000C_{45}^{24} 000 B_{40}^{20} 00000000001 B_{28}^{18} 00000 B_{21}^5 000000 B_{15}^3 000 B_{10}^{15} 01000000 C_2^{24} C_1^{24} B_0^3$$

$$c_1^{24} = 1 \oplus B_1^{13}, C_2^{24} = B_2^{23} \oplus B_2^{22} \oplus B_1^{14}, C_{45}^{24} = 1 \oplus B_{44}^0$$

Each lane is represented in 64 bits. Obviously, the above equations are constructed under the assumption that a single bit fault is injected on $A^{22}(0,0,0)$. On the other hand, the attacker needs to know the position of single bit fault to obtain these 22 bits of internal state. Next we describe the technique to know the fault location itself.

- **Determining the Fault Position:** Given a faulty output and a correct output of SHA3-384 (or SHA3-512) (say $H(M)$), it is possible to compute $(\chi^{23})^{-1}((\iota^{23})^{-1}(plane_{H(M)}(0)))$. In fact, the $plane(0)$ of differential output of π^{23} can be computed as follows:

$$(\chi^{23})^{-1}((\iota^{23})^{-1}(plane_{H(M)}(0))) \oplus (\chi^{23})^{-1}((\iota^{23})^{-1}(plane_{H(M)}(0'))). \quad (1)$$

Again suppose a single bit fault is injected on $A^{22}(0,0,0)$, the $plane(0)$ of differential output of π^{23} is shown as follow (Fig. 3):

Any bit in the above differential output could be 1, 0 or x. When the value of a bit is 1 or 0 then the difference in that position is deterministic. However, any bit that is marked as x could appear as 1 or 0 in the real differential output. It can be shown that the attacker can find the position of the fault by using this deterministic differential bits. We describe this method in Algorithm 1. Given the inverted $plane(0)$ of the output of the correct and faulty messages, based on a single bit fault on $A^{22}(x,y,z)$, Algorithm 1 can be used to determine the position of the fault.

We also verified this algorithm in simulations. For any single fault in $A^{22}(x,y,z)$, the algorithm returned the position of the fault correctly. In the off-line phase of the algorithm, for any possible positions of fault in $A^{22}(x,y,z)$ (i.e. all 1600 positions), related differential outputs up to the end of π^{23} are generated and $plane(0)$ of the state at the output of π^{23} is stored in a table $\mathcal{T}$. In Appendix A, the distribution of fault on $plane(0)$ after π^{23}, for injecting a single fault in any bits of $slice(0)$ are presented. Any other positions of fault is just a simple rotation of one of these patterns.

Since there is no collision between deterministic bits of stored values in $\mathcal{T}$, our Algorithm 1 will return the position of fault correctly, if the fault happened at the input of θ^{22}. Our simulations verify this claim and the success probability of the given algorithm to return the correct position of the injected single bit fault is '1'.

```
xx1-------xx---1-----x1-----x------------x--x1x----------------
-x------1-----------x1--xxx------------------11-x-------x1----x---
1-------x----------x---1x1-------------------xxx-----xx-----x----
--x------------------xx1-----x1----x1-----x-----xx----------1--
--------------xxx------1-x---x------x-----x1-----------x---x----
```

Each lane represents in 64 bits. We stress that the differential output happen with probability 1.

Fig. 3. The $plane(0)$ of differential output of π^{23}

Algorithm 1. Determining the position of a single bit fault, injected at the beginning of θ^{22}

---

off-line phase: For any possible position of injecting a single bit fault in $A^{22}(x, y, z)$, calculate related differential characteristics up to the output of π^{23} and store $plane(0)$ of the differential output of π^{23} in table $\mathcal{T}$.

on-line phase: Given a faulty output($plane_{H(M)}(0)'$)):
   S.1 Compute the expected value of the difference of $plane(0)$ at the output of π^{23}, using Eq. 1.
   S.2 For any entry in $\mathcal{T}$, compare it with the computed value in step **S.1** based on the deterministic bits only.
   S.3 If there is a match between the computed value in step **S.1** and an entry in $\mathcal{T}$, return the related position as the fault location; otherwise return unknown.

---

4   Theoretical Bound and Simulation Result

Following the discussion in Sect. 3, for any single bit fault injection, we can determine the position of fault uniquely and determine 22 bits of internal state. Given that we have inverted 5 lanes from output, the total number of required independent equations are $1600 - 5 \times 64 = 1280$. However, we have no control on the position of faults and therefore the extracted bits. Hence, we consider the upper-bound of required independent equations as 1600. The lower bound on the number of faulty messages is $\frac{1600}{22} \cong 73$ if all the faults produce independent equations. However, this lower bound is unlikely to be achieved in reality because of the possible injection of the fault in same locations or the overlap between extracted bits of the state by different position of faults. To provide a bound of the number of detected bits after injecting a single bit fault on N messages, one can argue that after the first fault one can extract 22 bits of internal state. On injecting the second fault, if it is in the same location as the first one then it does not provide any new information, otherwise it is possible to extract the information of any bit that is not extracted with the first fault. The same argument can be stated for other faults. Hence, a raw bound of the number of detected bits after injecting a single bit fault on N messages, denoted by $\mathcal{S}^N$, are as follows:

$$\mathcal{S}^N = \sum_{1}^{N} \mathcal{X}_i,$$ (2)

where $\mathcal{X}_1 = 22$ and $\mathcal{X}_i = \left(1 - \frac{i-1}{1600}\right)\left(22 \times \frac{1600-\mathcal{X}_{i-1}}{1600}\right)$. We evaluated this bound by simulations as well. The results of our simulations are shown in Fig. 4. For the experiments, for any value of N, we injected faults randomly in N positions of $A^{22}(x,y,z)$ and counted the number of extracted bits of internal state after these faults. We also repeated this simulation 100 times for any value of N and counted the average. The simulation results match the given theoretical bound.

Extracting all bits of the internal state only by injecting single bit faults requires few faults, because the number of the extracted bits decreases when N is increased. For example, based on the given theoretical bound, to extract 1480 bits of the internal state, we need 200 faulty messages. However, after extracting a part of the internal state, given that the only non-linear layer in Keccak-p round function is χ we can use Observation 2 to extract some of the unknown bits, without injecting extra faults. On the other hand, we know that the expected number of the unknown bits after injecting a single bit fault on N messages is $1600 - \mathcal{S}^N$. Hence, for any bit at the input of χ^{23} to be unknown, the probability is determined as follows, denoted by $Pr^N(Uk)$:

$$Pr^N(Uk) = \frac{1600 - \mathcal{S}^N}{1600}.$$

Based on Observation 3, we have non-linear bit at the output of χ^{22} if there are at least two consecutive unknown bits. We know that there are 1600 possibilities for two consecutive bits in the internal state, and any two consecutive bits are both unknown with probability $(Pr^N(Uk))^2$. Therefore the probability of any bit of the output of χ^{22} to be a non-linear function of the unknown bits is $(Pr^N(Uk))^2$ which we denote by $Pr^N(NL)$. Later, θ^{23} combines the state bits linearly. However, if any of bits that are XOR'ed together to generate a single bit of the internal state after θ^{23} is non-linear, the output bit will also be non-linear. Hence, we can state that the probability of any bit at the output of θ^{23} to be linear, denoted by $Pr^N(L)$, is as follows:

$$Pr^N(L) = (1 - Pr^N(NL))^{11}.$$

Therefore, the expected number of linear equations in the 320 bits that are extracted from $(\rho^{23})^{-1}((\pi^{23})^{-1}((\chi^{23})^{-1}((\iota^{23})^{-1}(plane_{H(M)}(0)))))$ are as follows, denoted by $\#L^N$:

$$\#L^N \cong 320 \times (1 - Pr^N(NL))^{11}.$$

Hence, we can rewrite $\mathcal{X}_i$ as follows:

$$\mathcal{X}_i = \left(1 - \frac{i-1}{1600}\right)\left((22 + \#L^{N-1}) \times \frac{1600 - \mathcal{X}_{i-1}}{1600}\right).$$ (3)

We simulated the stated theoretical bound $\mathcal{S}^N$ of the number of the extracted bits after injecting a single bit fault on N messages. The results of our experiments are shown in Fig. 4. It is clear from this figure that to recover the complete internal state of the last permutation of Keccak-p at the input of χ^{22}, we need at most 80 faulty messages and the related correct message. To be more precise, given 80 messages with single bit faults, we can retrieve 1592 out of 1600 input bits of χ^{22}. The remaining 8 bits can be found by exhaustive search. Given the internal state of the permutation, it would be straightforward to recover the secret key or forge MAC of any desired message.

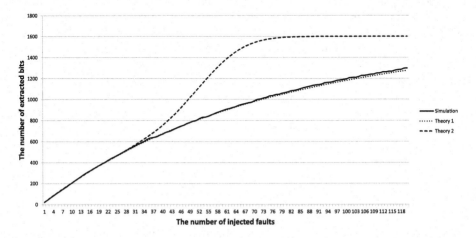

Fig. 4. The number of extracted bits of internal state after injecting a single bit fault on N messages. Here, Theory 1 and Theory 2 are the theoretical bounds that are given by Eqs. 2 and 3 respectively and Simulation is number of the extracted bits after simulating the attack without using information from Observations 2 and 3 which matches Theory 1.

5   Conclusions

In this work, we investigated the security of SHA-3 against DFA. The main idea was to extract the capacity of the sponge construction given a correct output and some related faulty outputs. Our study shows that by injecting around a random single bit faults on 80 messages, one can obtain the internal state of the compression function. Given that the primitives of SHA-3 and other sponge based schemes are permutations, it would be possible to do state/key recovery on such sponge based MACs/AE. This approach can also be applicable to perform DFA on sponge based CAESAR candidates, which is an on going work. Another direction to continue this study is to use some relaxation on the fault injection model (both the value of fault and its location) which is the subject for future works.

A   Appendix

We stress that all differential output happen with probability 1. The state is
described as a matrix of 5×5 lanes of 64 bits, ordered from right to left, where
each lane is given in bit using the little-endian format.
x represents a linear equation of χ input (B).

References

1. Ali, S., Mukhopadhyay, D., Tunstall, M.: Differential fault analysis of AES: towards reaching its limits. J. Crypt. Eng. **3**(2), 73–97 (2013)
2. AlTawy, R., Youssef, A.M.: Differential fault analysis of streebog. In: Lopez, J., Wu, Y. (eds.) ISPEC 2015. LNCS, vol. 9065, pp. 35–49. Springer, Heidelberg (2015)

3. Aumasson, J.-P., Meier, W.: Zero-sum distinguishers for reduced Keccak-f and for the core functions of Luffa and Hamsi. Rump session of Cryptographic Hardware and Embedded Systems-CHES 2009, p. 67 (2009)

4. Banik, S., Maitra, S.: A differential fault attack on MICKEY 2.0. In: Bertoni, G., Coron, J.-S. (eds.) CHES 2013. LNCS, vol. 8086, pp. 215–232. Springer, Heidelberg (2013)

5. Banik, S., Maitra, S., Sarkar, S.: A differential fault attack on the grain family of stream ciphers. In: Prouff, E., Schaumont, P. (eds.) CHES 2012. LNCS, vol. 7428, pp. 122–139. Springer, Heidelberg (2012)

6. Bellare, M., Canetti, R., Krawczyk, H.: Keying hash functions for message authentication. In: Koblitz, N. (ed.) CRYPTO 1996. LNCS, vol. 1109, pp. 1–15. Springer, Heidelberg (1996)

7. Benoît, O., Peyrin, T.: Side-channel analysis of six SHA-3 candidates. In: Mangard, S., Standaert, F.-X. (eds.) CHES 2010. LNCS, vol. 6225, pp. 140–157. Springer, Heidelberg (2010)

8. Bertoni, G., Daemen, J., Peeters, M.: Cryptographic sponge functions. Report, STMicroelectronics, Antwerp, Belgium, January 2011

9. Bertoni, G., Daemen, J., Peeters, M., Assche, G.V.: The Keccak SHA-3 submission (2009)

10. Biham, E., Shamir, A.: Differential fault analysis of secret key cryptosystems. In: Kaliski Jr., B.S. (ed.) CRYPTO 1997. LNCS, vol. 1294, pp. 513–525. Springer, Heidelberg (1997)

11. Boura, C., Canteaut, A.: A zero-sum property for the Keccak-f permutation with 18 rounds. In: ISIT 2010s, pp. 2488–2492. IEEE (2010)

12. Boura, C., Canteaut, A.: Zero-sum distinguishers for iterated permutations and application to KECCAK-f and Hamsi-256. In: Biryukov, A., Gong, G., Stinson, D.R. (eds.) SAC 2010. LNCS, vol. 6544, pp. 1–17. Springer, Heidelberg (2011)

13. Boura, C., Canteaut, A., De Cannière, C.: Higher-order differential properties of KECCAK and Luffa. In: Joux, A. (ed.) FSE 2011. LNCS, vol. 6733, pp. 252–269. Springer, Heidelberg (2011)

14. Chakraborti, A., Chang, D., Nandi, M.: Fault based forgeries on CLOC and SILC. In: Latincrypt, LNCS. Springer (2015). https://groups.google.com/forum/#!topic/crypto-competitions/_qxORmqcSrY

15. Das, S., Meier, W.: Differential biases in reduced-round Keccak. In: Pointcheval, D., Vergnaud, D. (eds.) AFRICACRYPT. LNCS, vol. 8469, pp. 69–87. Springer, Heidelberg (2014)

16. Dey, P., Chakraborty, A., Adhikari, A., Mukhopadhyay, D.: Improved practical differential fault analysis of Grain-128. DATE 2015, 459–464 (2015)

17. Dinur, I., Dunkelman, O., Shamir, A.: New attacks on Keccak-224 and Keccak-256. In: Canteaut, A. (ed.) FSE 2012. LNCS, vol. 7549, pp. 442–461. Springer, Heidelberg (2012)

18. Dinur, I., Dunkelman, O., Shamir, A.: Collision attacks on up to 5 rounds of SHA-3 using generalized internal differentials. In: Moriai, S. (ed.) FSE 2013. LNCS, vol. 8424, pp. 219–240. Springer, Heidelberg (2014)

19. Dinur, I., Dunkelman, O., Shamir, A.: Improved practical attacks on round-reduced Keccak. J. Crypt. 27(2), 183–209 (2014)

20. Dinur, I., Morawiecki, P., Pieprzyk, J., Srebrny, M., Straus, M.: Cube attacks and cube-attack-like cryptanalysis on the round-reduced Keccak sponge function. In: Oswald, E., Fischlin, M. (eds.) EUROCRYPT 2015. LNCS, vol. 9056, pp. 733–761. Springer, Heidelberg (2015)

21. Duc, A., Guo, J., Peyrin, T., Wei, L.: Unaligned rebound attack: application to Keccak. In: Canteaut, A. (ed.) FSE 2012. LNCS, vol. 7549, pp. 402–421. Springer, Heidelberg (2012)
22. Dusart, P., Letourneux, G., Vivolo, O.: Differential fault analysis on A.E.S. In: Zhou, J., Yung, M., Han, Y. (eds.) ACNS 2003. LNCS, vol. 2846, pp. 293–306. Springer, Heidelberg (2003)
23. FIPS-202. SHA-3 Standard: Permutation-Based Hash and Extendable-Output Functions. National Institute for Standards and Technology, pub-NIST, May 2014
24. Fischer, W., Reuter, C.A.: Differential fault analysis on Grøstl. In: Bertoni, G., Gierlichs, B. (eds.) FDTC 2012, pp. 44–54. IEEE Computer Society (2012)
25. Giraud, C.: DFA on AES. In: Dobbertin, H., Rijmen, V., Sowa, A. (eds.) AES 2005. LNCS, vol. 3373, pp. 27–41. Springer, Heidelberg (2005)
26. Giraud, C., Thillard, A.: Piret and quisquater's DFA on AES revisited. IACR Cryptology ePrint Archive, 2010:440 (2010)
27. Hemme, L.: A differential fault attack against early rounds of (Triple-)DES. In: Joye, M., Quisquater, J.-J. (eds.) CHES 2004. LNCS, vol. 3156, pp. 254–267. Springer, Heidelberg (2004)
28. Hemme, L., Hoffmann, L.: Differential fault analysis on the SHA1 compression function. In: Breveglieri, L., Guilley, S., Koren, I., Naccache, D., Takahashi, J. (eds.) FDTC 2011, pp. 54–62. IEEE Computer Society (2011)
29. Jean, J., Nikolic, I.: Internal differential boomerangs: practical analysis of the round-reduced Keccak-f permutation. IACR Cryptology ePrint Archive 2015:244 (2015)
30. Karmakar, S., Chowdhury, D.R.: Differential fault analysis of MICKEY-128 2.0. In: Fischer, W., Schmidt, J. (eds.) FDTC 2013, pp. 52–59. IEEE Computer Society (2013)
31. Kim, C.H.: Differential fault analysis of AES: toward reducing number of faults. Inf. Sci. **199**, 43–57 (2012)
32. Kim, C.H., Quisquater, J.-J.: New differential fault analysis on AES key schedule: two faults are enough. In: Grimaud, G., Standaert, F.-X. (eds.) CARDIS 2008. LNCS, vol. 5189, pp. 48–60. Springer, Heidelberg (2008)
33. Kölbl, S., Mendel, F., Nad, T., Schläffer, M.: Differential cryptanalysis of Keccak variants. In: Stam, M. (ed.) IMACC 2013. LNCS, vol. 8308, pp. 141–157. Springer, Heidelberg (2013)
34. Luo, P., Fei, Y., Fang, X., Ding, A.A., Kaeli, D.R., Leeser, M.: Side-channel analysis of MAC-Keccak hardware implementations. Cryptology ePrint Archive, Report 2015/411 (2015). http://eprint.iacr.org/
35. Morawiecki, P., Pieprzyk, J., Srebrny, M.: Rotational cryptanalysis of round-reduced Keccak. In: Moriai, S. (ed.) FSE 2013. LNCS, vol. 8424, pp. 241–262. Springer, Heidelberg (2014)
36. Naya-Plasencia, M., Röck, A., Meier, W.: Practical analysis of reduced-round Keccak. In: Bernstein, D.J., Chatterjee, S. (eds.) INDOCRYPT 2011. LNCS, vol. 7107, pp. 236–254. Springer, Heidelberg (2011)
37. Piret, G., Quisquater, J.-J.: A differential fault attack technique against SPN structures, with application to the AES and KHAZAD. In: Walter, C.D., Koç, Ç.K., Paar, C. (eds.) CHES 2003. LNCS, vol. 2779, pp. 77–88. Springer, Heidelberg (2003)
38. Preneel, B., van Oorschot, P.C.: On the security of iterated message authentication codes. IEEE Trans. Inf. Theory **45**(1), 188–199 (1999)
39. Rivain, M.: Differential fault analysis on DES middle rounds. In: Clavier, C., Gaj, K. (eds.) CHES 2009. LNCS, vol. 5747, pp. 457–469. Springer, Heidelberg (2009)

40. Saha, D., Kuila, S., Chowdhury, D.R.: EscApe: diagonal fault analysis of APE. In: Meier, W., Mukhopadhyay, D. (eds.) INDOCRYPT 2014. LNCS, vol. 8885, pp. 197–216. Springer, Heidelberg (2014)
41. Tunstall, M., Mukhopadhyay, D., Ali, S.: Differential fault analysis of the advanced encryption standard using a single fault. In: Ardagna, C.A., Zhou, J. (eds.) WISTP 2011. LNCS, vol. 6633, pp. 224–233. Springer, Heidelberg (2011)

A Key to Success
Success Exponents for Side-Channel Distinguishers

Sylvain Guilley[1,2], Annelie Heuser[1(✉)], and Olivier Rioul[1,3]

[1] Department Comelec, Télécom ParisTech, Institut Mines-Télécom, CNRS LTCI,
46 Rue Barrault, 75 634 Paris Cedex 13, France
{sylvain.guilley,annelie.heuser,olivier.rioul}@telecom-paristech.fr
[2] Secure-IC S.A.S., 15 Rue Claude Chappe, Bât. B, ZAC des Champs Blancs,
35 510 Cesson-Sévigné, France
[3] Applied Mathematics Department, École Polytechnique, Palaiseau, France

Abstract. The success rate is the classical metric for evaluating the performance of side-channel attacks. It is generally computed empirically from measurements for a particular device or using simulations. Closed-form expressions of success rate are desirable because they provide an explicit functional dependence on relevant parameters such as number of measurements and signal-to-noise ratio which help to understand the effectiveness of a given attack and how one can mitigate its threat by countermeasures. However, such closed-form expressions involve high-dimensional complex statistical functions that are hard to estimate.

In this paper, we define the success exponent (SE) of an arbitrary side-channel distinguisher as the first-order exponent of the success rate as the number of measurements increases. Under fairly general assumptions such as soundness, we give a general simple formula for any arbitrary distinguisher and derive closed-form expressions of it for DoM, CPA, MIA and the optimal distinguisher when the model is known (template attack). For DoM and CPA our results are in line with the literature. Experiments confirm that the theoretical closed-form expression of the SE coincides with the empirically computed one, even for reasonably small numbers of measurements. Finally, we highlight that our study raises many new perspectives for comparing and evaluating side-channel attacks, countermeasures and implementations.

Keywords: Side-Channel distinguisher · Evaluation metric · Success rate · Success exponent · Closed-form expressions

1 Introduction

Side-channel attacks analyse physical leakage that is unintentionally emitted during cryptographic operations in a device. This side-channel leakage is statistically dependent on intermediate processed values involving the secret key.

Annelie Heuser is a Google European fellow in the field of privacy and is partially founded by this fellowship.

© Springer International Publishing Switzerland 2015
A. Biryukov and V. Goyal (Eds.): INDOCRYPT 2015, LNCS 9462, pp. 270–290, 2015.
DOI: 10.1007/978-3-319-26617-6_15

It is then possible to retrieve the secret from the measured data by maximizing some statistical distinguisher. In the past decade, many distinguishers have been proposed: difference of means test [17] (DoM), Pearson correlation [4] (CPA), mutual information [12] (MIA), etc. Such distinguishers have different characteristics and performances, depending on the implementation, measurement noise, and assumed knowledge on how the device leaks.

To evaluate the performance of a given distinguisher for a limited number of measurements, the *average probability of success* a.k.a. *success rate* (SR) is the ideal and most common evaluation metric [30]. It provides everything one needs to know about the performance of a particular attack scenario. Ideally, one would exhibit an explicit functional relationship of the SR with the number of measurements, signal-to-noise ratio (SNR), and other important quantities determining the relationship between correct and false key hypotheses such as confusion coefficients [10]. The resulting closed-form expression would allow one to better understand how effective the attack can be under specific conditions and how one can mitigate it with appropriate countermeasures.

So far, however, it can be theoretically computed only for a very narrow range of distinguishers (DoM [10], CPA [18,29,31], Bayesian attacks [29]) and only under restrictive "ideal" scenarios (e.g., perfectly known leakage model in Gaussian noise). Moreover, the resulting exact expressions involve high dimensional functions whose dependency on the relevant parameters (such as confusion coefficients) can be very complex. For DoM and CPA under ideal scenarios, the resulting formulas involve a multivariate normal c.d.f. [28] for which no closed-form expression exists, while as was found in the case of CPA [29] the corresponding matrices are not of full rank and require heavy Monte-Carlo computation.

In this paper, we carry out a theoretical derivation of the SR for quite arbitrary distinguishers, at the first order of the exponent. More precisely, our computation yields closed-form expressions of the success exponent (SE) associated to the failure rate (1–SR) at first order as the number of measurements m increases:

$$1 - \mathrm{SR} \approx e^{-m \cdot \mathrm{SE}}. \tag{1}$$

(The precise mathematical meaning of the equivalence $\approx$ will be given in Definition 7.) Even though we obtain the derived expression for the SE under the asymptotic condition that m tends to infinity, simulations show that Eq. (1) is still accurate even for fairly small values of m.

Such an evaluation of the success rate, suitable even for a small number of traces, allows one to compare all possible distinguishers in any scenario (noise distribution, unprotected or protected implementation, etc.). A recent paper by Duc et al. [9, Theorem 2] tackles this problem and achieves a unilateral bound. Here we give both a lower and an upper bound, and as an illustration derive the exact expression of the SE for DoM, CPA, MIA and the optimal distinguisher when model is known (template attack) in terms of the appropriate relevant parameters.

The rest of this paper is organized as follows. Section 2 gives the necessary definitions about distinguishers, success and soundness. In Sect. 3, we examine the convergence of success rate and apply a central limit theorem to derive the SE (Theorem 1). Section 4 validates the SE even for relatively small number of traces, and Sect. 5 provides closed-form expressions of SE for some popular distinguishers. The conclusions and promising perspectives are in Sect. 6.

2   Preliminaries

In the sequel, we consider a standard univariate side-channel scenario as defined in [21]. Let k^* denote the secret cryptographic key, k any possible key hypothesis. Also let X be a random variable[1] representing the measured leakage and T be the (random) input or cipher text used for a given encryption request. The attacker knows some mapping f corresponding to an the internally processed variable $f(k, T)$. A common consideration is $f(T, k) = \text{Sbox}[T \oplus k]$ where Sbox is a substitution box. The measured leakage X can then be written as

$$X = \varphi(f(T, k^*)) + N, \tag{2}$$

where φ is a deterministic leakage function and where N is an independent—not necessarily Gaussian—additive noise with zero mean ($\mathbb{E}\{N\} = 0$). The device-specific deterministic function φ is normally unknown to the attacker but she may estimate it as $\hat{\varphi}$ and compute the *sensitive variable* $Y(k) = \hat{\varphi}(f(T, k))$ for each key hypothesis k. For later ease of notation we may drop the letter k and write $Y = Y(k)$ and $Y^* = Y(k^*)$. We do not make any particular assumption on φ or f so that our framework can be applied to any arbitrary scenario.

2.1   Distinguisher

In practice, the distinguisher is a function of m i.i.d. leakage measurements $X_1, X_2, \ldots, X_m$ and sensitive variables $Y_1(k), Y_2(k), \ldots, Y_m(k)$ whose maximization over the key hypothesis yields $\hat{k} = \arg\max_k \widehat{\mathcal{D}}(k)$, where

$$\widehat{\mathcal{D}}(k) = \widehat{\mathcal{D}}(X_1, X_2, \ldots, X_m; Y_1(k), Y_2(k), \ldots, Y_m(k)). \tag{3}$$

Definition 1 (Theoretical Distinguisher). *We assume that there is a "theoretical" value of the distinguisher*

$$\mathcal{D}(k) = \mathcal{D}(X, Y(k)) \tag{4}$$

for each k such that $\widehat{\mathcal{D}}(k)$ converges to $\mathcal{D}(k)$ as $m \to +\infty$ in the mean-squared sense, i.e., the mean-squared error

$$\text{MSE}_m = \mathbb{E}\left\{ \left(\widehat{\mathcal{D}}(k) - \mathcal{D}(k)\right)^2 \right\} \to 0 \text{ as } m \to +\infty. \tag{5}$$

---

[1] Capitals such as X denote random variables. The corresponding lowercase x denotes realizations of these random variables. We write $\mathbb{P}\{A\}$ for the probability of an event A and $\mathbb{E}\{X\}$ for the expectation of a random variable X.

This implies that $\widehat{\mathcal{D}}(k) \to \mathcal{D}(k)$ in probability. Thus we may consider the practical distinguisher $\widehat{\mathcal{D}}(k)$ as an *estimator* of the theoretical $\mathcal{D}(k)$. The corresponding *bias* and *variance* of $\widehat{\mathcal{D}}(k)$ are

$$B_m(k) = \mathbb{E}\{\widehat{\mathcal{D}}(k)\} - \mathcal{D}(k) \tag{6}$$

$$V_m(k) = \mathrm{Var}(\widehat{\mathcal{D}}(k)). \tag{7}$$

Example 1 (CPA [4]). For correlation analysis we have

$$\widehat{\mathcal{D}}(k) = \frac{m\sum_{i=1}^m X_i Y_i - \sum_{i=1}^m X_i \sum_{i=1}^m Y_i}{\sqrt{m\sum_{i=1}^m X_i^2 - (\sum_{i=1}^m X_i)^2}\sqrt{m\sum_{i=1}^m Y_i^2 - (\sum_{i=1}^m Y_i)^2}} \tag{8}$$

$$\mathcal{D}(k) = \rho(X,Y) = \frac{\mathrm{Cov}(X,Y)}{\sigma_X \sigma_Y} = \frac{\mathbb{E}\{(X - \mu_X)(Y - \mu_Y)\}}{\sigma_X \sigma_Y}. \tag{9}$$

Example 2 (MIA [12]). For mutual information

$$\mathcal{D}(k) = I(X,Y) = H(X) - H(X|Y) \tag{10}$$

can be estimated e.g. with histograms as

$$\widehat{\mathcal{D}}(k) = \sum_x \sum_y \hat{\mathbb{P}}(x,y) \log_2 \frac{\hat{\mathbb{P}}(x,y)}{\hat{\mathbb{P}}(x)\hat{\mathbb{P}}(y)}. \tag{11}$$

Lemma 1. *Bias $B_m(k)$ and variance $V_m(k)$ tend to zero as m increases.*

Proof. One has the well-known bias-variance compromise: $\mathrm{MSE}_m = \mathbb{E}\{(\widehat{\mathcal{D}}(k) - \mathbb{E}\{\widehat{\mathcal{D}}(k)\} + B_m(k))^2\} = V_m(k) + B_m(k)^2 + 0$ where the cross-term vanishes. Since $\mathrm{MSE}_m \to 0$ it follows that $V_m(k) \to 0$ and $B_m(k) \to 0$. $\qquad\square$

2.2 Success Rate

The success rate (SR) is the classical evaluation metric when comparing empirical side-channel distinguishers $\widehat{\mathcal{D}}(k)$. It is generally calculated empirically [8,19,21]. The exact (theoretical) value of SR [10,18,29,31] is as follows.

Definition 2 (Success Rate and Failure Rate). *The average success probability is defined by*

$$\mathrm{SR}(\widehat{\mathcal{D}}) = \mathbb{P}\{\forall k \neq k^*, \ \widehat{\mathcal{D}}(k^*) > \widehat{\mathcal{D}}(k)\}. \tag{12}$$

where k^* is the actual value of the secret key. It is sometimes convenient to consider the average failure rate *as the complementary probability*

$$\mathrm{FR}(\widehat{\mathcal{D}}) = 1 - \mathrm{SR}(\widehat{\mathcal{D}}) = \mathbb{P}\{\exists k \neq k^*, \ \widehat{\mathcal{D}}(k) \geq \widehat{\mathcal{D}}(k^*)\}. \tag{13}$$

Evaluating probabilities of events like $\{\exists k \neq k^*, \ \widehat{\mathcal{D}}(k) \geq \widehat{\mathcal{D}}(k^*)\}$ may be cumbersome. In order to pass from those to individual events $\{\widehat{\mathcal{D}}(k) \geq \widehat{\mathcal{D}}(k^*)\}$ for each k, the following lemma is convenient.

Lemma 2 (Squeezing the Failure Rate). *One can lower and upper bound the failure rate as follows:*

$$\max_{k \neq k^*} \mathbb{P}\{\widehat{\mathcal{D}}(k) \geq \widehat{\mathcal{D}}(k^*)\} \leq \mathrm{FR}(\widehat{\mathcal{D}}) \leq \sum_{k \neq k^*} \mathbb{P}\{\widehat{\mathcal{D}}(k) \geq \widehat{\mathcal{D}}(k^*)\}. \qquad (14)$$

Proof. We can write $\mathrm{FR}(\widehat{\mathcal{D}}) = \mathbb{P}\{\bigcup_{k \neq k^*}\{\widehat{\mathcal{D}}(k) \geq \widehat{\mathcal{D}}(k^*)\}\}$. The upper bound follows from the union bound $\mathbb{P}\{\bigcup_k A_k\} \leq \sum_k \mathbb{P}\{A_k\}$. Now the probability of the union is not less that of any individual event $\{\widehat{\mathcal{D}}(k) \geq \widehat{\mathcal{D}}(k^*)\}$. Choosing the one with maximal probability gives the lower bound. □

Remark 1. The lower bound approximation in Eq. (14) is reminiscent of ideas developed by Whitnall and Oswald in [33] where they define a framework for the theoretical evaluation of side-channel distinguishers. Their outcome is captured by the relative behavior of the distinguisher for the correct key and its nearest rival. We leverage on this idea to prove our Theorem 1 in Sect. 3.

Lemma 2 leads us to define pairwise quantities (see e.g., [29, Eq. (13)]).

Definition 3 (Pairwise Deltas). *For any function $f(k)$ define*

$$\Delta f(k^*, k) = f(k^*) - f(k). \qquad (15)$$

Thus $\Delta\widehat{\mathcal{D}}(k^*, k) = \widehat{\mathcal{D}}(k^*) - \widehat{\mathcal{D}}(k)$ and $\Delta\mathcal{D}(k^*, k) = \mathcal{D}(k^*) - \mathcal{D}(k)$. The pairwise error probability for the transition $k^* \to k$ is

$$\mathbb{P}\{\widehat{\mathcal{D}}(k) \geq \widehat{\mathcal{D}}(k^*)\} = \mathbb{P}\{\Delta\widehat{\mathcal{D}}(k^*, k) \leq 0\}. \qquad (16)$$

Lemma 3. *The difference $\Delta\widehat{\mathcal{D}}(k^*, k)$ estimates $\Delta\mathcal{D}(k^*, k)$ with bias and variance*

$$B_m(k^*, k) = B_m(k^*) - B_m(k) \qquad (17)$$

$$V_m(k^*, k) = \mathrm{Var}(\Delta\widehat{\mathcal{D}}(k^*, k)) \qquad (18)$$

tending to zero as $m \to +\infty$.

Proof. Since $\widehat{\mathcal{D}}(k) \to \mathcal{D}(k)$ and $\widehat{\mathcal{D}}(k^*) \to \mathcal{D}(k^*)$ in the mean-square sense (Definition 1) we can deduce that $\widehat{\mathcal{D}}(k^*) - \widehat{\mathcal{D}}(k) \to \mathcal{D}(k^*) - \mathcal{D}(k)$ also in the mean-square sense. This follows from Minkowski's inequality $\sqrt{\mathbb{E}\{(X \pm Y)^2\}} \leq \sqrt{\mathbb{E}\{X^2\}} + \sqrt{\mathbb{E}\{Y^2\}}$. The proof of Lemma 1 now applies verbatim to show that $B_m(k^*, k) \to 0$ and $V_m(k^*, k) \to 0$. □

2.3 Soundness

Definition 4 (Soundness Condition). *The attack using distinguisher $\widehat{\mathcal{D}}(k)$ is* sound *if the corresponding theoretical distinguisher's values satisfy the inequalities*

$$\mathcal{D}(k^*) > \mathcal{D}(k) \qquad for\ all\ k \neq k^*. \qquad (19)$$

In other words $\Delta\mathcal{D}(k^*, k) > 0$ for all bad key hypotheses k.

In [13] the authors give a proof of soundness for CPA. Note that, DoM can be seen as a special case of CPA (when $m \to \infty$) where $Y \in \{\pm 1\}$ and thus is all the more sound. MIA was proven sound for Gaussian noise in [23, 26].

Proposition 1 (Soundness). *When the attack is sound, the success eventually tends to 100 % as m increases:*

$$\text{SR}(\widehat{\mathcal{D}}) \to 1 \; as \; m \to +\infty. \tag{20}$$

This has been taken as a definition of soundness in [30, Sect. 5.1]. We provide an elegant proof.

Proof. By Lemma 2, $1 - \text{SR}(\widehat{\mathcal{D}}) \le \sum_{k \neq k^*} \mathbb{P}\{\Delta\widehat{\mathcal{D}}(k^*, k) \le 0\}$. It suffices to show that for each $k \neq k^*$, $\mathbb{P}\{\Delta\widehat{\mathcal{D}}(k^*, k) \le 0\} = \mathbb{P}\{\Delta\mathcal{D}(k^*, k) - \Delta\widehat{\mathcal{D}}(k^*, k) \ge \Delta\mathcal{D}(k^*, k)\}$ tends to zero. Now by the soundness assumption, $\Delta\mathcal{D} = \Delta\mathcal{D}(k^*, k) > 0$. Dropping the dependency on (k^*, k) for notational convenience, one obtains

$$\mathbb{P}\{\Delta\mathcal{D} - \Delta\widehat{\mathcal{D}} \ge \Delta\mathcal{D}\} \le \frac{\mathbb{E}\left\{(\Delta\mathcal{D} - \Delta\widehat{\mathcal{D}})^2\right\}}{\Delta\mathcal{D}^2} \to 0 \tag{21}$$

where we have used Chebyshev's inequality $\mathbb{P}\{X \ge \epsilon\} \le \frac{\mathbb{E}\{X^2\}}{\epsilon^2}$ and the fact that $\Delta\widehat{\mathcal{D}}(k^*, k) \to \Delta\mathcal{D}(k^*, k)$ in the mean-square sense (Lemma 3). $\qquad\square$

Since $\text{SR}(\widehat{\mathcal{D}}) \to 1$ as m increases we are led to investigate the rate of convergence of $\text{FR}(\widehat{\mathcal{D}}) = 1 - \text{SR}(\widehat{\mathcal{D}})$ toward zero. This is done next.

3 Derivation of Success Exponent

3.1 Normal Approximation and Assumption

We first prove some normal (Gaussian) behavior in the case of additive distinguishers and then generalize.

Definition 5 (Additive Distinguisher [18]). An additive distinguisher can be written in the form of a sum of i.i.d. terms:

$$\widehat{\mathcal{D}}(X_1, X_2, \ldots, X_m; Y_1(k), Y_2(k), \ldots, Y_m(k)) = \frac{1}{m} \sum_{i=1}^{m} \widehat{\mathcal{D}}(X_i; Y_i(k)). \tag{22}$$

Remark 2. DoM is additive (see e.g., [10]). Attacks maximizing scalar products $\sum_{i=1}^{m} X_i Y_i$ are clearly additive; they constitute a good approximation to CPA, and are even equivalent to CPA if one assumes that the first and second moments of $Y(k)$ are constant independent of k (see [14, 27, 29] for similar assumptions).

Lemma 4. *When the distinguisher is additive, the corresponding theoretical distinguisher is*

$$\mathcal{D}(X, Y(k)) = \mathbb{E}\{\widehat{\mathcal{D}}(X; Y(k))\}. \tag{23}$$

Thus $\varDelta\widehat{\mathcal{D}}(k^*, k)$ is an unbiased estimator of $\varDelta\mathcal{D}(k^*, k)$, whose variance is

$$V_m(k^*, k) = \frac{\mathrm{Var}\big(\widehat{\mathcal{D}}(X; Y(k^*)) - \widehat{\mathcal{D}}(X; Y(k))\big)}{m} \tag{24}$$

Proof. Letting $\mathbb{E}\{\widehat{\mathcal{D}}(X; Y(k))\} = \mathcal{D}(k)$, since the terms $\widehat{\mathcal{D}}(X_i; Y_i(k))$ are independent and identically distributed, one has

$$\mathbb{E}\Big\{\big(\widehat{\mathcal{D}}(k) - \mathcal{D}(k)\big)^2\Big\} = \tfrac{1}{m^2}\mathbb{E}\Big\{\textstyle\sum_{i=1}^{m}\big(\widehat{\mathcal{D}}(X_i; Y_i(k)) - \mathcal{D}(k)\big)^2\Big\} \tag{25}$$

$$= \tfrac{1}{m}\mathbb{E}\Big\{\big(\widehat{\mathcal{D}}(X; Y(k)) - \mathcal{D}(k)\big)^2\Big\} \to 0. \tag{26}$$

Therefore, $\frac{1}{m}\sum_{i=1}^{m}\widehat{\mathcal{D}}(X_i; Y_i(k)) \to \mathbb{E}\{\widehat{\mathcal{D}}(X; Y(k))\}$ in the mean-square sense. (This is actually an instance of the weak law of large numbers). The corresponding bias is zero: $\mathbb{E}\{\widehat{\mathcal{D}}(k)\} - \mathcal{D}(k) = 0$.

Taking differences, it follows from Lemma 3 that $\varDelta\widehat{\mathcal{D}}(k^*, k) \to \varDelta\mathcal{D}(k^*, k)$ in the mean-square sense with zero bias. The corresponding variance is computed as above as $\mathbb{E}\big\{\big(\varDelta\widehat{\mathcal{D}}(k^*, k) - \varDelta\mathcal{D}(k^*, k)\big)^2\big\} = \frac{1}{m}\mathbb{E}\big\{\big((\widehat{\mathcal{D}}(X; Y(k^*)) - \widehat{\mathcal{D}}(X; Y(k))) - (\mathcal{D}(X; Y(k^*)) - \mathcal{D}(X; Y(k)))\big)^2\big\} = \frac{1}{m}\mathrm{Var}\big(\widehat{\mathcal{D}}(X; Y(k^*)) - \widehat{\mathcal{D}}(X; Y(k))\big)$. $\square$

Proposition 2 (Normal Approximation). *When the distinguisher is additive, $\varDelta\widehat{\mathcal{D}}(k^*, k)$ follows the normal approximation*

$$\varDelta\widehat{\mathcal{D}}(k^*, k) \sim \mathcal{N}\big(\varDelta\mathcal{D}(k^*, k), V_m(k^*, k)\big) \tag{27}$$

as m increases. This means that

$$\frac{\varDelta\widehat{\mathcal{D}}(k^*, k) - \varDelta\mathcal{D}(k^*, k)}{\sqrt{V_m(k^*, k)}} \tag{28}$$

converges to the standard normal $\mathcal{N}(0, 1)$ in distribution.

Proof. Apply the central limit theorem to the sum of i.i.d. variables $m\varDelta\widehat{\mathcal{D}}(k^*, k) = \sum_{i=1}^{m}\widehat{\mathcal{D}}(X_i; Y_i(k^*)) - \widehat{\mathcal{D}}(X_i; Y_i(k))$. It follows that

$$\frac{m\varDelta\widehat{\mathcal{D}}(k^*, k) - m\varDelta\mathcal{D}(k^*, k)}{\sqrt{m \cdot \mathrm{Var}\big(\varDelta\widehat{\mathcal{D}}(k^*, k)\big)}} = \frac{\varDelta\widehat{\mathcal{D}}(k^*, k) - \varDelta\mathcal{D}(k^*, k)}{\sqrt{V_m(k^*, k)}} \tag{29}$$

tends in distribution to $\mathcal{N}(0, 1)$. $\square$

Remark 3. Notice that the normal approximation is *not* a consequence of a Gaussian noise assumption or anything actually related to the leakage model but is simply a genuine consequence of the central limit theorem.

The above result for additive distinguishers leads us to the following.

Definition 6 (Normal Assumption). *We say that a sound distinguisher follows the* normal assumption *if*

$$\Delta\widehat{D}(k^*, k) \sim \mathcal{N}\big(\mathbb{E}\{\Delta\widehat{D}(k^*, k)\}, V_m(k^*, k)\big) \tag{30}$$

as m increases.

Remark 4. We note that in general

$$\mathbb{E}\{\Delta\widehat{D}(k^*, k)\} = \Delta D(k^*, k) + \Delta B_m(k^*, k) \tag{31}$$

has a bias term (Lemma 3). By Proposition 2 any additive distinguisher follows the above normal assumption (with zero bias). We shall adopt the normal assumption even in situations where the distinguisher is not additive (as is the case of MIA) with possibly nonzero bias. The corresponding outcomes will be justified by simulations in Sect. 4.

3.2   The Main Result: Success Exponent

Recall a well-known mathematical definition that two functions are *equivalent*: $f(x) \sim g(x)$ if $f(x)/g(x) \to 1$ as $x \to +\infty$. The following defines a weaker type of equivalence $f(x) \approx g(x)$ at first order of exponent, which is required to derive the success exponent SE.

Definition 7 (First-Order Exponent [7, Chap. 11]). *We say that a function* $f(x)$ *has first order exponent* $\xi(x)$ *if* $\big(\ln f(x)\big) \sim \xi(x)$ *as* $x \to +\infty$, *in which case we write*

$$f(x) \approx \exp \xi(x). \tag{32}$$

Lemma 5. *Let* $Q(x) = \frac{1}{\sqrt{2\pi}} \int_x^{+\infty} e^{-t^2/2}\, dt$ *be the tail probability of the standard normal (a.k.a. Marcum function). Then as* $x \to +\infty$,

$$Q(x) \approx e^{-x^2/2}. \tag{33}$$

Proof. For $t > x$, we can write

$$\int_x^{+\infty} \frac{1 + 1/t^2}{1 + 1/x^2} \frac{e^{-t^2/2}}{\sqrt{2\pi}}\, dt \le Q(x) \le \int_x^{+\infty} \frac{t}{x} \frac{e^{-t^2/2}}{\sqrt{2\pi}}\, dt. \tag{34}$$

Taking antiderivative yields

$$\frac{1}{1 + 1/x^2} \frac{1}{\sqrt{2\pi}} \frac{e^{-x^2/2}}{x} \le Q(x) \le \frac{1}{x\sqrt{2\pi}} e^{-x^2/2}. \tag{35}$$

Taking the logarithm gives

$$-x^2/2 - \ln(x + 1/x) - \ln(2\pi)/2 \le \ln Q(x) \le -x^2/2 - \ln x - \ln(2\pi)/2 \tag{36}$$

which shows that $\ln Q(x) \sim -x^2/2$.   □

Lemma 6. *Under the normal assumption,*

$$\mathbb{P}\{\varDelta\widehat{\mathcal{D}}(k^*, k) \leq 0\} \approx \exp\left(-\frac{\left(\varDelta\mathcal{D}(k^*, k) + \varDelta B_m(k^*, k)\right)^2}{2\,V_m(k^*, k)}\right). \tag{37}$$

Proof. Noting that

$$\mathbb{P}\{\varDelta\widehat{\mathcal{D}}(k^*, k) \leq 0\} = \mathbb{P}\left\{\frac{\varDelta\widehat{\mathcal{D}}(k^*, k) - \mathbb{E}\{\varDelta\widehat{\mathcal{D}}(k^*, k)\}}{\sqrt{V_m(k^*, k)}} \leq \frac{-\mathbb{E}\{\varDelta\widehat{\mathcal{D}}(k^*, k)\}}{\sqrt{V_m(k^*, k)}}\right\} \tag{38}$$

and using the normal approximation it follows that

$$\mathbb{P}\{\varDelta\widehat{\mathcal{D}}(k^*, k) \leq 0\} \approx Q\left(\frac{\mathbb{E}\{\varDelta\widehat{\mathcal{D}}(k^*, k)\}}{\sqrt{V_m(k^*, k)}}\right) \tag{39}$$

where $\mathbb{E}\{\varDelta\widehat{\mathcal{D}}(k^*, k)\} = \varDelta\mathcal{D}(k^*, k) + \varDelta B_m(k^*, k)$. The assertion now follows from Lemma 5. $\qquad\square$

Theorem 1. *Under the normal assumption,*

$$\mathrm{FR}(\widehat{\mathcal{D}}) = 1 - \mathrm{SR}(\widehat{\mathcal{D}}) \approx \exp\left(-\min_{k \neq k^*} \frac{\left(\varDelta\mathcal{D}(k^*, k) + \varDelta B_m(k^*, k)\right)^2}{2\,V_m(k^*, k)}\right). \tag{40}$$

Proof. We combine Lemmas 2 and 6. The lower bound of $\mathrm{FR}(\widehat{\mathcal{D}})$ is

$$\approx \max_{k \neq k^*} \exp\left(-\frac{\left(\varDelta\mathcal{D}(k^*, k) + \varDelta B_m(k^*, k)\right)^2}{2\,V_m(k^*, k)}\right) \tag{41}$$

$$= \exp\left(-\min_{k \neq k^*} \frac{\left(\varDelta\mathcal{D}(k^*, k) + \varDelta B_m(k^*, k)\right)^2}{2\,V_m(k^*, k)}\right). \tag{42}$$

The upper bound is the sum of vanishing exponentials (for $k \neq k^*$) which is equivalent to the maximum of the vanishing exponentials, which yields the same expression. The result follows since the lower and upper bounds from Lemma 2 are equivalent as m increases. $\qquad\square$

Corollary 1. *For any additive distinguisher,*

$$\mathrm{FR}(\widehat{\mathcal{D}}) = 1 - \mathrm{SR}(\widehat{\mathcal{D}}) \approx e^{-m \cdot \mathrm{SE}(\widehat{\mathcal{D}})} \tag{43}$$

where

$$\mathrm{SE}(\widehat{\mathcal{D}}) = \min_{k \neq k^*} \frac{\varDelta\mathcal{D}(k^*, k)^2}{2\,\mathrm{Var}\left(\widehat{\mathcal{D}}(X; Y(k^*)) - \widehat{\mathcal{D}}(X; Y(k))\right)}. \tag{44}$$

Proof. Apply the above theorem using Lemma 4 and Proposition 2. $\qquad\square$

Remark 5. We show in Sect. 5 that for non-additive distinguisher such as MIA the closed-form expression for the first-order exponent is linear in the number of measurements m so that the expression $1 - \mathrm{SR} \approx e^{-m \cdot \mathrm{SE}}$ may be considered as fairly general for large m. Moreover, we experimentally show in the next section that this approximation already holds with excellent approximation for a relatively small number of measurements m.

4   Success Exponent for Few Measurements

Some devices such as unprotected 8-bit microprocessors require only a small number of measurements to reveal the secret key. As the SNR is relatively high, the targeted variable has the length of the full size, and on such processors, the pipeline is short or even completely absent. On such worst-case platforms, such as the AVR ATMega, the SNR can be has high as 7, for those instructions consisting in memory look-ups. A CPA requires $m = 12$ measurements (cf. DPA contest v4, for attacks reported in [2]).

In order to investigate the relation SR $\approx 1 - e^{-m\mathrm{SE}}$ for such small values of m, we target PRESENT [3], which is an SPN (Substitution Permutation Network) block cipher, with leakage model given by $Y(k) = HW(\mathtt{Sbox}(T \oplus k))$, where $\mathtt{Sbox} : \mathbb{F}_2^4 \to \mathbb{F}_2^4$ is the PRESENT substitution box and $k \in \mathbb{F}_2^4$. We considered $N \sim \mathcal{N}(0,1)$ in our simulations applied to the following distinguishers:

- optimal distinguisher (a.k.a. template attack [6], whose formal expression is given in [15] for Gaussian noise);
- DoM [17]^2 on bit #2;
- CPA (Example 1),
- MIA (Example 2), with three distinct bin widths of length $\Delta x \in \{1, 2, 4\}$, and two kinds of binning:
  - B1, which partitions $\mathbb{R}$ as $\bigcup_{i \in \mathbb{N}}[i\Delta x, (i+1)\Delta x[$, and
  - B2, which partitions $\mathbb{R}$ as $\bigcup_{i \in \mathbb{N}}[(i - \frac{1}{2})\Delta x, (i + \frac{1}{2})\Delta x[$.

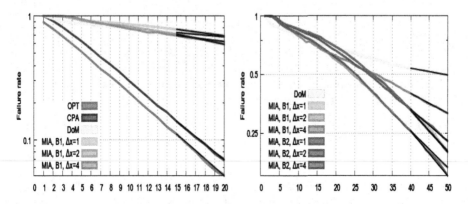

Fig. 1. Failure rate for few measurements. (a) Optimal distinguisher, CPA, DoM, and MIA. (b) Zoom out for less efficient attacks DoM and MIA.

Figure 1 shows the failure rate in a logarithmic scale for 10,000 simulations with additional error bars as described in [19]. To assess the linear dependence

---

^{2} It is known that for bit #1, the DoM is not sound: the same distinguisher value can be obtained for the correct key $k = k^*$ and for at least one incorrect key $k = k^* \oplus \mathtt{0x9}$.

$\log(1 - \text{SR}) = -m\text{SE}$ between the logarithm of the error rate and the number of traces, we have superimposed the linear slope $-\text{SE}$ in black. We find that CPA and the optimal distinguishers behave according to the law for m as small as 2! The error rate of MIA and DoM becomes linear for $m \geq 40$. Interestingly, for MIA, the binning size has an impact (see also [12,23]). The best parameterization of the MIA corresponds to $\Delta x = 2$, for both B1 and B2.

5   Closed-Form Expressions of SE

5.1   Success Exponents for DoM and CPA

We precise our side-channel model from Eq. (2) in case of additive distinguishers. As these distinguishers are most usually used when the leakage X is linearly depend on Y^*, we assume similar to previous works [10,31] $X = \alpha Y^* + N$. To simplify the derivation, we assume that the distribution of $Y(k)$ is identical for all k. In other words, knowing the distribution of $Y(k)$ does not give any evidence about the secret (see [14,27] for similar assumptions). In particular $\text{Var}\{Y(k)\}$ is constant for all k. Without loss of generality we may normalize the sensitive variable Y such that $\mathbb{E}\{Y(k)\} = 0$ and $\text{Var}\{Y(k)\} = \mathbb{E}\{Y(k)^2\} = 1$. The SNR is thus equal to α^2/σ^2.

We first extend the idea of confusion similar to [31], which we call *general 2-way confusion coefficients*.

Definition 8 (General 2-way Confusion Coefficients). *For $k \neq k^*$ we define*

$$\kappa(k^*, k) = \mathbb{E}\left\{ \left(\frac{Y(k^*) - Y(k)}{2} \right)^2 \right\}, \tag{45}$$

$$\kappa'(k^*, k) = \mathbb{E}\left\{ Y(k^*)^2 \left(\frac{Y(k^*) - Y(k)}{2} \right)^2 \right\}. \tag{46}$$

Remark 6. The authors of [10] defined the confusion coefficient as $\kappa(k^*, k) = \mathbb{P}\{Y(k^*) \neq Y(k)\}$. A straightforward computation gives

$$\mathbb{P}\{Y(k^*) \neq Y(k)\} = \mathbb{P}\{Y(k^*) = -1, Y(k) = 1\} + \mathbb{P}\{Y(k^*) = -1, Y(k) = 1\}$$

$$= \mathbb{E}\left\{ \left(\frac{Y(k^*) - Y(k)}{2} \right)^2 \right\}. \tag{47}$$

Thus our definition is consistent and a natural extension of the work in [10].

The alternative confusion coefficient introduced in [31] is defined as $\kappa^\circ(k^*, k) = \mathbb{E}\{Y(k^*)Y(k)\}$. The following relationship is easily obtained:

$$\kappa^\circ(k^*, k) = 1 - 2\kappa(k^*, k). \tag{48}$$

Proposition 3 (SE for CPA). *The success exponent for CPA takes the closed-form expression*

$$\text{SE} = \min_{k \neq k^*} \frac{\alpha^2 \kappa^2(k^*, k)}{2(\alpha^2(\kappa'(k^*, k) - \kappa^2(k^*, k)) + \sigma^2 \kappa(k^*, k))}. \tag{49}$$

Proof. Proposition 3 is an immediate consequence of the formula in Eq. (44) and the following lemma.   □

Lemma 7. *The first two moments of $\widehat{\Delta\mathcal{D}}(k^*, k)$ are given by*

$$\mathbb{E}\{\widehat{\Delta\mathcal{D}}(k^*, k)\} = 2\alpha\kappa(k^*, k), \tag{50}$$

$$\mathrm{Var}(\widehat{\Delta\mathcal{D}}(k^*, k)) = 4[\alpha^2(\kappa'(k^*, k) - \kappa^2(k^*, k)) + \sigma^2\kappa(k^*, k)]. \tag{51}$$

Proof. Recall from Remark 2 that $\widehat{\Delta\mathcal{D}}(k^*, k) = XY^* - XY = (\alpha Y^* + N)(Y^* - Y)$. On one hand, since we assumed that $\mathbb{E}\{(Y^*)^2\} = 1$, we obtain

$$\mathbb{E}\{Y^*(Y^* - Y)\} = 1 - \mathbb{E}\{Y^*Y\} = 2\mathbb{E}\left\{\left(\frac{Y^* - Y}{2}\right)^2\right\} = 2\kappa(k^*, k). \tag{52}$$

On the other hand, since N is independent of Y,

$$\mathbb{E}\{N(Y^* - Y)\} = \mathbb{E}\{N\} \cdot \mathbb{E}\{Y^* - Y\} = 0. \tag{53}$$

Combining we obtain $\mathbb{E}\{\widehat{\Delta\mathcal{D}}(k^*, k)\} = 2\alpha\kappa(k^*, k)$. For the variance we have

$$\mathbb{E}\{\widehat{\Delta\mathcal{D}}(k^*, k)^2\} = \mathbb{E}\{(XY^* - XY)^2\} \tag{54}$$

$$= \mathbb{E}\{N^2(Y^* - Y)^2\} + \alpha^2\mathbb{E}\{Y^{*2}(Y^* - Y)^2\} \tag{55}$$

$$= 4\sigma^2\kappa(k^*, k) + \alpha^2 4\kappa'(k^*, k), \tag{56}$$

since all cross terms with N vanish. It follows that

$$\mathrm{Var}(\widehat{\Delta\mathcal{D}}(k^*, k)) = \mathbb{E}\{\widehat{\Delta\mathcal{D}}(k^*, k)^2\} - \mathbb{E}\{\widehat{\Delta\mathcal{D}}(k^*, k)\}^2 \tag{57}$$

$$= 4[\alpha^2(\kappa'(k^*, k) - \kappa^2(k^*, k)) + \sigma^2\kappa(k^*, k)]. \tag{58}$$

as announced.   □

For DoM with one-bit variables $Y(k) \in \{\pm 1\}$ we can further simplify the success exponent such that it can be expressed directly through the SNR $= \alpha^2/\sigma^2$, number of measurements and 2-way confusion coefficient $\kappa(k^*, k)$:

Proposition 4 (SE for 1-bit DoM). *The success exponent for* DoM *takes the closed-form expression*

$$\mathrm{SE} = \frac{1}{\max\limits_{k \neq k^*}\left(\dfrac{2 - 2\kappa(k^*, k)}{\kappa(k^*, k)} + \dfrac{2}{\kappa(k^*, k)\,\mathrm{SNR}}\right)} \tag{59}$$

Proof. When $Y(k) \in \{\pm 1\}$ on has the additional simplification:

$$\kappa(k^*, k) = \mathbb{E}\left\{\left(\frac{Y(k^*) - Y(k)}{2}\right)^2\right\} = \mathbb{E}\left\{Y(k^*)^2\left(\frac{Y(k^*) - Y(k)}{2}\right)^2\right\} = \kappa'(k^*, k). \tag{60}$$

Now Proposition 4 follows directly from Proposition 3.   □

Remark 7. Estimating the success rate directly from confusion coefficients includes a computation of a multivariate normal cumulative distribution function [28] for which we have found that no closed-form expression exists. Moreover, the corresponding covariance matrices $[\kappa(k^*, i, j)]_{i,j}$ and $[\kappa(k^*, i) \times \kappa(k^*, j)]_{i,j}$ that depend on the confusion coefficients are not of full rank. This effect was similarly discovered for CPA by Rivain in [29], where the author propose to use Monte-Carlo simulation to overcome this problem.

Therefore, it is difficult to rederive the expressions above for the success exponent from the exact expressions of SR in [10, 29]. However, one clearly obtains the same exponential convergence behavior of SR toward 100 %.

As a result, we stress that the closed-form expressions of SE above are more convenient than the exact expressions for the SR for DoM and CPA, since in the SE, only 2-way confusion coefficients $\kappa(k^*, k), \kappa'(k^*, k)$ are involved without the need to compute multivariate distributions.

5.2   Success Exponent for the Optimal Distinguisher

Definition 9 (Optimal Distinguisher [15]). In case α is known and the noise is Gaussian the *optimal distinguisher* is additive and given by

$$\mathcal{D}(k) = -(X - \alpha Y)^2 \tag{61}$$

$$\widehat{\mathcal{D}}(X, Y(k)) = -(X - \alpha Y(k))^2. \tag{62}$$

Interestingly, as we show in the following proposition the optimal distinguisher involves the following confusion coefficient.

Definition 10 (Confusion Coefficient for the Optimal Distinguisher). For $k \neq k^*$ we define

$$\kappa''(k^*, k) = \mathbb{E}\left\{ \left(\frac{Y(k^*) - Y(k)}{2} \right)^4 \right\}. \tag{63}$$

Proposition 5 (SE for the Optimal Distinguisher). *The success exponent for the optimal distinguisher takes the closed-form expression*

$$\mathrm{SE} = \min_{k \neq k^*} \frac{\alpha^2 \kappa^2(k^*, k)}{2(\sigma^2 \kappa(k^*, k) + \alpha^2(\kappa''(k^*, k) - \kappa(k^*, k)))}. \tag{64}$$

Proof. Proposition 5 is an immediate consequence of the formula in Eq. (44) and the following lemma.   □

Lemma 8. *The first two moments of $\widehat{\Delta D}(k^*, k)$ are given by*

$$\mathbb{E}\{\widehat{\Delta D}(k^*, k)\} = 4\alpha^2 \kappa(k^*, k), \tag{65}$$

$$\mathrm{Var}(\widehat{\Delta D}(k^*, k)) = 16\alpha^2(\sigma^2 \kappa(k^*, k) + \alpha^2(\kappa(k^*, k)'' - \kappa(k^*, k))). \tag{66}$$

Proof. Recall that $\mathbb{E}\{N\}=0$. Straightforward calculation yields

$$\mathbb{E}\{\widehat{\Delta\mathcal{D}}(k^*,k)\} = \mathbb{E}\{-(X-\alpha Y^*)^2 + (X-\alpha Y)^2\} \tag{67}$$
$$= \mathbb{E}\{2N\alpha(Y^*-Y)\} + \mathbb{E}\{\alpha^2(Y^*-Y)^2\} \tag{68}$$
$$= 4\alpha^2\kappa(k^*,k). \tag{69}$$

Next we have

$$\mathbb{E}\{\widehat{\Delta\mathcal{D}}(k^*,k)^2\} = \mathbb{E}\{(2N\alpha(Y^*-Y) + \alpha^2(Y^*-Y)^2)^2\} \tag{70}$$
$$= \mathbb{E}\{4N^2\alpha^2(Y^*-2)^2\} + \mathbb{E}\{(Y^*-Y)^4\alpha^4\} \tag{71}$$
$$= 16\alpha^2\sigma^2\kappa(k^*,k) + 16\alpha^4\kappa''(k^*,k) \tag{72}$$

which yields the announced formula for the variance. $\qquad\square$

Corollary 2. *The closed-form expressions for* DoM, CPA *and for the optimal distinguisher simplify for high noise* $\sigma \gg \alpha$ *in a single equation:*

$$\mathrm{SE} \approx \min_{k\neq k^*} \frac{\alpha^2\kappa^2(k^*,k)}{2\sigma^2\kappa(k^*,k)} = \frac{1}{2}\cdot\mathrm{SNR}\cdot\min_{k\neq k^*}\kappa(k^*,k). \tag{73}$$

Proof. Trivial and left to the reader. $\qquad\square$

Remark 8. Corollary 2 is inline with the findings in [15], that CPA and the optimal distinguisher become closer the lower the SNR. However, note that, in [15] CPA is the correlation of the absolute value.

Remark 9. From Corollary 2 and the relationship $1 - \mathrm{SR} \approx e^{-m\cdot\mathrm{SE}}$ one can directly determine that if, e.g., the SNR is decreased by a factor of 2 the number of measurements m have to multiplied by 2 in order to achieve the same success. This verifies a well-known "rule of thumb" for side-channel attacks (see e.g., [20]).

5.3   Success Exponent for MIA

Unlike CPA or DoM, the estimation of the mutual information in MIA:

$$\mathcal{D}(k) = I(X,Y) = H(X) - H(X|Y) \tag{74}$$
$$= -\int p(x)\log p(x)\,\mathrm{d}x + \sum_y p(y)\int p(x|y)\log p(x|y)\,\mathrm{d}x \tag{75}$$

is a nontrivial problem. While Y is discrete, the computation of mutual information requires the estimation of the conditional pdfs $p(x|y)$. For a detailed evaluation of estimation methods for MIA we refer to [32].

In the following, we consider the estimation with histograms (H-MIA) in order to simplify the derivation of a closed-form expression for SE. One partitions the leakage X into h distinct bins b_i of width Δx with $i = 1,\dots,h$.

Definition 11. *Let $\hat{p}(x) = \frac{\#b_i}{m}$ where $\#b_i$ is the number of leakage values falling into bin b_i and let $\hat{p}(x|y)$ be the estimated probability knowing $Y = y$. Then*

$$\widehat{\mathcal{D}}(k) = -\sum_x \hat{p}(x) \log \hat{p}(x) + \sum_y \hat{p}(y) \sum_x \hat{p}(x|y) \log \hat{p}(x|y). \qquad (76)$$

To simplify the presentation that follows, we consider only the conditional negentropy $-\hat{H}(X|Y)$ as a distinguisher, since $\hat{H}(X)$ does not depend on the key hypothesis k. Additionally, we assume that the distribution of Y is known to the attacker so that she can use $p(y)$ instead of $\hat{p}(y)$. Now H-MIA simplifies to

$$\text{H-MIA}(X, Y) = \sum_y p(y) \sum_x \hat{p}(x|y) \log \hat{p}(x|y) + \log \Delta x. \qquad (77)$$

The additional term $\log \Delta x$ arises due to the fact that we have estimated the differential entropy $H(X)$. For more information on differential entropy and mutual information we refer to [7].

Proposition 6 (SE for H-MIA).

$$\text{SE} \approx \min_{k^* \neq k} \frac{\frac{1}{2}\left(\Delta\mathcal{D}(k^*, k) + \frac{\Delta x^2}{24}(\Delta J(k^*, k))\right)^2}{\sum_y p(y)\text{Var}\{-\log p(X|Y = y)\} + \sum_{y^*} p(y^*)\text{Var}\{-\log p(X|Y = y^*)\}}, \qquad (78)$$

where $\Delta\mathcal{D}(k^*, k) = H(X|Y) - H(X|Y^*)$, $\Delta J(k^*, k) = J(X|Y) - J(X|Y^*)$, $J(X|Y) = \sum_y p(y) J(X|Y = y)$ and $J(X|Y)$ is the Fisher information [11]:

$$J(X|Y = y) = \int_{-\infty}^{\infty} \frac{[\frac{d}{dx}p(x|y)]^2}{p(x|y)} \, dx. \qquad (79)$$

Proof. Since Y is discrete the bias only arise due to the discretization of X and the limited number of measurements m. Therefore, we use the approximations given for the bias of $\hat{H}(X)$ in [22] (3.14) to calculate $\mathbb{E}\{\widehat{\mathcal{D}}(k)\}$ and $\mathbb{E}\{\widehat{\Delta\mathcal{D}}(k^*, k)\}$ for H-MIA. To be specific, let h define the number of bins and Δx their width. Then

$$\mathbb{E}\{\widehat{\mathcal{D}}(k)\} = -\mathbb{E}\{\hat{H}(X|Y)\} = -\sum_y p(y)\mathbb{E}\{\hat{H}(X|Y = y)\}, \qquad (80)$$

$$\approx -\sum_y p(y)\left[H(X|Y = y) + \frac{\Delta x^2}{24}J(X|Y = y)\right] - \frac{h - 1}{2m}, \qquad (81)$$

$$\mathbb{E}\{\widehat{\Delta\mathcal{D}}(k^*, k)\} \approx \sum_y p(y)\left[H(X|Y = y) + \frac{\Delta x^2}{24}J(X|Y = y)\right]$$

$$- \left(\sum_{y^*} p(y^*)\left[H(X|Y^* = y^*) + \frac{\Delta x^2}{24}J(X|Y^* = y^*)\right]\right), \qquad (82)$$

with $J(X|Y) = \sum_y p(y) J(X|Y = y)$ and $J(X|Y = y)$ being the Fisher informa-
tion $\int_{-\infty}^{\infty} \frac{[\frac{d}{dx} p(x|y)]^2}{p(x|y)} dx$ [11].

To calculate $\mathrm{Var}\{\widehat{\mathcal{D}}(k)\}$ we use the law of total variance [16] and the approximations for the variance given in [22] (4.9):

$$\mathrm{Var}\{\widehat{\mathcal{D}}(k)\} = \mathrm{Var}\{\hat{H}(X|Y)\}\} = \mathrm{Var}\{\mathbb{E}\{\hat{H}(X|Y = y)\}\} \tag{83}$$

$$\approx \mathrm{Var}\{H(X)\} - \frac{1}{m} \sum_y p(y) \mathrm{Var}\{-\log p(x|y)\} \tag{84}$$

$$\mathrm{Var}\{\widehat{\Delta\mathcal{D}}(k^*, k)\} = \mathrm{Var}\{\mathbb{E}\{\hat{H}(X|Y = y)\}\} - \mathrm{Var}\{\mathbb{E}\{\hat{H}(X|Y^* = y^*)\}\} \tag{85}$$
$$- 2\mathrm{Cov}(\mathbb{E}\{\hat{H}(X|Y = y)\}, \mathbb{E}\{\hat{H}(X|Y^* = y^*)\})$$

$$\approx \frac{1}{m} \left(\sum_y p(y) \mathrm{Var}\{-\log p(x|y)\} + \sum_y p(y^*) \mathrm{Var}\{-\log p(x|y^*)\} \right)$$
$$\tag{86}$$

From Eqs. (82) and (86) Proposition 6 follows directly. □

Remark 10. Interestingly, even if MIA is not additive the SE is linear in the
number of measurements m just like for DoM and CPA. This is also confirmed
experimentally in the next subsection.

Remark 11. If N is normal distributed with variance σ^2 we can further simplify
$H(X|Y^* = y^*) = \frac{1}{2} \log(2\pi e \sigma^2)$ since $p(x|y^*) = p_N(x - y^*)$. Moreover, one has
$J(X|Y = y) = \frac{1}{\sigma^2}$ and $\mathrm{Var}\{-\log p(x|y^*)\} = \frac{1}{2m}$.

Remark 12. Remarkably, the variance term does not depend on the size of Δx
except in extreme cases like $\Delta x = 1$ and $\Delta x \to \infty$ – see [22] for more information.

5.4 Validation of the SE

To illustrate the validity of the success exponent and the derived closed-form
expressions, we choose the same scenario as in Sect. 4 (targeting the Sbox of
PRESENT) with a higher variance of the noise. We increased the bin width Δx
to 4 for MIA, which lead to the best success when comparing with other widths.
To be reliable we conducted 500 independent experiments in each setting.

With the appropriate parameters (confusion coefficients, SNR, etc.), we have
computed the exact values for the closed-form expressions in Eqs. (49), (59), (64),
and (78) for CPA, DoM, the optimal distinguisher, and MIA which are listed in
Table 1 with SE for several σ's. Additionally, we computed for CPA, DoM, and
the optimal distinguisher the SE in case of low noise from Eq. (73). To show
that these values are valid and reasonable, we estimated the success exponent
$\widehat{\mathrm{SE}}$ from the general theoretical formula in Eq. (44) using simulations. One can
observe that Corollary 2 is valid.

Moreover, we estimated the success exponent directly from the obtained suc-
cess rate as $-\log(1 - \mathrm{SR}(\widehat{\mathcal{D}}))/m$; this was done for limited values of m to avoid

Table 1. Experimental validation of SE for several σ (values $\times 10^{-3}$)

$\times 10^{-3}$	$\sigma = 5$				$\sigma = 7$				$\sigma = 10$			
	DPA	CPA	OPT	MIA	DPA	CPA	OPT	MIA	DPA	CPA	OPT	MIA
SE	0.2	4.5	4.8	1.4	0.1	2.3	2.4	0.8	0.01	1.2	1.2	0.4
SE (Eq. (73))	0.2	4.7	4.7	—	0.1	2.4	2.4	—	0.01	1.2	1.2	—
$\widehat{SE}$	0.3	4.7	4.6	1.4	0.1	2.3	2.3	0.8	0.1	1.1	1.2	0.2

the saturation effect of the $\mathrm{SR}(\widehat{\mathcal{D}}) = 1$. Figure 2b displays the theoretical value of SE along with the estimations as a function of the number of measurements for $\sigma = 5$. For comparison we plot the success rate in Fig. 2a.

Remarkably, one can see that for all distinguishers, the two estimated values are getting closer to the theoretical SE as m increases. This confirms our theoretical study in Sect. 3 and also demonstrates that the first-order exponent of MIA is indeed linear in the number of measurements as expected.

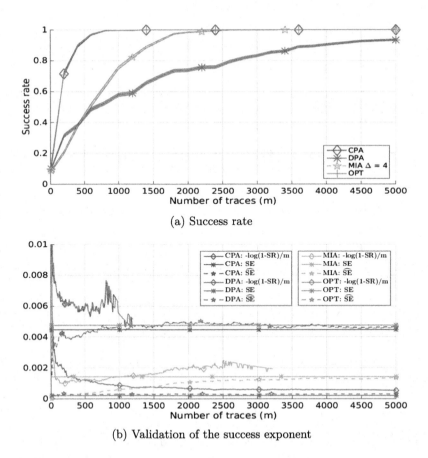

(a) Success rate

(b) Validation of the success exponent

Fig. 2. Success rate [top graph] and success exponent (SE) [bottom graph]

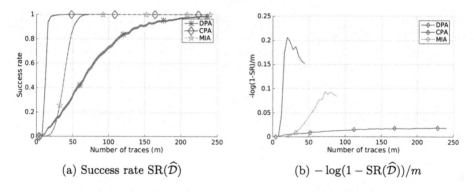

(a) Success rate SR($\widehat{\mathcal{D}}$)   (b) $-\log(1 - \mathrm{SR}(\widehat{\mathcal{D}}))/m$

Fig. 3. Empirical results using real traces (Arduino board)

Furthermore, for practical measurements we used an Arduino pro mini board with an AVR 328p micro-controller running at 16 MHz. We captured the operation of the AES Substitution box during the first round at 2 GSa/s using an EM probe. Figure 3a shows the success rate for DoM, CPA and MIA for 1600 independent retries. We plot $-\log(1 - \mathrm{SR}(\widehat{\mathcal{D}}))/m$ in Fig. 3b. One can observe that DoM converges to a constant. For CPA and MIA the saturation effect of $\mathrm{SR}(\widehat{\mathcal{D}}) = 1$ is disguising the convergence.

These results raise a lot of new perspectives which we discuss next.

6   Conclusion and Perspectives for Further Applications

In this work we investigated in the first-order exponent (success exponent SE) of the success rate for arbitrary sound distinguishers under a mild normal assumption as m increases. The resulting expressions were derived under the asymptotic condition that the number of measurements m tends to infinity, but already hold accurately for reasonable low values of m. More precisely, in the investigated scenarios the approximations for CPA hold for $m \geq 2$ whereas for MIA we have $m \geq 40$. As an illustration we derived the closed-form expressions of the SE for DoM, CPA, the optimal distinguisher, and MIA and showed that they agree theoretically and empirically.

This novel first-order exponent raises many new perspectives. In particular, the resulting closed-form expressions for the SE allows one to answer questions such as: *"How many more traces?"* for achieving a given goal. For example, suppose that one has obtained SE = 90 % after m measurements. To obtain 99 % success with the same distinguisher (hence the same SE), one should approximately square $(1 - \mathrm{SR})^2 = (0.1)^2 = 0.01$ which amounts to doubling m. Thus as a rule of thumb we may say that *"doubling the number of traces allows one to go from 90 % to 99 % chance of success"*.

Finally, we underline that the success exponent would constitute another approach to the question of comparing substitution boxes with respect to their exploitability in side-channel analysis. It can nicely complement methods like transparency order [25] (and variants thereof [5,24]). It can also characterize, in the same framework, various countermeasures such as no masking vs. masking.

The generality of the proposed approach to derive the success exponent allows one to investigate attack performance in many different scenarios, and we feel that for this reason it is a promising tool.

Acknowledgements. The authors are grateful to Darshana Jayasinghe for the real-world validation on traces taken from the Arduino board.

References

1. Batina, L., Robshaw, M. (eds.): CHES 2014. LNCS, vol. 8731. Springer, Heidelberg (2014)
2. Belgarric, P., Bhasin, S., Bruneau, N., Danger, J.-L., Debande, N., Guilley, S., Heuser, A., Najm, Z., Rioul, O.: Time-frequency analysis for second-order attacks. In: Francillon, A., Rohatgi, P. (eds.) CARDIS 2013. LNCS, vol. 8419, pp. 108–122. Springer, Heidelberg (2014)
3. Bogdanov, A.A., Knudsen, L.R., Leander, G., Paar, C., Poschmann, A., Robshaw, M., Seurin, Y., Vikkelsoe, C.: PRESENT: an ultra-lightweight block cipher. In: Paillier, P., Verbauwhede, I. (eds.) CHES 2007. LNCS, vol. 4727, pp. 450–466. Springer, Heidelberg (2007)
4. Brier, E., Clavier, C., Olivier, F.: Correlation power analysis with a leakage model. In: Joye, M., Quisquater, J.-J. (eds.) CHES 2004. LNCS, vol. 3156, pp. 16–29. Springer, Heidelberg (2004)
5. Chakraborty, K., Sarkar, S., Maitra, S., Mazumdar, B., Mukhopadhyay, D., Prouff, E.: Redefining the transparency order. In: The Ninth International Workshop on Coding and Cryptography, WCC 2015, Paris, France, April 13–17, 2015
6. Chari, S., Rao, J.R.: Rohatgi template attacks. In: Kaliski Jr., B.S., Koç, Ç.K., Paar, C. (eds.) CHES 2002. LNCS, vol. 2523, pp. 13–28. Springer, Heidelberg (2003)
7. Cover, T.M., Thomas, J.A.: Elements of Information Theory, 2nd edn. Wiley-Interscience, July 18, 2006. ISBN-10: ISBN-10: 0471241954, ISBN-13: 978–0471241959
8. Doget, J., Prouff, E., Rivain, M., Standaert, F.-X.: Univariate side channel attacks and leakage modeling. J. Cryptographic Eng. $1(2)$, 123–144 (2011)
9. Duc, A., Faust, S., Standaert, F.-X.: Making masking security proofs concrete. In: Oswald, E., Fischlin, M. (eds.) EUROCRYPT 2015. LNCS, vol. 9056, pp. 401–429. Springer, Heidelberg (2015)
10. Fei, Y., Luo, Q., Ding, A.A.: A statistical model for DPA with novel algorithmic confusion analysis. In: Prouff, E., Schaumont, P. (eds.) CHES 2012. LNCS, vol. 7428, pp. 233–250. Springer, Heidelberg (2012)
11. Ronald, A.: Statistical Methods for Research Workers. Oliver and Boyd, Edinburgh (1925)
12. Gierlichs, B., Batina, L., Tuyls, P., Preneel, B.: Mutual information analysis. In: Oswald, E., Rohatgi, P. (eds.) CHES 2008. LNCS, vol. 5154, pp. 426–442. Springer, Heidelberg (2008)

13. Guilley, S., Hoogvorst, P., Pacalet, R., Schmidt, J.: Improving side-channel attacks by exploiting substitution boxes properties. In: Presse Universitaire de Rouen et du Havre (ed.), BFCA, Paris, France, pp. 1–25, May 02–04, 2007. http://www.liafa.jussieu.fr/bfca/books/BFCA07.pdf

14. Heuser, A., Kasper, M., Schindler, W., Stottinger, M.: How a symmetry metric assists side-channel evaluation - a novel model verification method for power analysis. In: Proceedings of the 2011 14th Euromicro Conference on Digital System Design, DSD 2011, pp. 674–681. IEEE Computer Society, Washington DC (2011)

15. Heuser, A., Rioul, O., Guilley, S.: Good Is not good enough - deriving optimal distinguishers from communication theory. In: Batina and Robshaw [1], pp. 55–74

16. Kardaun, O.J.W.F.: Classical Methods of Statistics. Springer, Heidelberg (2005)

17. Kocher, P.C., Jaffe, J., Jun, B.: Differential power analysis. In: Wiener, M. (ed.) CRYPTO 1999. LNCS, vol. 1666, pp. 388–397. Springer, Heidelberg (1999)

18. Lomné, V., Prouff, E., Rivain, M., Roche, T., Thillard, A.: How to estimate the success rate of higher-order side-channel attacks. In: Batina and Robshaw [1], pp. 35–54

19. Maghrebi, H., Rioul, O., Guilley, S., Danger, J.-L.: Comparison between side-channel analysis distinguishers. In: Chim, T.W., Yuen, T.H. (eds.) ICICS 2012. LNCS, vol. 7618, pp. 331–340. Springer, Heidelberg (2012)

20. Mangard, S.: Hardware countermeasures against DPA – a statistical analysis of their effectiveness. In: Okamoto, T. (ed.) CT-RSA 2004. LNCS, vol. 2964, pp. 222–235. Springer, Heidelberg (2004)

21. Mangard, S., Oswald, E., Standaert, F.-X.: One for All - All for One: Unifying Standard DPA Attacks. IET Inf. Secur. 5(2), 100–111 (2011). doi:10.1049/iet-ifs.2010.0096. ISSN: 1751–8709

22. Moddemeijer, R.: On estimation of entropy and mutual information of continuous distributions. Signal Process. 16(3), 233–248 (1989)

23. Moradi, A., Mousavi, N., Paar, C., Salmasizadeh, M.: A comparative study of mutual information analysis under a gaussian assumption. In: Youm, H.Y., Yung, M. (eds.) WISA 2009. LNCS, vol. 5932, pp. 193–205. Springer, Heidelberg (2009)

24. Picek, S., Mazumdar, B., Mukhopadhyay, D., Batina, L.: Modified transparency order property: solution or just another attempt. In: Chakraborty, R.S., Schwabe, P., Solworth, J. (eds.) Security, Privacy, and Applied Cryptography Engineering. LNCS, vol. 9354, pp. 210–227. Springer, Heidelberg (2015)

25. Prouff, E.: DPA attacks and S-boxes. In: Gilbert, H., Handschuh, H. (eds.) FSE 2005. LNCS, vol. 3557, pp. 424–441. Springer, Heidelberg (2005)

26. Prouff, E., Rivain, M.: Theoretical and practical aspects of mutual information based side channel analysis. In: Abdalla, M., Pointcheval, D., Fouque, P.-A., Vergnaud, D. (eds.) ACNS 2009. LNCS, vol. 5536, pp. 499–518. Springer, Heidelberg (2009)

27. Prouff, E., Rivain, M., Bevan, R.: Statistical analysis of second order differential power analysis. IEEE Trans. Comput. 58(6), 799–811 (2009)

28. Rao, C.R.: Linear Statistical Inference and its Applications, 2nd edn. J. Wiley and Sons, New York (1973)

29. Rivain, M.: On the exact success rate of side channel analysis in the gaussian model. In: Avanzi, R.M., Keliher, L., Sica, F. (eds.) SAC 2008. LNCS, vol. 5381, pp. 165–183. Springer, Heidelberg (2009)

30. Standaert, F.-X., Malkin, T.G., Yung, M.: A unified framework for the analysis of side-channel key recovery attacks. In: Joux, A. (ed.) EUROCRYPT 2009. LNCS, vol. 5479, pp. 443–461. Springer, Heidelberg (2009)

31. Thillard, A., Prouff, E., Roche, T.: Success through confidence: evaluating the effectiveness of a side-channel attack. In: Bertoni, G., Coron, J.-S. (eds.) CHES 2013. LNCS, vol. 8086, pp. 21–36. Springer, Heidelberg (2013)
32. Veyrat-Charvillon, N., Standaert, F.-X.: Mutual information analysis: how, when and why? In: Clavier, C., Gaj, K. (eds.) CHES 2009. LNCS, vol. 5747, pp. 429–443. Springer, Heidelberg (2009)
33. Whitnall, C., Oswald, E.: A fair evaluation framework for comparing side-channel distinguishers. J. Cryptographic Eng. 1(2), 145–160 (2011)

Information Theoretic Cryptography

Non-malleable Extractors with Shorter Seeds and Their Applications

Yanqing Yao[1,2](✉) and Zhoujun Li[1,2]

[1] School of Computer Science and Engineering, Beihang University, Beijing, China
yaoyanqing1984@buaa.edu.cn
[2] Beijing Key Laboratory of Network Technology, Beihang University, Beijing, China

Abstract. Motivated by the problem of how to communicate over a public channel with an active adversary, Dodis and Wichs (STOC'09) introduced the notion of a non-malleable extractor. A non-malleable extractor $\mathsf{nmExt} : \{0,1\}^n \times \{0,1\}^d \to \{0,1\}^m$ takes two inputs, a weakly-random W and a uniformly random seed S, and outputs a string which is nearly uniform, given S as well as $\mathsf{nmExt}(W, \mathcal{A}(S))$, for an arbitrary function $\mathcal{A}$ with $\mathcal{A}(S) \neq S$.

In this paper, by developing the combination and permutation techniques, we improve the error estimation of the extractor of Raz (STOC'05), which plays an extremely important role in the constraints of the non-malleable extractor parameters including seed length. Then we present improved explicit construction of non-malleable extractors. Though our construction is the same as that given by Cohen, Raz and Segev (CCC'12), the parameters are improved. More precisely, we construct an explicit $(1016, \frac{1}{2})$-non-malleable extractor $\mathsf{nmExt} : \{0,1\}^n \times \{0,1\}^d \to \{0,1\}$ with $n = 2^{10}$ and seed length $d = 19$, while Cohen et al. showed that the seed length is no less than $\frac{46}{63} + 66$. Therefore, our method beats the condition "$2.01 \cdot \log n \leq d \leq n$" proposed by Cohen et al., since d is just $1.9 \cdot \log n$ in our construction. We also improve the parameters of the general explicit construction given by Cohen et al. Finally, we give their applications to privacy amplification.

Keywords: Extractors · Non-malleable extractors · Seed length · Privacy amplification protocol

1 Introduction

Randomness extractors are functions that convert weakly random sources into nearly uniform bits. Though the motivation of extractors is to simulate randomized algorithms with weak random sources as might arise in nature, randomness extractors have been successfully applied to coding theory, cryptography, complexity, etc. [12,14,22]. In this paper, we focus on the extractors that can be applied to privacy amplification. In this scenario, two parties Alice and Bob share

Y. Yao—Most of this work was done while the author visited New York University.

© Springer International Publishing Switzerland 2015
A. Biryukov and V. Goyal (Eds.): INDOCRYPT 2015, LNCS 9462, pp. 293–311, 2015.
DOI: 10.1007/978-3-319-26617-6_16

a weakly random secret $W \in \{0,1\}^n$. W may be a human-memorizable password, some biometric data, and physical sources, which are themselves weakly random, or a uniform secret which may have been partially leaked to an adversary Eve. Thus, only the min-entropy of W is guaranteed. Alice and Bob interact over a public communication channel in order to securely agree on a nearly uniform secret key $R \in \{0,1\}^m$ in the presence of the adversary, Eve, who can see every message transmitted in the public channel. The public seed length and min-entropy of W are two main measures of efficiency in this setting. If Eve is passive, a (strong) randomness extractor yields the following solution: Alice sends a uniformly random seed S to Bob, then they both compute $R = \mathsf{Ext}(W, S)$ as the nearly uniform secret key [18]. If Eve is active (i.e., it may change the messages in arbitrary ways), some protocols have been proposed to achieve this goal [4,6–9,13–15,21,23].

As a major progress, Dodis and Wichs [9] introduced non-malleable extractors to study privacy amplification protocols, where the attacker is active and computationally unbounded. If an attacker sees a random seed S and modifies it into an arbitrarily related seed S', then the relationship between $R = \mathsf{Ext}(W, S)$ and $R' = \mathsf{Ext}(W, S')$ is bounded to avoid related key attacks. More formally, a non-malleable extractor is a function $\mathsf{nmExt} : \{0,1\}^n \times \{0,1\}^d \to \{0,1\}^m$ that takes two inputs, a weakly-random secret source[1] W with min-entropy α and uniformly random seed S, and outputs a string which is γ-close to uniform (see Definition 1), given S as well as $\mathsf{nmExt}(W, \mathcal{A}(S))$, for an arbitrary function $\mathcal{A}$ with $\mathcal{A}(S) \neq S$. They proved that $(\alpha, 2\gamma)$-non-malleable extractors exist as long as $\alpha > 2m + 3\log\frac{1}{\gamma} + \log d + 9$ and $d > \log(n - \alpha + 1) + 2\log\frac{1}{\gamma} + 7$. The first explicit non-malleable extractor was constructed by Dodis, Li, Wooley and Zuckerman [8]. It works for any weakly random input source with the min-entropy $\alpha > \frac{n}{2}$ and uniformly random seed of length $d = n$ (It works even if the seed has entropy only $\Theta(m + \log n)$). However, when outputting more than a logarithmic number of bits, its efficiency relies on a longstanding conjecture on the distribution of prime numbers.

Li [14] proposed that $(\alpha, 2\gamma)$-non-malleable extractor $\mathsf{nmExt} : \{0,1\}^n \times \{0,1\}^d \to \{0,1\}$, where $\alpha = (\frac{1}{2}-\delta)\cdot n$ and $d = O(\log n+\log(1/\gamma))$ for any constant $\delta > 0$, can be constructed as follows: the seed S is encoded using the parity check matrix of a BCH code, and then the output is the inner product function of the encoded source and the encoded seed over $\mathbb{F}_2$. Dodis and Yu [11] observed that for 4-wise independent hash function family $\{h_w : \{0,1\}^d \to \{0,1\}^m \mid w \in \{0,1\}^n\}$, $\mathsf{nmExt}(w, s) = h_w(s)$ is a $(\alpha, 2\sqrt{2^{n-\alpha-d}})$-non-malleable extractor. In 2012, an alternative explicit construction based on the extractor of Raz [20] was given by Cohen et al. [6]. Without using any conjecture, their construction works for any weakly random source with the min-entropy $\alpha = (\frac{1}{2} + \delta) \cdot n$ and uniformly random seed of length $d \geq \frac{23}{\delta} \cdot m + 2\log n$ (see Theorem 1 for details). However, their result suffers from some drawbacks: The non-malleable extractor is constructed based on the explicit seeded extractor of Raz [20], while the error[2] estimation in

---

[1] When we say a source in this paper, we mean a random variable.

[2] The concept of the error of seeded extractor can be seen in Definition 1.

that construction is too rough. Furthermore, though one main purpose of [6] is to shorten the length of the seed, the lower bound on the seed length is still not optimal.

Our Contributions and Techniques

- By developing the combination and permutation techniques, we improve the error estimation of Raz's extractor in STOC'05 [20], a special case of which was used by Cohen et al. in CCC'12 [6]. For simplicity, denote γ_1 as the error of the extractor in [6], and γ_2 as the counterpart in this paper. Recall that $\gamma_1 = 2^{\frac{(\frac{1}{2}-\delta)n}{k}} \cdot (2\epsilon)^{\frac{1}{k}}$ in [6] under the assumption that $\epsilon \geq 2^{-\frac{dk}{2}} \cdot k^k$ and $0 < \delta \leq \frac{1}{2}$ (see Lemma 1). If $\epsilon \geq \frac{1}{2^{(\frac{1}{2}-\delta)n+1}}$, then $\gamma_1 = 2^{\frac{(\frac{1}{2}-\delta)n}{k}} \cdot (2\epsilon)^{\frac{1}{k}} \geq 1$. In this case, the error estimation is meaningless. One main reason is that in those proofs, the partition method about the sum [6,20] which bounds the error didn't capture the essence of the biased sequence for linear tests (see Definition 2). In this paper, we propose another partition method and give a better bound on the sum by employing the combination and permutation formulas. In particular, the combination and permutation techniques (see Proposition 1) may be useful in future works. Correspondingly, the error is $\gamma_2 = 2^{\frac{(\frac{1}{2}-\delta)n}{k}} \cdot [2^{-\frac{dk}{2}} \cdot (k-1) \cdot (k-3) \cdots 1 \cdot (1-\epsilon) + \epsilon]^{\frac{1}{k}}$ (see Theorem 2). Since $\epsilon \geq 2^{-\frac{dk}{2}} \cdot k^k$ and $2^{-\frac{dk}{2}} \cdot k^k > 2^{-\frac{dk}{2}} \cdot (k-1) \cdot (k-3) \cdots 1$ for any even integer k, we get $\gamma_1 > \gamma_2$. To simplify this bound, let k be a specific value. For instance, let $k = 4$, then the error $\gamma_2 = 2^{\frac{(\frac{1}{2}-\delta)n}{4}} \cdot [2^{-2d} \cdot 3 \cdot (1-\epsilon) + \epsilon]^{\frac{1}{4}}$.
- Note that the error estimation of the Raz's extractor impacts greatly on the constraints of the parameters including the seed length, the weak source's min-entropy and the error[3] of the non-malleable extractor. Based on the above improvement of the error estimation, we present an explicit construction of non-malleable extractors, which is an improvement of the construction of Cohen et al. in CCC'12 [6] in the sense that the seed length is shorter. More concretely, we present an explicit $(1016, \frac{1}{2})$-non-malleable extractor nmExt : $\{0,1\}^n \times \{0,1\}^d \rightarrow \{0,1\}$ with $n = 1024$ and $d = 19$, which beats the condition "$2.01 \cdot \log n \leq d \leq n$" in [6], since seed length d is just $1.9 \cdot \log n$ in our construction while it is no less than $\frac{46}{63} + 66$ according to [6]. Moreover, we improve the parameters of the general explicit construction given by Cohen et al.
- We show how our non-malleable extractors are applied to privacy amplification.

Organization. The remainder of the paper is organized as follows. In Sect. 2, we review some notations, concepts, and results. In Sect. 2, we show an existing central lemma about the error estimation of Raz's Extractor and improve it by proposing a new partition method. In Sect. 4, we propose the explicit construction of non-malleable extractors with shorter seed length compared with that in [6]. In Sect. 5, we show how the non-malleable extractors are applied to privacy amplification. Section 6 concludes the paper.

---

[3] The concept of the error of non-malleable extractor can be seen in Definition 3.

2   Preliminaries

For any positive integer n, denote $[n] = \{1, 2, \ldots, n\}$. Denote U_m as the uniformly random distribution over $\{0,1\}^m$. We measure the distance between two distributions by the $\mathcal{L}_1$ norm in order to be consistent with [6]. The statistical distance of X and Y is defined as $\mathsf{SD}(X, Y) = \frac{1}{2}\|X - Y\|_1$. It's well known that for any function f, $\mathsf{SD}(f(X), f(Y)) \leq \mathsf{SD}(X, Y)$. Denote $\mathsf{SD}((X_1, X_2), (Y_1, Y_2) \mid Z)$ as the abbreviation of $\mathsf{SD}((X_1, X_2, Z), (Y_1, Y_2, Z))$.

The *min-entropy* of variable W is $H_\infty(W) = -\log\max_w Pr(W = w)$. W over $\{0,1\}^n$ is called an (n, α)it-source if $H_\infty(W) \geq \alpha$. We say that a source (i.e., a random variable) is a *weak source* if its distribution is not uniform. We say W is a *flat source* if it is a uniform distribution over some subset $S \subseteq \{0,1\}^n$. Chor and Goldreich [5] observed that the distribution of any (n, α)-source is a convex combination of distributions of flat (n, b)-sources. Therefore, for general weak sources, it will be enough to consider flat sources instead in most cases.

Definition 1. *We say that the distribution X is ϵ-close to the distribution Y if $\|X - Y\|_1 = \sum_s |\Pr[X = s] - \Pr[Y = s]| \leq \epsilon^4$. A function $\mathsf{Ext} : \{0,1\}^n \times \{0,1\}^d \rightarrow \{0,1\}^m$ is an (α, γ)-seeded extractor if for every (n, α)-source W and an independent uniformly random variable S (called seed) over $\{0,1\}^d$, the distribution of $\mathsf{Ext}(W, S)$ is γ-close to U_m. γ is called the error of the seeded extractor. A seeded extractor is a strong (α, γ)-extractor if for W and S as above, $(\mathsf{Ext}(W, S), S)$ is γ-close to (U_m, U_d).*

Definition 2. *A random variable Z over $\{0,1\}$ is ϵ-biased if $\mathrm{bias}(Z) = |\Pr[Z = 0] - \Pr[Z = 1]| \leq \epsilon$ (i.e., Z is ϵ-close to uniform). A sequence of 0-1 random variables $Z_1, Z_2, \ldots, Z_N$ is ϵ-biased for linear tests of size k if for any nonempty $\tau \subseteq [N]$ with $|\tau| \leq k$, the random variable $Z_\tau = \oplus_{i \in \tau} Z_i$ is ϵ-biased. We also say that the sequence $Z_1, Z_2, \ldots, Z_N$ ϵ-fools linear tests of size k.*

For every k', $N \geq 2$, variables $Z_1, \cdots, Z_N$ as above can be explicitly constructed using $2 \cdot \lceil \log(1/\epsilon) + \log k' + \log\log N \rceil$ random bits [1].

The Extractor of Raz. Raz [20] constructed an extractor based on a sequence of 0-1 random variables that have small bias for linear tests of a certain size. Let $Z_1, \cdots, Z_{m \cdot 2^d}$ be 0-1 random variables that are ϵ-biased for linear tests of size k' that are constructed using n random bits. The set of indices $[m \cdot 2^d]$ can be considered as the set $\{(i, s) : i \in [m], s \in \{0,1\}^d\}$. Define $\mathsf{Ext} : \{0,1\}^n \times \{0,1\}^d \rightarrow \{0,1\}^m$ by $\mathsf{Ext}(w, s) = Z_{(1,s)}(w)\|Z_{(2,s)}(w)\ldots\|Z_{(m,s)}(w)$, where "$\|$" is the concatenation operator. Raz proposed that Ext is a seeded extractor with good parameters [20].

Cohen et al. [6] proved that the above extractor is in fact non-malleable. We'll also construct non-malleable extractors based on it. The formal definition of non-malleable extractors is as follows.

---

[4] In other papers (e.g., [9,11,14,24]), X is ϵ-close to Y if $\frac{1}{2}\|X - Y\|_1 = \frac{1}{2}\sum_s |\Pr[X = s] - \Pr[Y = s]| \leq \epsilon$. To keep consistency, Definition 1 holds throughout this paper.

Definition 3. *(see [6]) We say that a function* $\mathcal{A} : \{0,1\}^d \rightarrow \{0,1\}^d$ *is an adversarial function, if for every* $s \in \{0,1\}^d$, $f(s) \neq s$ *holds. A function* $nmExt : \{0,1\}^n \times \{0,1\}^d \rightarrow \{0,1\}^m$ *is a* (α, γ)-*non-malleable extractor if for every* (n, α)-*source* W, *independent uniformly random variable* S, *and every adversarial function* $\mathcal{A}$,

$$\|(nmExt(W,S), nmExt(W, \mathcal{A}(S)), S) - (U_m, nmExt(W, \mathcal{A}(S)), S)\|_1 \leq \gamma.$$

γ *is called the error of the non-malleable extractor.*

One-time message authentication code (MAC) is used to guarantee that the received message is sent by a specified legitimate sender in an unauthenticated channel. Formally,

Definition 4. *A family of functions* $\{MAC_r : \{0,1\}^v \rightarrow \{0,1\}^\tau\}_{r \in \{0,1\}^m}$ *is a* ε-*secure (one-time) message authentication code (MAC) if for any* μ *and any function* $f : \{0,1\}^\tau \rightarrow \{0,1\}^v \times \{0,1\}^\tau$, *it holds that,*

$$\Pr_{r \leftarrow U_m} [MAC_r(\mu') = \sigma' \wedge \mu' \neq \mu \mid (\mu', \sigma') = f(MAC_r(\mu))] \leq \varepsilon.$$

Recall that the main theorem about the explicit construction of non-malleable extractors proposed in [6] is as follows.

Theorem 1. *(see [6]) For any integers* n, d, *and* m, *and for any* $0 < \delta \leq \frac{1}{2}$ *such that* $d \geq \frac{23}{\delta} \cdot m + 2 \log n$, $n \geq \frac{160}{\delta} \cdot m$, *and* $\delta \geq 10 \cdot \frac{\log(nd)}{n}$, *there exists an explicit* $((\frac{1}{2} + \delta) \cdot n, 2^{-m})$-*non-malleable extractor* $nmExt : \{0,1\}^n \times \{0,1\}^d \rightarrow \{0,1\}^m$.

3 Error Estimation of Raz's Extractor and Its Improvement

In this section, we first recall the central lemma used in [6], which is a special case about the error estimation of Raz's Extractor [20]. Then we point out the flaw in the proof and improve its error estimation. Afterwards, we compare our result with the original one and roughly show the role of the improvement.

3.1 A Special Case of Raz's Extractor

The central lemma used in [6] is below, the proof of which is essentially the same as that in [20]. It can be considered as a special case of Raz's Extractor [20].

Lemma 1. *Let* $D = 2^d$. *Let* $Z_1, \ldots, Z_D$ *be 0-1 random variables that are* ϵ-*biased for linear tests of size* k' *that are constructed using* n *random bits. Define* $Ext^{(1)} : \{0,1\}^n \times \{0,1\}^d \rightarrow \{0,1\}$ *by* $Ext^{(1)}(w,s) = Z_s(w)$, *that is,* $Ext^{(1)}(w,s)$ *is the random variable* Z_s, *when using* w *as the value of the* n *bits needed to produce* $Z_1, \ldots, Z_D$. *Then, for any* $0 < \delta \leq \frac{1}{2}$ *and even integer* $k \leq k'$ *s.t.* $k \cdot (\frac{1}{\epsilon})^{\frac{1}{k}} \leq D^{\frac{1}{2}}$, $Ext^{(1)}$ *is a* $((\frac{1}{2} + \delta) \cdot n, \gamma_1)$-*seeded-extractor, with* $\gamma_1 = (\epsilon \cdot 2^{(\frac{1}{2} - \delta)n + 1})^{\frac{1}{k}}$.

Proof. Let W be a $(n, (\frac{1}{2} + \delta) \cdot n)$-source. Let S be a random variable that is uniformly distributed over $\{0,1\}^d$ and is independent of W. We will show that the distribution of $\mathsf{Ext}^{(1)}(W, S)$ is γ_1-close to uniform. As in [5], it is enough to consider the case where W is uniformly distributed over a set $W' \subseteq \{0,1\}^n$ of size $2^{(1/2+\delta)n}$. For every $w \in \{0,1\}^n$ and $s \in \{0,1\}^d$ denote $e(w, s) = (-1)^{Z_s(w)}$.

Claim 1. *For any $r \in [k]$ and any different $s_1, \ldots, s_r \in \{0,1\}^d$,*

$$\sum_{w \in \{0,1\}^n} \prod_{j=1}^{r} e(w, s_j) \le \epsilon \cdot 2^n.$$

Proof.

$$\sum_{w \in \{0,1\}^n} \prod_{j=1}^{r} e(w, s_j) = \sum_{w \in \{0,1\}^n} \prod_{j=1}^{r} (-1)^{Z_{s_j}(w)} = \sum_{w \in \{0,1\}^n} (-1)^{Z_{s_1}(w) \oplus \cdots \oplus Z_{s_r}(w)},$$

and since $Z_{s_1}(w) \oplus \cdots \oplus Z_{s_r}(w)$ is ϵ-biased, the last sum is at most $\epsilon \cdot 2^n$.   □

The $\mathcal{L}_1$ distance of $\mathsf{Ext}^{(1)}(W, S)$ and U is

$$\|\mathsf{Ext}^{(1)}(W, S) - U\|_1$$
$$= |\Pr[\mathsf{Ext}^{(1)}(W, S) = 0] - \Pr[\mathsf{Ext}^{(1)}(W, S) = 1]|$$
$$= |\frac{1}{2^{(\frac{1}{2}+\delta)n}} \cdot \frac{1}{2^d} (\sum_{w \in W'} \sum_{s \in \{0,1\}^d} e(w, s))|.$$

Denote $\gamma(W, S) = \frac{1}{2^{(\frac{1}{2}+\delta)n}} \cdot \frac{1}{2^d} (\sum_{w \in W'} \sum_{s \in \{0,1\}^d} e(w, s))$.

Define $f : [-1, 1] \to [-1, 1]$ by $f(z) = z^k$, then f is a convex function for any even positive integer k.

Thus, by a convexity argument, we have

$$2^{(\frac{1}{2}+\delta)n} \cdot (2^d \cdot \gamma(W, S))^k = 2^{(\frac{1}{2}+\delta)n} \cdot \{\sum_{w \in W'} [\frac{1}{2^{(1/2+\delta)n}} \sum_{s \in \{0,1\}^d} e(w, s)]\}^k$$

$$\le 2^{(\frac{1}{2}+\delta)n} \cdot \{\sum_{w \in W'} \frac{1}{2^{(1/2+\delta)n}} [\sum_{s \in \{0,1\}^d} e(w, s)]^k\}$$

$$\le \sum_{w \in \{0,1\}^n} [\sum_{s \in \{0,1\}^d} e(w, s)]^k$$

$$= \sum_{w \in \{0,1\}^n} \sum_{s_1, \ldots, s_k \in \{0,1\}^d} \prod_{j=1}^{k} e(w, s_j)$$

$$= \sum_{s_1, \ldots, s_k \in \{0,1\}^d} \sum_{w \in \{0,1\}^n} \prod_{j=1}^{k} e(w, s_j).$$

The sum over $s_1, \ldots, s_k \in \{0,1\}^d$ is broken into two sums. The first sum is over $s_1, \ldots, s_k \in \{0,1\}^d$ such that in each summand, at least one s_j is different than all other elements in the sequence $s_1, \ldots, s_k$[5], and the second sum is over $s_1, \ldots, s_k \in \{0,1\}^d$ such that in each summand every s_j is identical to at least one other element in the sequence $s_1, \ldots, s_k$. The number of summands in the first sum is trivially bounded by $2^{d \cdot k}$, and by Claim 1 each summand is bounded by $2^n \cdot \epsilon$. The number of summands in the second sum is bounded by $2^{d \cdot \frac{k}{2}} \cdot (\frac{k}{2})^k$, and each summand is trivially bounded by 2^n. Therefore,

$$2^{(\frac{1}{2}+\delta)n} \cdot 2^{d \cdot k} \cdot \gamma(W,S)^k \leq 2^n \cdot \epsilon \cdot 2^{d \cdot k} + 2^n \cdot 2^{d \cdot \frac{k}{2}} \cdot (\frac{k}{2})^k \leq 2 \cdot 2^n \cdot \epsilon \cdot 2^{d \cdot k},$$

where the last inequality follows by the assumption that $k \cdot (1/\epsilon)^{1/k} \leq D^{\frac{1}{2}}$. That is, $\gamma(W,S) \leq (\epsilon \cdot 2^{(\frac{1}{2}-\delta)n+1})^{\frac{1}{k}}$. □

The above partition method about the sum over $s_1, \ldots, s_k \in \{0,1\}^d$ is not optimal, since it doesn't capture the essence of random variable sequence that is biased for linear tests (i.e., $Z_1, \ldots, Z_{2^d}$ is called ϵ-biased for linear tests of size k if for any nonempty $\tau \subseteq [2^d]$ with $|\tau| \leq k$, the random variable $Z_\tau = \oplus_{i \in \tau} Z_i$ is ϵ-biased). Moreover, the bounds on the number of summands in the two sums are too large. The same problem exists in [20].

In fact, when every s_j is identical to at least one other element in the sequence $s_1, \ldots, s_k$ under the assumption that at least one s_j appears odd times in the sequence $s_1, \ldots, s_k$, the summand $\sum_{w \in \{0,1\}^n} \prod_{j=1}^{k} e(w, s_j)$ is still upper bounded by $2^n \cdot \epsilon$, since $\sum_{w \in \{0,1\}^n} \prod_{j=1}^{k} e(w, s_j) = \sum_{w \in \{0,1\}^n} \prod_{j=1}^{k} (-1)^{Z_{s_j}(w)} = \sum_{w \in \{0,1\}^n} (-1)^{Z_{s_1}(w) \oplus \cdots \oplus Z_{s_k}(w)}$ and $Z_1, \ldots, Z_D$ are 0-1 random variables that are ϵ-biased for linear tests of size k'. However, in this case the upper bound on the summand $\sum_{w \in \{0,1\}^n} \prod_{j=1}^{k} e(w, s_j)$ was considered to be 2^n in [6,20].

3.2   Improvement for the Error Estimation of Raz's Extractor

We improve the error estimation of Raz's extractor as follows. Unlike [6,20], we present another partition method of the sum in the following proof. The combination and permutation formulas are exploited to show a tight bound on the sum. Correspondingly, the error can be reduced.

Proposition 1. *Consider fixed positive numbers k and d. Assume that a sequence $s_1, \ldots, s_k$ satisfies the following two conditions: (1) for every $i \in [k]$, $s_i \in \{0,1\}^d$, and (2) for every $j \in [k]$, s_j appears even times in the sequence $s_1, \ldots, s_k$. Then the number of such sequences $s_1, \ldots, s_k$ is $2^{\frac{dk}{2}} \cdot (k-1) \cdot (k-3) \cdot \ldots \cdot 1$.*

---

[5] In this paper, two elements s_i and s_j in the sequence $s_1, \ldots, s_k$, where $i \neq j$, might represent the same string.

Proof. Denote C_r^l as the number of possible combinations of r objects from a set of l objects. Then $C_r^l = \frac{l!}{r!(l-r)!} = \frac{l(l-1)(l-2)\cdots(l-r+1)}{r!}$. Denote P_r^l as the number of possible permutations of r objects from a set of l objects. Then $P_r^l = \frac{l!}{(l-r)!} = l(l-1)(l-2)\cdots(l-r+1)$. Hence the number of the corresponding sequences is

$$\frac{C_2^k \cdot C_2^{k-2} \cdot \cdots \cdot C_2^2}{P_{\frac{k}{2}}^{\frac{k}{2}}} \cdot 2^{\frac{dk}{2}} = \frac{k! \cdot \frac{1}{2^{\frac{k}{2}}}}{(\frac{k}{2})!} \cdot 2^{\frac{dk}{2}} = \frac{k!}{(\frac{k}{2})! \cdot 2^{\frac{k}{2}}} \cdot 2^{\frac{dk}{2}} = 2^{\frac{dk}{2}} \cdot (k-1) \cdot (k-3) \cdots \cdots 1.$$

Theorem 2. *Let $D = 2^d$. Let $Z_1, \ldots, Z_D$ be 0-1 random variables that are ϵ-biased for linear tests of size k' that are constructed using n random bits. Define $\mathsf{Ext}^{(1)} \colon \{0,1\}^n \times \{0,1\}^d \to \{0,1\}$ by $\mathsf{Ext}^{(1)}(w,s) = Z_s(w)$, that is, $\mathsf{Ext}^{(1)}(w,s)$ is the random variable Z_s, when using w as the value of the n bits needed to produce $Z_1, \ldots, Z_D$. Then, for any $0 < \delta \le \frac{1}{2}$ and any even integer $k \le k'$, $\mathsf{Ext}^{(1)}$ is a $((\frac{1}{2}+\delta) \cdot n, \gamma_2)$-seeded-extractor, where $\gamma_2 = 2^{\frac{(\frac{1}{2}-\delta) \cdot n}{k}} \cdot [2^{-\frac{dk}{2}} \cdot (k-1) \cdot (k-3) \cdots \cdots 1 \cdot (1-\epsilon) + \epsilon]^{\frac{1}{k}}$.*

Proof. We improve the proof by proposing another method for partitioning the sum $\sum_{s_1, \ldots, s_k \in \{0,1\}^d} \sum_{w \in \{0,1\}^n} \prod_{j=1}^{k} e(w, s_j)$ into two sums. The first sum is over $s_1, \ldots, s_k \in \{0,1\}^d$ such that in each summand, at least one s_j appears odd times in the sequence $s_1, \ldots, s_k$, and the second sum is over $s_1, \ldots, s_k \in \{0,1\}^d$ such that in each summand every s_j appears even times in the sequence $s_1, \ldots, s_k$. By Proposition 1, the number of summands in the second sum is $2^{\frac{dk}{2}} \cdot (k-1) \cdot (k-3) \cdots \cdots 1$, and each summand is 2^n. Therefore, the number of summands in the first sum is $2^{dk} - 2^{\frac{dk}{2}} \cdot (k-1) \cdot (k-3) \cdots \cdots 1$, and by Claim 1 each summand is bounded by $2^n \cdot \epsilon$. Hence, $2^{(\frac{1}{2}+\delta) \cdot n} \cdot 2^{d \cdot k} \cdot \gamma(W, S)^k \le 2^n \cdot [2^{\frac{dk}{2}} \cdot (k-1) \cdot (k-3) \cdots \cdots 1] + 2^n \cdot \epsilon \cdot [2^{dk} - 2^{\frac{dk}{2}} \cdot (k-1) \cdot (k-3) \cdots \cdots 1]$. Correspondingly,

$$\gamma(W,S)^k \le \frac{2^n \cdot 2^{dk}}{2^{(\frac{1}{2}+\delta) \cdot n} \cdot 2^{d \cdot k}} \cdot [2^{-\frac{dk}{2}} \cdot (k-1) \cdot (k-3) \cdots \cdots 1 \cdot (1-\epsilon) + \epsilon]$$
$$= 2^{(\frac{1}{2}-\delta) \cdot n} \cdot [2^{-\frac{dk}{2}} \cdot (k-1) \cdot (k-3) \cdots \cdots 1 \cdot (1-\epsilon) + \epsilon]$$

That is, $\gamma(W,S) \le 2^{\frac{(\frac{1}{2}-\delta) \cdot n}{k}} \cdot [2^{-\frac{dk}{2}} \cdot (k-1) \cdot (k-3) \cdots \cdots 1 \cdot (1-\epsilon) + \epsilon]^{\frac{1}{k}}$. $\square$

3.3   Comparison

For simplicity, in the rest of the paper, denote γ_1 as the error of the extractor in Lemma 1, and γ_2 as the counterpart in Theorem 2.

Proposition 2. $(k-1) \cdot (k-3) \cdots \cdots 1 \le (\frac{k}{2})^k$ *for any positive even integer k, and "$=$" holds iff $k = 2$. Furthermore,* $\lim_{k \to \infty} \frac{(k-1) \cdot (k-3) \cdots \cdots 1}{2^{\frac{1}{2}} \cdot (\frac{k}{e})^{\frac{k}{2}}} = 1$.

Proof. When $k = 2$, it's trivial that $(k-1) \cdot (k-3) \cdots \cdots 1 = (\frac{k}{2})^k$. In the following, we only consider any positive even integer k with $k > 2$.

Since $\frac{k!}{(\frac{k}{2})!} < \frac{k^k}{2^{\frac{k}{2}}}$, we have $\frac{k!}{(\frac{k}{2})! \cdot 2^{\frac{k}{2}}} < \frac{k^k}{2^k}$. Hence,

$$(k-1) \cdot (k-3) \cdots \cdots 1 = \frac{k!}{(\frac{k}{2})! \cdot 2^{\frac{k}{2}}} < \frac{k^k}{2^k}.$$

From the Stirling's Formula, we have $\lim_{k \to \infty} \frac{k!}{\sqrt{2\pi k}(\frac{k}{e})^k} = 1$. Therefore,

$$\lim_{k \to \infty} \frac{(k-1) \cdot (k-3) \cdots \cdots 1}{2^{\frac{1}{2}} \cdot (\frac{k}{e})^{\frac{k}{2}}} = \lim_{k \to \infty} [\frac{k!}{\sqrt{2\pi k} \cdot (\frac{k}{e})^k} \cdot \frac{\sqrt{2\pi \cdot \frac{k}{2}} \cdot (\frac{k}{2e})^{\frac{k}{2}}}{(\frac{k}{2})!}] = 1. \qquad \square$$

The error estimation of the extractor in Theorem 1 is better than that in Lemma 1. Recall that in Theorem 1, we have

$$\gamma_2 = 2^{\frac{(\frac{1}{2}-\delta)n}{k}} \cdot [2^{-\frac{dk}{2}} \cdot (k-1) \cdot (k-3) \cdots \cdots 1 \cdot (1 - \epsilon) + \epsilon]^{\frac{1}{k}}$$

$$= 2^{\frac{(\frac{1}{2}-\delta)n}{k}} \cdot \{2^{-\frac{dk}{2}} \cdot (k-1) \cdot (k-3) \cdots \cdots 1 + [1 - 2^{-\frac{dk}{2}} \cdot (k-1) \cdot (k-3) \cdots \cdots 1] \cdot \epsilon\}^{\frac{1}{k}},$$

while in Lemma 1, we have $\gamma_1 = 2^{\frac{(\frac{1}{2}-\delta)n}{k}} \cdot (2\epsilon)^{\frac{1}{k}}$ in [6] under the assumption that $\epsilon \geq 2^{-\frac{dk}{2}} \cdot k^k$ and $0 < \delta \leq \frac{1}{2}$.

In general, since $\epsilon \geq 2^{-\frac{dk}{2}} \cdot k^k$ and $2^{-\frac{dk}{2}} \cdot k^k > 2^{-\frac{dk}{2}} \cdot (k-1) \cdot (k-3) \cdots \cdots 1$ for any even integer k, we get $\gamma_1 > \gamma_2$. In particular, when k is large enough, from Proposition 2, we get that $(k-1) \cdot (k-3) \cdots \cdots 1 \approx 2^{\frac{1}{2}} \cdot (\frac{k}{e})^{\frac{k}{2}}$. Therefore,

$$\gamma_2 \approx 2^{\frac{(\frac{1}{2}-\delta)n}{k}} \cdot \{2^{-\frac{dk}{2}} \cdot 2^{\frac{1}{2}} \cdot (\frac{k}{e})^{\frac{k}{2}} + [1 - 2^{-\frac{dk}{2}} \cdot 2^{\frac{1}{2}} \cdot (\frac{k}{e})^{\frac{k}{2}}] \cdot \epsilon\}^{\frac{1}{k}}.$$

Correspondingly, $\epsilon \geq 2^{-\frac{dk}{2}} \cdot k^k > 2^{-\frac{dk}{2}} \cdot 2^{\frac{1}{2}} \cdot (\frac{k}{e})^{\frac{k}{2}}$. Hence, $\gamma_1 > \gamma_2$.

Remark 1. To simplify γ_2, let k be a specific value. For instance, let $k = 4$, then the error $\gamma_1 = 2^{\frac{(\frac{1}{2}-\delta)n}{4}} \cdot (2\epsilon)^{\frac{1}{4}}$ and $\gamma_2 = 2^{\frac{(\frac{1}{2}-\delta)n}{4}} \cdot [2^{-2d} \cdot 3 \cdot (1 - \epsilon) + \epsilon]^{\frac{1}{4}}$.

Remark 2. Noted that when k is large enough, $(\frac{k}{2})^k$ is much greater than $(k-1) \cdot (k-3) \cdots \cdots 1$. For instance, when $k = 6$, we have $(\frac{k}{2})^k = 729$ and $(k-1) \cdot (k-3) \cdots \cdots 1 = 15$. Therefore, "The number of summands in the second sum is $2^{\frac{dk}{2}} \cdot (k-1) \cdot (k-3) \cdots \cdots 1$, and each summand is 2^n." in the proof of Theorem 2 is a great improvement on "The number of summands in the second sum is bounded by $2^{d \cdot \frac{k}{2}} \cdot (\frac{k}{2})^k$, and each summand is trivially bounded by 2^n." in the proof of Lemma 1.

Remark 3. If $\epsilon \geq \frac{1}{2^{(\frac{1}{2}-\delta)n+1}}$, then $\gamma_1 = 2^{\frac{(\frac{1}{2}-\delta)n}{k}} \cdot (2\epsilon)^{\frac{1}{k}} \geq 1$. In this case, the error estimation is meaningless.

3.4   Important Role in Improving the Seed Length of Non-malleable Extractors

It should be noticed that the error of the non-malleable extractor in Theorem 1 given by Cohen et al. [6] relies on some constrained parameters. The main idea of the proof about Theorem 1 given by Cohen et al. [6] is as follows. Assume for contradiction that Ext is not a non-malleable extractor, then after some steps, an inequality $\gamma_1 > A$ is deduced, where A denotes a certain formula. On the other hand, from the assumption of Theorem 1, $\gamma_1 < A$ should hold. Thus Ext is a non-malleable extractor. Essentially, the constraints on the parameters in Theorem 1 are chosen according to the inequality $\gamma_1 < A$. From Proposition 2, we have $\gamma_1 > \gamma_2$ for any positive even integer $k \geq 4$. Therefore, we may relax the constraints on the parameters in Theorem 1 according to $\gamma_2 < A$. See the proofs of Theorems 3 and 4 below for details. Correspondingly, the seed length may be further shortened.

4   Explicit Construction of Non-malleable Extractors with Shorter Seed Length

In this section, we improve the parameters of the explicit construction of non-malleable extractors by Cohen et al. in [6]. The seed length here is shorter than that in Theorem 1.

We first review two lemmas that will be used later.

Lemma 2. (see [6]) *Let X be a random variable over $\{0,1\}^m$. Let Y, S be two random variables. Then,*

$$\|(X,Y,S) - (U_m,Y,S)\|_1 = \mathbb{E}_{s \sim S}[\|(X,Y,S)|_{S=s} - (U_m,Y,S)|_{S=s}\|_1].$$

Lemma 3. (see [6]) *Let X, Y be random variables over $\{0,1\}^m$ and $\{0,1\}^n$ respectively. Then $\|(X,Y) - (U_m,Y)\|_1 \leq \sum\limits_{\emptyset \neq \sigma \subseteq [m], \tau \subseteq [n]} bias(X_\sigma \oplus Y_\tau)$, where X_i is the i-th bit of X, Y_j is the jth bit of Y, $X_\sigma = \oplus_{i \in \sigma} X_i$, and $Y_\tau = \oplus_{j \in \tau} Y_j$.*

In what follows, we show a specific explicit construction of a non-malleable extractor such that it is an improvement of [6] in the sense that the seed length is shorter.

Theorem 3. *There exists an explicit $(1016, \frac{1}{2})$-non-malleable extractor Ext : $\{0,1\}^{1024} \times \{0,1\}^{19} \to \{0,1\}$.*

   Proof Idea. *We borrow the reductio ad absurdum approach in the proof of Theorem 1. The proof sketch is as follows. Assume by contradiction that Ext is not non-malleable. Then*

   Phase 1: There must exist a weak source W with min-entropy at least α and an adversarial function $\mathcal{A}$ such that the statistical distance between $(\mathsf{Ext}(W,S), \mathsf{Ext}(W,\mathcal{A}(S)), S)$ and $(U_1, \mathsf{Ext}(W,\mathcal{A}(S)), S)$ has a certain lower bound. Then there exists $S \subseteq \{0,1\}^d$ s.t. for every $s \in S$, $Y_s = \mathsf{Ext}(W,s) \oplus$

$\mathsf{Ext}(W, \mathcal{A}(s))$ *is biased. Consider the directed graph* $G = (S \cup \mathcal{A}(s), E)$ *with* $E = \{(s, \mathcal{A}(s) : s \in S\}$, *where* G *might contains cycles. By employing a lemma about graph as shown in [6], we can find a subset* $S' \subseteq S$ *s.t. the induced graph of* G *by* $S' \cup \mathcal{A}(S')$ *is acyclic.*

Phase 2: *We prove that the set of variables* $\{Y_s\}_{s\in S'}$ *is ϵ-biased for linear tests of size at most* $k/2$. *Consider the extractor of Raz built on the variables* $\{Y_s\}_{s\in S'}$. *It's a good seeded-extractor, which yields a contradiction.*

Phase 1 of the proof is almost the same as that in [6]. Phase 2 jumps out of the idea in [6]. We exploit the error estimation of the extractor in Theorem 2 instead of Lemma 1. We use a trick such that the even integer k is just 4 instead of the largest even integer that is not larger than $\frac{\lceil 128\delta \rceil}{2}$, *where δ can be seen in Theorem 1. Therefore the extractor error can be simplified and we don't need to prove* $k \cdot (\frac{1}{\epsilon})^{\frac{1}{k}} \leq (2^d)^{\frac{1}{2}}$ *as shown in Lemma 1.*

Proof. The explicit construction we present is the extractor constructed in [20]. Alon et al. [1] observed that for every k', $N \geq 2$, the sequence of 0-1 random variables $Z_1, \ldots, Z_N$ that is ϵ-biased for linear tests of size k' can be explicitly constructed using $2 \cdot \lceil \log(1/\epsilon) + \log k' + \log \log N \rceil$ random bits. Therefore, let $D = 2^{19}$ and $\epsilon = 2^{-\frac{1024}{2}+r}$ with $r = 1 + \log k' + \log 19$, then we can construct a sequence of 0-1 random variables $Z_1, \ldots, Z_{2^{19}}$ that is ϵ-biased for linear tests of size k' using n random bits. Let $k' = 8$. Define $\mathsf{Ext} : \{0,1\}^{1024} \times \{0,1\}^{19} \to \{0,1\}$ by $\mathsf{Ext}(w, s) = Z_s(w)$.

Let S be a random variable uniformly distributed over $\{0,1\}^{19}$.

Assume for contradiction that Ext is not a $(1016, \frac{1}{2})$-non-malleable-extractor. Then there exists a source W of length 1024 with min-entropy 1016, and an adversarial function $\mathcal{A} : \{0,1\}^{19} \to \{0,1\}^{19}$ such that

$$\|(\mathsf{Ext}(W, S), \mathsf{Ext}(W, \mathcal{A}(S)), S) - (U_1, \mathsf{Ext}(W, \mathcal{A}(S)), S)\|_1 > \frac{1}{2}.$$

As in [5], suppose W is uniformly distributed over a set $W' \subseteq \{0,1\}^{1024}$ of size 2^{1016}.

For every $s \in \{0,1\}^{19}$, let X_s be the random variable $\mathsf{Ext}(W, s)$. By Lemmas 2 and 3, we have

$$\sum_{\emptyset \neq \sigma \subseteq [1], \tau \subseteq [1]} \mathbb{E}_{s\sim S}[bias((X_s)_\sigma \oplus (X_{\mathcal{A}(s)})_\tau)] > \frac{1}{2}.$$

Let $\sigma^*, \tau^* \subseteq [1]$ be the indices of (one of) the largest summands in the above sum. For every $s \in \{0,1\}^{19}$, let $Y_s = (X_s)_{\sigma^*} \oplus (X_{\mathcal{A}(s)})_{\tau^*}$.

There is a set $S'' \subseteq \{0,1\}^{19}$ satisfying that

$$|S''| > \frac{\xi \cdot 2^{19-2}}{2(1+1)^2} = 2^{13}.$$

The S'' here is the same as that in the proof of Theorem 1 by replacing t there with 1 and the error 2^{-m} there with $\frac{1}{2}$. Please see [6] for details.

Define a random variable $Y_{S''}$ over $\{0, 1\}$ as follows: To sample a bit from $Y_{S''}$, uniformly sample a string s from S'', and then independently sample a string w uniformly from W'. The sampled value is $Y_s(w)$. We have that $bias(Y_{S''}) > \frac{\frac{1}{2}}{2^{1+1}(2-1)(1+1)} = \frac{1}{2^4}$. For every $s \in S''$, let $Y'_s = Z_{(1,s)} \oplus (\oplus_{j \in T^*} Z_{(j, \mathcal{A}(s))})$, where $Z_{(1,s)} = Z_s$.

Let $t = 1$ and $m = 1$ in Claim 7.2 of [6], we get the following claim.

Claim 2. *The set of random variables* $\{Y'_s\}_{s \in S''}$ *ϵ-fools linear tests of size 4.*

We apply Theorem 2 on the random variables $\{Y'_s\}_{s \in S''}$. For simplicity of presentation we assume $|S''| = 2^{d'}$. By Theorem 2, the distribution of $\mathsf{Ext}^{(1)}(W, S'')$ is γ_2-biased for $\gamma_2 = 2^{\frac{8}{k}} \cdot [2^{-\frac{d'k}{2}} \cdot (k-1) \cdot (k-3) \cdots 1 \cdot (1 - \epsilon) + \epsilon]^{\frac{1}{k}}$. Let $k = \frac{k'}{2} = 4$, then $\gamma_2 = 2^{\frac{8}{4}} \cdot [2^{-2d'} \cdot 3 \cdot (1 - \epsilon) + \epsilon]^{\frac{1}{4}}$. We note that $\mathsf{Ext}^{(1)}(W, S'')$ has the same distribution as $Y_{S''}$. In particular, both random variables have the same bias. Therefore, we get

$$2^{\frac{8}{4}} \cdot [2^{-2d'} \cdot 3 \cdot (1 - \epsilon) + \epsilon]^{\frac{1}{4}} \geq bias(Y_{S''}) > \frac{1}{2^4},$$

Moreover, since $2^{d'} = |S''| > 2^{13}$, we have

$$2^2 \cdot [4 \cdot 2^{-28} \cdot 3 \cdot (1 - \epsilon) + \epsilon]^{\frac{1}{4}} > 2^2 \cdot [2^{-2d'} \cdot 3 \cdot (1 - \epsilon) + \epsilon]^{\frac{1}{4}} > \frac{1}{2^4}.$$

That is,

$$2^{-38} > \frac{2^{-4} \cdot 2^{-20} - \epsilon}{3(1 - \epsilon) \cdot 2^{12}}, \tag{a}$$

where $\epsilon = 2^{-516+r}$ and $r = 4 + \log 19$.

On the other hand, we have $2^{-38} < \frac{2^{-4} \cdot 2^{-20} - \epsilon}{3(1-\epsilon) \cdot 2^{10} \cdot 2^2}$, which is in contradiction to the inequality (a).   □

Comparison. In Theorem 1, the seed length d and the source length n should satisfy $d \geq \frac{23}{\delta} m + 2 \log n$ with $0 < \delta \leq \frac{1}{2}$. However, in the above construction, we have $d = 1.9 \log n$. We compare them in detail as follows.

Let $n = 2^{10}$, $m = 1$, and $\delta = \frac{63}{128}$ in Theorem 1, then it can be easily verified that $n \geq \frac{160}{\delta} \cdot m$. To construct an explicit $((\frac{1}{2} + \delta) \cdot n, 2^{-m})$-non-malleable extractor $\mathsf{nmExt} : \{0, 1\}^n \times \{0, 1\}^d \to \{0, 1\}^m$ (that is, an explicit $(1016, \frac{1}{2})$-non-malleable extractor nmExt), by Theorem 1, the seed length d should satisfy $d \geq \frac{23}{\delta} \cdot m + 2 \log n = \frac{46}{63} + 66$. Moreover, when $d \leq 2^{41}$, the precondition $\delta \geq 10 \cdot \frac{\log(nd)}{n}$ in Theorem 1 is satisfied. Meanwhile, by Theorem 3, the seed length d can just be 19. In this sense, our construction is much better than that of [6].

Using the extractor with improved error estimation (see Theorem 2), we can also improve the parameters of the explicit non-malleable extractor $\mathsf{nmExt} : \{0, 1\}^n \times \{0, 1\}^d \to \{0, 1\}^m$ constructed by Cohen et al. [6] below.

Theorem 4. *Assume that*

$$0 < 2^{\log 3 - 2\theta + 4m + 8} - 2^{\log 3 - \frac{n}{2} + 4 + \log d - 2\theta + 4m + 8} \leq 2^{2d + 4\theta - 8m - 8 - n + \alpha} - 2^{2d - \frac{n}{2} + 4 + \log d}.$$

Then there exists an explicit $(\alpha, 2^{\theta})$-*non-malleable extractor* nmExt : $\{0,1\}^n \times \{0,1\}^d \rightarrow \{0,1\}^m$.

The proof is similar to that of Theorem 3. Please see Appendix A for details.

Due to the analysis of Sect. 3.4, we conclude that the above theorem is really an improvement in the sense that the seed length here is shorter. Though the constrains on the parameters in Theorem 4 are complex, we show some simplification in Appendix B. How to further simplify the constraints is an open problem.

5   Application to Privacy Amplification

In this section, we show how the non-malleable extractor is applied to the privacy amplification protocol [8,9] (also known as an information-theoretic key agreement protocol), the formal concept of which can be seen in Appendix C.

Roughly speaking, in this scenario, Alice and Bob share a shared weak secret W, the min-entropy of which is only guaranteed. They communicate over a public and unauthenticated channel to securely agree on a nearly uniform secret key R, where the attacker Eve is active and computationally unbounded. To achieve this goal, the protocol is designed as follows.

Table 1. The Dodis-Wichs privacy amplification protocol.

Alice: W	Eve	Bob: W
Sample random S.		
	$S \longrightarrow S'$	
		Sample random S_0.
		$R' = \mathsf{nmExt}(W, S')$.
		$T_0 = \mathsf{MAC}_{R'}(S_0)$.
		Reach **KeyDerived** state.
		Output $R_B = \mathsf{Ext}(W, S_0)$.
	$(S_0', T_0') \longleftarrow (S_0, T_0)$	
$R = \mathsf{nmExt}(W, S)$.		
If $T_0' \neq \mathsf{MAC}_R(S_0')$, output $R_A = \bot$.		
Otherwise, reach **KeyConfirmed** state,		
and output $R_A = \mathsf{Ext}(W, S_0')$.		

Assume that we'll authenticate the seed S_0. Alice initiates the conversation by transmitting a uniformly random seed S to Bob. During this transmission, S may be modified by Eve into any value S'. Then Bob samples a uniform seed S_0, computes the authentication key $R' = \mathsf{nmExt}(W, S')$, and sends S_0 together with the authentication tag $T_0 = \mathsf{MAC}_{R'}(S_0)$ to Alice. At this point, Bob reaches the **KeyDerived** state and outputs $R_B = \mathsf{Ext}(W, S_0)$. During this transmission, (S_0, T_0) may be modified by Eve into any pair (S_0', T_0'). Alice computes the authentication key $R = \mathsf{nmExt}(W, S)$ and verifies that $T_0' = \mathsf{MAC}_R(S_0')$. If the verification fails then Alice rejects and outputs $R_A = \bot$. Otherwise, Alice reaches the KeyConfirmed state and outputs $R_A = \mathsf{nmExt}(W, S_0')$.

The security can be analyzed in two cases [6,8]. Case 1: Eve does not modified the seed S in the first round. Then Alice and Bob share the same authentication key (i.e., $R' = R$), which is statistically close to a uniform distribution. Therefore, Eve has only a negligible probability of getting a valid authentication tag T_0' for any seed $S_0' \neq S_0$. Case 2: Eve does modify the seed S to a different seed S'. Since T_0 is a deterministic function of S_0 and R', Eve may guess R'. According to the definition of non-malleable extractors, the authentication key R computed by Alice is still statistically close to a uniform distribution. Thus, again, the adversary has only a negligible probability of computing a valid authentication T_0' for any seed S_0' with respect to the authentication key R. Consequently, the above protocol is secure.

Theorem 5. (see [6,9]) *Assume* $\mathsf{nmExt} : \{0,1\}^n \times \{0,1\}^{d_1} \rightarrow \{0,1\}^{m_1}$ *is a* (α, γ_{nmExt})-*non-malleable extractor,* $\mathsf{Ext} : \{0,1\}^n \times \{0,1\}^{d_2} \rightarrow \{0,1\}^{m_2}$ *is a strong* $(\alpha - (d_1 + m_1) - \log \frac{1}{\epsilon'}, \gamma_{Ext})$-*extractor, and* $\{\mathsf{MAC}_r : \{0,1\}^{d_2} \rightarrow \{0,1\}^{\tau}\}_{r \in \{0,1\}^{m_1}}$ *is a* ε_{MAC}-*secure message authentication code. Then for any integers n and $\alpha \leq n$, the protocol in Table 1 is a 2-round (n, α, m, η)-privacy amplification protocol, with communication complexity $d_1 + d_2 + \tau$ and $\eta = \max\{\epsilon' + \gamma_{Ext}, \gamma_{nmExt} + \varepsilon_{MAC}\}$.*

The explicit non-malleable extractor in this paper can be applied to construct the above privacy amplification protocol with low communication complexity.

6 Conclusion

Non-malleable extractor is a powerful theoretical tool to study privacy amplification protocols, where the attacker is active and computationally unbounded. In this paper, we improved the error estimation of Raz's extractor using the combination and permutation techniques. Based on the improvement, we presented an improved explicit construction of non-malleable extractors with shorter seed length. Similar to [6], our construction is also based on biased variable sequence for linear tests. However, our parameters are improved. More precisely, we presented an explicit $(1016, \frac{1}{2})$-non-malleable extractor $\mathsf{nmExt} : \{0,1\}^{1024} \times \{0,1\}^d \rightarrow \{0,1\}$ with seed length 19, while it is no less than $\frac{46}{63} + 66$ according to Cohen et al. in CCC'12 [6]. We also improved the parameters of the general explicit construction of non-malleable extractors proposed by Cohen et al. and analyzed the simplification of the constraints on the parameters (see Appendix B for details). How to further simplify the constraints is an open problem. Finally, we showed their applications to privacy amplification protocol (or information-theoretic key agreement protocol).

Acknowledgments. We would like to thank Divesh Aggraval, Yevgeniy Dodis, Feng-Hao Liu, and Xin Li for helpful discussions. This work is supported in part by High Technology Research and Development Program of China under grant No.

2015AA016004, NSFC (Nos. 61170189, 61370126, 61202239), the Fund of the State Key Laboratory of Software Development Environment (No. SKLSDE-2015ZX-16), the Fund of the China Scholarship Council under grant No. 201206020063, and the Fundamental Research Funds for the Central Universities.

A   Proof of Theorem 4

Proof. The explicit construction we present is the extractor constructed in [20]. Alon et al. [1] observed that for every k', $N \geq 2$, the sequence of 0-1 random variables $Z_1, \ldots, Z_N$ that is ϵ-biased for linear tests of size k' can be explicitly constructed using $2 \cdot \lceil \log(1/\epsilon) + \log k' + \log \log N \rceil$ random bits. Therefore, let $D = m \cdot 2^d$ and $\epsilon = 2^{-\frac{n}{2}+r}$ with $r = 1 + \log k' + \log \log D$, then we can construct a sequence of 0-1 random variables $Z_1, \ldots, Z_D$ that is ϵ-biased for linear tests of size k' using n random bits. Let $k' = 8m$. We interpret the set of indices $[D]$ as the set $\{(i,s) : i \in [m], s \in \{0,1\}^d\}$. Define $\mathsf{Ext} : \{0,1\}^n \times \{0,1\}^d \to \{0,1\}^m$ by $\mathsf{Ext}(w,s) = Z_{(1,s)}(w) \cdots \| Z_{(m,s)}(w)$, where "$\|$" is the concatenation operator.

Let S be a random variable uniformly distributed over $\{0,1\}^d$.

Assume for contradiction that Ext is not a $(\alpha, 2^\theta)$-non-malleable-extractor. Then there exists a source W of length n with min-entropy α, and an adversarial-function $\mathcal{A} : \{0,1\}^d \to \{0,1\}^d$ such that

$$\|(\mathsf{Ext}(W,S), \mathsf{Ext}(W, \mathcal{A}(S)), S) - (U_m, \mathsf{Ext}(W, \mathcal{A}(S)), S)\|_1 > 2^\theta.$$

As in [5], suppose W is uniformly distributed over $W' \subseteq \{0,1\}^n$ of size 2^α.

For every $s \in \{0,1\}^d$, let X_s be the random variable $\mathsf{Ext}(W,s)$. By Lemmas 2 and 3, we have $\sum_{\emptyset \neq \sigma \subseteq [m], \tau \subseteq [m]} \mathbb{E}_{s \sim S}[bias((X_s)_\sigma \oplus (X_{\mathcal{A}(s)})_\tau)] > 2^\theta$. Let $\sigma^*, \tau^* \subseteq [m]$ be the indices of (one of) the largest summands in the above sum. For every $s \in \{0,1\}^d$, let $Y_s = (X_s)_{\sigma^*} \oplus (X_{\mathcal{A}(s)})_{\tau^*}$. There is a set $S'' \subseteq \{0,1\}^d$ satisfying that

$$|S''| > \frac{2^\theta \cdot 2^{d-2}}{2^m t(2^m - 1)(t+1)^2} = \frac{2^\theta \cdot 2^{d-2}}{2^{m+2}(2^m - 1)}.$$

The S'' here is the same as that in the proof of Theorem 1 by replacing t there with 1 and the error 2^{-m} there with 2^θ. Please see [6] for details.

Define a random variable $Y_{S''}$ over $\{0,1\}$ as follows: To sample a bit from $Y_{S''}$, uniformly sample a string s from S'', and then independently sample a string w uniformly from W'. The sampled value is $Y_s(w)$. We have that

$$bias(Y_{S''}) > \frac{2^\theta}{2^m t + 1(2^m - 1)(t+1)} = \frac{2^\theta}{2^{m+2}(2^m - 1)}.$$

For every $s \in S''$, let $Y'_s = \oplus_{i \in \sigma^*} Z_{(i,s)} \oplus (\oplus_{j \in \tau^*} Z_{(j, \mathcal{A}(s))})$.

Let $t = 1$ in Claim 7.2 of [6], we get the following claim.

Claim 2'. *The set $\{Y'_s\}_{s \in S''}$ ϵ-fools linear tests of size $\frac{k'}{(t+1)m} = 4$.*

We apply Theorem 2 on the random variables $\{Y'_s\}_{s \in S''}$. For simplicity of presentation, we assume $|S''| = 2^{d'}$. By Theorem 2, the distribution of $\mathsf{Ext}^{(1)}(W, S'')$

is γ_2-biased for $\gamma_2 = 2^{\frac{n-\alpha}{k}} \cdot [2^{-\frac{d'k}{2}} \cdot (k-1) \cdot (k-3) \cdots 1 \cdot (1-\epsilon) + \epsilon]^{\frac{1}{k}}$. Let $k = 4$, then $\gamma_2 = 2^{\frac{n-\alpha}{4}} \cdot [2^{-2d'} \cdot 3 \cdot (1-\epsilon) + \epsilon]^{\frac{1}{4}}$. We note that $\mathsf{Ext}^{(1)}(W, S'')$ has the same distribution as $Y_{S''}$. In particular, both random variables have the same bias. Therefore, we get

$$2^{\frac{n-\alpha}{4}} \cdot [2^{-2d'} \cdot 3 \cdot (1-\epsilon) + \epsilon]^{\frac{1}{4}} \geq bias(Y_{S''}) > \frac{2^\theta}{2^{m+2}(2^m - 1)}.$$

Moreover, since $2^{d'} = |S''| > \frac{2^\theta \cdot 2^{d-2}}{2^{m+2}(2^m-1)}$, we have

$$2^{\frac{n-\alpha}{4}} \cdot [(2^\theta)^{-2} \cdot 2^{-2d+2m+8} \cdot (2^m - 1)^2 \cdot 3 \cdot (1-\epsilon) + \epsilon]^{\frac{1}{4}} > \frac{2^\theta}{2^{m+2} \cdot (2^m - 1)}.$$

Hence, $2^{n-\alpha} \cdot [2^{-2\theta} \cdot 2^{-2d+4m+8} \cdot 3 \cdot (1-\epsilon) + \epsilon] > \frac{2^{4\theta}}{2^{8m+8}}$. That is,

$$2^{-2d} > \frac{2^{4\theta-8m-8-n+\alpha} - \epsilon}{3(1-\epsilon)2^{-2\theta+4m+8}}$$

with $\epsilon = 2^{-\frac{n}{2}+4+\log d}$, which is in contradiction to the assumption of the theorem. $\qquad\square$

B   Analysis of the Assumption in Theorem 4

In order to construct an explicit non-malleable extractor, it's enough to guarantee that the parameters satisfies

$$0 < 2^{\log 3} \cdot (1 - 2^{-\frac{n}{2}+4+\log d}) \cdot 2^{-2\theta+4m+8} \leq 2^{2d+4\theta-8m-8-n+\alpha} - 2^{2d-\frac{n}{2}+4+\log d}. \quad (b)$$

For simplicity, denote

$$A' = \log 3 - 2\theta + 4m + 8, \quad B' = \log 3 - \frac{n}{2} + 4 + \log d - 2\theta + 4m + 8,$$

$$C' = 2d + 4\theta - 8m - 8 - n + \alpha, \quad D' = 2d - \frac{n}{2} + 4 + \log d.$$

then (b) $holds$ $\Leftrightarrow$ $0 < 2^{A'} - 2^{B'} \leq 2^{C'} - 2^{D'}$. We discuss what happens under the assumption (b) in three cases as follows.

Case 1. Assume that $A' \geq C'$ and $B' \geq D'$. Since "$B' \geq D'$" implies "$A' \geq C'$", we only need to consider $B' \geq D'$ (i.e., $\log 3 - 2\theta + 4m + 8 \geq 2d$). Let $1 - \epsilon = 1 - 2^{-\frac{n}{2}+4+\log d} = 2^{\rho'}$.

From $\log 3 + 8 + 4m \geq 2d + 2\theta$, $\alpha \leq n$, $m \geq 1$, and $\theta < 0$, we get

$$-16 > -8m - 8 + 4\theta - n + \alpha$$
$$= (\log 3 + 8 + 4m) + 4\theta - 12m - 16 - \log 3 - n + \alpha$$
$$\geq 2d + 2\theta + 4\theta - 12m - 16 - \log 3 - n + \alpha.$$

Let $\rho' \geq -16$. Then we have $\rho' > 2d + 2\theta + 4\theta - 12m - 16 - \log 3 - n + \alpha$.

Therefore, $\log 3 + \rho' - 2\theta + 4m + 8 > 2d + 4\theta - 8m - 8 - n + \alpha$, which is in contradiction to the inequality (b).

Consequently, when $\epsilon \in (0, 1-2^{-16}]$, $A' \geq C'$, and $B' \geq D'$, (b) does not hold. From Theorem 2, only if ϵ is small enough, the corresponding seeded extractor is useful. Therefore, we assume that $\epsilon \in (0, 1 - 2^{-16}]$.

Case 2. Assume that $A' \geq C'$ and $B' < D'$, then it's in contradiction to the inequality (b).

Case 3. Assume that $A' < C'$, then it's trivial that $B' < D'$. Thus, we only need to consider $A' < C'$. Since $A' > B'$, we have $C' > D'$, that is, $4\theta - 8m - 12 - \frac{n}{2} + \alpha > \log d$.

Therefore, we obtain the following corollary.

Corollary. *Assume that $\epsilon = 2^{-\frac{n}{2}+4+\log d} \in (0, 1 - 2^{-16}]$ and*

$$2^{\log 3} \cdot \left(1 - 2^{-\frac{n}{2}+4+\log d}\right) \cdot 2^{-2\theta+4m+8} \leq 2^{2d+4\theta-8m-8-n+\alpha} - 2^{2d-\frac{n}{2}+4+\log d}.$$

Then there exists an explicit $(\alpha, 2^{\theta})$-non-malleable extractor $\mathsf{nmExt} : \{0,1\}^n \times \{0,1\}^d \to \{0,1\}^m$.

In particular, the parameters of the non-malleable extractor can be chosen according to the inequality system

$$\begin{cases} \log 3 - 6\theta + 16 + 12m + n - \alpha < 2d \\ 4\theta - 8m - 12 - \frac{n}{2} + \alpha > \log d \\ 2^{-\frac{n}{2}+4+\log d} \leq 1 - 2^{-16} \end{cases} \quad (1)$$

then check whether they satisfy the inequality

$$2^{\log 3 - 2\theta + 4m + 8} - 2^{\log 3 - \frac{n}{2}+4+\log d - 2\theta + 4m + 8} \leq 2^{2d+4\theta-8m-8-n+\alpha} - 2^{2d-\frac{n}{2}+4+\log d}.$$

Remark. α *can't be less than* $\frac{n}{2}$, *since* $4\theta - 8m - 12 - \frac{n}{2} + \alpha > \log d$.

C   The Concept of Privacy Amplification Protocol

Definition 5. (see [6,9]) *In an (n, α, m, η)-privacy amplification protocol (or information-theoretic key agreement protocol), Alice and Bob share a weak secret W, and have two candidate keys $r_A, r_B \in \{0,1\}^m \cup \bot$, respectively. For any adversarial strategy employed by Eve, denote two random variables R_A, R_B as the values of the candidate keys r_A, r_B at the conclusion of the protocol execution, and random variable T_E as the transcript of the (entire) protocol execution as seen by Eve. We require that for any weak secret W with min-entropy at least α the protocol satisfies the following three properties:*

- **Correctness:** *If Eve is passive, then one party reaches the state, the other party reaches the* KeyConfirmed *state, and $R_A = R_B$.*

- **Privacy:** *Denote KeyDerived$_A$ and KeyDerived$_B$ as the indicators of the events in which Alice and Bob reach the KeyDerived state, respectively. Then during the protocol execution, for any adversarial strategy employed by Eve, if Bob reaches the KeyDerived$_B$ state then $SD((R_B, T_E), (U_m, T_E)) \leq \eta$; if Alice reaches the KeyDerived$_A$ state, then $SD((R_A, T_E), (U_m, T_E)) \leq \eta$.*

- **Authenticity:** *Denote KeyConfirmed$_A$ and KeyConfirmed$_B$ as the indicators of the events in which Alice and Bob reach the KeyConfirmed state, respectively. Then, for any adversarial strategy employed by Eve, it holds that*

$$\Pr[(\textit{KeyConfirmed}_A \vee \textit{KeyConfirmed}_B) \wedge R_A \neq R_B] \leq \eta.$$

References

1. Alon, N., Goldreich, O., Håstad, J., Peralta, R.: Simple construction of almost k-wise independent random variables. Random Struct. Algorithms **3**(3), 289–304 (1992)
2. Bourgain, J.: More on the sum-product phenomenon in prime fields and its applications. Int. J. Number Theory **1**, 1–32 (2005)
3. Cheraghchi, M., Guruswami, V.: Non-malleable coding against bit-wise and split-state tampering. In: Lindell, Y. (ed.) TCC 2014. LNCS, vol. 8349, pp. 440–464. Springer, Heidelberg (2014)
4. Chandran, N., Kanukurthi, B., Ostrovsky, R., Reyzin, L.: Privacy amplification with asymptotically optimal entropy loss. In: STOC 2010, pp. 785–794 (2010)
5. Chor, B., Goldreich, O.: Unbiased bits from sources of weak randomness and probabilistic communication complexity. SIAM J. Comput. **17**(2), 230–261 (1988)
6. Cohen, G., Raz, R., Segev, G.: Non-malleable extractors with short seeds and applications to privacy amplification. In: CCC 2012, pp. 298–308 (2012)
7. Dodis, Y., Katz, J., Reyzin, L., Smith, A.: Robust fuzzy extractors and authenticated key agreement from close secrets. In: Dwork, C. (ed.) CRYPTO 2006. LNCS, vol. 4117, pp. 232–250. Springer, Heidelberg (2006)
8. Dodis, Y., Li, X., Wooley, T.D., Zuckerman, D.: Privacy amplification and non-malleable extractors via character sums. In: FOCS 2011, pp. 668–677 (2011)
9. Dodis, Y., Wichs, D.: Non-malleable extractors and symmetric key cryptography from weak secrets. In: STOC 2009, pp. 601–610 (2009)
10. Dziembowski, S., Pietrzak, K., Wichs, D.: Non-malleable codes. In: Proceedings of Innovations in Computer Science (ICS 2010), pp. 434–452 (2010)
11. Dodis, Y., Yu, Y.: Overcoming weak expectations. In: Sahai, A. (ed.) TCC 2013. LNCS, vol. 7785, pp. 1–22. Springer, Heidelberg (2013)
12. Fortnow, L., Shaltiel, R.: Recent developments in explicit constructions of extractors, 2002. Bull. EATCS **77**, 67–95 (2002)
13. Kanukurthi, B., Reyzin, L.: Key agreement from close secrets over unsecured channels. In: Joux, A. (ed.) EUROCRYPT 2009. LNCS, vol. 5479, pp. 206–223. Springer, Heidelberg (2009)
14. Li, X.: Non-malleable extractors, two-source extractors and privacy amplification. In: FOCS 2012, pp. 688–697 (2012)
15. Maurer, U.M., Wolf, S.: Privacy amplification secure against active adversaries. In: Kaliski Jr., B.S. (ed.) CRYPTO 1997. LNCS, vol. 1294, pp. 307–321. Springer, Heidelberg (1997)

16. Maurer, U.M., Wolf, S.: Secret-key agreement over unauthenticated public channels III: privacy amplification. IEEE Trans. Inf. Theory **49**(4), 839–851 (2003)
17. Naor, J., Naor, M.: Small-bias probability spaces: efficient constructions and applications. SIAM J. Comput. **22**(4), 838–856 (1993)
18. Nisan, N., Zuckerman, D.: Randomness is linear in space. J. Comput. Syst. Sci. **52**(1), 43–52 (1996)
19. Rao, A.: An exposition of Bourgain's 2-source extractor. Technical report TR07-34, ECCC (2007). http://eccc.hpi-web.de/eccc-reports/2007/TR07-034/index.html
20. Raz, R.: Extractors with weak random seeds. In: STOC 2005, pp. 11–20 (2005)
21. Renner, R.S., Wolf, S.: Unconditional authenticity and privacy from an arbitrarily weak secret. In: Boneh, D. (ed.) CRYPTO 2003. LNCS, vol. 2729, pp. 78–95. Springer, Heidelberg (2003)
22. Vadhan, S.: Randomness extractors and their many guises: invited tutorial. In: FOCS 2002, p. 9 (2002)
23. Wolf, S.: Strong security against active attacks in information-theoretic secret-key agreement. In: Ohta, K., Pei, D. (eds.) ASIACRYPT 1998. LNCS, vol. 1514, pp. 405–419. Springer, Heidelberg (1998)
24. Zuckerman, D.: Linear degree extractors and the inapproximability of max clique and chromatic number. In: Theory of Computing 2007, pp. 103–128 (2007)

Efficiently Simulating High Min-entropy Sources in the Presence of Side Information

Maciej Skórski[(✉)]

University of Warsaw, Warsaw, Poland
maciej.skorski@mimuw.edu.pl

Abstract. In this paper we show that every source X having very high min-entropy conditioned on side information Z, can be efficiently simulated from Z. That is, there exists a simulator $\mathsf{Sim}(\cdot)$ such that (a) it is efficient, (b) $(\mathsf{Sim}(Z), Z)$ and (X, Z) are indistinguishable and (c) the min-entropy of $\mathsf{Sim}(Z)$ and X given Z is (almost, up to a few bits) the same. Concretely, the simulator achieves (s, ϵ)-indistinguishability running in time $s \cdot \operatorname{poly}\left(\frac{1}{\epsilon}, 2^{\Delta}, |Z|\right)$, where Δ is the entropy deficiency and $|Z|$ is the length of Z.

This extends the result of Trevisan, Tulsani and Vadhan (CCC '09), who proved a special case when Z is empty.

Our technique is based on the standard min-max theorem combined with a new convex L_2-approximation result resembling the well known result due to Maurey-Jones-Barron.

Keywords: Simulating high entropy · Min-entropy · Convex L_2-approximation

1 Introduction

Simulatability. Entropy sources we have in practice are far from ideal and suffer from many limitations, in particular exact sampling may be too expensive or time consuming. In this paper we study the possibility of sampling a given entropy source X only approximately, but in an efficient way and preserving its entropy; everything against computationally bounded adversaries. Such a feature is called *simulatability*. In a more realistic setting, the simulator may know some information Z about X and the entropy is conditioned on Z. The existence of such a simulator for a given distribution (and possibly side information) is not only an elegant foundational question. Simulatable high entropy sources are important in memory delegation [CKLR11], and the (more general) problem of simulating a given random variable (not necessarily of high entropy) is important in leakage-resilient cryptography [JP14, VZ13, SPY13] and even in the theory of zero-knowledge [CLP13].

This work was partly supported by the WELCOME/2010-4/2 grant founded within the framework of the EU Innovative Economy Operational Programme.

© Springer International Publishing Switzerland 2015
A. Biryukov and V. Goyal (Eds.): INDOCRYPT 2015, LNCS 9462, pp. 312–325, 2015.
DOI: 10.1007/978-3-319-26617-6_17

1.1   Problem Statement

Informal statement. The discution above motivates the following question

> **Problem:** Given a possibly inefficient entropy source X and (possibly
> long) side information Z, can one design a simulator Sim, which takes Z
> on its input and satisfy the following postulates:
> (a) *Efficiency*: the simulator Sim is efficient.
> (b) *Accuracy*: the distributions (X, Z) and $(\text{Sim}(Z), Z)$ are close
> (c) *Entropy Preserving*: the conditional min-entropy of $\text{Sim}(Z)$ and X
>   given Z are the same

Precise requirements. To have a meaningful statement, we want the simula-
tion cost to be at most polynomial in $|Z|$, which is motivated by applications
where side information might be *noisy* (i.e. very long but still entropy is high);
such settings are of theoretical and practical concern [FH]. We also require the
simulator to be at most polynomial in the accuracy (which is parametrized by
computational indistinguishability). Finally, we allow the entropy of the simulator output to be smaller by a constant than the entropy of X which doesn't
matter in most of applications. Generally speaking, we want to construct a simulator by *trading accuracy for efficiency*. Note however, that the simulator has
to be more complicated than the required indistinguishability, see discussions in
[TTV09, JP14].

1.2   Some Technical Issues

Approximating distributions in the total variation. Our problem regarded as
a two player game has a very common form: the first player plays with functions (distinguishers) and the second player plays with distributions. It might be
tempting then to think that the application of the min-max theorem solves the
problem as in similar cases, except maybe not best possible complexity. However, in our case the set of pure strategies corresponding to the distributions is
not convex! Indeed, we not only want to have a high entropy distribution, but
also make it efficiently samplable. Thus, at some point in the proof, we need
to approximate combinations of efficiently samplable distributions in the total
variation (statistical distance). Doing this by a standard application of the Chernoff Bound leads to an unacceptable cost equal the size of the domain! This is
because to make the variation distance smaller than ϵ, we need to approximate
every point mass up to $\epsilon/2^n$ where n is the length of X. Therefore, more clever
approximation techniques are required.

1.3   Related Works

Trevisan, Tulsani and Vadhan [TTV09] gave a bunch of "decomposition" results,
in particular they studied the problem of efficiently simulating a high min-
entropy source X. Their analysis however does not cover side information.

Indeed, from their result one can only derive a simulator for some (X', Z') which is close (X, Z) but the marginal distribution Z' is not guaranteed to be identical to Z. This distinction is important in settings where Z is information that an adversary might have learned.

1.4   Our Contribution

Summary. We prove that every distribution with high conditional min-entropy is efficiently simulatable (indistinguishable from a samplable distribution having essentially the same amount of entropy.

Simulating high min-entropy entropy sources with side information. We give the following positive answer to the posted question

Theorem 1 (Simulating High Min-entropy Sources with Side Information, Informal). *Let $X \in \{0,1\}^n$ be an min-entropy source and $Z \in \{0,1\}^m$ ba a side information, such that X conditioned on Z has $n - \Delta$ bits of min-entropy[1]. Let $\mathcal{D}$ be a class of $[0,1]$-valued functions on n-bits. Then there exists a random variable $Y \in \{0,1\}^n$ such that*

(a) $(Y, Z) = (\mathsf{Sim}(Z), Z)$, where Sim is a randomized function $2^{O(\Delta)}\epsilon^{-O(1)}$ times more complex than $\mathcal{D}$

(b) Y and X are ϵ-indistinguishable given Z by functions in $\mathcal{D}$

(c) the conditional min-entropy of Y and X given Z differ by at most $O(1)$ bits.

The exact parameters are given in Theorem 2.

It would be nice to have a boosting argument, which could perhaps reduce the complexity by a factor of $\epsilon^{\Omega(1)}$. However, it might be very complicated because of the consistency issue. These problems appear in [JP14] and lead to severe complications in the proof. In fact, we do not expect a boosting proof to be simpler than the min-max theorem based argument, which is already involved.

Proof techniques of independent interests. To get rid of the exponential dependency on the domain, we replace the popular argument based on the Chernoff Bound [BSW03] (see also Sect. 1.2) by a more delicate approximation technique. Our auxiliary result on L_2-convex approximation is actually an extension of the classical theorem attributed to Maurey, Jones and Barron. This is the more tricky part of our proof. See Sect. 3 for more details.

Impact and possible applications. The problem we solved is very foundational and certainly deserves explanation. But we also believe that the result we present improves the understanding of high entropy sources and might be used elsewhere, in particular in cryptography. In fact, simulatability condition already appeared in works on memory delegation [CKLR11]. Moreover, what we proved suggests

---

[1] There are two ways to define conditional min-entropy as explained in Sect. 2. Our result holds in any case.

that it might be possible to extend the other results in [TTV09, JP14, VZ13] to the conditional case, gaining a huge improvement in interesting entropy regimes where the deficiency is small. Replacing the dimension by the deficiency in these results would be a remarkable and important result.

2   Preliminaries

Entropy. In information theory the notion of Shannon entropy is a fundamental measure of randomness. In cryptography we use min-entropy.

Definition 1. *The* **min-entropy** *of a variable X is*

$$\mathbf{H}_\infty(X) = -\log \max_x \Pr[X = x]$$

More generally, for a joint distribution (X, Z), the **average min-entropy** *of X conditioned on Z is*

$$\tilde{\mathbf{H}}_\infty(X|Z) = -\log \mathbb{E}_{z \leftarrow Z} \max_x \Pr[X = x|Z = z]$$

$$= -\log \mathbb{E}_{z \leftarrow Z} 2^{-H_\infty(X|Z=z)} .$$

and **worst-case min-entropy** *of X conditioned on Z is*

$$\mathbf{H}_\infty(X|Z) = -\log \max z \max_x \Pr[X = x|Z = z]$$

Sometimes the second conditional notion is more convenient to work with. We can go back to average entropy losing only $\log(1/\epsilon)$ in the amount

Lemma 1 (From Average to Worst-Case Conditional Min-entropy). *For any pair (X, Z) there exists a pair (X', Z) such that*

(a) (X, Z) and (X', Z) are ϵ-close
(b) $\tilde{\mathbf{H}}_\infty(X'|Z) \geqslant \mathbf{H}_\infty(X|Z) - \log(1/\epsilon)$.

Probabilities. By U_S we denote the uniform distribution over the given set S, we omit the subscript if it is clear from the context. When $S = \{0, 1\}^n$ we also use the shortcut $U_S = U_n$. By $\mathbf{P}_Y$ we denote the probability mass function of the given distribution Y. By $Y_{Z|=z}$ we denote the conditional distribution $\Pr[Y = x|Z = z]$. Sometimes we slightly abuse the notation and write $Y \overset{d}{=} p_1 \cdot Y_1 + p_2 \cdot Y_2$ by which we man that the distribution of Y is a convex combination of distributions of Y_1 and Y_2 with coefficients p_1, p_2, that is $\mathbf{P}_Y = p_1 \cdot \mathbf{P}_{Y_1} + p_2 \cdot \mathbf{P}_{Y_2}$.

Computational and variational distance. Given a class $\mathcal{D}$ of functions (typically taking values in $[0, 1]$ or $[-1, 1]$) we say that the distributions X_1, X_2 are $(\mathcal{D}, \epsilon)$-indistnguishable if for every $D \in \mathcal{D}$ we have $|\mathbb{E}D(X_1) - \mathbb{E}D(X_2)| \leqslant \epsilon$. If $\mathcal{D}$ is the class of all circuits of size s then we also say about (s, ϵ)-indistinguishability. If $\mathcal{D}$ consists of all possible boolean functions then we get the $\max_D |\mathbb{E}D(X_1) - \mathbb{E}D(X_2)| = d_{TV}(X_1; X_2)$ where $d_{TV}(X_1; X_2)$ is the variation distance, defined as $d_{TV}(X_1; X_2) = \frac{1}{2} \sum_x |\mathbf{P}_{X_1}(x) - \mathbf{P}_{X_2}(x)|$.

Vectors and scalar products. Sometimes we slightly abuse the notation and understand $\mathbf{P}_Y$ as a vector which assigns the probability mass to each point in the range of Y. Also, we can think of a function $D : \{0,1\}^d \to \mathbb{R}$ as a vector with 2^d real coefficients. By '·' we denote not only the multiplication sign, but also the scalar product. Given a vector v, we denote $v^2 = v \cdot v$. In particular, $(\mathbf{P}_Y)^2 = \sum_x \mathbf{P}_Y(x)^2$ and $\mathbb{E}D(Y) = D \cdot \mathbf{P}_Y$.

3   A New Result on Convex L_2-approximation

We need the following technical result, which can be understand as a conditional version of the famous Maurey-Jones-Barron Theorem (see Lemma 1 in [Bar93]) on approximating convex hulls in Hilbert spaces. It shows how to approximate long convex combinations in L_2-norm by much shorter combinations.

Lemma 2 (Convex Approximation). *Let* (Y,Z), $\{(Y_i, Z)\}_{i \in I}$ *be random variables such that the distribution of* (Y,Z) *is a convex combination of distributions* (Y_i, Z). *Then for any t there are indexes $i_1, \ldots, i_t \in I$ such that if we define* $\mathbf{P}_{Y'|Z} = \frac{1}{t} \sum_{j=1}^t \mathbf{P}_{Y_{i_j}|Z}$ *then*

$$\mathbb{E}_{z \leftarrow Z} \left(\mathbf{P}_{Y'|Z=z} - \mathbf{P}_{Y|Z=z} \right)^2 \leqslant \frac{\max_i \mathbb{E}_{z \leftarrow Z} \mathrm{Var}_i \left(\mathbf{P}_{Y_i|Z=z} \right)}{t} \tag{1}$$

The proof appears in Appendix A.

4   Main Result

4.1   Statement

Below we present our main result with some discussion given in the remarks.

Theorem 2 (High Conditional Min-entropy is Simulatable). *Let* $X \in \{0,1\}^n$ *and* $Z \in \{0,1\}^m$ *be correlated random variables and* $\mathbf{H}_\infty(X|Z) = n - \Delta$. *Then there exists a distribution Y, Z such that*

(a) *There is a circuit* Sim *of complexity* $O\left(n(n+m)2^{2\Delta}\epsilon^{-5}\right)$ *and such that* $\mathsf{Sim}(Z) \overset{d}{=} Y$

(b) (X, Z) *and* (Y, Z) *are* (s, ϵ)-*indistinguishable*

(c) *We have* $\mathbf{H}_\infty(Y|Z) \geqslant n - \Delta - 6$.

Remark 1 (A version for average conditional entropy). We stress that it holds also for average min-entropy, that is when $\widetilde{\mathbf{H}}_\infty(X|Z) = n - \Delta$, with the simulating complexity $O\left(n(n+m)2^{2\Delta}\epsilon^{-7}\right)$.

Remark 2 (Independency on m). With a little bit of more effort one can replace m by $\log(1/\epsilon)$ and make the result completely independent on m. Even stated as above, the factor $n + m$ does not matter in asymptotic settings. Indeed, the circuits which computes Sim must be of size at least $m = |Z|$ in order to read its input.

4.2   Proof of Theorem 2

How to fool a boolean distinguisher. We will show that for any fixed boolean D there exists an efficiently samplable source Y such that $\mathbf{ED}(X) \leqslant \mathbf{ED}(Y) + \epsilon$ and $\mathbf{H}_\infty(Y) \geqslant \mathbf{H}_\infty(X) - 3$. The idea is very simple: if we don't care about computational issues, we just define $p = \min(2^{n-k}\mathbf{ED}(U), 1)$ and set

$$Y \stackrel{d}{=} p \cdot U_{\{x:\ D(x)=1\}} + (1-p) \cdot U_n$$

It is easy to check that $\mathbf{H}_\infty(Y) \geqslant k-1$ and that $\mathbf{ED}(X) \leqslant \min(2^{-k}|D|, 1) = p \leqslant \mathbf{ED}(Y)$. To make this distribution efficiently samplable, we need to compute the weight p and sample from the set $\{x :\ D(x) = 1\}$. The sampling is relatively easy when the set is dense in the domain but the efficiency decreases exponentially in the density, see Proposition 1. However, it turns out that this is the case when p is very small, so that Y is close to U_n. This observation is the heart of the proof for the computational case. First, we give a formal proof of the simple observation about sampling from the dense subset

Proposition 1. *Let* $\mathrm{D} :\ \{0,1\}^n \rightarrow \{0,1\}$ *be a function of size s and let* $\delta = \mathbf{ED}(U)$ *and* $\epsilon > 0$. *Then there is a distribution* $\widetilde{U}_D$ *samplable in time* $O\left(\log(1/\epsilon) \cdot \delta^{-1}\right)$ *which is ϵ-close to* $U_{\{x:\ D(x)=1\}}$ *and has at least the same min-entropy.*

Proof. We define $\widetilde{U}_D$ as follows: we sample x at random, reject and repeat if $\mathrm{D}(x) = 0$, and return x if $D(x) = 1$ or the number of rejections exceeds $t = \log(1/\epsilon) \cdot \delta^{-1}$ times. This distribution is given by

$$\widetilde{U}_D \stackrel{d}{=} (1 - (1-\delta)^t) \cdot U_{\{x:\ D(x)=1\}} + (1-\delta)^t \cdot U \tag{2}$$

Note that $d_{TV}(\widetilde{U}_D, U_{\{x:\ D(x)=1\}}) \leqslant (1 - \delta)^t \leqslant \epsilon$ and that we also have $\mathbf{H}_\infty\left(\widetilde{U}_D\right) \geqslant \mathbf{H}_\infty\left(U_{\{x:\ D(x)=1\}}\right)$ □

Now we are in position to prove the main lemma:

Lemma 3. *Let* $X \in \{0,1\}^n$ *be a random variable of min-entropy at least $k = n - \Delta$. There exists a uniform probabilistic algorithm $\mathcal{S} = \mathcal{S}\left(1^n, \mathrm{D}, \Delta, \epsilon\right)$ that for any boolean D, using only $O\left(2^\Delta/\epsilon\right)$ queries to D and in time $O\left(n \cdot 2^\Delta/\epsilon\right)$, outputs a random variable $Y \in \{0,1\}^n$ that has min-entropy $k - 3$ and satisfies*

$$\mathbf{ED}(X) \leqslant \mathbf{ED}(Y) + \epsilon \tag{3}$$

Proof. Define $\delta = 2^{-(\Delta+2)}\epsilon$. Let $\mu = \mathbf{ED}(U)$ and $\bar{\mu} = \frac{1}{\ell}\sum_{i=1}^{\ell} \mathrm{D}(x_i)$ where x_i are sampled at random and independently. If $\bar{\mu} < 2\delta$ then we output random x. Otherwise if $\bar{\mu} > 2\delta$, we consider the distribution Y_0 defined as follows: we sample $x \leftarrow \{0,1\}^n$ and check if $\mathrm{D}(x) = 1$; if this is true we output x otherwise we reject it and repeat - but no more than t times. We define Y to be Y_0 with probability $p = \min(2 \cdot 2^\Delta \bar{\mu}, 1)$ and U with probability $1 - p$ (see Algorithm -). The values for t and ℓ we set so that $t > n/\delta$ and $\ell > n/\delta$; now observe that, by definition

---

Function FOOLBOOLEAN(D)

Input : accuracy ϵ, deficiency Δ, $\{0,1\}$-valued D
Output: $x \in \{0,1\}^n$

1 $\delta \leftarrow 2^{-2-\Delta}\epsilon$, $\ell = \lceil 2n/\delta \rceil$, $t = \lceil 2n/\delta \rceil$
2 **for** $i = 1$ **to** ℓ **do**
3 $\quad \mid \quad x_i \leftarrow \{0,1\}^n$
4 **end**
5 $\bar{\mu} = \frac{1}{\ell}\sum_{i=1}^{\ell} D(x_i)$ /* estimating $\bar{\mu} \approx \mathbb{ED}(U)$ */
6 $p = \min(2 \cdot 2^{\Delta}\bar{\mu}, 1)$
7 $x \leftarrow \{0,1\}^n$
8 **if** $\bar{\mu} < 2\delta$ **then**
9 $\quad \mid \quad$ **return** x /* w.h.p. $\mathbb{ED}(U) < \epsilon$ */
10 **else**
11 $\quad \mid \quad i \leftarrow 0, b \leftarrow 0$
12 $\quad \mid \quad$ **while** $b = 0$ **and** $i < t$ **do**
13 $\quad \mid \quad \quad \mid \quad x_0 \leftarrow \{0,1\}^n$
14 $\quad \mid \quad \quad \mid \quad$ **if** $D(x_0) = 0$ **then**
15 $\quad \mid \quad \quad \mid \quad \quad \mid \quad i \leftarrow i+1$
16 $\quad \mid \quad \quad \mid \quad$ **else**
17 $\quad \mid \quad \quad \mid \quad \quad \mid \quad b \leftarrow 1$ /* success: a point x_0 s.t. $D(x_0) = 1$ */
18 $\quad \mid \quad \quad \mid \quad$ **end**
19 $\quad \mid \quad$ **end**
20 $\quad \mid \quad r \leftarrow [0,1]$ /* random real number */
21 $\quad \mid \quad$ **if** $r \leqslant p$ **then**
22 $\quad \mid \quad \quad \mid \quad$ **return** x_0
23 $\quad \mid \quad$ **else**
24 $\quad \mid \quad \quad \mid \quad$ **return** x
25 $\quad \mid \quad$ **end**
26 **end**

(a) if $\bar{\mu} < 2\delta$ then $Y \overset{d}{=} U$
(b) if $\bar{\mu} > 2\delta$ then Y_0 is $(1-\mu)^t$-close to the distribution uniform over the set $\{x : D(x) = 1\}$.

By the Multiplicative Chernoff Bound[2] we obtain the following properties

Claim. If $\mu < \delta$ then with probability $1 - \exp(-\ell\delta)$

$$\bar{\mu} \leqslant (1 + \delta/\mu) \cdot \mu \leqslant \mu + \delta \tag{4}$$

Claim. If $\mu > 4\delta$ then with probability $1 - \exp(-\ell\delta)$

$$\frac{\mu}{2} \leqslant \bar{\mu} \leqslant 2\mu \tag{5}$$

---

[2] Here we slightly abuse the notation and for simplicity write $\exp(\cdot)$ meaning the exponential function at such a base such that the written inequalities are valid.

We chose $\ell > n/\delta$. Consider the case $\mu > 4\delta$, when we are with high probability in case (b). Observe that for $t > n/\delta$ the distribution Y_0 is 2^{-n}-close to the distribution uniform over $\{x : \mathrm{D}(x) = 1\}$. Therefore, for $\epsilon > 2 \cdot 2^{-n}$ we obtain

$$
\begin{aligned}
\mathbf{ED}(Y) &\geqslant \mathbf{E}_{x_1,\ldots,x_\ell}[\mathrm{D}(Y)|\bar{\mu} > 2\delta] - 2^{-n} \\
&\geqslant \mathbf{E}_{x_1,\ldots,x_\ell}[p\mathbf{ED}(Y_0) + (1-p)\mathbf{ED}(U)] \\
&\geqslant \min(2^{\Delta}\mu, 1) + (1-p)\mathbf{ED}(U) - 2 \cdot 2^{-n} \\
&> \min(2^{\Delta}\mathbf{ED}(U), 1) - \epsilon \\
&\geqslant \mathbf{ED}(X) - \epsilon
\end{aligned}
\tag{6}
$$

In turn the min-entropy of Y can be estimated as follows

$$
\begin{aligned}
\Pr[Y = x] &\leqslant \Pr_{x_1,\ldots,x_\ell}[Y = x|\bar{\mu} > 2\delta] + 2^{-n} \\
&\leqslant \mathbf{E}_{x_1,\ldots,x_\ell}[p\Pr[Y_0 = x] + (1-p)\Pr[U = x]] + 2^{-n} \\
&\leqslant 2^{\Delta+2}\mu \cdot /|\mathrm{D}| + 4 \cdot 2^{-n} \\
&= 2^{\Delta+2} \cdot 2^{-n} + 4 \cdot 2^{-n} \\
&\leqslant 2^{\Delta+3} \cdot 2^{-n} = 2^{-k+3}
\end{aligned}
\tag{7}
$$

The case $\mu < \delta$ is much simpler because then trivially

$$
\mathbf{ED}(X) \leqslant 2^{\Delta}\mathbf{ED}(U) < \epsilon < \mathbf{ED}(Y) + \epsilon,
\tag{8}
$$

and since (a) holds with probability $1 - 2^{-n}$,

$$
\Pr[Y = x] \leqslant \Pr_{x_1,\ldots,x_\ell}[Y = x|\bar{\mu} < 2\delta] + 2^{-n} = 2 \cdot 2^{-n}.
\tag{9}
$$

It remains to consider the case when $\delta < \mu < 4\delta$. It is easy to observe that, as before

$$
\mathbf{E}_{x_1,\ldots,x_\ell}[\mathrm{D}(Y)|\bar{\mu} > 2\delta] \geqslant \mathbf{ED}(X) - \epsilon
\tag{10}
$$

and, trivially

$$
\mathbf{E}_{x_1,\ldots,x_\ell}[\mathrm{D}(Y)|\bar{\mu} < 2\delta] \geqslant \mathbf{ED}(X) - \epsilon
\tag{11}
$$

Therefore, no matter what the probabilities of $\bar{\mu} > 2\delta$ and $\bar{\mu} < 2\delta$ are, we obtain $\mathbf{ED}(X) \leqslant \mathbf{ED}(Y) + \epsilon$. Similarly, we obtain

$$
\Pr[Y = x|\bar{\mu} > 2\delta] \leqslant 2^{-k+3}
\tag{12}
$$

and

$$
\Pr[Y = x|\bar{\mu} < 2\delta] \leqslant 2^{-n}
\tag{13}
$$

and again, $\mathbf{H}_\infty(Y) \geqslant k - 3$ independently of the probabilities of $\bar{\mu} > 2\delta$ and $\bar{\mu} < 2\delta$. $\qquad\square$

It is easy to see that what we have proven applies also to high-min entropy distributions in the conditional case. We simply apply the last result to $D(\cdot, z)$.

Corollary 1. *Let $X \in \{0,1\}^n$ and $Z \in \{0,1\}^m$ be random variables such that $\mathbf{H}_\infty(X|Z) \geqslant n - \Delta$. There exists a uniform probabilistic algorithm $S = S(1^n, D, \Delta, \epsilon, Z)$ that for any boolean D on $\{0,1\}^{n+m}$ and distribution Z, using only $O(2^\Delta/\epsilon)$ queries to D and in time $O(n \cdot 2^\Delta/\epsilon)$, outputs a random variable $Y(Z) \in \{0,1\}^n$ that satisfies*

$$\mathbf{H}_\infty(Y(Z)|Z) \geqslant n - \Delta - 3 \tag{14}$$

and

$$\mathbf{ED}(X, Z) \leqslant \mathbf{ED}(Y(Z), Z) + \epsilon \tag{15}$$

How to fool a real-valued distinguisher. The statement and the efficiency is similar for the real-valued case, however the proof becomes a bit complicated and is referred to Appendix B. The main idea is to discretize the distinguisher D and approximate by boolean distinguishers.

Lemma 4. *Let $X \in \{0,1\}^n$ be a random variable of min-entropy at least $n - \Delta$. There exists a uniform probabilistic algorithm $S = S(1^n, D, \Delta, \epsilon)$ that for any $D : \{0,1\}^n \to [0,1]$, using only $\mathcal{O}(2^\Delta/\epsilon)$ queries to $\mathcal{D}$ and in time $\mathcal{O}(n \cdot 2^\Delta/\epsilon)$, outputs a random variable $Y \in \{0,1\}^n$ that satisfies*

$$\mathbf{ED}(X) \leqslant \mathbf{ED}(Y) + \epsilon \tag{16}$$

and[3]

$$\mathbf{H}_\infty(Y) \geqslant n - \Delta - 6 \tag{17}$$

How to fool a real conditional distinguisher. Again, it is easy to see that the result above applies also to high-min entropy distributions in the conditional case. This is because $\mathbf{H}_\infty(Y|Z) \geqslant k$ means $\mathbf{H}_\infty(Y|Z = z) \geqslant k$ for every z.

Corollary 2 (Fooling Real Distinguisher, Conditional Case). *Let $X \in \{0,1\}^n$ and $Z \in \{0,1\}^m$ be random variables such that $\mathbf{H}_\infty(X|Z) \geqslant n - \Delta$. There exists a uniform probabilistic algorithm $S = S(1^n, D, \Delta, \epsilon, Z)$ that for any boolean D on $\{0,1\}^{n+m}$ and distribution Z, using only $O(2^\Delta/\epsilon)$ queries to D and in time $O(n \cdot 2^\Delta/\epsilon)$, outputs a random variable $Y(Z) \in \{0,1\}^n$ that satisfies*

$$\mathbf{ED}(X) \leqslant \mathbf{ED}(Y) + \epsilon \tag{18}$$

and[4]

$$\mathbf{H}_\infty(Y) \geqslant n - \Delta - 6 \tag{19}$$

---

[3] The constant 6 can be replaced by 1, and even by any arbitrary small number, at the price of increasing the constant hidden under the asymptotic notation.

[4] The constant 6 can be replaced by 1, and even by any arbitrary small number, at the price of increasing the constant hidden under the asymptotic notation.

How to fool the whole class of distinguishers. We have shown that one can simulate against every single distinguisher. By the min-max theorem and a non-trivial approximation argument we switch the order of quantifiers, obtaining one simulator for the entire class of distinguishers. Indeed, let $\mathcal{D}'$ be a class of distinguishers of size $s' = se^2/(n+m)$, $\mathcal{D}$ be the class of distinguishers of size s and let $\mathcal{Y}$ be the set of all distributions of the pairs (Y', Z) such that $\mathbf{H}_\infty(Y'|Z) \geqslant n - \Delta - 6$ and Y' is simulatable in time $O\left(s \cdot 2^\Delta \epsilon^{-1}\right)$ from Z. Directly from Lemma 4 we obtain

$$\min_{D \in \mathcal{D}} \max_{(Y,Z) \in \mathcal{Y}} \left(\mathbb{E}D(X, Z) - \mathbb{E}D(Y, Z) \right) \leqslant \epsilon$$

By the Chernoff Bound it's easy to see that for every $D \in \text{conv}(\mathcal{D}')$ there exists D' of size only $s' \cdot (m+n)\epsilon^{-2} = s$ such that $|D(x, z) - D'(x, z)|$ for all x, z. This means that

$$\min_{D \in \text{conv}(\mathcal{D}')} \max_{(Y,Z) \in \mathcal{Y}} \left(\mathbb{E}D(X, Z) - \mathbb{E}D(Y, Z) \right) \leqslant 2\epsilon.$$

By the min-max theorem it is equivalent to

$$\max_{(Y,Z) \in \text{conv}(\mathcal{Y})} \min_{D \in \mathcal{D}'} \left(\mathbb{E}D(X, Z) - \mathbb{E}D(Y, Z) \right) \leqslant 2\epsilon. \tag{20}$$

This is not what we really want, because long convex combinations are inefficient to compute. We will remove the convex hull over $\mathcal{Y}$ by finding a shorter combination which satisfies similar inequality. The standard Chernoff Bound is not enough, as we need not the uniform error but the error in the variational norm. Consequently, the use of Chernoff Bound would lead to an unacceptable blowup of $2^{|Z|}$ in complexity. Therefore, we will use the L_2-norm approximation result in Lemma 2. Namely, for every $(Y, Z) \in \text{conv}(\mathcal{Y})$ and ℓ there is $(Y', Z) \in \text{conv}(\mathcal{Y})$ being a convex combination of only t terms and such that

$$\mathop{\mathbb{E}}_{z \leftarrow Z} \left(\mathbf{P}_{Y|Z=z} - \mathbf{P}_{Y'|Z=z} \right)^2 \leqslant \frac{\max_{(Y,Z) \in \mathcal{Y}} \mathbb{E}_{z \leftarrow Z} \mathbf{P}_{Y|Z=z}^2}{t}.$$

Since $(Y, Z) \in \mathcal{Y}$ we have $\mathbf{P}_{Y|Z=z}^2 \leqslant 2^{-(n-\Delta-6)}$ for every z, which means

$$\mathop{\mathbb{E}}_{z \leftarrow Z} \left(\mathbf{P}_{Y|Z=z} - \mathbf{P}_{Y'|Z=z} \right)^2 \leqslant t^{-1} 2^{-(n-\Delta-6)}$$

where squares are interpreted as scalar products. Now the above bound, the Cauchy-Schwarz Inequality, and the assumption

$$|\mathbb{E}D(Y', Z) - \mathbb{E}D(Y, Z)| = \left| \mathop{\mathbb{E}}_{z \leftarrow Z} D(\cdot, z) \cdot \left(\mathbf{P}_{Y'|Z=z} - \mathbf{P}_{Y|Z=z} \right) \right|$$

$$\leqslant \mathop{\mathbb{E}}_{z \leftarrow Z} \sqrt{D(\cdot, z)^2} \cdot \sqrt{\left(\mathbf{P}_{Y'|Z=z} - \mathbf{P}_{Y|Z=z} \right)^2}$$

$$\leqslant \sqrt{2^n} \cdot \sqrt{t^{-1} 2^{-(n-\Delta-6)}} = 8 \cdot 2^{-\frac{\Delta}{2}} t^{\frac{1}{2}}$$

and choosing $t = 2^\Delta \epsilon^{-2}$ we obtain from Eq. 20 that

$$\min_{D \in \mathcal{D}'} (\mathbb{E}D(X, Z) - \mathbb{E}D(Y', Z)) \leqslant 10\epsilon.$$

Since $\mathcal{D}$ can be assumed to be closed under complements (that is if $D \in \mathcal{D}$ then also $1 - D \in \mathcal{D}$ (with an additive loss of 1 in complexity) we get

$$\min_{D \in \mathcal{D}'} |\mathbb{E}D(X, Z) - \mathbb{E}D(Y', Z)| \leqslant 10\epsilon, \tag{21}$$

and the result follows. We only note that we lose uniformity due to the nonconstructive min-max theorem.

5   Conclusion

We proved that high conditional entropy sources can be efficiently simulated. We believe that the simulator costs can be improved much by a clever boosting argument, and that using our ideas (especially the L_2-approximation trick) one can prove other strong generalizations of different results in the literature. We leave this as a problem for further studies.

A   Proof of Lemma 2

Proof (Proof of Lemma 2). Suppose that $\mathbf{P}_{Y',Z} = \sum_j \theta_j \mathbf{P}_{Y_j,Z}$ is a convex combination of distributions $\{\mathbf{P}_{Y_j,Z}\}_i$. Let $i_1, \dots, i_t$ be sampled independently according to the probability measure given by $\{\theta_j\}_j$, that is $i_1 = j$ with probability θ_j. Define

$$\mathbf{P}_{Y',Z} = \frac{1}{t} \sum_{j=1}^{t} \mathbf{P}_{Y_{i_j},Z}$$

Using independency of i_j and the fact that $\mathbb{E}_{j_i} \mathbf{P}_{Y'_{i_j},Z} = \mathbf{P}_{Y_{i_j},Z}$ we obtain

$$\mathbb{E}_{i_1,\dots,i_t} \mathbb{E}_{z \leftarrow Z} (\mathbf{P}_{Y'|Z=z} - \mathbf{P}_{Y|Z=z})^2 = \frac{1}{t^2} \mathbb{E}_{z \leftarrow Z} \mathbb{E}_{i_1,\dots,i_t} \left(\sum_{j=1}^{t} \left(\mathbf{P}_{Y'_{i_j}|Z=z} - \mathbf{P}_{Y_{i_j}|Z=z} \right) \right)^2$$

$$= \frac{1}{t^2} \mathbb{E}_{z \leftarrow Z} \left[\sum_{j=1}^{t} \left(\mathbb{E}_{ij} \mathbf{P}^2_{Y'_{i_j}|Z=z} - \left(\mathbb{E}_{ij} \mathbf{P}_{Y'_{i_j}|Z=z} \right)^2 \right) \right]$$

which means that there is a choice of $i_1, \dots, i_j$ such that

$$\mathbb{E}_{z \leftarrow Z} (\mathbf{P}_{Y'|Z=z} - \mathbf{P}_{Y|Z=z})^2 \leqslant \frac{1}{t^2} \sum_{j=1}^{t} \mathrm{Var}_{i_j} \left(\mathbf{P}_{Y'_{i_j}|Z=z} \right)$$

and the result follows.   □

B Proof of Lemma 4

Proof Let $\Delta = n - k$. By replacing ϵ with 2ϵ we can assume that $\mathcal{D} = \sum_{i=1}^{j} \alpha_i \mathcal{D}_i$ where $\alpha_i = 1 - (i - 1)\epsilon$ for $i = 1, \ldots, \lceil 1/\epsilon \rceil$ and $\mathcal{D}_i$ are boolean such that $1 = \sum_i \mathcal{D}_i$. Define

$$d(i) = \Pr[\mathrm{D}(U) \geqslant \alpha_i]. \tag{22}$$

and let M be the smallest number i such that $d(i) \geqslant 2^{-\Delta}$. Note that if we didn't care about computational efficiency then the best answer would be

$$Y^+ \stackrel{d}{=} \frac{d(M-1)}{2^{-\Delta}} \cdot U_{\mathcal{D}_1 + \ldots + \mathcal{D}_{M-1}} + \frac{2^{-\Delta} - d(M-1)}{2^{-\Delta}} \cdot U_{\mathcal{D}_M} \tag{23}$$

because then

$$\mathbf{ED}(Y^+) = \frac{\sum_{i=1}^{M-1} \alpha_i |\mathrm{D}_i| + \left(2^k - \sum_{i=1}^{M-1} \alpha_i |\mathrm{D}_i|\right) \alpha_M}{2^k}$$

$$= \max_{Y:\, \mathbf{H}_\infty(Y) \geqslant k} \mathbf{ED}(Y) \tag{24}$$

The approach we chose is quite obvious - we efficiently approximate the distribution Y^+. For any i, sample $x_1, \ldots, x_\ell$ where $\ell > 2^\Delta n \log(1/\epsilon)/\epsilon$ and let

$$\tilde{d}(i) = \ell^{-1} \sum_{j=1}^{\ell} \mathbf{1}_{\{\mathrm{D}(x_i) \geqslant \alpha_i\}} \tag{25}$$

Now let M' be the smallest number such that $\tilde{d}(M') > \frac{3}{4} \cdot 2^{-\Delta}$. Note that that M' is well defined with probability $1 - 2^{-n}$, and then we have

$$\tilde{d}(M' - 1) < \frac{3}{4} \cdot 2^{-\Delta} < \tilde{d}(M') \tag{26}$$

Now we define Y as follows:

$$Y \stackrel{d}{=} \begin{cases} \frac{\tilde{d}(M'-1)}{2^{-\Delta}} \cdot \tilde{U}_{\mathcal{D}_1 + \ldots + \mathcal{D}_{M'-1}} + \left(1 - \frac{\tilde{d}(M'-1)}{2^{-\Delta}}\right) \cdot \tilde{U}_{\mathcal{D}_{M'}}, & 2^{-\Delta}\epsilon < \tilde{d}(M'-1) < 2^{-\Delta}/16 \\ \tilde{U}_{\mathcal{D}_1 + \ldots + \mathcal{D}_{M'-1}}, & 2^{-\Delta}/16 < \tilde{d}(M'-1) \\ \tilde{U}_{\mathcal{D}_{M'}}, & 2^{-\Delta}\epsilon > \tilde{d}(M'-1) \end{cases} \tag{27}$$

Observe that if $d(i) < 2^{-\Delta}/4$ then with probability $1 - 2^{-n}$ we get $\tilde{d}(i) < 2^{-\Delta}/2$. Thus, the probability that $d(M') < 2^{-\Delta}/4$ and $\tilde{d}(M') > 2^{-\Delta}/2$ is at most $2^{-n} \log(1/\epsilon)$ and we can assume that $d(M') > 2^{-\Delta}/4$. Similarly, if $d(i) > 2^{-\Delta}$ then with probability $1 - 2^{-n}$ we have $\tilde{d}(i) > \frac{3}{4} \cdot 2^{-\Delta}$ which means $M' \leqslant i$. Therefore, with probability $1 - 2^{-n} \log(1/\epsilon)$ we can assume that $d(M' - 1) < 2^{-\Delta}$. Now we split the analysis into the following cases

(a) $\tilde{d}(M'-1) < 2^{-\Delta}\epsilon$ and $d(M'-1) < 2 \cdot 2^{-\Delta}\epsilon$. Since $|D_{M'}| = 2^n(d(M') - d(M'-1)) \geqslant 2^{n-\Delta}/8$, we see that $U_{\mathcal{D}_{M'}}$ is samp「able in time $\mathcal{O}\left(2^{\Delta}\log(1/\epsilon)\right)$ and that $\mathbf{H}_\infty(U_{\mathcal{D}_{M'}}) \geqslant k-3$. Note that

$$
\begin{aligned}
\mathbf{ED}(Y^+) &= \mathbf{ED}(Y^+)\mathbf{1}_{\mathrm{D}(Y^+) \geqslant \alpha_{M'-1}} + \mathbf{ED}(Y^+)\mathbf{1}_{\mathrm{D}(Y^+) \leqslant \alpha_{M'}} \\
&\leqslant 2\epsilon + \alpha_{M'} \\
&\leqslant 3\epsilon + \mathbf{ED}(Y)
\end{aligned}
\tag{28}
$$

(b) $\tilde{d}(M'-1) > 2^{-\Delta}/16$ and $2^{-\Delta} > d(M'-1) > 2^{-\Delta}/32$. Then we have $|D_1| + \ldots + |D_{M'-1}| \geqslant 2^{n-\Delta-5}$ and thus $\mathbf{H}_\infty(\widetilde{U}_{\mathcal{D}_1+\ldots+\mathcal{D}_{M'-1}}) \geqslant n - \Delta - 5$ and $\widetilde{U}_{\mathcal{D}_1+\ldots+\mathcal{D}_{M'-1}}$ is samplable in time $\mathcal{O}\left(2^{\Delta}\log(1/\epsilon)\right)$. Since $|D_1| + \ldots + |D_{M'-1}| \leqslant 2^{n-\Delta}$, we have

$$
\begin{aligned}
\mathbf{ED}(Y^+) &\leqslant \mathbf{ED}(U_{\mathcal{D}_1+\ldots+\mathcal{D}_{M'-1}}) \\
&\leqslant \mathbf{ED}(\widetilde{U}_{\mathcal{D}_1+\ldots+\mathcal{D}_{M'-1}}) + \epsilon
\end{aligned}
\tag{29}
$$

(c) $2^{-\Delta}\epsilon < \tilde{d}(M'-1) < 2^{-\Delta}/16$ and $2^{-\Delta}\epsilon/2 < d(M'-1) < 2^{-\Delta}/8$ and $\tilde{d}(M'-1) \leqslant 2d(M'-1)$. We have $|D_1| + \ldots + |D_{M'-1}| = 2^n d(M'-1) > 2^{n-\Delta}\epsilon/2$ and $|D_{M'}| = 2^n(d(M') - d(M'-1)) \geqslant 2^{n-\Delta}/8$, therefore Y is samplable in time $\mathcal{O}\left(2^{\Delta}\log(1/\epsilon)/\epsilon\right)$. Moreover, we have $\mathbf{H}_\infty(\widetilde{U}_{\mathcal{D}_1+\ldots+\mathcal{D}_{M'-1}}) \geqslant \log(|D_1| + \ldots + |D_{M'-1}|)$ and $\mathbf{H}_\infty(\widetilde{U}_{\mathcal{D}_{M'}}) \geqslant \log|D_{M'}|$. Hence $\mathbf{H}_\infty(\widetilde{U}_{\mathcal{D}_1+\ldots+\mathcal{D}_{M'-1}}) \geqslant n + \log d(M'-1)$ and $\mathbf{H}_\infty(\widetilde{U}_{\mathcal{D}_{M'}}) \geqslant n - \Delta - 3$ and

$$
\begin{aligned}
\Pr[Y = x] &\leqslant \frac{\tilde{d}(M'-1)}{d(M'-1)} \cdot 2^{-n+\Delta} + 2^{-n+\Delta+3} \\
&\leqslant 2^{-n+\Delta+4}
\end{aligned}
\tag{30}
$$

Suppose now that $d(M'-1) < 2^{-\Delta}\epsilon/2$. Then, by the Chernoff Bound with probability $1 - 2^{-n}$ we have $\tilde{d}(M'-1) < 2^{-\Delta}\epsilon/2 + d(M'-1) < 2^{-\Delta}\epsilon$ and we are in case (a). If $2^{-\Delta}\epsilon/2 < d(M'-1) < 2^{-\Delta}/32$ then with probability $1 - 2^{-n}$ we have $\frac{1}{2} < \frac{\tilde{d}(M'-1)}{d(M'-1)} < 2$ and it is easy to check that we can be either in (a) or in (c), depending on $\tilde{d}(M'-1)$. If $2^{-\Delta}/32 < d(M'-1) < 2^{-\Delta}/8$ then with probability $1 - 2^{-n}$ we are either in (c) or in (b). If $2^{-\Delta}/8 < d(M'-1) < 2^{-\Delta}$ then with probability $1 - 2^{-n}$ we can be only in (b). $\qquad\square$

References

[Bar93] Barron, A.R.: Universal approximation bounds for superpositions of a sigmoidal function. IEEE Trans. Inf. Theory **39**(3), 930–945 (1993)

[BSW03] Barak, B., Shaltiel, R., Wigderson, A.: Computational analogues of entropy. In: Arora, S., Jansen, K., Rolim, J.D.P., Sahai, A. (eds.) RANDOM 2003 and APPROX 2003. LNCS, vol. 2764, pp. 200–215. Springer, Heidelberg (2003)

[CKLR11] Chung, K.-M., Kalai, Y.T., Liu, F.-H., Raz, R.: Memory delegation. In: Rogaway, P. (ed.) CRYPTO 2011. LNCS, vol. 6841, pp. 151–168. Springer, Heidelberg (2011)

[CLP13] Chung, K.-M., Lui, E., Pass, R.: From weak to strong zero-knowledge and applications. In: Dodis, Y., Nielsen, J.B. (eds.) TCC 2015, Part I. LNCS, vol. 9014, pp. 66–92. Springer, Heidelberg (2015)

[FH] Fuller, B., Hamlin, A.: Unifying leakage classes: simulatable leakage and pseudoentropy. In: Lehmann, A., Wolf, S. (eds.) Information Theoretic Security. LNCS, vol. 9063, pp. 69–86. Springer, Heidelberg (2015)

[JP14] Jetchev, D., Pietrzak, K.: How to fake auxiliary input. In: Lindell, Y. (ed.) TCC 2014. LNCS, vol. 8349, pp. 566–590. Springer, Heidelberg (2014)

[SPY13] Standaert, F.-X., Pereira, O., Yu, Y.: Leakage-resilient symmetric cryptography under empirically verifiable assumptions. In: Canetti, R., Garay, J.A. (eds.) CRYPTO 2013, Part I. LNCS, vol. 8042, pp. 335–352. Springer, Heidelberg (2013)

[TTV09] Trevisan, L., Tulsiani, M., Vadhan, S.: Regularity, boosting, and efficiently simulating every high-entropy distribution. In: CCC (2009)

[VZ13] Vadhan, S., Zheng, C.J.: A uniform min-max theorem with applications in cryptography. In: Canetti, R., Garay, J.A. (eds.) CRYPTO 2013, Part I. LNCS, vol. 8042, pp. 93–110. Springer, Heidelberg (2013)

Lightweight Cryptography

BitCryptor: Bit-Serialized Flexible Crypto Engine for Lightweight Applications

Ege Gulcan, Aydin Aysu[✉], and Patrick Schaumont

Secure Embedded Systems, Center for Embedded Systems for Critical Applications,
Bradley Department of ECE, Virginia Tech, Blacksburg, VA 24061, USA
{egulcan,aydinay,schaum}@vt.edu

Abstract. There is a significant effort in building lightweight cryptographic operations, yet the proposed solutions are typically single-purpose modules that can implement a single functionality. In contrast, we propose BitCryptor, a multi-purpose, compact processor for cryptographic applications on reconfigurable hardware. The proposed crypto engine can perform pseudo-random number generation, strong collision-resistant hashing and variable-key block cipher encryption. The hardware architecture utilizes SIMON, a recent lightweight block cipher, as its core. The complete engine uses a bit-serial design methodology to minimize the area. Implementation results on the Xilinx Spartan-3 s50 FPGA show that the proposed architecture occupies 95 slices (187 LUTs, 102 registers), which is 10× smaller than the nearest comparable multi-purpose design. BitCryptor is also smaller than the majority of recently proposed lightweight single-purpose designs. Therefore, it is a very efficient cryptographic IP block for resource-constrained domains, providing a good performance at a minimal area overhead.

Keywords: Lightweight cryptography · Bit-serialization · Hardware architecture · Crypto engine · SIMON · FPGA

1 Introduction

Lightweight cryptography studies the challenges of enabling security services on resource-constrained platforms. Typical applications on such devices require a protocol execution for secure key exchange or entity authentication. For example, the protocol with non-reversible functions (Sect. 6.1.5. of [12]) uses a PRNG, hash function and encryption, all within a single protocol run. Yet, most compact implementations in the literature are single-purpose building blocks that can perform only one of these three operations. How should a designer combine a multi-purpose requirement with an area resource-constraint? Figure 1 shows the advantage of our proposal compared to the traditional approaches. The straightforward approach is to include optimized single-purpose hardware blocks and to glue them with a finite-state machine wrapper. Clearly, a solution that uses these disjoint kernels (like PRESENT [7] for encryption, PHOTON [19] for hashing, and TRIVIUM [13] for PRNG) yields a design that is larger than

© Springer International Publishing Switzerland 2015
A. Biryukov and V. Goyal (Eds.): INDOCRYPT 2015, LNCS 9462, pp. 329–346, 2015.
DOI: 10.1007/978-3-319-26617-6_18

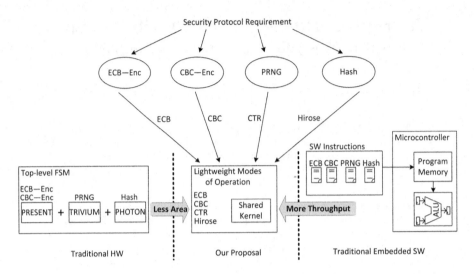

Fig. 1. Comparison of design methods for flexibility

the sum of its composing kernels. It also ignores the opportunity to share the internal designs for each kernel. Another solution would be to use embedded software on a compact microcontroller. The program memory of the microcontroller can store an instruction-level description of each operation, and can configure the small microarchitecture. But such a solution is not ideal either, because the instruction-set of the microcontroller is generic, and not optimized for the multi-purpose kernel which we have in mind. Therefore, we will evaluate a third option: the design of a flexible yet specialized crypto-engine.

In this paper, we propose BitCryptor, a bit-serialized compact crypto engine that can execute fundamental cryptographic operations. BitCryptor is a multi-purpose design that can perform PRNG, encryption and hash operations. Therefore, we are promoting BitCryptor as a generic lightweight crypto engine upon which protocols can be built as a sequence of BitCryptor commands. We show that the BitCryptor is significantly smaller than competing crypto engines and it has a better performance than low-cost microcontrollers.

1.1 Compact and Efficient Crypto Engine on FPGAs

ASIC technology offers a high integration density and a low per-unit price, yet there exist a myriad of applications where FPGAs are preferred over ASICs due to their lower NRE cost and reconfigurable nature. Wireless sensor nodes (WSN) [30], wearable computers (WC) [31], and Internet-of-Things (IoT) [25] are amongst such application domains that require compact solutions and still incorporate FPGAs. In addition to their primary functionality, secure systems in FPGAs need a method to perform cryptographic operations. Thus, the resource-constrained device should embody this method with low operational and area costs.

We are not the first to propose a multi-purpose design in FGPA, but our proposal is the smallest so far. BitCryptor occupies 95 slices, 12 % of available resources of a Spartan-3 s50 FPGA whereas the nearest competitor with similar functionalities [24] occupies 916 slices and cannot even fit into the same device. Hence, a system using [24] must migrate to a larger device (eg. Spartan-3 s200), effectively increasing the component cost. A larger device also increases the system cost, as it increases static power dissipation, and possibly PCB cost. So, the argument that it is always possible to use a larger FPGA, and thus that FPGA area optimization has little value, is not correct in the context of IoT, WSN, and so on.

One can argue the use of embedded microcontrollers for low-cost reconfigurable systems. However, these platforms are at a disadvantage in terms of operational costs: A recent work [14] shows that, compared to BitCryptor, encryption on a 16-bit MSP430 microcontoller needs 4.8× more clock cycles, 70.8× execution time, and 15.2× energy[1]. Alternatively, to achieve a higher operating frequency, the same general purpose MSP430 microcontroller can also be configured as a soft-core processor on FPGAs. However, this trivial approach is problematic as the resulting hardware occupies a very large area, requiring the system to again move to an expensive board. Therefore, a designer has to find the delicate balance between the area-cost, performance, and flexibility. BitCryptor is such a solution that offers multiple cryptographic operations at minimal area-cost and performance hit on reconfigurable hardware.

1.2 Novelty

Achieving the combination of area resource constraints with multi-purpose functionality requires sound cryptographic engineering. It requires picking a lightweight crypto kernel, applying specific functionalities with a careful analysis of modes-of-operations, selecting proper configuration parameters, employing an appropriate design methodology, and back-end engineering for EDA tool optimizations. In this paper, we guide through these steps to reveal how to realize a compact and multi-purpose crypto-engine on FPGA. We also provide detailed analysis on the trade-offs within the design space.

The major contributions of this work are as follows

- We demonstrate a multi-purpose design that is 10× smaller than the nearest comparable crypto-engine [24] and even smaller than the majority of single-purpose encryption and all hash function implementations.
- We develop a systematic design approach with optimizations at several abstraction levels.
- We show area-performance trade-offs between different serialization methods and on multiple platforms.

---

[1] The previous work implements a 128-bit security encryption with a fixed key, results section elaborates on comparisons.

Table 1. BitCryptor construction

Operation	Kernel and configuration	Modes-of-operation	Security-level
Encryption	SIMON 96/96	ECB, CBC	96-bits
Hash function	SIMON 96/144	Hirose [20]	96-bits[a]
PRNG	SIMON 96/96	CTR	96-bits

[a] SIMON 96/144 generates a digest of 192-bits which has 96-bits of strong collision resistance.

- We present a comparison with 8-bit, 16-bit, and 32-bit microcontroller designs and quantify the performance improvement of our solutions.
- We provide a small isolated security module that is easier to validate and certify.

1.3 Organization

The rest of the paper is organized as follows. Section 2 explains SIMON, the lightweight core of the crypto engine, and discusses high-level design parameters. Section 3 illustrates the bit-serial design methodology with a simple example. Section 4 describes the hardware architecture of BitCryptor. Section 5 shows the implementation results and its comparison to previous work and Sect. 6 concludes the paper.

2 High-Level Description of BitCryptor

Table 1 summarizes the design of BitCryptor. The heart of BitCryptor is a flexible block cipher, SIMON [6]. The flexibility of SIMON allows multiple key and block lengths. The choice of security-level (96-bits, corresponding to ECRYPT-II Level 5 or 'legacy standard-level' [36]) is a trade-off between selecting the shortest key length possible while offering reasonable security for the intended application domains. Using SIMON as the kernel, we then configure different mode-of-operations to achieve message confidentiality (encryption), message integrity (hashing), and pseudo-random number generation. Each row in Table 1 describes such a mode-of-operation. In all of these configurations, we maintain the selected 96-bit security level.

2.1 SIMON Block Cipher

The lightweight block cipher SIMON is developed by NSA, targeting compact hardware implementations [6]. So far, conventional cryptanalytic techniques against SIMON did not demonstrate any weaknesses [1,3,38]. Equations 1 and 2 formulate the SIMON round and key expansion functions respectively, and Fig. 2 illustrates them. SIMON has ten configurations with different block and key sizes, giving users the flexibility to choose the best one that fits into their applications requirements. Block size indicates the bit length of the input message

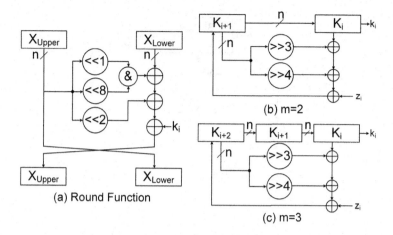

Fig. 2. (a) SIMON round function, (b) SIMON key expansion function for $m = 2$, (c) SIMON key expansion function for $m = 3$

to the block cipher while the key size is the bit length of the key. SIMON is a Feistel-based block cipher and the input block $(2n)$ is divided into two words, shown as the word size (n). The key is formed by (m) words making the key size (mn). SIMON using a block size $2n$ and key size mn is denoted as SIMON $2n/mn$.

$$R(X_u, X_l) = (X_u \lll 1) \wedge (X_u \lll 8) \oplus (X_u \lll 2) \oplus X_l \oplus k_i \qquad (1)$$

$$K(i + m) = \begin{cases} k_i \oplus (k_{i+1} \ggg 3) \oplus (k_{i+1} \ggg 4) \oplus z_i \; for \, m = 1 \\ k_i \oplus (k_{i+2} \ggg 3) \oplus (k_{i+2} \ggg 4) \oplus z_i \; for \, m = 2 \end{cases} \qquad (2)$$

2.2   Parameter Selection

The parameters we select directly affect the area and performance of the crypto engine. Typically, to reduce the area, lightweight cryptographic systems utilize shorter keys (80-bits). In our design, we aim to find the best configuration that will at least meet this security level while minimizing the area. We utilize SIMON 96/96 for symmetric key encryption and PRNG, and SIMON 96/144 for hashing.

One of the challenges in selecting the parameters of the crypto engine is to satisfy the security needs of the hash function. The security level of a hash is determined by the size of the output digest and the probability of a collision on the value of a digest. We choose the most stringent security constraint of strong collision resistance [27] which requires that a hash at a k-bit security level provides a 2k-bit digest. A common practice in building hash functions is to use a block cipher with single-block-length (SBL) constructions like Davies-Meyer [39] or double-block-length (DBL) constructions like Hirose [20]. In SBL, the output size of the hash function is equal to the block size of the underlying block cipher, while in DBL it is twice the block size. For lightweight applications, DBL is more

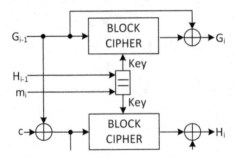

Fig. 3. Hirose double-block-length hash function

advantageous than SBL because it requires a smaller block cipher. To have a strong collision resistance of minimum 80-bits with SBL, the underlying block cipher must have a block and key size of at least 160-bits, which is in fact an undefined configuration for lightweight ciphers like SIMON. On the other hand, DBL can achieve the same level of security with a block size of only 80-bits, which will result in a smaller architecture.

Figure 3 shows the DBL Hirose construction. The input message m_i is concatenated with the chaining value H_{i-1} and fed into the key input. Both block ciphers use the same key generated by a single key expansion function. The Hirose construction requires a block cipher with a key size that is larger than the block size. The digest is the concatenation of the last two chaining values H_i and G_i. The computation equations of H_i and G_i are as follows.

$$H_i = E(G_{i-1} \oplus c, m_i \| H_{i-1}) \oplus G_{i-1} \oplus c \tag{3}$$

$$G_i = E(G_{i-1}, m_i \| H_{i-1}) \oplus G_{i-1} \tag{4}$$

The Hirose construction is regarded as secure if the underlying block cipher is secure. However, the security of SIMON is not yet evaluated for such a usecase. For the scope of this paper, rather than focusing on security proofs, we study practical lightweight instantiations. The configuration of SIMON that will be used in Hirose construction must have a block size that is at least 80-bits for strong collision resistance and it must have a key size that is larger than the block size. Therefore we select SIMON 96/144 because it gives us the most compact solution and provides a security level even stronger than the minimum requirements. The resulting hash function reads messages in 48-bit blocks and produces a 192-bit digest.

To minimize the area, the crypto engine shares the SIMON block cipher used in hash function to implement symmetric key encryption and PRNG. However, having a 144-bit key is unnecessary in both operations since it is beyond our security requirements. Therefore, the performance of the system improves if we use SIMON 96/96 which has the same block size but a shorter key. In [18], Gulcan et al. show that the flexible architecture of SIMON with all block and key sizes is still very compact. So, the crypto engine uses a flexible SIMON architecture with 96-bit key size for symmetric key encryption and PRNG, and 144-bit key size for hash function. Since only the key size is flexible, the number of words in the key expansion function changes while the datapath remains exactly the same.

For the implementation of the PRNG, the crypto engine uses SIMON 96/96 in counter mode of operation. The host system provides a 96-bit key (seed) as the source of entropy for the PRNG and is responsible to reseed the PRNG when necessary. In [15] authors suggest that a single key be used to generate at most 2^{16} blocks of random data. For a block size of 96-bits, this corresponds to approximately 2^{22} bits hence the PRNG module uses a 22-bit counter.

3 Design Methodology

The way to systematically reduce the area of a circuit is through sequential-ization; dividing operations in time and reusing the same resources for similar computations. In our design, we have applied bit-serialization [4], a sequentializa-tion methodology that processes one output bit at a time. We have adapted and applied this methodology with an architecture optimization using shift register logic (SRL-16) for the target FPGA technology.

3.1 Datapath

Figure 4 illustrates an example where the datapath computes $c = a \oplus b$ by XORing two 16-bit registers a and b, and generates the 16-bit output c. In this example, the datapath uses the same value of a multiple times while the value of b changes. If all the bits are processed in the same clock cycle (Fig. 4(a)), the datapath produces all bits of c in parallel. This datapath utilizes 48 registers (to store a, b, and c) and 16 LUTs (to compute 16 XOR operations of $c = a \oplus b$). We can map these elements to 24 slices.

If we bit-serialize the entire datapath (Fig. 4(b)), the resulting hardware will produce one output bit in one clock cycle. The 16-bit register blocks can now

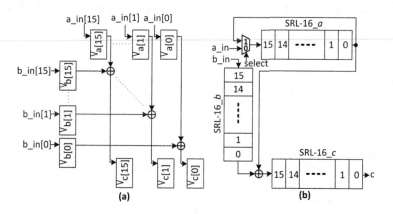

Fig. 4. (a) Bit-parallel datapath (b) Bit-serial datapath

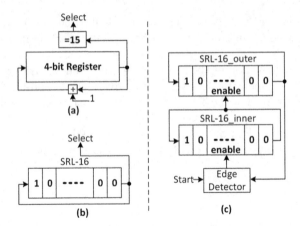

Fig. 5. (a) Control with up-counters (b) Control with ring counters (c) Control of nested loops

be mapped to SRL-16 logic and the output of a and b can be XORed using a single XOR gate. To keep the value of a, SRL-16_a should have a feedback from its output to input. Thus, the resulting hardware architecture will consist of 5 LUTs (3 SRL-16 to store a, b, and c, 1 LUT to compute the XOR operation and 1 LUT to apply the feedback via a multiplexer). Now, the datapath can be mapped to a total of 3 slices, which is one-eight of the size of the bit-parallel implementation.

3.2 Control

Bit-serialization comes with control overhead. If not dealt carefully, this can counteract the area gain of the datapath. In bit-serial designs, to identify when to start and end loading shift registers, and when to finish operations, we need to keep track of the bit positions during computations. In the example, since the value of a is fixed for a number of $c = a \oplus b$ executions, the control needs to determine the value of the select signal at the input multiplexer of SRL-16_a. It will select 0 while a_in is loaded, otherwise it will select 1. Usually, this is implemented with counters and comparators. Figure 5(a) shows a 4-bit counter with a corresponding comparator. In each clock cycle, the counter value increments by one and four registers update their values in parallel. A comparator checks the counter value and returns 1 when the check condition occurs. This architecture consists of 5 LUTs (4 LUTs for counter and 1 LUT for comparator) and 4 registers.

Instead of using an up-counter, the same functionality can be realized using a ring counter. Ring counters consist of circular shift registers. Figure 5(b) shows a 16-bit ring counter. After 16 clock cycles, the output of this counter will return 1 indicating that 16 cycles have passed. The control unit can use a single LUT (SRL-16) to implement the ring counter which is less than one-fifth of a counter-based control mechanism. If the control signal has to remain 1 after 16 clock cycles, the controller can use an edge detector which costs an extra LUT and register, to check when a transition from 1 to 0 occurs.

Managing the hierarchy of control is also simpler using ring counters and edge detectors. Consider an example with two nested loops both counting up to 16. Figure 5(c) shows the implementation of this nested loop with two ring counters and an edge detector. The outer (*SRL-16_outer*) loop may count the number of rounds while the inner (*SRL-16_inner*) loop counts the number of bits. The *Edge Detector* will convert the *start* pulse into a continuous *enable* signal which will keep *SRL-16_inner* active until a positive edge is detected at the output of *SRL-16_outer*. Once the *SRL-16_inner* is active, its output will be 1 every 16 clock cycles and enable the *SRL-16_outer* for a single clock cycle. This control unit can be realized with 4 LUTs and 3 register (2 LUTs for SRL-16, 2 LUT and 3 register for the Edge Detector).

3.3   Bit-Serializing BitCryptor

The datapath of BitCryptor is serialized similar to the example. The bit-parallel operations are converted into bit-serial ones and the necessary data elements are stored in SRL-16. The sequentialization of the control flow is achieved by using ring counters and edge detectors. The ring counters control the internal signals when there is a data transmission with the host system. The I/O structure of BitCryptor is also simplified using bit-serial design methodology. The data input and output of the BitCryptor are single bit ports which makes it very suitable for standard serial communication interfaces.

4   Hardware Implementation

Figure 6 shows the block diagram of BitCryptor. The host system indicates the operation *mode* as 1, 2 or 3 for hash, encryption and PRNG respectively. It also provides the *input data, key/IV* (Initialization Vector) and the *start* signal. There are two output signals showing the current status of the engine. The first status signal *Next Block* indicates that a new block of input data can be hashed while the second signal *Done* states that the operation is completed and the output can be sampled. All the data interfaces (*Data In, Key/IV, Data Out*) are realized as serial ports and the control signals (*Start, Mode, Next Block, Done*) are synchronized with the corresponding data.

BitCryptor is an autonomous module and it does not reveal any internal state to outside. To have a secure mode switching, the crypto engine requires the host system to provide a *key/IV* at the start of each operation. This process overwrites the residues of the *key/IV* from a previous execution and ensures that no secret information is leaked between two consecutive operations. Output data is revealed together with the done signal if and only if the operation is completed. Hence, an adversary abusing the input control signals cannot dump out the internal states of the engine.

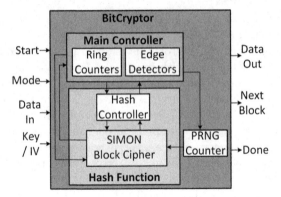

Fig. 6. Block diagram of the BitCryptor

The main controller of BitCryptor handles the selection of operation modes, starting the functions and reading the output values. Ring counters and edge detectors are used to manage the control hierarchy of modes following the methodology in Sect. 3.2. The hash function encapsulates the block cipher module and controls it during the hashing operation. Also, the main controller has direct access to the block cipher for encryption and PRNG modes, bypassing the hash controller. Next, we describe the details of the individual operations.

4.1 Hash Function

In the Hirose construction, we can use two block ciphers to compute the two halves of the digest. However, this does not necessarily mean that there have to be two full block cipher engines. Since both encryption engines use the same key, they can share a single key expansion function. Moreover, the internal control signals of both round functions are the same so they can share the same control logic. We call this architecture the Double-Datapath (DDP) SIMON with a master round function, slave round function and a shared key expansion function. The master round function is the full version that is capable of running on its own, independently. On the other hand, the slave round function gets the internal control signals from the master so it can only run while the master is running.

Figure 7 shows the DDP SIMON architecture following a master/slave configuration. The architecture is bit-serialized using the design methodology of Sect. 3. The hash function has two 96-bit chaining variables G_i and H_i, which are produced by the master and slave round functions respectively. These two variables are loaded with the IV value at the beginning of each operation. A 96-bit shift register (6 SRL-16) stores the G_i value while the shift registers of the key expansion function store H_i. When the hash function is completed, it returns G_i and H_i as the lower and upper 96-bits of the digest respectively.

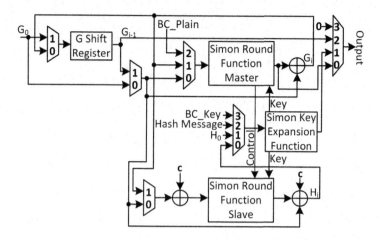

Fig. 7. Architecture of the Double-Datapath SIMON with Hirose construction

4.2 Symmetric Key Encryption

At the core, the crypto engine uses the SIMON block cipher with a 96-bit block and key size. In [5], Aysu *et al.* implement the bit-serial version of SIMON 128/128 and show that it is an extremely compact design. To adapt the bit-serial SIMON block cipher to our crypto engine, we modify the implementation in [5] and convert it into SIMON 96/96. We also extend it to perform Cipher-block-chaining (CBC) mode as well as Electronic-code-book (ECB).

Figure 7 shows the hardware architecture of the hash function, which also includes the SIMON 96/96 block cipher. When the crypto engine is in encryption mode, it only uses the master round function while the slave round function is inactive. The key expansion function uses the 96-bit key configuration. The input data *BC_plain* and key *BC_key* come directly from the host system through the main controller, bypassing the hash function. When the block cipher completes encryption, it gives the output from the same data output port that is shared with the hash function.

4.3 PRNG

The PRNG uses the SIMON 96/96 in counter mode of operation. When the host system requests a random number, it provides the key as the source of entropy and the PRNG module feeds the 22-bit *PRNG counter* value to the block cipher padded with zeros. The host system is also responsible to change the key after receiving 2^{22} bits of random data. After the block cipher generates the random number, the PRNG module increments the counter value. We verified that the output of the PRNG passes the NIST statistical test suite [32].

Table 2. Area-cost breakdown of BitCryptor

Block	LUTs	Registers
Round-function master	15	29
Round-function slave	9	15
Key expansion	16	53
Control	62	91

5 Results

In this section, we first focus on BitCryptor, the lightweight bit-serialized implementation. Then, we show the trade-off between the area and performance on a round-serial variant of BitCryptor.

5.1 Smallest Area – BitCryptor

The proposed hardware architecture is written in Verilog HDL and synthesized in Xilinx ISE 14.7 for the target Spartan-3 XC3S50-4 FPGA as well as a more recent Spartan-6 XC3S50-4 FPGA. In order to minimize the slice count, the synthesized design is manually mapped to the FPGA resources using Xilinx PlanAhead and finally the design is placed and routed. The power consumptions are measured using Xilinx XPower. BitCryptor occupies 95 slices (187 LUTs, 102 Registers) and the area consumption details are given in Table 2. On the target FPGA, BitCryptor achieves a throughput of 4 Mbps for encryption and PRNG, and 1.91 Mbps for hashing at 118 MHz.

For classic hardware design, efficiency (area-per-throughput) is the typical comparison metric. However, for lightweight applications, area-cost is the most important evaluation criterion. Indeed, even reducing the area-cost of a design by 2.3 % is accepted as a significant contribution [42]. The improvement we achieve in this paper is much more significant. Comparing hardware implementations is a hard task due to the differences in EDA tools, target technology, optimization heuristics, and the security level of the underlying primitives. In Fig. 8, we demonstrate the area comparison of the designs that have the same practical functionality for lightweight applications, which is having at least 80-bits of security. The figure shows a detailed area comparison of BitCryptor with the smallest previous multi-purpose engine [24] and with various standalone area-optimized block ciphers [5,10,11,22,26,28,37,41], hash functions [2,23,28], and PRNGs [21]. The results show how small a flexible solution can become with sound cryptographic engineering. Next, we discuss the details of area comparisons and the compromise in performance and efficiency.

Migrating to More Recent Xilinx FPGAs. For comparison fairness with the previous work, we implement our hardware architecture on a Spartan-3 family FPGA (XC3S50-4TQG144C). In addition, we also map our design on a more

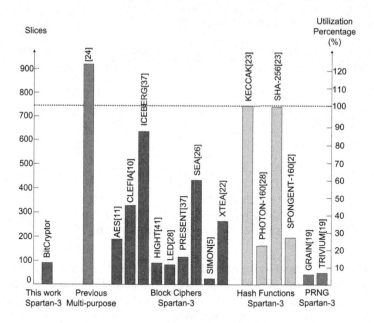

Fig. 8. Implementation results and comparison with previous work. For comparison fairness with the previous work, we map our architecture on an older Spartan-3 FPGA but we also provide the result on a recent Spartan-6 FPGA.

recent Spartan-6 device. On a Spartan-6 XC6SLX4-2 FPGA, BitCryptor occupies 5 % of available resources which corresponds to only 35 slices (136 LUTs, 103 Registers) with a maximum frequency of 172 MHz.

Comparison with Single-Purpose Designs. The results show that our design is more compact than the sum of implementing these functionalities individually. Moreover, it is even smaller than the majority of the lightweight block ciphers and all hash functions. Standalone PRNGs are usually based on simple stream cipher constructions thus making them very compact. Table 2 reveals that the increase of BitCryptor is largely due to the control overhead of the flexible engine.

Comparison with Other Multi-purpose Designs. Most of the previous multi-purpose designs do not target lightweight applications [29, 33–35]. They are mapped on ASIC technology and optimized primarily for performance to provide a throughput at the orders of Gbps. Bossuet *et al.* survey a number of multi-purpose designs and document the smallest to be 847 slices [8]. In [24], Laue *et al.* propose a compact hardware engine on FPGAs that offers the closest functionality to our design. However, they do not apply our design and optimization methods. Therefore, it requires 916 slices on a Virtex-II family FPGA (which has the same slice structure with Spartan-3). Compared to this design, our architecture has an area improvement of almost 10×. The net area reduction of 800

Table 3. Comparison of encryption performance with low-cost microcontrollers

Platform	Clock cycles	Max. frequency (MHz)	Throughput (Kbps)
ATmega128 [14]	24369	16	82.07
MSP430F1611 [14]	12902	8	77.50
This work (bit-serial) XC3S50-4TQG144C	2685	118	4120
This work (bit-serial) XC6SLX4-2TQG144C	2685	172	6005

slices is not achievable by simply plugging in SIMON instead of AES because the area difference of these kernels is 150 slices. Instead, the improvement is due to the combination of our design and optimization methods.

Comparison with Soft-core and Embedded Processors. We also compare our results with the software implementations on actual microcontrollers and on FPGAs using soft-core processors. Good *et al.* provide the smallest soft-core processor in the literature that is capable of running only a single-purpose AES encryption [17]. This design utilizes the 8-bit PicoBlaze processor [9], achieves 0.71 Mbps, and occupies 119 slices and a BRAM (≈ 452 slice equivalent), making it larger and slower than BitCryptor. Likewise, the 16-bit MSP430 softcore processor [16] on FPGAs occupies more than 10× of BitCryptor and it can not even fit into the same device.

Table 3 shows the comparison of a SIMON block cipher encryption on FPGAs vs. low-cost 8-bit and 16-bit microcontrollers. BitCryptor is two orders of magnitude better than ATmega128 and MSP430 based microcontroller implementations. Note that the previous work [14] uses a fixed-key implementation that requires fewer operations and we provide throughput results to compensate for different SIMON configurations. Unfortunately, the power and energy results of Dinu *et al.* is not available, but we can make a rough estimation on TI MSP430F1611. The typical energy consumption of this microcontroller at an energy optimized configuration of 2.2 V and 1 MHz is 330μJ. A SIMON execution with this setting takes 1.3 ms and consumes 4.26×10^{-6} J of energy which is 15.2× of our bit-serial compact design. Table 4 shows the details of the performance figures.

5.2   Relaxing Area – Round-Serial Variant

Area-Performance Tradeoff. A bit-serial design exchanges performance for area savings. Typically, by extending the width of the datapath, it is possible to achieve a more area- and energy-efficient design, because the control and the storage do not scale with the datapath. Therefore, bit-serial architectures are less efficient than 4-bit, 8-bit, or round-serial designs. We have evaluated the relative impact of this trade-off, by comparing a bit-serial implementation of BitCryptor with a round-serial version of BitCryptor. The area improvement comes at

Table 4. Area-performance tradeoff (@100 MHz XC3S50-4)

	Bit-serial	Round-serial[a]	Unit
Block cipher & PRNG	3.41	169.54	Mbps
Hash short block[b]	1.64	83.23	Mbps
Hash long block[b]	1.80	86.37	Mbps
Static power[c]	3.24	14.31	mW
Dynamic power	7	38	mW
Total power	10.24	52.31	mW
Average energy[d]	2.80×10^{-7}	2.85×10^{-8}	J
Energy-delay	7.79×10^{-12}	1.57×10^{-14}	J-s
Area	95	500	Slice

[a] The Round-Serial results are estimated from a simulation of SIMON 96/96 hardware
[b] Short block is one 48-bit input block, long block is 1000 48-bit input blocks
[c] Static power is scaled with respect to the resource utilization ratio
[d] Average energy refers to the averaged energy consumption of three modes

the expense of throughput and energy-efficiency. Compared to bit-serial architectures, round-serial designs have simpler control and a faster execution time, resulting in a reduced energy consumption and a higher throughput. Table 4 quantifies these trade-offs. The round-serial design is approximately two orders of magnitude faster and more energy efficient, but it occupies 5 times the area compared to the bit-serial. However, the power requirement of the bit-serial design is lower due to sequentialization (dynamic) and reduced total area (static). The area-efficiency (defined as Mbps/slice) of the flexible bit-serial and round serial designs are 0.036 and 0.339. Normalized at the same frequency (to compensate for FPGA difference), the efficiency of the previous compact flexible engine is 0.311. The compact single-purpose designs like HIGHT [40] has an efficiency of 0.71 which is, as expected, much better than the flexible engines.

The round-serial variant of BitCryptor is still smaller than previous multipurpose implementations and can also fit into the same Spartan-3 and Spartan-6 FPGA with the bit-serial design. Table 5 shows that this architecture can achieve a two orders of magnitude performance improvement compared to a capable 32-bit ARM microcontroller.

6   Conclusion

The key contribution of this work is to provide a flexible engine with a minimal area overhead. We showed that selecting the optimum encryption kernel and parameters, using a bit-serial design methodology, targeting the architecture optimization for the shift register logic (SRL-16), and manual placement of LUTs

Table 5. Comparison of encryption performance with moderate microcontrollers

Platform	Clock cycles	Max. frequency (MHz)	Throughput (Mbps)
ATSAM3A8 ARM-CORTEX-M3 [14]	1406	84	7.29
This work (round-serial) XC3S50-4TQG144C	54	112	189.88
This work (round-serial) XC6S50-2TQG144C	54	162	274.66

and registers can significantly minimize the area. The resulting hardware architecture is 10× smaller than a previous multi-purpose design and smaller than majority of single-purpose crypto modules. BitCryptor can fit into the smallest FPGA of Spartan-3 and Spartan-6 family with only 12 % and 5 % resource utilization respectively, leaving a large amount of logic for other embedded functionalities. Hence, the proposed hardware architecture is a promising IP block for system designers who seek compact and efficient solutions on reconfigurable hardware.

Acknowledgements. This project was supported in part by the National Science Foundation grant no 1115839 and 1314598.

References

1. Abed, F., List, E., Lucks, S., Wenzel, J.: Differential and linear cryptanalysis of reduced-round SIMON. Cryptology ePrint Archive, Report 2013/526 (2013). http://eprint.iacr.org/
2. Adas, M.: On the FPGA based implementation of SPONGENT (2011)
3. Alkhzaimi, H.A., Lauridsen, M.M.: Cryptanalysis of the SIMON family of block ciphers. Cryptology ePrint Archive, Report 2013/543 (2013)
4. Andraka, R.J.: Building a high performance bit-serial processor in an FPGA. In: Proceedings of Design SuperCon., vol. 96, pp. 1–5 (1996)
5. Aysu, A., Gulcan, E., Schaumont, P.: SIMON says: break area records of block ciphers on FPGAs. IEEE Embed. Syst. Lett. **6**(2), 37–40 (2014)
6. Beaulieu, R., Shors, D., Smith, J., Treatman-Clark, S., Weeks, B., Wingers, L.: The SIMON and SPECK families of lightweight block ciphers. Cryptology ePrint Archive, Report 2013/404 (2013). http://eprint.iacr.org/
7. Bogdanov, A.A., Knudsen, L.R., Leander, G., Paar, C., Poschmann, A., Robshaw, M., Seurin, Y., Vikkelsoe, C.: PRESENT: an ultra-lightweight block cipher. In: Paillier, P., Verbauwhede, I. (eds.) CHES 2007. LNCS, vol. 4727, pp. 450–466. Springer, Heidelberg (2007). http://dx.doi.org/10.1007/978-3-540-74735-2_31
8. Bossuet, L., Grand, M., Gaspar, L., Fischer, V., Gogniat, G.: Architectures of flexible symmetric key crypto engines-a survey: from hardware coprocessor to multi-crypto-processor system on chip. ACM Comput. Surv. **45**(4), 41:1–41:32 (2013). http://doi.acm.org/10.1145/2501654.2501655

9. Chapman, K.: Picoblaze 8-bit microcontroller for virtex-e and spartan-ii/iie devices. Xilinx Application Notes (2003)

10. Chaves, R.: Compact CLEFIA implementation on FPGAs. In: Athanas, P., Pnevmatikatos, D., Sklavos, N. (eds.) Embedded Systems Design with FPGAs, pp. 225–243. Springer, New York (2013). http://dx.doi.org/10.1007/978-1-4614-1362-2_10

11. Chu, J., Benaissa, M.: Low area memory-free FPGA implementation of the AES algorithm. In: 2012 22nd International Conference on Field Programmable Logic and Applications (FPL), pp. 623–626, August 2012

12. Clark, J.A., Jacob, J.L.: A survey of authentication protocol literature: Version 1.0 (1997)

13. De Cannière, C.: TRIVIUM: a stream cipher construction inspired by block cipher design principles. In: Katsikas, S.K., López, J., Backes, M., Gritzalis, S., Preneel, B. (eds.) ISC 2006. LNCS, vol. 4176, pp. 171–186. Springer, Heidelberg (2006). http://dx.doi.org/10.1007/11836810_13

14. Dinu, D., Corre, Y.L., Khovratovich, D., Perrin, L., Groschdl, J., Biryukov, A.: Triathlon of lightweight block ciphers for the internet of things. Cryptology ePrint Archive, Report 2015/209 (2015). http://eprint.iacr.org/

15. Ferguson, N., Schneier, B.: Practical Cryptography. Wiley, New York (2003). http://books.google.com/books?id=7SiKtxPrrRMC

16. Girard, O.: openmsp430 (2009)

17. Good, T., Benaissa, M.: AES on FPGA from the fastest to the smallest. In: Rao, J.R., Sunar, B. (eds.) CHES 2005. LNCS, vol. 3659, pp. 427–440. Springer, Heidelberg (2005)

18. Gulcan, E., Aysu, A., Schaumont, P.: A flexible and compact hardware architecture for the SIMON block cipher. In: Eisenbarth, T., Öztürk, E. (eds.) LightSec 2014. LNCS, vol. 8898, pp. 34–50. Springer, Heidelberg (2015)

19. Guo, J., Peyrin, T., Poschmann, A.: The PHOTON family of lightweight hash functions. In: Rogaway, P. (ed.) CRYPTO 2011. LNCS, vol. 6841, pp. 222–239. Springer, Heidelberg (2011). http://dx.doi.org/10.1007/978-3-642-22792-9_13

20. Hirose, S.: Some plausible constructions of double-block-length hash functions. In: Robshaw, M. (ed.) FSE 2006. LNCS, vol. 4047, pp. 210–225. Springer, Heidelberg (2006)

21. Hwang, D., Chaney, M., Karanam, S., Ton, N., Gaj, K.: Comparison of FPGA-targeted hardware implementations of eSTREAM stream cipher candidates. In: State of the Art of Stream Ciphers Workshop, SASC 2008, Lausanne, Switzerland, pp. 151–162, February 2008

22. Kaps, J.-P.: Chai-tea, cryptographic hardware implementations of xTEA. In: Chowdhury, D.R., Rijmen, V., Das, A. (eds.) INDOCRYPT 2008. LNCS, vol. 5365, pp. 363–375. Springer, Heidelberg (2008)

23. Kaps, J., Yalla, P., Surapathi, K.K., Habib, B., Vadlamudi, S., Gurung, S.: Lightweight implementations of SHA-3 finalists on FPGAs. In: The Third SHA-3 Candidate Conference (2012)

24. Laue, R., Kelm, O., Schipp, S., Shoufan, A., Huss, S.: Compact AES-based architecture for symmetric encryption, hash function, and random number generation. In: International Conference on Field Programmable Logic and Applications, FPL 2007, pp. 480–484, August 2007

25. Liu, S., Xiang, L., Xu, J., Li, X.: Intelligent engine room IoT system based on multi-processors. Microelectron. Comput. **9**, 049 (2011)

26. Mace, F., Standaert, F.X., Quisquater, J.J.: FPGA implementation(s) of a scalable encryption algorithm. IEEE Trans. Very Large Scale Integr. (VLSI) Syst. **16**(2), 212–216 (2008)
27. Menezes, A.J., Van Oorschot, P.C., Vanstone, S.A.: Handbook of Applied Cryptography. CRC Press, Boca Raton (2010)
28. Nalla-Anandakumar, N., Peyrin, T., Poschmann, A.: A very compact FPGA implementation of LED and PHOTON. In: Meier, W., Mukhopadhyay, D. (eds.) INDOCRYPT 2014. LNCS, vol. 8885, pp. 304–321. Springer, Heidelberg (2014). http://dx.doi.org/10.1007/978-3-319-13039-2_18
29. Paul, G., Chattopadhyay, A.: Three snakes in one hole: the first systematic hardware accelerator design for sosemanuk with optional serpent and snow 2.0 modes. IEEE Trans. Comput. **PP**(99), 1–1 (2015)
30. De la Piedra, A., Braeken, A., Touhafi, A.: Sensor systems based on FPGAs and their applications: a survey. Sensors **12**(9), 12235–12264 (2012)
31. Plessl, C., Enzler, R., Walder, H., Beutel, J., Platzner, M., Thiele, L.: Reconfigurable hardware in wearable computing nodes. In: Proceedings of the Sixth International Symposium on Wearable Computers, ISWC 2002, pp. 215–222. IEEE (2002)
32. Rukhin, A., Soto, J., Nechvatal, J., Smid, M., Barker, E.: A statistical test suite for random and pseudorandom number generators for cryptographic applications. Technical report, DTIC Document (2001)
33. Sayilar, G., Chiou, D.: Cryptoraptor: high throughput reconfigurable cryptographic processor. In: Proceedings of the 2014 IEEE/ACM International Conference on Computer-Aided Design, ICCAD 2014, pp. 154–161. IEEE Press, Piscataway (2014). http://dl.acm.org/citation.cfm?id=2691365.2691398
34. Sen Gupta, S., Chattopadhyay, A., Khalid, A.: Designing integrated accelerator for stream ciphers with structural similarities. Crypt. Commun. **5**(1), 19–47 (2013). http://dx.doi.org/10.1007/s12095-012-0074-6
35. Shahzad, K., Khalid, A., Rakossy, Z., Paul, G., Chattopadhyay, A.: Coarx: a coprocessor for arx-based cryptographic algorithms. In: 2013 50th ACM/EDAC/IEEE Design Automation Conference (DAC), pp. 1–10, May 2013
36. Smart, N., Babbage, S., Catalano, D., Cid, C., de Weger, B., Dunkelman, O., Ward, M.: ECRYPT II yearly report on algorithms and keysizes (2011–2012). European Network of Excellence in Cryptology (ECRYPT II), September 2012
37. Standaert, F.X., Piret, G., Rouvroy, G., Quisquater, J.J.: FPGA implementations of the ICEBERG block cipher. In: International Conference on Information Technology: Coding and Computing, ITCC 2005, vol. 1, pp. 556–561 (2005)
38. Wang, Q., Liu, Z., Varıcı, K., Sasaki, Y., Rijmen, V., Todo, Y.: Cryptanalysis of reduced-round SIMON32 and SIMON48. In: Meier, W., Mukhopadhyay, D. (eds.) INDOCRYPT 2014. LNCS, vol. 8885, pp. 143–160. Springer, Heidelberg (2014)
39. Winternitz, R.S.: A secure one-way hash function built from DES. In: 2012 IEEE Symposium on Security and Privacy, p. 88. IEEE Computer Society (1984)
40. Yalla, P., Kaps, J.: Compact FPGA implementation of CAMELLIA. In: International Conference on Field Programmable Logic and Applications, FPL 2009, pp. 658–661 (2009)
41. Yalla, P., Kaps, J.: Lightweight cryptography for FPGAs. In: International Conference on Reconfigurable Computing and FPGAs, ReConFig 2009, pp. 225–230 (2009)
42. Yang, G., Zhu, B., Suder, V., Aagaard, M.D., Gong, G.: The SIMECK family of lightweight block ciphers. In: Güneysu, T., Handschuh, H. (eds.) CHES 2015. LNCS, vol. 9293, pp. 307–329. Springer, Heidelberg (2015). http://dx.doi.org/10.1007/978-3-662-48324-4_16

Low-Resource and Fast Binary Edwards Curves Cryptography

Brian Koziel[1]([✉]), Reza Azarderakhsh[1], and Mehran Mozaffari-Kermani[2]

[1] Computer Engineering Department,
Rochester Institute of Technology, Rochester, NY 14623, USA
{bck6520,rxaeec}@rit.edu
[2] Electrical and Microelectronic Engineering Department,
Rochester Institute of Technology, Rochester, NY 14623, USA
mmkeme@rit.edu

Abstract. Elliptic curve cryptography (ECC) is an ideal choice for low-resource applications because it provides the same level of security with smaller key sizes than other existing public key encryption schemes. For low-resource applications, designing efficient functional units for elliptic curve computations over binary fields results in an effective platform for an embedded co-processor. This paper proposes such a co-processor designed for area-constrained devices by utilizing state of the art binary Edwards curve equations over mixed point addition and doubling. The binary Edwards curve offers the security advantage that it is complete and is, therefore, immune to the exceptional points attack. In conjunction with Montgomery Ladder, such a curve is naturally immune to most types of simple power and timing attacks. The recently presented formulas for mixed point addition in [1] were found to be invalid, but were corrected such that the speed and register usage were maintained. We utilize corrected mixed point addition and doubling formulas to achieve a secure, but still fast implementation of a point multiplication on binary Edwards curves. Our synthesis results over NIST recommended fields for ECC indicate that the proposed co-processor requires about 50 % fewer clock cycles for point multiplication and occupies a similar silicon area when compared to the most recent in literature.

Keywords: Crypto-processor · Binary edwards curves · Gaussian normal basis · Point multiplication · Low-resource devices

1 Introduction

Deeply-embedded computing systems, nowadays, are essential parts of emerging, sensitive applications. With the transition to the Internet of Things (IoTs), where all tools and electronics will be linked wirelessly, there is a need to secure these devices from malicious intent. However, these devices are mainly designed in such a way that the functionality and connectivity monopolize the device's area and power. Little power and area are allocated for the establishment of

A. Biryukov and V. Goyal (Eds.): INDOCRYPT 2015, LNCS 9462, pp. 347–369, 2015.
DOI: 10.1007/978-3-319-26617-6_19

security. Therefore, a secure co-processor that can fill this niche in the current technological world is necessary for the evolution of achieving security for IoTs in the near future.

Elliptic Curve Cryptography (ECC) is the ideal implementation for this application because it provides a secure application for far fewer bits than RSA and other public key encryption schemes. ECC provides key exchange ECDH, authentication ECDSA, and encryption ECIES protocols. An elliptic curve is composed of all points that satisfy an elliptic curve equation as well as a point at infinity. This forms an Abelian group, E, over addition, where the point at infinity represents the zero element or identity of the group. The most basic operations over this Abelian group are point addition and point doubling. Using a double-and-add method, a point multiplication, $Q = kP$, where $k \in \mathbb{Z}$ and $Q, P \in E$, can be computed quickly and efficiently. Protocols implemented over ECC rely on the difficulty to solve the elliptic curve discrete logarithm problem (ECDLP), that given Q and P in $Q = kP$, it is infeasible to solve for k [2]. For the computations of ECC, several parameters should be considered including representation of field elements and underlying curve, choosing point addition and doubling method, selecting coordinate systems such as affine, projective, Jacobian, and mixed, and finally arithmetic (addition, inversion, multiplication, squaring) on finite field. Field multiplication determines the efficiency of point multiplication on elliptic curves as its computation is complex and point multiplication requires many field multiplications. IEEE and NIST recommended the usage of both binary and prime fields for the computation of ECC [3,4]. However, in hardware implementations and more specifically for area-constrained applications, binary fields outperform prime fields, as shown in [5]. Therefore, a lot of research in the literature has been focused on investigating the efficiency of computing point multiplication on elliptic curves over binary fields. For instance, one can refer to [6–9] to name a few, covering a wide variety of cases including different curve forms, e.g., generic and Edwards, and different coordinate systems, e.g., affine, projective, and mixed. The formulas for point addition and point addition can be determined by using geometric properties. In [10], binary Edwards curves are presented for the first time for ECC and their low-resource implementations appeared in [9]. It has been shown that a binary Edwards curves (BEC) is isomorphic to a general elliptic curve if the singularities are resolved [10]. Based on the implementations provided in [9], it has been observed that their implementations are not as efficient as other standardized curves. Recently, in [1], the authors revisited the original equations for point addition and doubling and provided competitive formulas. We observed that the revisited formulas for mixed point addition in [1] are invalid. After modifying their formulas, we employed them for the computation of point multiplication using a mixed coordinate system and proposed an efficient crypto-processor for low-resource devices. The main contributions of this paper can be summarized as follows:

- We propose an efficient hardware architecture for point multiplication on binary Edwards curves. We employed Gaussian normal basis (GNB) for representing field elements and curves as the computation of squaring, inversion, and trace function can be done very efficiently over GNB in hardware.
- We modified and corrected the w-coordinates differential point addition formulas presented in [1]. We provide explicit formulas over binary Edwards curves that maintain the speed and register usage provided in [1] and employed the formulas in steps on the Montgomery Ladder [11]. This is the first time this double-and-add algorithm has appeared in literature. This implementation was competitive with many of the area-efficient elliptic curve crypto-processors found in literature, but adds the additional security benefit of completeness.
- We implemented and synthesized our proposed algorithms and architectures for the computation of point multiplication on binary Edwards curves and compared our results to the leading ones available in the literature.

This paper is organized as follows. In Sect. 2, the binary Edwards curve is introduced and proper mixed coordinate addition formulas are presented. Section 3 details the area-efficient architecture used for this ECC co-processor. Section 4 compares this work to other ECC crypto-processors in terms of area, latency, computation time, and innate security. Section 5 concludes the paper with takeaways and the future of area efficient implementations of point multiplication.

2   Point Multiplication on Binary Edwards Curves

ECC cryptosystems can be implemented over a variety of curves. Some curves have more inherent properties than others. Table 1 contains a comparison of point addition and doubling formulas presented in literature. Completeness means that there are no exceptional cases to addition or doubling (e.g., adding the neutral point). From this table, the choice was to apply the new mixed coordinate addition and doubling formulas over new binary Edwards curves presented in [1].

2.1   Binary Edwards Curve

Definition 1. *Consider a finite field of characteristic two, K. Let $d_1, d_2 \in K$ such that $d_1 \neq 0$ and $d_2 \neq d_1^2 + d_1$. Then the binary Edwards curve with coefficients d_1 and d_2 is the affine curve [10]:*

$$E_{\mathbb{F}_{2^m}, d_1, d_2} : d_1(x + y) + d_2(x^2 + y^2) = xy + xy(x + y) + x^2 y^2 \qquad (1)$$

This curve is symmetric in that if (x, y) is on the curve, then (y, x) is also on the curve. In fact, these points are additive inverses over the Edwards addition law. The point $(0, 0)$ is isomorphic to the point at infinity in a binary generic curve. This represents the neutral point in the binary Edwards curve. The point

Table 1. Cost of point operations on binary generic curves (BGCs) [12], binary Edwards curves (BECs) [10], binary Edwards curves revisited [1], and generalized Hessian curves (GHC) [13] over $GF(2^m)$.

Curve	Coordinate System	Differential PA and PD	Completeness
BGC	Projective	$6M + 1D + 5S$	✗
	Mixed	$5M + 1D + 4S$	✗
BEC ($d_1 = d_2$)	Projective	$7M + 2D + 4S$	✓
	Mixed	$5M + 2D + 4S$	✓
BEC-R ($d_1 = d_2$)	Projective	$7M + 2D + 4S$	✓
	Mixed	$5M + 1D + 4S$	✓
GHC	Projective	$7M + 2D + 4S$	✓
	Mixed	$5M + 2D + 4S$	✓

(1,1) is also on every binary Edwards curve, and has order 2. The curve is complete if there is no element $t \in K$ that satisfies the relation $t^2 + t + d_2 = 0$ [10]. Alternatively, this means that if $\text{Tr}(d_2) = 1$, then the curve is complete [14].

Point addition and point doubling do not have the same representation or equations as standard generic curves. The Edwards Addition Law is presented below. The sum of any two points $(x_1, y_1), (x_2, y_2)$ on the curve defined by $E_{\mathbb{F}_{2^m}, d_1, d_2}$ to (x_3, y_3) is defined as [10]

$$x_3 = \frac{d_1(x_1+x_2)+d_2(x_1+y_1)(x_2+y_2)+(x_1+x_1^2)(x_2(y_1+y_2+1)+y_1y_2)}{d_1+(x_1+x_1^2)(x_2+y_2)}$$

$$y_3 = \frac{d_1(y_1+y_2)+d_2(x_1+y_1)(x_2+y_2)+(y_1+y_1^2)(y_2(x_1+x_2+1)+x_1x_2)}{d_1+(y_1+y_1^2)(x_2+y_2)}$$

(2)

We note that [10] uses this addition law to prove that ordinary elliptic curves over binary fields are birationally equivalent to binary Edwards curves.

2.2   Revised Differential Addition and Doubling Formulas

Point multiplication utilizes point doubling and point addition to quickly generate large multiples of a point. As a deterrent to timing and other side channel attacks, the Montgomery Ladder [11] is used as a method to generate multiplications efficiently and securely. Montgomery Ladder is shown in Algorithm 1. At each step of the ladder, there is an addition and doubling. The current bit of a key determines which point is doubled and where the point addition and doubling reside. For standard point additions and point doublings, the finite field inversion dominates the computation. To reduce this impact, the typical convention is to use projective coordinates, $(x, y) \rightarrow (X, Y, Z)$, where $x = \frac{X}{Z}$ and $y = \frac{Y}{Z}$. $X, Y,$ and Z are updated at each step of the ladder and a single inversion is performed at the end. An additional improvement to this convention is to use w-coordinates, $(x, y) \rightarrow (w)$, where $w = x+y$. Mixed coordinates is the combination

---

Algorithm 1. Montgomery algorithm [11] for point multiplication using w-coordinates.

Inputs: A point $P = (x_0, y_0) \in E(\mathbb{F}_{2^m})$ on a
binary curve and an integer $k = (k_{l-1}, \cdots, k_1, k_0)_2$.
Output: $w(Q) = w(kP) \in E(\mathbb{F}_{2^m})$.
1: **set** : $w_0 \leftarrow x_0 + y_0$ and initialize
   a: $W_1 \leftarrow w_0$ and $Z_1 \leftarrow 1$ and $c = \frac{1}{w_0}$ (inversion)
   b: $(W_2, Z_2) = \texttt{DiffDBL}(W_1, Z_1)$
2: **for** i **from** $l - 2$ down to 0 **do**
   a: **if** $k_i = 1$ **then**
       i): $(W_1, Z_1) = \texttt{MDiffADD}(W_1, Z_1, W_2, Z_2, c)$
       ii): $(W_2, Z_2) = \texttt{DiffDBL}(W_2, Z_2)$
   b: **else**
       i): $(W_1, Z_1) = \texttt{DiffDBL}(W_1, Z_1)$
       ii): $(W_2, Z_2) = \texttt{MDiffADD}(W_1, Z_1, W_2, Z_2, c)$
   end if
   end for
3: **return** $w(kP) \leftarrow (W_1, Z_1)$ and $w((k+1)P) \leftarrow (W_2, Z_2)$

---

of w-coordinates and the projective coordinates. Hence, $(x, y) \rightarrow w \rightarrow (W, Z)$, where $w = x + y = \frac{W}{Z}$. For computing point multiplication, let P be a point on a binary Edwards curve $E_{\mathbb{F}_{2^m}, d_1, d_2}$ and let us assume $w(nP)$ and $w((n+1)P)$, $0 < n < k$ are known. Therefore, one can use the w-coordinate differential addition and doubling formulas to compute their sum as $w((2n+1)P)$ and double of $w(nP)$ as $w(2nP)$ [10].

2.3 Fixed w-Coordinate Differential Addition

In [1], the authors present faster equations for w-coordinates and mixed coordinates addition than those presented in [10]. This equation makes the assumption that $d_1 = d_2$. An analysis of the formula, however, shows that they do not properly produce the correct w-coordinates. The authors correctly identify the relation, $\frac{w_3 w_0}{d_1(w_1^2 + w_2^2)} = \frac{w_3 + w_0 + 1}{d_1}$, but incorrectly solve for w_3. We observe that the final equation for differential point addition that is presented in subsection (3.19) of [1] is faulty. Therefore, we wrote a sage script to verify this claim[1]. This algebra was performed correctly and here we present the revised formulas. The incorrect formula presented in [1] is in (3) and the corrected formula is shown in (4). This formula defines the addition of $w_1 + w_2 = w_3$, given that $w_i = x_i + y_i$ and $w_0 = w_2 - w_1$.

Proposition 1. *The w-coordinate differential addition formula over binary Edwards curves with $d_1 = d_2$ proposed in [1] does not provide correct formulation based on the following equation:*

---

[1] http://github.com/briankoziel/BEC_Small.

$$w_3 = 1 + \frac{\frac{1}{w_0}(w_1^2 + w_2^2)}{\frac{1}{w_0}(w_1^2 + w_2^2) + 1} \tag{3}$$

Proof. In the following equations, the correct w-coordinate differential addition formula over binary Edwards curves with $d_1 = d_2$ is discovered from the starting relation in [1].

$$\frac{w_3 w_0}{d_1(w_1^2 + w_2^2)} = \frac{w_3 + w_0 + 1}{d_1}$$

$$\frac{w_3 w_0}{(w_1^2 + w_2^2)} = w_3 + w_0 + 1$$

$$\frac{w_3 w_0}{(w_1^2 + w_2^2)} + w_3 = w_0 + 1$$

$$w_3\left(\frac{w_0}{(w_1^2 + w_2^2)} + 1\right) = w_0 + 1$$

$$w_3(w_0 + w_1^2 + w_2^2) = (w_0 + 1)(w_1^2 + w_2^2)$$

$$w_3 = \frac{(w_0 + 1)(w_1^2 + w_2^2)}{w_0 + w_1^2 + w_2^2}$$

$$w_3 = \frac{(1 + \frac{1}{w_0})(w_1^2 + w_2^2)}{\frac{1}{w_0}(w_1^2 + w_2^2) + 1}$$

Corrected w-Coordinate Differential Addition

$$w_3 = \frac{w_1^2 + w_2^2 + \frac{1}{w_0}(w_1^2 + w_2^2)}{\frac{1}{w_0}(w_1^2 + w_2^2) + 1}. \tag{4}$$

The explicit affine w-coordinate differential addition is

$$A = (w_1 + w_2)^2, \quad B = A \cdot \frac{1}{w_0}, \quad N = A + B, \tag{5}$$
$$D = B + 1, \quad E = \frac{1}{D}, \quad w_3 = N \cdot E.$$

The total cost of this corrected formula is still $1I + 1M + 1D + 1S$, but now the differential addition functions as intended. Assuming that inversion requires at least two registers, a total of three registers are required. $\frac{1}{w_0}$ is the inverse of the difference between the points and will not be updated in each step of the point multiplication algorithm. For the application in Montgomery Ladder [11], the difference between the two points is always P (specifically $w(P)$). Therefore, this value can be determined at the start of the ladder and used throughout to cut down on each step.

[1] uses the faulty formula (3) for determining explicit formulas in mixed w-coordinates, but also gives a faster and correct formula for affine w-coordinate differential addition which requires $1I + 1M + 2S$, so long as the values $\frac{1}{w_0 + w_0^2}$ and w_0 are known. This formula is shown below.

$$w_3 = w_0 + 1 + \frac{1}{\frac{1}{w_0 + w_0^2}(w_1^2 + w_2^2 + w_0)} \tag{6}$$

The explicit affine w-coordinate differential addition is

$$A = (w_1 + w_2)^2, \quad B = A + w_0, \quad D = B \cdot \frac{1}{w_0 + w_0^2}, \tag{7}$$

$$E = \frac{1}{D}, w_3 = E + 1 + w_0$$

Assuming that w_0 and $\frac{1}{w_0 + w_0^2}$ are known, the actual cost for w-coordinate differential addition can be reduced down to $1I + 1D + 1S$. This method requires two registers for w_1 and w_2, and the storage of w_0 and $\frac{1}{w_0 + w_0^2}$.

Mixed w-Coordinate Differential Addition and Doubling. Equation (4) can be applied to mixed w-coordinate differential addition and doubling. The general formula and explicit formula are shown below. This formula defines the addition of $\frac{W_1}{Z_1} + \frac{W_2}{Z_2} = \frac{W_3}{Z_3}$, given that $w_0 = w_2 - w_1$.

$$\frac{W_3}{Z_3} = \frac{(W_1 Z_2 + W_2 Z_1)^2 + \frac{1}{w_0}(W_1 Z_2 + W_2 Z_1)^2}{Z_1^2 Z_2^2 + \frac{1}{w_0}(W_1 Z_2 + W_2 Z_1)^2} \tag{8}$$

$$C = (W_1 Z_2 + W_2 Z_1)^2, \, D = (Z_1 Z_2)^2, \, E = \frac{1}{w_0} \cdot C, \tag{9}$$

$$W_3 = E + C, \; Z_3 = E + D$$

Thus, mixed w-coordinate differential addition requires $3M + 1D + 2S$. From a simple analysis of the formula, four registers are needed.

For mixed w-coordinate differential addition and doubling, the doubling formula from [10] can be used in conjunction with this corrected differential addition formula, with the assumption that $d_1 = d_2$. This formula defines the addition of $\frac{W_1}{Z_1} + \frac{W_2}{Z_2} = \frac{W_3}{Z_3}$ and doubling of $2 \times \frac{W_1}{Z_1} = \frac{W_4}{Z_4}$ given that $w_0 = w_2 - w_1$.

$$\frac{W_4}{Z_4} = \frac{(W_1(W_1 + Z_1))^2}{d_1 \cdot Z_1^4 + (W_1(W_1 + Z_1))^2} \tag{10}$$

$$C = (W_1 Z_2 + W_2 Z_1)^2, \quad D = (Z_1 Z_2)^2, \quad E = \frac{1}{w_0} \cdot C, \tag{11}$$

$$W_3 = E + C, \quad Z_3 = E + D \quad W_4 = (W_1(W_1 + Z_1))^2,$$

$$Z_4 = W_4 + d_1 \cdot Z_1^4$$

Thus, mixed w-coordinate differential addition and doubling requires $5M + 1D + 5S$. From an analysis of the formula, five registers are needed.

Mixed w-Coordinate Differential Addition and Doubling with the Co-Z Trick. We note that in [15] the common-Z trick is proposed. This method to reduces the number of registers required per step of the Montgomery Ladder [11] and simplifies the number of operations per step. Each step of the Montgomery Ladder is a point doubling and addition. By using a common-Z coordinate system, one less register is required for a step on the ladder, and the method becomes more efficient, requiring one less squaring operation. The doubling formula was obtained from [10] and it is assumed that $d_1 = d_2$. The general formula and explicit formulas are shown below. This formula defines the addition of $\frac{W_1}{Z} + \frac{W_2}{Z} = \frac{W_3}{Z'}$ and doubling of $2 \times \frac{W_1}{Z} = \frac{W_4}{Z'}$ given that $w_0 = w_2 - w_1$.

$$\frac{W_3}{Z'} = \frac{(W_1 + W_2)^2 + \frac{1}{w_0}(W_1 + W_2)^2}{Z^2 + \frac{1}{w_0}(W_1 + W_2)^2} \tag{12}$$

$$\frac{W_4}{Z'} = \frac{(W_1(W_1 + Z))^2}{d_1 \cdot Z^4 + (W_1(W_1 + Z))^2} \tag{13}$$

$$C = (W_1 + W_2)^2, \quad D = Z^2, \quad E = \frac{1}{w_0} \cdot C, \tag{14}$$
$$U = E + C, V = E + D, S = (W_1(W_1 + Z))^2,$$
$$T = S + d_1 \cdot D^2, W_3 = U \cdot T, W_4 = V \cdot S,$$
$$Z' = V \cdot T$$

Thus, the mixed w-coordinate differential addition and doubling formula requires $5M+1D+4S$. An analysis of this formula shows that it requires only four registers. As will be discussed later, this implementation incorporates shifting for the multiplication within the register file, forcing the need for an additional register. This formula requires one less squaring than that provided in [10], and also uses registers much more efficiently. Table 2 shows a comparison of differential point addition schemes for BEC with $d_1 = d_2$.

Table 2. Comparison of Differential Point Addition Schemes for BEC with $d_1 = d_2$.

Operation	Formula	Complexity	#Registers
Affine w-coordinate Differential Addition	(5)	$1I + 1M + 1D + 2S$	3
Affine w-coordinate Differential Addition	(7)	$1I + 1D + 1S$	2
Mixed Differential Addition	(9)	$3M + 1D + 1S$	4
Mixed Differential Addition and Doubling	(11)	$5M + 1D + 5S$	5
Mixed Differential Addition and Doubling w/Co-Z	(14)	$5M + 1D + 4S$	4

Algorithm 2. Retrieving x and y from w-coordinates

Inputs: A point $P = (x_0, y_0) \in E(\mathbb{F}_{2^m})$ on a
binary curve and an integer $k = (k_{l-1}, \cdots, k_1, k_0)_2$.
Output: $Q = kP \in E(\mathbb{F}_{2^m})$.
1: set: $w_0 \leftarrow x_0 + y_0$ and initialize
2: compute: $w_2 \leftarrow w(kP)$, $w_3 \leftarrow w(kP + 1)$
3: solve (15) for $x_2 + x_2^2$
4: **if** $\text{Tr}(x_2 + x_2^2) = 0$ **then**
    a: $x_2 =$ **half-trace**$(x_2 + x_2^2)$
end if
3: solve (16) for $y_2 + y_2^2$
4: **if** $\text{Tr}(y_2 + y_2^2) = 0$ **then**
    a: $y_2 =$ **half-trace**$(y_2 + y_2^2)$
end if
5: **return** $Q = (x_2, y_2) = kP \in E(\mathbb{F}_{2^m})$

Retrieving x and y from w-Coordinates. The formula to retrieve the x-coordinate from w-coordinates is presented in [10]. This formula requires P, $w(kP)$, and $w(kP + 1)$. Again, relating back to the application of Montgomery Ladder [11], each consecutive step produces $w(mP)$ and $w(mP + 1)$, where m represents the scalar multiplication over each steps. The formula to solve for the x-coordinate of mP is shown below [10]. In this formula, $P = (x_1, y_1)$, $w_0 = x_1 + y_1$, $w_2 = w(kP)$, and $w_3 = w(kP + 1)$.

$$x_2^2 + x_2 = \frac{w_3(d_1 + w_0 w_2(1 + w_0 + w_2) + \frac{d_2}{d_1} w_0^2 w_2^2) + d_1(w_0 + w_2) + (y_1^2 + y_1)(w_0^2 + w_2)}{w_0^2 + w_0} \tag{15}$$

This formula requires $1I + 4M + 4S$ if $d_2 = d_1$. After solving for $x_2^2 + x_2 = A$, if $\text{Tr}(A) = 0$, then the value of x_2 or $x_2 + 1$ can be recovered by using the half-trace.

After the value of x_2 has been found, y_2 can be retrieved by solving the curve equation for $y_2^2 + y_2$ (16) and also using the half-trace to solve for y_2 or $y_2 + 1$.

$$y_2^2 + y_2 = \frac{d(x_2 + x_2^2)}{d + x_2 + x_2^2} \tag{16}$$

Therefore, recovering y_2 requires $1I + 2M + S$, and the total cost of recovering points from w-coordinates is $2I + 6M + 5S$. Even though the point $(x_2 + 1, y_2 + 1)$ is not the same as (x_2, y_2), both points will produce the same value in standard ECC applications. Algorithm 2 summarizes how to retrieve the x and y-coordinates.

The implementation of the algorithms noted in this section require a binary Edwards curve with $d_1 = d_2$. The standardized NIST curves over binary generic curves [3] could be converted to binary Edwards curves. However, there is no guarantee that these isomorphic binary Edwards curves would satisfy $d_1 = d_2$.

Therefore, values for x and d were randomly picked and used in conjunction with (16) to solve for y. If the point (x,y) was on the curve, then the point and corresponding binary Edwards curve were valid and could be used with the above algorithms. It can also be noted that there are no restrictions on d, so it could be chosen to be small for faster arithmetic.

2.4   Resistance Against Side-Channel Attacks

The binary Edwards curve features the unique properties that its addition formula is unified and complete. Unified implies that the addition and doubling formulas are the same. This gives the advantage that no checking is required for the points to differentiate if an addition or doubling needs to take place. Complete implies that the addition formula works for any two input points, including the neutral point. Therefore, as long as two points are on the curve, no checking is needed for the addition formula, as it will always produce a point for a complete binary Edwards Curve [10].

One common attack to reveal bits of an ECC system's key is to use the exceptional points attack [16]. This attacks the common projective coordinate system. For the point at infinity in a non-binary Edwards curve system, the point is often represented as $(X_k, Y_k, 0)$. Hence, a conversion back to the (x_k, y_k) coordinate system would attempt to divide by zero, causing an error or revealing a point that is not on the curve [16]. In either case, an adversary could detect that the point at infinity was attempted to be retrieved. The attack relies on picking different base points, which after multiplied by the hidden key, reveal that the point of infinity was retrieved.

The binary Edwards curve's completeness property and coordinate system make the curve immune to this form of attack. For a complete binary Edwards curve, the projective coordinate system representation for the neutral point, which is isomorphic to the point at infinity of other curves, is $(X_k, Y_k, 1)$. Furthermore, the completeness also ensures that no other sets of points can be used to break the system and reveal critical information about the key. The mixed w-coordinates that are used for their speed in the binary Edwards curve are also invulnerable to this attack as long as $w_0 \neq 0, 1$, since the denominator will never be 0 [1]. With the Montgomery Ladder [11], a proper curve and starting point will never violate this condition.

Montgomery Ladder [11] is a secure way to perform repeated point addition and point doublings to thwart side channel attacks. The ladder provides a point addition and point doubling for each step, with each step taking the same amount of time. Therefore, this application provides an extremely powerful defense against power analysis attacks and timing attacks. Power analysis attacks identify characteristics of the power consumption of a device to reveal bits of the key and timing attacks identify characteristics of the timing as the point multiplication is performed. By application of the binary Edwards curve with Montgomery Ladder, the binary Edwards curve features an innate defense against many of the most common attacks on ECC systems today.

It should be noted that this work does not investigate resistance against differential power analysis (DPA) [17] or electromagnetic (EM) radiation leaks. These will be investigated in detail in a future work.

3   Architecture

The architecture of the ECC co-processor that was implemented resembles that of [6]. However, there are several major differences. An analysis of the explicit formula presented for mixed w-coordinate addition and doubling revealed that five registers $(T_0, T_1, R_0, R_1, R_2)$ and four constants $(\frac{1}{w_0}, d_1, x_1, y_1)$ were required. Additionally, it was deemed that the neutral element in GNB multiplication (all '1's) was not required for any part of the multiplication, which reduced the size of the 4:2 output multiplexer to a 3:2 multiplexer. These following sections will explain the design in more detail. Architectures for each of the components can be found in Fig. 1.

3.1   Field Arithmetic Unit

The field arithmetic unit is designed to incorporate the critical finite field operations in as small of a place as possible. In particular, this requires multiplication, squaring, and addition. The XOR gate to add two elements was reused in the multiplication and addition to reduce the total size of the FAU. Since the neutral element was not necessary for this point multiplier, the neutral element select from the output multiplexer in [6] was removed to save area. Swap functionality was added to incorporate quick register file swap operations. The field arithmetic unit incorporates the GNB multiplier from [18]. The operations are as follows:

- **Addition** $C = A + B$: Addition is a simple XOR of two inputs. The first input is loaded to Z by selecting the first input in the register file, and setting $s_1 = $ "01" and $s_2 = $ "00". The next cycle, the second operand is selected from the register file, and $s_2 = $ "01" so that the output register has the addition of the two input elements. The output is written on the third cycle. This operation requires three clock cycles.
- **Squaring** $C = A \gg 1$: Squaring is a right circular shift of the input. The input is loaded to Z by selecting the input in the register file, and setting $s_1 = $ "01" and $s_2 = $ "00". The next cycle, $s_1 = $ "10" and $s_2 = $ "10" so that the output register has been shifted. The output is written on the third cycle. This operation requires three clock cycles.
- **Multiplication** $C = T_0 \times T_1$: Multiplication is a series of shifted additions. For the first cycle, $s_1 = $ "00", $s_2 = $ "00", and $s_{T_0} = s_{T_1} = $ "1". The next cycle, $s_2 = $ "01". After m cycles of shifts and addition, $s_{T_0} = s_{T_1} = $ "0", and the output is ready. The output is written on the mth cycle. This operation requires m clock cycles.

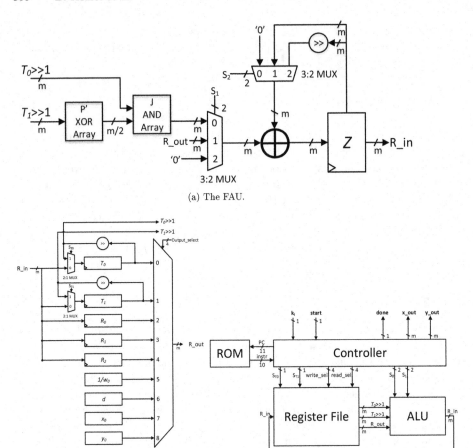

(a) The FAU.

(b) The register file and top-level control unit.

Fig. 1. Architecture of the Proposed Co-processor for Point Multiplication on Binary Edwards Curve. This includes (a) Field Arithmetic Unit, (b) Register File, and Top-Level Control Unit.

- **Swapping** $A, B = B, A$: Swapping is a switch of two registers within the register file. The first register is loaded to Z by selecting the input in the register file, and setting $s_1 =$ "01" and $s_2 =$ "00". The next cycle, the first register is written to the second register's location as it is being loaded to Z. The second register's value is written to the first register's place on the third cycle. This operation requires three clock cycles.

3.2   Register File

Similar to [6,9], the register file was designed to contain registers, with two particular registers that perform special shifting for the finite field multiplication. An analysis of the formulas used in this ECC unit revealed that four registers and four constants were required. However, with two registers being designated as multiplication registers, an extra register is needed for swapping in the value of d_1 for a multiplication with D^2. The other three registers would be holding (U, V, S). Thus, the formulas require five registers with the Co-Z trick implementation.

For unified access to constants and not impact the retrieval, the registers and constants are co-located in the register file. However, since this implementation targets a future standardization of a binary Edwards curve, the idea was that a starting point and curve parameters would be strictly defined. Therefore, there is no reason to add flexibility to the parameters of the base point or d_1. Hardwiring these coordinates to the register file provides the advantage that they can be used on-the-fly and that no extra control is necessary to bring these into the register file. For instance, [9] uses a small and external RAM chip to hold these constants. Such a design requires extra interfacing and extra cycles to load the value into the register file. After NIST standards for ECC are revised, hardwiring the constants in a place close to the register file is the best solution to save power and area.

The register file is random access to values including the constants. A register is written to when write is enabled and the multiplexer for writing selects that register.

3.3   Control Unit

The control unit handles the multiplexers for reading, writing, and performing operations. The four operations are ADD, SQ, MULT, and SWAP. The control unit uses a Finite State Machine to switch between these operations. A program counter is sent to an external ROM device that feeds in the current instruction. Instructions are ten bits long. The first two bits indicate which instruction is being used. The next four bits indicate the input register. This value does not matter for multiplication. The last four bits indicate the output register.

The key is never stored in the control unit, such as how it was in [6]. The controller signals the master device to provide the next bit as the Montgomery Ladder [11] is being performed. Special SWAP instructions that depend on the key were left inside the controller to handle each step of the ladder, depending on the provided bit of the key. The subroutine for a step on the Montgomery Ladder with the corresponding register usage is shown below in Table 3. Table 3 shows the registers after each instruction. Six multiplications are required for each step.

To save area, the half-trace functionality was left as a series of squarings and additions. Adding additional area to handle the half-trace saves a relatively small fraction of instructions but adds an additional multiplexer select in the FAU.

Table 3. Point Addition and Doubling Register Usage

#	Op	T_0	T_1	R_0	R_1	R_2
1	ADD T0 T1	W_1	$W_1 + W_2$	Z		
2	SQ T1 R1	W_1	$W_1 + W_2$	Z	C	
3	SWAP T1 R0	W_1	Z	$W_1 + W_2$	C	
4	SQ T1 R0	W_1	Z	D	C	
5	ADD T0 T1	W_1	$W_1 + Z$	D	C	
6	MULT T1 T0	$W_1 \cdot (W_1 + Z)$	$W_1 + Z$	D	C	
7	SQ T0 R2	$W_1 \cdot (W_1 + Z)$	$W_1 + Z$	D	C	S
8	SWAP R1 T1	$W_1 \cdot (W_1 + Z)$	C	D	$W_1 + Z$	S
9	SWAP R3 T0	$\frac{1}{w_0}$	C	D	$W_1 + Z$	S
10	MULT T0 R1	$\frac{1}{w_0}$	C	D	E	S
11	ADD R1 T1	$\frac{1}{w_0}$	U	D	E	S
12	ADD R0 R1	$\frac{1}{w_0}$	U	D	E	S
13	SQ R0 R0	$\frac{1}{w_0}$	U	D^2	V	S
14	SWAP R0 T1	$\frac{1}{w_0}$	D^2	U	V	S
15	SWAP R4 T0	d	D^2	U	V	S
16	MULT T1 T0	$d \cdot D^2$	D^2	U	V	S
17	ADD R2 T0	T	D^2	U	V	S
18	SWAP R0 T1	T	U	D^2	V	S
19	MULT T0 T1	T	W_3	D^2	V	S
20	SWAP T1 R1	T	V	D^2	W_3	S
21	MULT T0 T0	Z'	V	D^2	W_3	S
22	SWAP T0 R2	S	V	D^2	W_3	Z'
23	MULT T0 T1	S	W_4	D^2	W_3	Z'
24	SWAP R0 R2	S	W_4	Z'	W_3	D^2
25	SWAP T0 R1	W_3	W_4	Z'	S	D^2

Inversion and the half-trace were implemented as subroutines within the ROM for instructions. The half traces uses a repetitive combination of double SQ then ADD. This was used to recover the x and y-coordinates of the final point. Inversion was used to obtain $w_i = \frac{W_i}{Z_i}$, recover the x-coordinate, and recover the y-coordinate. Itoh-Tsujii inversion algorithm [19] was used to reduce the number of multiplications. For $\mathbb{F}_{2^{283}}$, the addition chain (1,2,4,8,16,17,34,35,70,140,141,282) was used. By implementing these repeated functionalities as subroutines, the number of instructions in the ROM is dramatically reduced. The main program is shown in Fig. 2. The subroutines for inversion in $\mathbb{F}_{2^{283}}$ and the half-trace are shown in Fig. 3. The total instruction count of the point multiplier for $\mathbb{F}_{2^{283}}$ is shown in Table 4. Approximately 132, 10-bit instructions were needed.

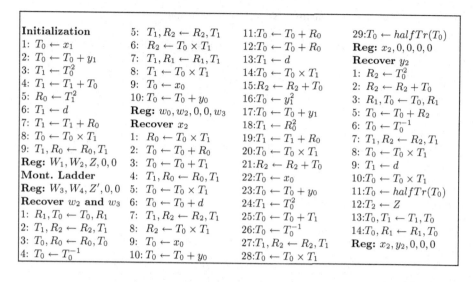

Fig. 2. Main Program Listing for Point Multiplication using Binary Edwards Curves [10]

Inversion			
1: $T_1 \leftarrow T_0^2$	12: $T_0 \leftarrow T_0 \times T_1$	24: $T_0 \leftarrow T_1^2$	36: $T_0 \leftarrow T_0^2$
2: $T_1 \leftarrow T_0 \times T_1$	13: $T_1, R_0 \leftarrow R_0, T_1$	25: $T_0 \leftarrow T_0^{2^{34}}$	**Half-Trace**
3: $T_0, R_0 \leftarrow R_0, T_0$	14: $T_0 \leftarrow T_0^2$	26: $T_0 \leftarrow T_0 \times T_1$	1: $T_1 \leftarrow T_0^2$
4: $T_0 \leftarrow T_1^2$	15: $T_0 \leftarrow T_0 \times T_1$	27: $T_0 \leftarrow T_1^2$	2: $T_1 \leftarrow T_1^2$
5: $T_0 \leftarrow T_0^2$	16: $T_1, R_0 \leftarrow R_0, T_1$	28: $T_0 \leftarrow T_0^{2^{69}}$	3: $T_0 \leftarrow T_0 + T_1$
6: $T_0 \leftarrow T_0 \times T_1$	17: $T_0 \leftarrow T_1^2$	29: $T_0 \leftarrow T_0 \times T_1$	4: $T_1 \leftarrow T_1^2$
7: $T_0 \leftarrow T_1^2$	18: $T_0 \leftarrow T_0^{2^{16}}$	30: $T_1, R_0 \leftarrow R_0, T_1$	5: $T_1 \leftarrow T_1^2$
8: $T_0 \leftarrow T_0^{2^3}$	19: $T_0 \leftarrow T_0 \times T_1$	31: $T_0 \leftarrow T_0^2$	6: $T_0 \leftarrow T_0 + T_1$
9: $T_0 \leftarrow T_0 \times T_1$	20: $T_1, R_0 \leftarrow R_0, T_1$	32: $T_0 \leftarrow T_0 \times T_1$	*Repeat steps 4-6 $\frac{m-2}{2}$ times
10: $T_0 \leftarrow T_1^2$	21: $T_0 \leftarrow T_0^2$	33: $T_0 \leftarrow T_1^2$	
11: $T_0 \leftarrow T_0^{2^7}$	22: $T_0 \leftarrow T_0 \times T_1$	34: $T_0 \leftarrow T_0^{2^{140}}$	
	23: $T_1, R_0 \leftarrow R_0, T_1$	35: $T_0 \leftarrow T_0 \times T_1$	

Fig. 3. Itoh-Tsujii [19] Inversion ($\mathbb{F}_{2^{283}}$) and Half-Trace Subroutines

4 Comparison and Discussion

This design was synthesized using Synopsys Design Compiler in $\mathbb{F}_{2^{283}}$, $\mathbb{F}_{2^{233}}$, and $\mathbb{F}_{2^{163}}$, each a different standardized binary field size by NIST [3]. The TSMC 65-nm CMOS standard technology and CORE65LPSVT standard cell library were used for results. This implementation was optimized for area.

The area was converted to Gate Equivalent (GE), where the size of a single NAND gate is considered 1 GE. For our particular technology library, the size of a synthesized NAND gate was $1.4\,\mu m^2$, so this was used as the conversion factor. Latency reports the total number of cycles to compute the final coordinates of a point multiplication. Parameters such as the type of curve used and if

Table 4. Necessary Subroutines.

Subroutine	Iterations	#ADD	#SQ	#MULT	#SWAP	Latency (cycles)
Init	1	3	2	1	4	310
Step	281	5	4	6	$10+2^a$	494,841
x Recovery, no HT and Inv	1	16	5	9	13	2,649
y Recovery, no HT and Inv	1	3	2	3	6	882
Half Trace	2×141	1	2	0	0	$2 \times 1,269$
Inversion	3×1	0	282	11	6	$3 \times 3,977$
Total		1,705	2,540	1,730	3,410	512,555

^{a} Special SWAP's that the controller handles.

Table 5. Comparison of Different Point Bit-Level Multiplications Targeted for ASIC

Work	Curve	Ladder?	Field Size	Tech (nm)	Mult	# of clock Cycles	Coord	Area (GE)
[15], 2007	BGC	✓	$\mathbb{F}_{2^{163}}$	180	Bit-serial	313,901	Projective	13,182
[8], 2008	BGC	✓	$\mathbb{F}_{2^{163}}$	130	Bit-serial	275,816	Mixed	12,506
[9], 2010	BEC	✓	$\mathbb{F}_{2^{163}}$	130	Bit-serial	219,148	Mixed	11,720
[20], 2011	BGC	✓	$\mathbb{F}_{2^{163}}$	130	Comb-serial	286,000	Projective	8,958
[6], 2014	BKC	✗	$\mathbb{F}_{2^{163}}$	65	Bit-serial	106,700	Affine	10,299
[21], 2014	BGC	✗	$\mathbb{F}_{p^{160}}$	130	Comb-serial	139,930	Projective	$12,448^a$
[7], 2015	BKC	✗	$\mathbb{F}_{2^{283}}$	130	Comb-serial	1,566,000	Projective	$10,204^b$ $(4,323)^c$
This work	BEC	✓	$\mathbb{F}_{2^{163}}$	65	Bit-serial	177,707	Mixed	$10,945^d$
		✓	$\mathbb{F}_{2^{233}}$			351,856		$14,903^d$
		✓	$\mathbb{F}_{2^{283}}$			512,555		$19,058^d$

^{a} Includes a Keccak module to perform ECDSA.
^{b} RAM results were not synthesized, but extrapolated from a different implementation.
^{c} Area excluding RAM.
^{d} Area excluding ROM. Approximately 274 GE more with ROM.

Montgomery Ladder were used to indicate some innate security properties of the curve. Power and energy results were not included as a comparison because they are dependent on the underlying technology, frequency of the processor, and testing methodology. The comparison results are shown in Table 5.

This ECC implementation over BEC does make a few assumptions that not necessarily each of these other implementations make. This architecture's area does not include the ROM to hold the instructions. The ROM was not synthesized, but approximately 165 bytes of ROM were required. By the estimate that 1,426 bytes is equivalent to 2,635 GE in [22], 165 bytes of ROM is roughly equivalent to 274 GE. This architecture assumes that each bit of the key will be fed into the co-processor. These assumptions are explained in previous sections. The areas of the implementation for $\mathbb{F}_{2^{163}}$, $\mathbb{F}_{2^{233}}$, and $\mathbb{F}_{2^{283}}$ excluding the register file and program ROM are 3,248 GE, 3,788 GE, and 5,566 GE, respectively.

Looking at timing for these implementation, the number of clock cycles appears to rise quadratically when comparing $\mathbb{F}_{2^{163}}$ to $\mathbb{F}_{2^{283}}$. This is to be expected, as the Montgomery Ladder performs 6 multiplications each step. A multiplication takes m clock cycles and there are $m - 2$ steps.

The area appears to have a linear relationship. This is also to be expected, as the register file's size increases linearly. The area of the FAU depends on the underlying finite field and the area of the controller is fairly constant. The area of the FAU and controller for $\mathbb{F}_{2^{233}}$ is only a slight increase over the area of the FAU and controller for $\mathbb{F}_{2^{163}}$ because the $\mathbb{F}_{2^{233}}$ is type II GNB, in contrast to $\mathbb{F}_{2^{163}}$ and $\mathbb{F}_{2^{283}}$ are type IV GNB. Therefore the p' block in $\mathbb{F}_{2^{233}}$ requires much fewer XOR gates.

The underlying architecture of this implementation was similar to [6]. This implementation uses more area because an additional register and two additional constants were used in the register file. However, one less multiplexer was required in the FAU since the neutral element in GNB was not required in any formulas. Other than that, the implementation in [6] does not use the Montgomery Ladder and performs over Koblitz curves, which speeds up the point multiplication at the cost of some security.

The only other light-weight implementation of BEC point multiplication is found in [9]. Many of the internals of our point multiplier are different. For instance, this implementation uses a circular register structure, and also a different bit-serial multiplier in Polynomial Basis. A Polynomial Basis parallel squaring unit was used in this implementation, which is costly when compared to the GNB. This implementation uses Common-Z differential coordinate system for the Montgomery Ladder, but each step requires 8 multiplications. Our implementation requires only 6 multiplications, representing a reduction of latency in the Montgomery Ladder by approximately 25 %. Lastly, this implementation requires a register file to hold 6 registers, whereas our register file only requires 5 registers. Hence, our implementation features a smaller and faster point multiplication scheme than that in [9].

The introduction of extremely area-efficient crypto-processors with comb-serial multiplication schemes [23] like the one proposed in [7] indicates that there is a need for new trade-off for future implementations of these ECC targeted at RFID chips. Bit-parallel multiplication architectures are among the fastest approaches to perform finite field multiplications, but this requires a tremendous amount of area. Digit-serial schemes require a factor more of cycles, but use less area. The most popular scheme for RFID chip point multiplication is bit-serial, which requires a fraction of the area of digit-serial and requires m cycles to perform a multiplication. Comb-serial multiplication takes this a step further by performing small multiplications over many small combs. Depending on the multiplication scheme, this could require more than m cycles but holds new records for area-efficiency. The work presented in [7] is among the smallest ECC co-processors, even in $\mathbb{F}_{2^{283}}$. It was designed as a drop-in concept, such that the co-processor can share RAM blocks with a microcontroller. This implementation utilizes a comb-serial multiplicationscheme in polynomial basis over Koblitz curves. As such, the latency of each operation is larger than that of this work. Field addition, squaring, and multiplication require 60, 200, and 829 cycles, respectively. This implementation needs space to hold 14 intermediate elements throughout the point multiplication operation. Including the constants, our implementation requires 9 intermediate values. The area of the co-processor without the RAM for the register file is 4,323 GE. Moreover, in [7], the RAM results that were included were extrapolated from a different implementation of ECC appeared in [22]. With these extrapolated results, the total area of the co-processor would be 10,204 GE. Our crypto-processor with the register file uses 87 % more area, but performs the point multiplication approximately three times faster, reducing the need to run at higher speeds to meet timing requirements in a device. Further, [7] utilizes zero-free tau-adic expansion to enforce a constant pattern of operations, similar to the Montgomery ladder [11], to protect against timing and power analysis attacks. However, this new technique has not been thoroughly explored like the Montgomery ladder. Furthermore, the co-processor does not have any protection against exceptional points attacks such as the ones presented in [16]. In summary, for higher levels of security as was implemented in [7], the time complexity was several factors higher, but the area was comparable to an implementation of a smaller finite field. As there is a push for larger field sizes for higher security levels, the time complexity of the comb-serial method of multiplication and other operations becomes inefficient.

5   Conclusion

In this paper, it is shown that new mixed w-coordinate differential addition and doubling formulas for binary Edwards curve produce a fast, small, and secure implementation of point multiplication. Corrected formulas for addition in this coordinate system have been provided and proven. Binary Edwards curves feature a complete and unified addition formula. The future of point multipliers targeted at RFID technology depends on the trade-offs among area, latency, and security. The binary Edwards curves implementation presented in this paper has demonstrated that BEC is highly-competitive with the dominant elliptic curve systems standardized by NIST and IEEE. As such, new standardizations that include binary Edwards curves are necessary for the future of elliptic curve cryptography. The detailed analysis in this paper also suggests that binary Edwards curves are among the fastest and most secure curves for point multiplication targeting resource-constrained devices.

Acknowledgments. The authors would like to thank the reviewers for their constructive comments. This material is based upon work supported by the National Science Foundation under Award No. CNS-1464118 to Reza Azarderakhsh.

A   Appendix

A.1   Subroutines

This contains a code listing of the program in assembly.

Algorithm 3 shows the Itoh-Tsujii [19] inversion subroutine for $\mathbb{F}_{2^{283}}$. This follows the addition chain (1,2,4,8,16,17,34,35,70,140,141,282). Eleven multiplications are required for this binary field. A similar approach was done for $\mathbb{F}_{2^{163}}$ and $\mathbb{F}_{2^{233}}$.

Algorithm 3. Itoh-Tsujii [19] Inversion Subroutine for $GF(2^{283})$

```
SQ T0 T1
MULT T1 T1 –2^2-1
SWAP T0 R0
SQ T1 T0
SQ T0 T0
MULT T1 T0 –2^4-1
SQ T0 T1
SQ T1 T1 3 Times
MULT T1 T0 –2^8-1
SQ T0 T1
SQ T1 T1 7 Times
MULT T1 T0 –2^16-1
SWAP T1 R0
SQ T0 T0
MULT T1 T0 –2^17-1
SWAP T1 R0
SQ T0 T1
SQ T1 T1 16 Times
MULT T1 T0 –2^34-1
SWAP T1 R0
SQ T0 T0
MULT T1 T0 –2^35-1
SWAP T1 R0
SQ T0 T1
SQ T1 T1 34 Times
MULT T1 T0 –2^70-1
SQ T0 T1
SQ T1 T1 69 Times
MULT T1 T0 –2^140-1
SWAP T1 R0
SQ T0 T0
MULT T1 T0 –2^141-1
SQ T0 T1
SQ T1 T1 140 Times
MULT T1 T0 –2^282-1
SQ T0 T0
```

Algorithm 4. Half-Trace Subroutine

```
SQ T0 T1
SQ T1 T1
ADD T1 T0
{SQ T1 T1
SQ T1 T1
ADD T1 T0} loop for (m-2)/2 times
```

Algorithm 4 shows the half-trace subroutine. This is a simple double square and add routine that produces the result after $\frac{m-1}{2}$ iterations.

Algorithm 5 shows the beginning of the main program that was used. This includes the initialization of the point and the repeated step of the Montgomery ladder [11].

Algorithm 5. General Program Flow

INIT
SWAP R5 T0
ADD R6 T0
SQ T0 T1
ADD T0 T1
SQ T1 R0 –W4
SWAP R4 T1
ADD R0 T1 –Z4
MULT T1 T0 –W1 revised
SWAP T1 R0 –W1 W4 Z4
STEP
SWAP T0 T0 –OUTPUT register selected by k bit
ADD T0 T1
SQ T1 R1
SWAP T1 R0
SQ T1 R0
ADD T0 T1
MULT T1 T0
SQ T0 R2 –S
SWAP R1 T1
SWAP R3 T0 –1/w0
MULT T0 R1 –E
ADD R1 T1 –U
ADD R0 R1 –V
SQ R0 R0
SWAP R0 T1
SWAP R4 T0 –d1
MULT T1 T0
ADD R2 T0 –T
SWAP R0 T1
MULT T0 T1 –W3
SWAP T1 R1
MULT T0 T0 –Z'
SWAP T0 R2
MULT T0 T1 –W4
SWAP R0 R2
SWAP T0 R1
SWAP T0 T0 –Output register selected by k bit. Repeat for every step

---

Algorithm 6 shows the end of the main program that was used. This includes the recovery of w_2, w_3, x_2, y_2.

Algorithm 6. General Program Flow (cont.)

RECOVER X
SWAP T0 R1
SWAP T1 R2
SWAP R0 T0
Invert T0 –1/Z
SWAP R2 T1
MULT T1 R2 –w3
SWAP R1 T1
MULT T1 T1 –w2
SWAP R5 T0
ADD R6 T0
MULT T1 R0
ADD R0 T0
ADD T1 T0
SWAP R0 T1 –(w1w2+w1+w2) w1w2 w2 0 w3
MULT T1 T0
ADD T1 T0
ADD R4 T0 –d1+w1w2+w1w2*(w1+w2+w1w2) w1w2 w2 0 w3
SWAP T1 R2
MULT T1 R2 –1st part of the numerator – 0 0 w2 0 1st
SWAP R5 T0
ADD R6 T0
ADD R0 T0 –w1+w2
SWAP R4 T1
MULT T1 T0
ADD T0 R2 –0 0 w2 0 1st+2nd
SQ R6 T0
ADD R6 T0
SQ R0 T1
ADD R0 T1
MULT T1 T0
ADD T0 R2 –Numerator complete, compute inversion now
SWAP R5 T0
ADD R6 T0 –w1 0 0 0 Numerator
SQ T0 T1
ADD T1 T0 –w1^2+w1 0 0 0 Numerator, now inversion
Invert T0 –1/(w0^2+w0)
SWAP R2 T1
MULT T1 T0
T0 = HalfTrace(T0) –x2 or x2+1
RECOVER Y
SQ T0 R2
ADD T0 R2
SWAP T0 R1
SWAP R4 T0
ADD R2 T0
Invert T0 –1/(d+x+x^2)
SWAP T1 R2
MULT T1 T0
SWAP R4 T1
MULT T1 T0
T0 = HalfTrace(T0) –y2 or y2+1
SWAP T0 T1
SWAP R1 T0 –Solution is x, y in T0 T1

References

1. Kim, K., Lee, C., Negre, C.: Binary edwards curves revisited. In: Meier, W., Mukhopadhyay, D. (eds.) INDOCRYPT 2014. LNCS, vol. 8885, pp. 393–408. Springer, Heidelberg (2014)
2. Hankerson, D.R., Vanstone, S.A., Menezes, A.J.: Guide to Elliptic Curve Cryptography. Springer-Verlag New York Inc., New York (2004)
3. U.S. Department of Commerce/NIST: National Institute of Standards and Technology. Digital Signature Standard, FIPS Publications 186–2, January 2000
4. IEEE Std 1363–2000: IEEE Standard Specifications for Public-Key Cryptography, January 2000
5. Wenger, E., Hutter, M.: Exploring the design space of prime field vs. binary field ECC-hardware implementations. In: Laud, P. (ed.) NordSec 2011. LNCS, vol. 7161, pp. 256–271. Springer, Heidelberg (2012)
6. Azarderakhsh, R., Jarvinen, K.U., Mozaffari Kermani, M.: Efficient algorithm and architecture for elliptic curve cryptography for extremely constrained secure applications. IEEE Trans. Circuits Syst. **61**(4), 1144–1155 (2014)
7. Roy, S.S., Jarvinen, K., Verbauwhede, I.: Lightweight coprocessor for Koblitz curves: 283-bit ECC including scalar conversion with only 4300 gates. Cryptology ePrint Archive, Report 2015/556 (2015). http://eprint.iacr.org/
8. Lee, Y.K., Sakiyama, K., Batina, L., Verbauwhede, I.: Elliptic-curve-based security processor for RFID. IEEE Trans. Comput. **57**(11), 1514–1527 (2008)
9. Kocabas, U., Fan, J., Verbauwhede, I.: Implementation of binary edwards curves for very-constrained devices. In: Proceedings of 21st International Conference on Application-Specific Systems Architectures and Processors (ASAP 2010), pp. 185–191 (2010)
10. Bernstein, D.J., Lange, T., Rezaeian Farashahi, R.: Binary edwards curves. In: Oswald, E., Rohatgi, P. (eds.) CHES 2008. LNCS, vol. 5154, pp. 244–265. Springer, Heidelberg (2008)
11. Montgomery, P.L.: Speeding the pollard and elliptic curve methods of factorization. Math. Comput. **48**, 243–264 (1987)
12. Lopez, J., Dahab, R.: Fast multiplication on elliptic curves over $GF(2^m)$ without precomputation. In: Proceedings of Workshop on Cryptographic Hardware and Embedded Systems (CHES 1999), pp. 316–327 (1999)
13. Farashahi, R.R., Joye, M.: Efficient arithmetic on Hessian curves. In: Nguyen, P.Q., Pointcheval, D. (eds.) PKC 2010. LNCS, vol. 6056, pp. 243–260. Springer, Heidelberg (2010)
14. Azarderakhsh, R., Reyhani-Masoleh, A.: Efficient FPGA implementations of point multiplication on binary Edwards and generalized Hessian curves using Gaussian normal basis. IEEE Trans. Very Large Scale Integr. Syst. **20**(8), 1453–1466 (2012)
15. Lee, Y.K., Verbauwhede, I.: A compact architecture for montgomery elliptic curve scalar multiplication processor. In: Kim, S., Yung, M., Lee, H.-W. (eds.) WISA 2007. LNCS, vol. 4867, pp. 115–127. Springer, Heidelberg (2008)
16. Izu, T., Takagi, T.: Exceptional procedure attack on elliptic curve cryptosystems. In: Desmedt, Y.G. (ed.) PKC 2003. LNCS, vol. 2567, pp. 224–239. Springer, Heidelberg (2002)
17. Kocher, P.C., Jaffe, J., Jun, B.: Differential power analysis. In: Wiener, M. (ed.) CRYPTO 1999. LNCS, vol. 1666, pp. 388–397. Springer, Heidelberg (1999)
18. Azarderakhsh, R., Jao, D., Lee, H.: Common subexpression algorithms for space-complexity reduction of Gaussian normal basis multiplication. IEEE Trans. Inf. Theory **61**(5), 2357–2369 (2015)

19. Itoh, T., Tsujii, S.: A fast algorithm for computing multiplicative inverses in $GF(2^m)$ using normal bases. Inf. Comput. **78**(3), 171–177 (1988)
20. Wenger, E., Hutter, M.: A hardware processor supporting elliptic curve cryptography for less than 9 kGEs. In: Prouff, E. (ed.) CARDIS 2011. LNCS, vol. 7079, pp. 182–198. Springer, Heidelberg (2011)
21. Pessl, P., Hutter, M.: Curved tags — a low-resource ECDSA implementation tailored for RFID. In: Sadeghi, A.-R., Saxena, N. (eds.) RFIDSec 2014. LNCS, vol. 8651, pp. 156–172. Springer, Heidelberg (2014)
22. Wenger, E.: Hardware architectures for MSP430-based wireless sensor nodes performing elliptic curve cryptography. In: Jacobson, M., Locasto, M., Mohassel, P., Safavi-Naini, R. (eds.) ACNS 2013. LNCS, vol. 7954, pp. 290–306. Springer, Heidelberg (2013)
23. Menezes, A.J., Vanstone, S.A., Oorschot, P.C.V.: Handbook of Applied Cryptography, 1st edn. CRC Press Inc., Boca Raton (1996)

Author Index

Printed in the United States
By Bookmasters